T0320400

FUNCTIONAL ANALYSIS

FUNCTIONAL ANALYSIS

PETER D. LAX
Courant Institute
New York University

WILEY-
INTERSCIENCE

A JOHN WILEY & SONS, INC., PUBLICATION

For ordering and customer service, call 1-800-CALL-WILEY.

Library of Congress Cataloging-in-Publication Data
Lax, Peter D.
    Functional analysis / Peter D. Lax.
       p.   cm.
    Includes bibliographical references and index.
    ISBN 0-471-55604-1 (cloth : alk. paper)
    1. Functional analysis. I. Title.

  QA320.L345 2002
  515—dc21   2001046547

Printed in the United States of America
30

CONTENTS

FOREWORD

This book grew out of a course of lectures on functional analysis taught over many years to second-year graduate students at the Courant Institute of New York University. It is a graduate text, not a treatise or a monograph. Most of the chapters are short, for it is easier to digest material in small chunks. Not all topics can be presented briefly, so some of the chapters are longer. Theorems and lemmas, as well as equations, are numbered consecutively in each chapter.

The first 23 chapters make only a modest technical demand on the reader; this material would serve very well as text for an introductory graduate course on functional analysis. The rest of the material is well suited as text for a more advanced graduate course on functional analysis in general, or on Hilbert space in particular.

When I was a student, the only text on functional analysis was Banach's original classic, written in 1932; Hille's book appeared in time to serve as my graduation present. For Hilbert space there was Stone's Colloquium publication, also from 1932, and Sz.-Nagy's Ergebnisse volume. Since then, our cup hath run over; first came Riesz and Sz.-Nagy, then Dunford and Schwartz, Yosida, later Reed and Simon, and Rudin. For Hilbert space, there was Halmos's elegant slender volume, and Achiezer and Glazman, all of which I read with pleasure and profit. Many, many more good texts have appeared since. Yet I believe that my book offers something new: the order in which the material is arranged, the interspersing of chapters on theory with chapters on applications, so that cold abstractions are made flesh and blood, and the inclusion of a very rich fare of mathematical problems that can be clarified and solved from the functional analytic point of view.

In choosing topics I heeded the warning of my teacher Friedrichs: "It is easy to write a book if you are willing to put into it everything you know about the subject." I present the basic structure of the subject, and those more advanced topics that loom large in the body of mathematics. Among these are the spectral resolution and spectral representation of self-adjoint operators, the theory of compact operators, the Krein-Milman theorem, Gelfand's theory of commutative Banach algebras, invariant subspaces, strongly continuous one-parameter semigroups. I discuss the index of operators, so important in calculating topological invariants; the celebrated trace formula of Lidskii, a powerful tool in analysis; the Fredholm determinant and its generalizations, rising again after almost a hundred years of hibernation; and scattering theory, another gift from physics to mathematics. I have also included some (but not all) special topics close to my heart.

What has been omitted? All of nonlinear functional analysis, for which I recommend the four-volume treatise by Zeidler. Operator algebras, except for Gelfand's theory of commutative Banach algebras. I slight the geometric theory of Banach spaces; happily a handbook on this subject, edited by Bill Johnson and Joram Lindenstrauss is about to be published by North Holland.

What are the prerequisites? What every second-year graduate student—and many undergraduates—knows:

- *Naive set theory.* Denumerable sets, the continuum, Zorn's lemma.

- *Linear algebra.* The alternative for linear maps, trace and determinant of a matrix, the spectral theory of general and symmetric matrices, functions of a matrix.

- *Point set topology.* Complete metric spaces, the Baire category principle, Hausdorff spaces, compact sets, Tychonov's theorem.

- *Basic theory of functions of a complex variable.*

- *Real variables.* The Arzela-Ascoli theorem, the Lebesgue decomposition of measures on $\mathbb{R}$, Borel measure on compact spaces.

It is an accident of history that measure theory was invented before functional analysis. The usual presentations of measure theory fail to take advantage of the concepts and constructions of functional analysis. In an appendix on the Riesz-Kakutani representation theorem I show how to use the tools of functional analysis in measure theory. Another appendix summarizes the basic facts of Laurent Schwartz's theory of distributions.

Many of the applications are to problems of partial differential equations. Here a nodding acquaintance with the Laplace and the wave equation would help, although an alert uninformed reader could pick up some of the basic results from these pages.

Like most mathematicians, I am no historian. Yet I have included historical remarks in some of the chapters, mainly where I had some firsthand knowledge, or where conventional history has been blatantly silent concerning the tragic fate of many of the founding fathers of functional analysis during the European horrors of the 1930s and 1940s.

I am indebted to many. I learned the rudiments of functional analysis, and how to apply them, from my teacher Friedrichs. Subsequently my views were shaped by the work of Tosio Kato, who has brought the power of functional analysis to bear on an astonishing range of problems. My happy and long collaboration with Ralph Phillips has led to some unusual uses of functional analysis. I learned much from Israel Gohberg, especially about the index of Toeplitz operators, from Bill Johnson about the fine points of the geometry of Banach spaces, and from Bob Phelps about Choquet's theorem. I thank Reuben Hersh and Louise Raphael for their critique of the appendix on distributions, and Jerry Goldstein for his expert comments on the material on semigroups and scattering theory. To all of them, as well as to Gábor Francsics, my thanks.

Jerry Berkowitz and I alternated teaching functional analysis at the Courant Institute. This would be a better book had he lived and looked the manuscript over critically.

I thank Jeff Rosenbluth and Paul Chernoff for a careful reading of the early chapters, and Keisha Grady for TEXing the manuscript, and cheerfully making subsequent changes and corrections.

The lecture course was popular and successful with graduate students of the Courant Institute. I hope this printed version retains the spirit of the lectures.

PETER D. LAX

New York, NY
November 2001

1

LINEAR SPACES

A *linear space X* over a *field* **F** is a mathematical object in which two operations are defined: addition and multiplication by scalars.

Addition, denoted by +, as in

$$x + y \tag{1}$$

is assumed to be *commutative*,

$$x + y = y + x, \tag{2}$$

associative,

$$x + (y + z) = (x + y) + z, \tag{3}$$

and to form a *group*, with the neutral element denoted as 0:

$$x + 0 = x. \tag{4}$$

The inverse of addition is denoted by $-$:

$$x + (-x) \equiv x - x = 0. \tag{5}$$

The second operation is the *multiplication* of elements of X *by elements k of the field* **F**:

$$kx.$$

The result of this multiplication is again an element of X. Multiplication by elements of **F** is assumed to be *associative*,

$$k(ax) = (ka)x, \tag{6}$$

and *distributive*,

$$k(x + y) = kx + ky \tag{7}$$

1

as well as

$$(a + b)x = ax + bx. \tag{8}$$

We assume that multiplication by the *unit* of $\mathbf{F}$, denoted as 1, acts as the identity:

$$1x = x. \tag{9}$$

These are the *axioms* of linear algebra. From them proceed to draw some deductions.

Set $b = 0$ in (8). It follows that for all x,

$$0x = 0. \tag{10}$$

Set $a = 1, b = -1$ in (8). Using (9) and (10), we deduce that for all x,

$$(-1)x = -x. \tag{11}$$

The finite-dimensional linear spaces are dealt with in courses on linear algebra. In this book the emphasis is on the infinite-dimensional ones—those that are not finite-dimensional. The field $\mathbf{F}$ will be either the real numbers $\mathbb{R}$ or the complex numbers $\mathbb{C}$. Here are some examples.

Example 1. X is the space of all polynomials in a single variable s, with real coefficients, here $\mathbf{F} = \mathbb{R}$.

Example 2. X is the space of all polynomials in N variables $s_1, \ldots, s_N$, with real coefficients, here $\mathbf{F} = \mathbb{R}$.

Example 3. G is a domain in the complex plane, and X the space of all functions complex analytic in G, here $\mathbf{F} = \mathbb{C}$.

Example 4. $X =$ space of all vectors

$$x = (a_1, a_2, \ldots)$$

with infinitely many real components, here $\mathbf{F} = \mathbb{R}$.

Example 5. Q is a Hausdorff space, X the space of all continuous real-valued functions on Q, here $\mathbf{F} = \mathbb{R}$.

Example 6. M is a C^∞ differentiable manifold, $X = C^\infty(M)$, the space of all differentiable functions on M.

Example 7. Q is a measure space with measure m, $X = L^1(Q, m)$.

Example 8. $X = L^p(Q, m)$.

Example 9. X = harmonic functions in the upper half-plane.

Example 10. X = all solutions of a linear partial differential equation in a given domain.

Example 11. All meromorphic functions on a given Riemann surface; $\mathbf{F} = \mathbb{C}$.

We start the development of the theory by giving the basic *constructions* and *concepts*. Given two subsets S and T of a linear space X, we define their *sum*, denoted as $S + T$ to be the set of all points x of the form $x = y + z$, y in S, z in T. The *negative* of a set S, denoted as $-S$, consists of all points x of the form $x = -y$, y in S.

Given two linear spaces Z and U over the same field, their *direct sum* is a linear space denoted as $Z \oplus U$, consisting of *ordered pairs* (z, u), z in Z, u in U. Addition and multiplication by scalars is componentwise.

Definition. A subset Y of a linear space X is called a *linear subspace* of X if sums and scalar multiples of Y belong to Y.

Theorem 1.

 (i) *The sets $\{0\}$ and X are linear subspaces of X.*
 (ii) *The sum of any collection of subspaces is a subspace.*
 (iii) *The intersection of any collection of subspaces is a subspace.*
 (iv) *The union of a collection of subspaces totally ordered by inclusion is a subspace.*

Exercise 1. Prove theorem 1.

Let S be some subset of the linear space X. Consider the collection $\{Y_\sigma\}$ of all linear subspaces that contain the set S. This collection is not empty, since it certainly contains X.

Definition. The intersection $\cap Y_\sigma$ of all linear subspaces Y_σ containing the set S is called the *linear span* of the set S.

Theorem 2.

 (i) *The linear span of a set S is the smallest linear subspace containing S.*
 (ii) *The linear span of S consists of all elements x of the form*

$$x = \sum_1^n a_i x_i, \qquad x_i \in S, a_i \in \mathbf{F}, n \text{ any natural number.} \tag{12}$$

Proof. Part (i) is merely a rephrasing of the definition of linear span. To prove part (ii), we remark that on the one hand, the elements of the form (12) form a linear subspace of X; on the other hand, every x of form (12) is contained in any subspace Y containing S.

REMARK 1. An element x of form (12) is called a *linear combination* of the points $x_1, \ldots, x_n$. So theorem 1 can be restated as follows:

The linear span of a subset S of a linear space consists of all linear combinations of elements of S.

Definition. X a linear space, Y a linear subspace of X. Two points x_1 and x_2 of X are called *equivalent* modulo Y, denoted as $x_1 \equiv x_2 \pmod Y$, if $x_1 - x_2$ belongs to Y.

It follows from the properties of addition that equivalence mod Y is an equivalence relation, meaning that it is symmetric, reflexive, and transitive. That being the case, we can divide X into distinct equivalence classes mod Y. We denote the set of equivalence classes as X/Y. The set X/Y has a natural linear structure; the sum of two equivalence classes is defined by choosing arbitrary points in each equivalence class, adding them and forming the equivalence class of the sum. It is easy to check that the last equivalence class is independent of the representatives we picked; put differently, if $x_1 \equiv z_1, x_2 \equiv z_2$, then $x_1 + x_2 \equiv z_1 + z_2$ mod Y. Similarly we define multiplication by a scalar by picking arbitrary elements in the equivalence class. The resulting operation does not depend on the choice, since, if $x_1 \equiv z_1$, then $kx_1 \equiv kz_1$ mod Y. The quotient set X/Y endowed with this natural linear structure is called the *quotient space* of X mod Y. We define codim $Y = \dim X/Y$.

Exercise 2. Verify the assertions made above.

As with all algebraic structures, so with linear structures we have the concept of *isomorphism*.

Definition. Two linear spaces X and Z over the same field are *isomorphic* if there is a one-to-one correspondence $\mathbf{T}$ carrying one into the other that maps sums into sums, scalar multiples into scalar multiples; that is,

$$\mathbf{T}(x_1 + x_2) = \mathbf{T}(x_1) + \mathbf{T}(x_2),$$
$$\mathbf{T}(kx) = k\mathbf{T}(x). \tag{13}$$

We define similarly *homomorphism*, called in this context a *linear map*.

Definition. X and U are linear spaces over the same field. A *mapping* $\mathbf{M} : X \longrightarrow U$ is called *linear* if it carries sums into sums, and scalar multiples into scalar multiples;

that is, if for all x, y in X and all k in $\mathbf{F}$

$$\mathbf{M}(x + y) = \mathbf{M}(x) + \mathbf{M}(y),$$
$$\mathbf{M}(kx) = k\mathbf{M}(x). \tag{14}$$

X is called the domain of $\mathbf{M}$, U its target.

REMARK 2. An isomorphism of linear spaces is a linear map that is one-to-one and onto.

Theorem 3.

(i) *The image of a linear subspace Y of X under a linear map $\mathbf{M} : X \to U$ is a linear subspace of U.*

(ii) *The inverse image under $\mathbf{M}$ of a linear subspace V of U is a linear subspace of X.*

Exercise 3. Prove theorem 3.

A very important concept in a linear space over the *reals* is *convexity*:

Definition. X is a linear space over the reals; a subset K of X is called *convex* if, whenever x and y belong to K, the whole segment with endpoints x, y, meaning all points of the form

$$ax + (1 - a)y, \qquad 0 \le a \le 1, \tag{15}$$

also belong to K.

Examples of convex sets in the plane are the circular disk, triangle, and semicircular disk. The following property of convex sets is an immediate consequence of the definition:

Theorem 4. *Let K be a convex subset of a linear space X over the reals. Suppose that $x_1, \ldots, x_n$ belong to K; then so does every x of the form*

$$x = \sum_1^n a_j x_j, \qquad a_j \ge 0, \tag{16}$$

$$\sum_1^n a_j = 1.$$

Exercise 4. Prove theorem 4.

An x of form (16) is called a *convex combination* of $x_1, x_2, \ldots, x_n$.

Theorem 5. *Let X be a linear space over the reals.*

(i) *The empty set is convex.*

(ii) *A subset consisting of a single point is convex.*

(iii) *Every linear subspace of X is convex.*

(iv) *The sum of two convex subsets is convex.*

(v) *If K is convex, so is* $-K$.

(vi) *The intersection of an arbitrary collection of convex sets is convex.*

(vii) *Let* $\{K_j\}$ *be a collection of convex subsets that is totally ordered by inclusion. Then their union* $\cup K_j$ *is convex.*

(viii) *The image of a convex set under a linear map is convex.*

(ix) *The inverse image of a convex set under a linear map is convex.*

Exercise 5. Prove theorem 5.

Definition. Let S be any subset of a linear space X over the reals. The *convex hull* of S is defined as the *intersection* of all convex sets containing S. The hull is denoted as $\hat{S}$.

Theorem 6.

(i) *The convex hull of S is the smallest convex set containing S.*

(ii) *The convex hull of S consists of all convex combinations (16) of points of S.*

Exercise 6. Prove theorem 6.

Definition. A subset E of a convex set K is called an *extreme subset* of K if:

(i) E is convex and nonempty.

(ii) whenever a point x of E is expressed as

$$x = \frac{y+z}{2}, \qquad y, z \text{ in } K,$$

then both y and z belong to E.

An extreme subset consisting of a single point is called an *extreme point* of K.

Example 1. K is the interval $0 \le x \le 1$; the two endpoints are extreme points.

Example 2. K is the closed disk

$$x^2 + y^2 \le 1.$$

Every point on the circle $x^2 + y^2 = 1$ is an extreme point.

Example 3. The open disk

$$x^2 + y^2 < 1$$

has no extreme points.

Example 4. K a polyhedron, including faces. Its extreme subsets are its faces, edges, vertices, and of course K itself.

Theorem 7. *Let K be a convex set, E an extreme subset of K, and F an extreme subset of E. Then F is an extreme subset of K.*

Exercise 7. Prove theorem 7.

Theorem 8. *Let M be a linear map of the linear space X into the linear space U. Let K be a convex subset of U, E an extreme subset of K. Then the inverse image of E is either empty or an extreme subset of the inverse image of K.*

Exercise 8. Prove theorem 8.

Exercise 9. Give an example to show that the image of an extreme subset under a linear map need not be an extreme subset of the image.

Taking U to be one dimensional, we get

Corollary 8′. *Denote by H a convex subset of a linear space X, ℓ a linear map of X into $\mathbb{R}$, $H_{\min}$ and $H_{\max}$ the subsets of H, where ℓ achieves its minimum and maximum, respectively.*

Assertion. *When nonempty, $H_{\min}$ and $H_{\max}$ are extreme subsets of H.*

2

LINEAR MAPS

2.1 ALGEBRA OF LINEAR MAPS

We recall from chapter 1 that a *linear map* from one linear space X into another, U, both over the same field of scalars, is a mapping of X into U,

$$\mathbf{M} : X \longrightarrow U,$$

that is an algebraic homomorphism:

$$\mathbf{M}(x + y) = \mathbf{M}(x) + \mathbf{M}(y),$$
$$\mathbf{M}(kx) = k\mathbf{M}(x). \tag{1}$$

In this section we explore those properties of linear maps that depend on the purely algebraic properties (1), without any topological restrictions imposed on the spaces X, U.

The *sum* of two linear maps $\mathbf{M}$ and $\mathbf{N}$ of X into U, and the *scalar multiple* is defined as

$$(\mathbf{M} + \mathbf{N})(x) = \mathbf{M}(x) + \mathbf{N}(x), \tag{2}$$

$$(k\mathbf{M})(x) = k\mathbf{M}(x). \tag{3}$$

This makes a *linear space* out of the set of linear maps of X into U. The space is denoted as $\mathcal{L}(X, U)$. Given two linear maps, one, $\mathbf{M}$ from $X \to U$, the other, $\mathbf{N}$ from $U \to W$, we can define their *product* as the *composite map*

$$(\mathbf{NM})(x) = \mathbf{N}(\mathbf{M}(x)). \tag{4}$$

Since compositon of maps in general is *associative*, so is in particular the composition of linear maps. As we will see, composition is far from being commutative.

From now on we omit the bracket and denote the action of a linear map on x as

$$\mathbf{M}(x) = \mathbf{M}x.$$

This notation suggests that the action of $\mathbf{M}$ on x is a kind of multiplication; indeed (1) and (2) give the distributive property of this kind of multiplication.

Exercise 1. Verify that the composite of two linear maps is linear, and that the distributive law holds:

$$M(N + K) = MN + MK,$$
$$(M + K)N = MN + KN.$$

Definition. A mapping is *invertible* if it maps X one-to-one and onto U.

If M is invertible, it has an inverse, denoted as M^{-1}, that satisfies

$$M^{-1}M = I, \quad MM^{-1} = I,$$

where I on the left is the *identity mapping* in X, on the right on U. If M is linear, so is M^{-1}.

Definition. The *nullspace* of M, denoted by N_M, is the set of points mapped into zero.

The *range* of M, denoted by R_M, is the image of X under M in U.

Theorem 1. *Let M be a linear map of $X \to U$.*

  (i) The nullspace N_M is a linear subspace of X, the range R_M a linear subspace of U.
 (ii) M is invertible iff $N_M = \{0\}$ and $R_M = U$.
(iii) M maps the quotient space X/N_M one-to-one onto R_M.
 (iv) If $M : X \to U$ and $K : U \to W$ are both invertible, so is their product, and

$$(KM)^{-1} = M^{-1}K^{-1}.$$

  (v) If KM is invertible, then

$$N_M = \{0\}, \quad R_K = W.$$

Exercise 2. Prove theorem 1.

We remark that when $x = U = W$ are finite dimensional, then the invertibility of the product NM implies that N and M separately are invertible. This is not so in the infinite-dimensional case; take, for instance, X to be the space of infinite sequences

$$x = (a_1, a_2, \ldots)$$

and define R and L to be right and left shift: $Rx = (0, a_1, a_2, \ldots)$, $Lx = (a_2, a_3, \ldots)$. Clearly, LR is the identity map, but neither R nor L are invertible; nor is RL the identity.

We formulate now a number of useful notions and results concerning mappings of a linear space *into itself*:

$$\mathbf{M} : X \longrightarrow X.$$

We denote by N_j the *nullspace* of the jth power of $\mathbf{M}$:

$$N_j = N_{\mathbf{M}^j}. \tag{5}$$

Theorem 2. *The subspaces N_j defined in (5) have these properties:*

$$N_j \subset N_{j+1} \quad \text{for all } j \tag{6}$$

and

$$\dim\left(\frac{N_j}{N_{j-1}}\right) \geq \dim\left(\frac{N_{j+1}}{N_j}\right) \quad \text{for all } j. \tag{7}$$

Proof. Equation (6) is an immediate consequence of (5). To show (7), we claim that $\mathbf{M}$ maps N_{j+1}/N_j into N_j/N_{j-1} in a one-to-one fashion. To see this, note that a nonzero element of N_{j+1}/N_j is represented by a point z in N_{j+1} that does not lie in N_j. Clearly, $\mathbf{M}z$ lies in N_j but not in N_{j-1}; this shows the one-to-oneness. It follows that N_{j+1}/N_j is isomorphic to a subspace of N_j/N_{j-1}, from which the statement (7) about dimension follows. When N_{j+1}/N_j is infinite-dimensional, so is N_j/N_{j-1}. □

The following is an immediate corollary of equation (7):

Theorem 2′. *Suppose that for some i the subspaces defined by (5) satisfy*

$$N_i = N_{i+1}; \tag{8}$$

then

$$N_i = N_k \quad \text{for all } k > i. \tag{8′}$$

Definition. A subspace Y of X is called an *invariant subspace* of a linear map $\mathbf{M}$: $X \to X$ if $\mathbf{M}$ maps Y into Y.

Theorem 3. *Suppose that Y is an invariant subspace of X for a mapping $\mathbf{M}$: $X \to X$. Then*

  (i) there is a natural interpretation of $\mathbf{M}$ as a mapping $X/Y \to X/Y$.
  (ii) if both maps

$$\mathbf{M} : Y \longrightarrow Y \text{ and } \mathbf{M} : X/Y \longrightarrow X/Y$$

are invertible, so is $\mathbf{M} : X \to X$.

Proof. We leave part (i) to the reader. In (ii) we show first that the null space of $\mathbf{M}$ on X is trivial. To see this, suppose that

$$\mathbf{M}z = 0;$$

then, since the nullspace of $\mathbf{M}$ on X/Y is assumed to be trivial, it follows that z belongs to Y. But since the nullspace of $\mathbf{M}$ on Y also is trivial, it follows that $z = 0$.

Next we show that $\mathbf{M} : X \to X$ is onto, meaning that

$$\mathbf{M}x_0 = u_0 \qquad (9)$$

has a solution x_0 for every u_0 in X. To this end we solve equation (9) in two stages. First we solve the congruence

$$\mathbf{M}x \equiv u_0 (\bmod Y),$$

which is possible since $\mathbf{M}$ maps X/Y onto itself. Let x_1 be an element of the solution class; then x_1 satisfies

$$\mathbf{M}x_1 = u_0 + z, \qquad z \text{ in } Y.$$

Therefore the solution x_0 of (9) is

$$x_0 = x_1 - y,$$

where y is the solution in Y of

$$\mathbf{M}y = z.$$

Such a solution exists since $\mathbf{M}$ is assumed to map Y onto Y. $\qquad \square$

We remark that whereas invertibility of $\mathbf{M}$ on Y and X/Y guarantees the invertibility of $\mathbf{M}$ on X, the converse by no means holds in spaces of infinite dimension. For example, let X be the space of all bounded continuous functions on $\mathbb{R}$, $\mathbf{S}$ the shift operator

$$(\mathbf{S}x)(t) = x(t - 1),$$

and Y the subspace of functions $x(t)$ that vanish on the negative axis. Clearly, Y is shift invariant, and equally clearly, $\mathbf{S}$ is invertible on X, its inverse being the left unit shift. But $\mathbf{S}$ is not invertible on either Y or X/Y; on Y its range consists of functions $x(t)$ that are zero for $t \leq 1$, and on X/Y it has a nontrivial nullspace.

Exercise 3. What is the nullspace of $\mathbf{S}$ on X/Y?

The construction of invariant subspaces will be taken up in chapter 25. Here we gather the following useful observations:

Theorem 4. *Let M be a linear map:* $X \to X$.

(i) *For any* y *in* X, *the set* $\{p(\mathbf{M})y\}$, *where* p *represents any polynomial, is an invariant subspace of* $\mathbf{M}$.

(ii) *Let* T *be a linear map:* $X \to X$ *that commutes with* $\mathbf{M} : \mathbf{TM} = \mathbf{MT}$. *Then the nullspace of* T *is an invariant subspace of* $\mathbf{M}$.

Proof. Part (i) rests on the observation that if $p(\mathbf{M})$ is a polynomial, so is $\mathbf{M}p(\mathbf{M})$. Part (ii) follows from the observation that if $\mathbf{M}$ and $\mathbf{T}$ commute, and if z is in the nullspace of $\mathbf{T} : \mathbf{T}z = 0$, then $\mathbf{TM}z = \mathbf{MT}z = \mathbf{M}0 = 0$. $\qquad\qquad$ $\square$

2.2 INDEX OF A LINEAR MAP

The next group of theorems describe an important special class of mappings.

Definition. A linear map $\mathbf{G}$ is called *degenerate* if its range is finite dimensional:

$$\dim R_{\mathbf{G}} < \infty. \tag{10}$$

Theorem 5. *The degenerate maps form an ideal in the following sense:*

(i) *The sum of two degenerate maps is degenerate.*

(ii) *The product of a degenerate map with any linear map, in either order, is degenerate; that is, if* $\mathbf{G}$ *is degenerate, so are* $\mathbf{MG}$ *and* $\mathbf{GN}$, *provided of course that the products can be defined.*

Exercise 4. Prove theorem 5.

Definition. The linear maps $\mathbf{M} : X \to U$ and $\mathbf{L} : U \to X$ are *pseudoinverse* to each other if

$$\mathbf{LM} = \mathbf{I} + \mathbf{G}, \quad \mathbf{ML} = \mathbf{I} + \mathbf{G}, \tag{11}$$

where $\mathbf{I}$ denotes the identity, $\mathbf{G}$ degenerate maps of $X \to X$, and $U \to U$, respectively.

Exercise 5. Prove that the right shift and the left shift described after theorem 1 are pseudoinverses of each other on the space of all sequences.

Theorem 6.

(i) *If* $\mathbf{L}$ *and* $\mathbf{M}$ *are pseudoinverses of each other, so are* $\mathbf{L} + \mathbf{G}_1$ *and* $\mathbf{M} + \mathbf{G}_2$, *where* $\mathbf{G}_1, \mathbf{G}_2$ *are arbitrary degenerate maps.*

(ii) *Suppose that* $\mathbf{M} : X \to U$ *and* $\mathbf{A}: U \to W$ *have pseudoinverses* $\mathbf{L}$ *and* $\mathbf{B}$, *respectively. Then* $\mathbf{AM}$ *and* $\mathbf{LB}$ *are pseudoinverse to each other.*

Exercise 6. Prove theorem 6.

We recall the definition of codimension of a subspace R of a linear space U:

$$\text{codim } R = \dim(U/R).$$

Theorem 7. *A linear map* $\mathbf{M} : X \to U$ *has a pseudoinverse if and only if*

$$\dim N_\mathbf{M} < \infty, \quad \text{codim } R_\mathbf{M} < \infty. \tag{12}$$

Proof. For the "only if" part we use a lemma:

Lemma 8. *If* $\mathbf{G}$ *is a degenerate map of* $X \to X$, *then*

$$\dim N_{\mathbf{I}+\mathbf{G}} < \infty, \quad \text{codim } R_{\mathbf{I}+\mathbf{G}} < \infty. \tag{13}$$

Proof. For x in $N_{\mathbf{I}+\mathbf{G}}$,

$$x + \mathbf{G}x = 0.$$

This shows that

$$N_{\mathbf{I}+\mathbf{G}} \subset R_\mathbf{G};$$

combined with (10) this shows the first part of (13).

According to theorem 1 (iii), $\mathbf{G}$ maps $X/N_\mathbf{G}$ one-to-one onto $R_\mathbf{G}$; so

$$\text{codim } N_\mathbf{G} = \dim R_\mathbf{G}. \tag{14}$$

Obviously $\mathbf{I} + \mathbf{G}$ maps every x in $N_\mathbf{G}$ into itself; this shows that $R_{\mathbf{I}+\mathbf{G}} \supset N_\mathbf{G}$. It follows from this relation that

$$\text{codim } R_{\mathbf{I}+\mathbf{G}} \leq \text{codim } N_\mathbf{G}. \tag{14'}$$

Combining (14) and (14'), we conclude that codim $R_{\mathbf{I}+\mathbf{G}} \leq \dim R_\mathbf{G}$; using (10), we deduce the second part of (13). $\qquad\square$

Suppose now that $\mathbf{M}$ has a pseudoinverse; then (11) holds. From the first relation in (11) we deduce that $N_\mathbf{M} \subset N_{\mathbf{I}+\mathbf{G}}$ and therefore $\dim N_\mathbf{M} \leq \dim N_{\mathbf{I}+\mathbf{G}}$; combining this with the first part of (13), we obtain the first part of (12). It follows from the second relation in (11) that $R_\mathbf{M} \supset R_{\mathbf{I}+\mathbf{G}}$. Therefore

$$\text{codim } R_\mathbf{M} \leq \text{codim } R_{\mathbf{I}+\mathbf{G}}.$$

Combining this with the second relation in (13), we deduce the second part of (12).

For the "if" part we need:

Lemma 9. *Every subspace N of a linear space has a complementary subspace Y, namely a linear subspace Y of X such that*

$$X = N \oplus Y,$$

meaning that every x in X can be decomposed uniquely as

$$x = n + y, \qquad n \in N, y \in Y. \tag{15}$$

Proof. Consider all subspaces Y of X whose intersection with N is $\{0\}$, partially ordered by inclusion. Every totally ordered collection of Y_j has as upper bound the union of the Y_j. Zorn's lemma shows that there is a maximal Y; this Y clearly has the property stated in the lemma. Now, if some x cannot be expressed of form (15), we could enlarge Y by adjoining x, contradicting the maximality of Y. $\qquad\square$

Note that the complementary subspace Y is in no way uniquely determined. Having determined a particular Y, we define the *projection* $\mathbf{P}$ onto N from the decomposition (15):

$$\mathbf{P}x = n.$$

Exercise 7. Prove that $\mathbf{P}$ is a linear map.

Exercise 8. Show that when N has finite codimension, $\dim Y = \operatorname{codim} N$.

We return now to the proof of the "if" part of theorem 7: it follows from (15) that every equivalence class of $X \mod N$ contains exactly one element belonging to Y, and that this correspondence is an isomorphism:

$$Y \leftrightarrow X/N.$$

Suppose that $\mathbf{M} : X \to U$ satisfies conditions (12); we choose complementary subspaces Y and V for the nullspace and range of $\mathbf{M}$:

$$X = N_{\mathbf{M}} \oplus Y, \quad U = R_{\mathbf{M}} \oplus V. \tag{16}$$

According to theorem 1 (iii), $\mathbf{M}$ maps $X/N_{\mathbf{M}}$ one-to-one onto $R_{\mathbf{M}}$. Since X/N_M is isomorphic with Y, we conclude that

$$\mathbf{M} : Y \longrightarrow R_{\mathbf{M}}$$

is invertible. Denote its inverse by $\mathbf{M}^{-1}$ and define the map $\mathbf{K}$ as follows:

$$\mathbf{K} = \mathbf{M}^{-1} \text{ on } R_{\mathbf{M}}, \quad \mathbf{K} = 0 \text{ on } V. \tag{17}$$

Using (16), we can extend $\mathbf{K}$ to all of U. Clearly,

$$\mathbf{KM} = \left\{ \begin{array}{ll} \mathbf{I} & \text{on } Y \\ \mathbf{0} & \text{on } N_{\mathbf{M}} \end{array} \right., \qquad \mathbf{MK} = \left\{ \begin{array}{ll} \mathbf{I} & \text{on } R_{\mathbf{M}} \\ \mathbf{0} & \text{on } V \end{array} \right. . \tag{17'}$$

We can rewrite (17') as follows:

$$\mathbf{KM} = \mathbf{I} - \mathbf{P}, \quad \mathbf{MK} = \mathbf{I} - \mathbf{Q},$$

where $\mathbf{P}$ is projection onto N, $\mathbf{Q}$ projection onto V. It follows from this that $\mathbf{K}$ and $\mathbf{M}$ are pseudoinverse to each other in the sense of (11). Since $\mathbf{P}$ and $\mathbf{Q}$ are degenerate, the proof of theorem 7 is complete. □

Definition. Let $\mathbf{M} : X \to U$ be a linear map with a pseudoinverse. We define the *index* of such an $\mathbf{M}$ as

$$\text{ind } \mathbf{M} = \dim N_{\mathbf{M}} - \text{codim } R_{\mathbf{M}}. \tag{18}$$

It follows from theorem 7 that this definition makes sense.

Theorem 10. $\mathbf{M} : X \to U$ *and* $\mathbf{L} : U \to W$ *are linear maps with pseudoinverse. Then the product* $\mathbf{LM}$ *has pseudoinverse, and*

$$\text{ind } (\mathbf{LM}) = \text{ind } \mathbf{L} + \text{ind } \mathbf{M}. \tag{19}$$

Proof. By theorem 6 (ii), $\mathbf{LM}$ has a pseudoinverse. To prove (19), we want to use as a counting device the notion of an exact sequence:

Definition. A sequence of linear spaces $V_0, V_1, \ldots, V_n$ and a sequence of linear maps $\mathbf{T}_j : V_j \to V_{j+1}$,

$$V_0 \xrightarrow{\mathbf{T}_0} V_1 \xrightarrow{\mathbf{T}_1} \ldots \xrightarrow{\mathbf{T}_{n-1}} V_n,$$

is called *exact* if the range of $\mathbf{T}_j$ is the nullspace of $\mathbf{T}_{j+1}$.

Lemma 11. *Suppose that all the V_j in the exact sequence above are finite dimensional and that*

$$\dim V_0 = 0 = \dim V_n. \tag{20}$$

Then

$$\sum_j (-1)^j \dim V_j = 0. \tag{20'}$$

Proof. Decompose V_j as

$$V_j = N_j \oplus Y_j,$$

where N_j is the nullspace of $\mathbf{T}_j$ and Y_j complementary to N_j. The condition of exactness requires that $\mathbf{T}_j$ be an isomorphism of Y_j with N_{j+1}. Since $\dim V_j = \dim N_j + \dim Y_j$, it follows that

$$\dim V_j = \dim N_j + \dim N_{j+1}, \qquad 0 \le j < n - 1. \tag{21}$$

By (20),

$$\dim N_0 = 0 \quad \text{and} \quad \dim V_{n-1} = \dim N_{n-1}. \tag{21'}$$

Setting (21) and (21′) in the left side of (20′) shows that the alternating sum is zero.
$\square$

To prove theorem 10, we construct the following exact sequence:

$$0 \to N_\mathbf{M} \xrightarrow{\mathbf{I}_0} N_\mathbf{LM} \xrightarrow{\mathbf{M}} N_\mathbf{L} \xrightarrow{\mathbf{Q}} U/R_\mathbf{M} \xrightarrow{\mathbf{L}} W/R_\mathbf{LM} \xrightarrow{\mathbf{E}} W/R_\mathbf{L} \to 0. \tag{22}$$

The mapping $\mathbf{I}_0$ identifies $N_\mathbf{M}$ as a subspace of $N_\mathbf{LM}$. $\mathbf{Q}$ is the natural map of points of U into the equivalence classes of $U \bmod R_\mathbf{M}$ containing them. $\mathbf{E}$ is the mapping of equivalence classes of $W \bmod R_\mathbf{LM}$ into equivalence classes $\bmod R_\mathbf{L}$.

Exercise 9. Verify that (22) is an exact sequence.

We apply relation (20′) to the exact sequence (22), with

$$V_0 = 0, \quad V_1 = N_\mathbf{M}, \quad V_2 = N_\mathbf{LM}, \quad V_3 = N_\mathbf{L},$$
$$V_4 = U/R_\mathbf{M}, \quad V_5 = W/R_\mathbf{LM}, \quad V_6 = W/R_\mathbf{L}, \quad V_7 = 0.$$

Using the definition of codimension, we can write (20′) as follows:

$$\dim N_\mathbf{M} - \dim N_\mathbf{LM} + \dim N_\mathbf{L} - \operatorname{codim} R_\mathbf{M} + \operatorname{codim} R_\mathbf{LM} - \operatorname{codim} R_\mathbf{L} = 0.$$

Using the definition (18) of the index, we deduce the *product formula* (19) for the index.
$\square$

The next result is called the *stability* of index:

Theorem 12. *Let* $\mathbf{M} : X \to U$ *be a linear map with a pseudoinverse, and* $\mathbf{G} : X \to U$ *a degenerate linear map. Then* $\mathbf{M} + \mathbf{G}$ *has a pseudoinverse, and*

$$\operatorname{ind}(\mathbf{M} + \mathbf{G}) = \operatorname{ind}\mathbf{M}. \tag{23}$$

Proof. We first verify (23) for $U = X$ and $\mathbf{M} = \mathbf{I}$. For this we need a lemma:

Lemma 13. *Let* X *be a linear space, and* $\mathbf{K} : X \to U$ *a linear map of* X *into* U *that has a pseudoinverse. Let* X_0 *be a linear subspace of* X *that has finite codimension.*

Then $\mathbf{K}_0 : X_0 \to U$, the restriction of $\mathbf{K}$ to X_0, has a pseudoinverse, and

$$\text{ind}\,\mathbf{K}_0 = \text{ind}\,\mathbf{K} - \text{codim}\,X_0. \tag{24}$$

Proof. Factor $\mathbf{K}_0$ as

$$\mathbf{K}_0 = \mathbf{K}\mathbf{I}_0, \tag{24'}$$

where $\mathbf{I}_0 : X_0 \to X$ is the identification map. Clearly $N_{\mathbf{I}_0} = \{0\}$, $R_{\mathbf{I}_0} = X_0$, so

$$\text{ind}\,\mathbf{I}_0 = -\text{codim}\,X_0. \tag{25}$$

Now we apply the product formula (19) to (24') and deduce (24). □

Let $\mathbf{G} : X \to X$ be a degenerate map; take $\mathbf{K} : X \to X$ to be

$$\mathbf{K} = \mathbf{I} + \mathbf{G}. \tag{26}$$

Clearly, $\mathbf{I}$ is a pseudoinverse to $\mathbf{K}$. Take X_0 to be the nullspace of $\mathbf{G}$:

$$X_0 = N_{\mathbf{G}}. \tag{27}$$

By (14), X_0 has finite codimension. Since $\mathbf{G}$ is zero on X_0, $\mathbf{K}_0$, the restriction of $\mathbf{K}$ to X_0, is the identification map $\mathbf{I}_0$. So by (25),

$$\text{ind}\,\mathbf{K}_0 = \text{ind}\,\mathbf{I}_0 = -\text{codim}\,X_0.$$

We apply now lemma 13 to $\mathbf{K}$. By (24),

$$\text{ind}\,\mathbf{K}_0 = \text{ind}\,\mathbf{K} - \text{codim}\,X_0.$$

We deduce from the last two relations that

$$\text{ind}\,\mathbf{K} = 0 \tag{28}$$

for every $\mathbf{K}$ of form (26). This proves (23) for $\mathbf{M} = \mathbf{I}$.

We take now $\mathbf{M}$ as any map with a pseudoinverse; denote by $\mathbf{L} : U \to X$ a pseudoinverse of $\mathbf{M}$. By definition,

$$\mathbf{L}\mathbf{M} = \mathbf{K} = \mathbf{I} + \mathbf{G}',$$

$\mathbf{G}'$ degenerate. So by (28),

$$\text{ind}\,(\mathbf{L}\mathbf{M}) = \text{ind}\,(\mathbf{I} + \mathbf{G}') = 0. \tag{29}$$

Using the product formula (19), we get from (29) that

$$\text{ind}\,\mathbf{L} = -\text{ind}\,\mathbf{M}. \tag{30}$$

As we saw in theorem 6 (i), for degenerate $\mathbf{G}$, $\mathbf{L}$ is also a pseudoinverse of $\mathbf{M} + \mathbf{G}$. Therefore, using (30), once more we deduce that

$$\operatorname{ind} \mathbf{L} = -\operatorname{ind}(\mathbf{M} + \mathbf{G}). \tag{30'}$$

Combining (30) and (30'), we get (23).

Notes

The first part of this chapter is standard fare. The nonstandard items are as follows:

(i) The notion of the index of linear maps that have a pseudoinverse, theorem 7.

(ii) The product formula for the index, theorem 10.

(iii) The invariance of the index under perturbation by degenerate maps, theorem 12.

Strange to say, these results of linear algebra were first discovered in the setting of bounded maps of normed linear spaces. That they hold without any topological assumptions has remained a folk theorem. The first statement and proof of the multiplicative property in print is due to Donald Sarason. The proof presented here, using exact sequences, is due to Sergiu Klainerman.

BIBLIOGRAPHY

Sarason, D. The multiplication theorem for Fredholm operators. *Am. Math. Monthly*, **94** (1987): 68–70.

3

THE HAHN-BANACH THEOREM

3.1 THE EXTENSION THEOREM

The result named in the title of this chapter is remarkable for its simplicity and for its far-reaching consequences. It deals with the extension of linear functionals.

Definition. A linear functional ℓ is a mapping of a linear space X over a field $\mathbf{F}$ into $\mathbf{F}$, that is linear:

$$\ell(x + y) = \ell(x) + \ell(y)$$

for all x, y in X and

$$\ell(kx) = k\ell(x)$$

for all k in $\mathbf{F}$.

In this section we will mainly deal with linear spaces over the field of reals, and real number valued linear functionals.

Theorem 1 (Hahn-Banach Theorem). *Let X be a linear space over the reals, and p a real-valued function defined on X, which has the following properties:*

(i) Positive homogeneity,

$$p(ax) = ap(x) \qquad \text{for all } a > 0 \tag{1}$$

for every x in X.

(ii) Subadditivity,

$$p(x + y) \leq p(x) + p(y) \tag{2}$$

for all x, y in X.

Y denotes a linear subspace of X on which a linear functional ℓ is defined that is dominated by p:

$$\ell(y) \leq p(y) \qquad \text{for all } y \text{ in } Y. \tag{3}$$

Assertion. *ℓ can be extended to all of X as a linear functional dominated by p:*

$$\ell(x) \leq p(x) \qquad \text{for all } x \text{ in } X. \tag{3'}$$

Proof. Suppose that Y is not all of X; then there is some z in X that is not in Y. Denote by Z the linear span of Y and z, meaning all points of the form

$$y + az, \qquad y \text{ in } Y; a \text{ in } \mathbb{R}.$$

Our aim is to extend ℓ as a linear functional to Z so that (3') is satisfied for x in Z. that is,

$$\ell(y + az) = \ell(y) + a\ell(z) \leq p(y + az)$$

holds for all y in Y and all real a. By (3), the inequality holds for $a = 0$. Since p is positive homogeneous, it suffices to verify it for $a = \pm 1$:

$$\ell(y) + \ell(z) \leq p(y + z), \quad \ell(y') - \ell(z) \leq p(y' - z).$$

Thus for all y, y' in Y,

$$\ell(y') - p(y' - z) \leq \ell(z) \leq p(y + z) - \ell(y) \tag{4}$$

must hold. Such an $\ell(z)$ exists iff for all pairs y, y',

$$\ell(y') - p(y' - z) \leq p(y + z) - \ell(y). \tag{5}$$

This is the same as

$$\ell(y') + \ell(y) = \ell(y' + y) \leq p(y + z) + p(y' - z). \tag{5'}$$

Since $y + y'$ lies in Y, (3) holds:

$$\ell(y' + y) \leq p(y + y'). \tag{6}$$

By subadditivity,

$$p(y + y') = p(y + z + y' - z) \leq p(y + z) + p(y' - z). \tag{7}$$

Combining (6) and (7) gives (5'), proving the possibility of extending ℓ from Y to Z. So (3') remains satisfied.

Consider all extensions of ℓ to linear spaces Z containing Y on which inequality (3') continues to hold. We order these extensions by defining

$$(Z, \ell) \leq (Z', \ell')$$

to mean that Z' contains Z, and that ℓ' agrees with ℓ on Z.

Let $\{Z_\nu, \ell_\nu\}$ be a totally ordered collection of extensions of ℓ. Then we can define ℓ on the union $Z = \cup Z_\nu$ as being ℓ_ν on Z_ν. Clearly, ℓ on Z satisfies (3′); equally clearly, $(Z_\nu, \ell_\nu) \leq (Z, \ell)$ for all ν. This shows that every totally ordered collection of extensions of ℓ has an upper bound. So the hypothesis of Zorn's lemma is satisfied, and we conclude that there exists a maximal extension. But according to the foregoing, a maximal extension must be to the whole space X. $\qquad\square$

3.2 GEOMETRIC HAHN-BANACH THEOREM

In spite (or perhaps because) of its nonconstructive proof, the HB theorem has plenty of very concrete applications. One of the most important is to separation theorems concerning convex sets; these are sometimes called *geometric Hahn-Banach theorems*.

Definition. X is a linear space over the reals, S a subset of X. A point x_0 is called an *interior point* of S if for any y in X there is an ϵ, depending on y, such that

$$x_0 + ty \in S \qquad \text{for all real } t, |t| < \epsilon.$$

Let K be a convex set that has an interior point, which we take to be the origin. We denote the *gauge* p_K of K with respect to the origin as follows:

$$p_K(x) = \inf a \qquad a > 0, \frac{x}{a} \in K. \tag{8}$$

Since the origin is assumed to be an interior point of K,

$$p_K(x) < \infty$$

for every x.

Theorem 2. *The gauge p_K of a convex set K in a linear space over the reals is positive homogeneous and subadditive.*

Proof. Positive homogeneity follows from the definition (8), even when K is not convex. To prove subadditivity, let x and y be any pair of points in X, a and b positive numbers such that

$$\frac{x}{a} \in K, \qquad \frac{y}{b} \in K. \tag{9}$$

Convexity, as defined in chapter 1, means that any convex combination of points of K belongs to K. We take the convex combination of x/a and y/b with weights $a/(a+b)$ and $b/(a + b)$. These are nonnegative numbers whose sum is 1. We conclude that

$$\frac{a}{a+b}\frac{x}{a} + \frac{b}{a+b}\frac{y}{b} = \frac{x+y}{a+b} \in K.$$

Since $(x+y)/(a+b)$ is in K, by definition (8), $p_K(x+y) \le a+b$. Since this holds for all a and b satisfying (9),

$$p_K(x+y) \le \inf(a+b) = \inf a + \inf b = p_K(x) + p_K(y),$$

where in the last step we have again used (8). This proves subadditivity of p_K. □

Theorem 3. *For any convex set K,*

$$p_K(x) \le 1 \qquad \textit{if } x \in K, \tag{10}$$

$$p_K(x) < 1 \textit{ iff } x \textit{ is an interior point of } K. \tag{10'}$$

Proof. (10) is an immediate consequence of definition (8) of p_K. □

Exercise 1. Prove (10′).

The converse of theorem 3 also is true:

Theorem 4. *Let p denote a positive homogeneous, subadditive function defined on a linear space X over the reals.*

  (i) The set of points x satisfying

$$p(x) < 1$$

  is a convex subset of X, and 0 is an interior point of it.
  (ii) The set of points x satisfying

$$p(x) \le 1$$

  is a convex subset of X.

Exercise 2. Prove theorem 4.

We turn now to the notion of a *hyperplane*. Suppose that ℓ is a linear functional not $\equiv 0$; for any real c, all points of X belong to one, and only one, of the following three sets:

$$\ell(x) < c, \quad \ell(x) = c, \quad \ell(x) > c.$$

The set of x that satisfies

$$\ell(x) = c$$

is called a *hyperplane*; the sets where $\ell(x) < c$, respectively $\ell(x) > c$ are called open *halfspaces*. The sets where

$$\ell(x) \geq c, \quad \text{or} \quad \ell(x) \leq c,$$

are called *closed halfspaces*.

Theorem 5 (Hyperplane Separation Theorem). *Let K be a nonempty convex subset of a linear space X over the reals; suppose that all points of K are interior. Any point y not in K can be separated from K by a hyperplane $\ell(x) = c$; that is, there is a linear functional ℓ, depending on y, such that*

$$\ell(x) < c \quad \text{for all } x \text{ in } K; \quad \ell(y) = c. \tag{11}$$

Proof. Assume that $0 \in K$, and denote by p_K the gauge of K. Since all points of K are interior, it follows from theorem 3 that $p_K(x) < 1$ for every x in K. We set

$$\ell(y) = 1. \tag{12}$$

Then ℓ is defined for all z of the form ay,

$$\ell(ay) = a. \tag{12'}$$

We claim that for all such z,

$$\ell(z) \leq p_K(z).$$

This is obvious for $a \leq 0$, for then $\ell(z) \leq 0$ while $p_K \geq 0$. Since y is not in K, by (8), $p_K(y) \geq 1$. So, by positive homogeneity, $p_K(ay) \geq a$ for $a > 0$.

Having shown that ℓ, as defined on the above one-dimensional subspace, is dominated by p_K, we conclude from the HB theorem that ℓ can be so extended to all of X. We deduce from this and ($10'$) that for any x in K,

$$\ell(x) \leq p_K(x) < 1$$

This gives the first part of (11), with $c = 1$; the second part is (12). $\qquad\square$

Corollary 5'. *Let K denote a convex set with at least one interior point. For any y not in K there is a nonzero linear functional ℓ that satisfies*

$$\ell(x) \leq \ell(y) \quad \text{for all } x \text{ in } K. \tag{13}$$

Theorem 6 (Extended Hyperplane Separation). *X is a linear space over $\mathbb{R}$, H, and M disjoint convex subsets of X, at least one of which has an interior point. Then H and M can be separated by a hyperplane $\ell(x) = c$; that is, there is a nonzero linear functional ℓ, and a number c, such that*

$$\ell(u) \leq c \leq \ell(v) \tag{14}$$

for all u in H, all v in M.

Proof. According to theorem 5 of chapter 1, the difference set $H - M = K$ is convex; since either H or M contains an interior point, so does K.

Since H and M are disjoint, $0 \notin K$; according to (13) of corollary $5'$ applied to $y = 0$, there is a linear functional ℓ such that

$$\ell(x) \leq \ell(0) = 0 \text{ for all } x \text{ in } K. \tag{15}$$

Since all x in $K = H - M$ is of the form $x = u - v$, u in H, v in M, (15) means that

$$\ell(u) \leq \ell(v);$$

(14) follows from this, with $c = \sup_{u \in H} \ell(u)$. $\qquad\qquad\square$

3.3 EXTENSIONS OF THE HAHN-BANACH THEOREM

The following extension of the H-B theorem, due to R. P. Agnew and A. P. Morse, is both useful and beautiful:

Theorem 7. *Let X denote a linear space over the reals and $\mathcal{A}$ be a collection of linear maps $\mathbf{A}_v : X \to X$ that commute; that is,*

$$\mathbf{A}_v \mathbf{A}_\mu = \mathbf{A}_\mu \mathbf{A}_v \tag{16}$$

for all pairs in the collection. Let p denote a real-valued, positive homogeneous, subadditive function on X—see (1) and (2)—that is invariant under each $\mathbf{A}_v$:

$$p(\mathbf{A}_v x) = p(x). \tag{17}$$

Let Y denote a linear subspace of X on which a linear functional ℓ is defined, with the following properties:

(i) ℓ is dominated by p, namely

$$\ell(y) \leq p(y) \tag{18}$$

for every y in Y.

(ii) Y is invariant under each mapping $\mathbf{A}$, namely

$$\text{for } y \text{ in } Y, \quad \mathbf{A}y \text{ in } Y. \tag{19}$$

(iii) ℓ is invariant under each mapping $\mathbf{A}$, namely

$$\ell(\mathbf{A}y) = \ell(y) \quad \text{for } y \text{ in } Y. \tag{19'}$$

Assertion. *ℓ can be extended to all of X so that ℓ is dominated by p in the sense of (18), and is invariant under each mapping $\mathbf{A}_v$.*

Proof. If (17) holds for two mappings **A** and **B** of the collection $\mathcal{A}$, it also holds for their product **AB**, defined as their composite. Similarly, if (19) and (19′) hold for **A** and **B**, they hold for the product **AB**. Likewise, if **A** and **B** commute with all $\mathbf{A}_\nu$, so does their product. Thus we may adjoin to the collection $\mathcal{A}$ any finite products and the identity **I**. This enlarged collection will now form a *semigroup*. Then, if **A** and **B** belong to it, so does their product **AB**. From now on we assume that the collection $\mathcal{A}$ is a semigroup under multiplication.

We define a new function g on X as follows:

$$g(x) = \inf p(\mathbf{C}x), \tag{20}$$

with **C** a convex combination of mappings in $\mathcal{A}$, namely maps of the form

$$\mathbf{C} = \sum a_j \mathbf{A}_j, \qquad a_j \geq 0, \sum a_j = 1, \mathbf{A}_j \text{ in } \mathcal{A}.$$

Since $\mathcal{A}$ is a semigroup, the product of two convex combinations of mappings in $\mathcal{A}$ is also a convex combination.

Using subadditivity, homogeneity, and invariance (17), we deduce that

$$p(\mathbf{C}x) = p\left(\sum a_j \mathbf{A}_j x\right) \leq \sum a_j p(\mathbf{A}_j x) = p(x). \tag{21}$$

Since in (20) we may take **C** to be the identity, it follows that

$$g(x) \leq p(x). \tag{21′}$$

Since p is positive homogeneous, it follows from (20) that so is g. We show next that g is subadditive.

Let x and y be arbitrary elements of X. By definition (20), for any $\epsilon > 0$ there are maps **C** and **D** in the convex hull of $\mathcal{A}$ such that

$$p(\mathbf{C}x) \leq g(x) + \epsilon, \qquad p(\mathbf{D}y) \leq g(y) + \epsilon. \tag{22}$$

Applying (20) to the map **CD**, we get, since **C** and **D** commute, that

$$g(x + y) \leq p(\mathbf{CD}(x + y)) = p(\mathbf{DC}x + \mathbf{CD}y). \tag{23}$$

Using subadditivity, and (21), the right side of (23) is seen to be less than

$$p(\mathbf{DC}x) + p(\mathbf{CD}y) \leq p(\mathbf{C}x) + p(\mathbf{D}y). \tag{24}$$

Using (22) to estimate (24), we conclude that

$$g(x + y) \leq g(x) + g(y) + 2\epsilon;$$

since ϵ is arbitrary, subadditivity of g follows.

Since, by (19′), ℓ on Y is invariant under each **A**, for any convex combination **C** of mappings in $\mathcal{A}$ and for any y in Y,

$$\ell(\mathbf{C}y) = \ell\left(\sum a_j \mathbf{A}_j y\right) = \sum a_j \ell(\mathbf{A}_j y) = \sum a_j \ell(y) = \ell(y).$$

It follows from (19) that if y belongs to Y, so does $\mathbf{C}y$. Applying (18) to $\mathbf{C}y$, we get that for y in Y,

$$\ell(\mathbf{C}y) \le p(\mathbf{C}y).$$

Since we have shown that $\ell(\mathbf{C}y) = \ell(y)$,

$$\ell(y) \le p(\mathbf{C}y);$$

by definition (20) of g, it follows from this that for all y in Y,

$$\ell(y) \le g(y). \tag{25}$$

We apply now the Hahn-Banach theorem to conclude that ℓ can be extended to all of X so that (25) holds. We claim that ℓ thus extended is invariant under all mappings $\mathbf{A}$ in $\mathcal{A}$ in the sense of (19). For any $\mathbf{A}$ in $\mathcal{A}$ and any natural number n, we define $\mathbf{C}_n$ by $\mathbf{C}_n = \frac{1}{n}\sum_0^{n-1} \mathbf{A}^j$. Since $\mathcal{A}$ is a semigroup, $\mathbf{C}_n$ belongs to the convex hull of $\mathcal{A}$. According to the basic formula for geometric series, $\mathbf{C}_n(\mathbf{I} - \mathbf{A}) = \frac{1}{n}(\mathbf{I} - \mathbf{A}^n)$.

Let x be any point in X; by definition (20) of g,

$$g(x - \mathbf{A}x) \le p(\mathbf{C}_n(x - \mathbf{A}x)) = p(\mathbf{C}_n(\mathbf{I} - \mathbf{A})x) = \frac{1}{n}p(x - \mathbf{A}^n x). \tag{26}$$

In the last step we used the formula for geometric series, and the positive homogeneity of p. Using subadditivity and (17), we deduce that

$$\frac{1}{n}p(x - \mathbf{A}^n x) \le \frac{1}{n}[p(x) + p(-\mathbf{A}^n x)] = \frac{1}{n}[p(x) + p(-x)].$$

Combining this with (26), we get

$$g(x - \mathbf{A}x) \le \frac{1}{n}[p(x) + p(-x)]. \tag{26'}$$

Now we let $n \to \infty$; since the right side of (26') tends to 0,

$$g(x - \mathbf{A}x) \le 0. \tag{27}$$

Since g dominates ℓ, we deduce from (27) that

$$\ell(x - \mathbf{A}x) \le 0.$$

Since ℓ is linear, this implies that for all x,

$$\ell(x) \le \ell(\mathbf{A}x). \tag{27'}$$

Replacing x by $-x$, we get

$$\ell(-x) \le \ell(-\mathbf{A}x),$$

which is the opposite of inequality (27′). So equality must hold, meaning that ℓ is invariant under each $\mathbf{A}$.

By construction, ℓ is dominated by g. It follows then from (21′) that it is dominated by p. □

Exercise 3. Show that theorem 7 remains true if condition (17) is replaced by $p(\mathbf{A}x) \leq p(x)$.

We conclude by a version of HB for complex linear space due to Bohnenblust and Sobczyk, and Soukhomlinoff:

Theorem 8. *Let X be a linear space over $\mathbb{C}$, and p a real valued function that satisfies*

(i)

$$p(ax) = |a|p(x) \tag{28}$$

for all complex a, all x in X;

(ii) subadditivity,

$$p(x + y) \leq p(x) + p(y).$$

Let Y be a linear subspace of X over $\mathbb{C}$, and let ℓ be a linear functional on Y that satisfies

$$|\ell(y)| \leq p(y) \quad \text{for } y \text{ in } Y. \tag{29}$$

Assertion. *ℓ can be extended to all of X so that (29) holds over X.*

Proof. Split ℓ into its real and imaginary part:

$$\ell(y) = \ell_1(y) + i\ell_2(y). \tag{30}$$

Clearly, ℓ_1 and ℓ_2 are linear over $\mathbb{R}$, and are related by

$$\ell_1(iy) = -\ell_2(y). \tag{31}$$

Conversely, if ℓ_1 is a linear functional over $\mathbb{R}$,

$$\ell(x) = \ell_1(x) - i\ell_1(ix) \tag{31′}$$

is linear over $\mathbb{C}$.

We turn now to the task of extending ℓ. It follows from (29) and (30) that

$$\ell_1(y) \leq p(y). \tag{32}$$

Therefore by the real H-B theorem, ℓ_1 can be extended to all of X so that (32) holds. We define ℓ on X by (31). Clearly, ℓ is linear over $\mathbb{C}$ and we claim that (29) holds. To see this, write

$$\ell(x) = \alpha r, \quad r \text{ real}, \quad |\alpha| = 1.$$

Then

$$|\ell(x)| = r = \alpha^{-1}\ell(x) = \ell(\alpha^{-1}x) = \ell_1(\alpha^{-1}x) \leq p(\alpha^{-1}x) = p(x).$$

This completes the proof of the complex H-B theorem.   □

A historical review and a modern update is given by Gerard Buskes in his survey article.

BIBLIOGRAPHY

Agnew, R. P. and Morse, A. P. Extension of linear functionals, with application to limits, integrals, measures, and densities, *An. Math.*, **39** (1938): 20–30.

Banach, S. Sur les fonctionelles linéaires. *Studia Math.*, **1** (1929): 211–216, 223–229.

Bohnenblust, H. F. and Sobczyk, A. Extension of functionals on complex linear spaces. *Bull. AMS*, **44** (1938): 91–93.

Buskes, G. The Hahn-Banach Theorem Surveyed. Dissertationes Mathematicae, **327**. 1993.

Hahn, H. Über lineare Gleichungssysteme in linearen Räumen. *J. Reine Angew. Math.*, **157** (1927): 214–229.

Soukhomlinoff, G. A. Über Fortsetzung von linearen Funktionalen in linearen komplexen Räumen und linearen Quaternion-räumen. *Sbornik, N.S.*, **3** (1938): 353–358.

4

APPLICATIONS OF THE
HAHN-BANACH THEOREM

4.1 EXTENSION OF POSITIVE LINEAR FUNCTIONALS

S denotes any abstract set, and $B = B(S)$ the collection of all real-valued functions x on S that are *bounded*, that is, satisfy

$$|x(s)| \leq c. \tag{1}$$

B is a linear space over the reals.

There is a natural *partial order* for the elements of B : $x \leq y$ means that $x(s) \leq y(s)$ for all s in S. A function x satisfying $0 \leq x$ is called *nonnegative*.

Let Y be a linear subspace of B that contains some nonnegative functions. A *linear functional* ℓ defined on Y is called *positive* on Y if $\ell(y) \geq 0$ for all nonnegative y in Y. Every positive linear functional ℓ is *monotone*:

$$y_1 \leq y_2 \quad \text{implies} \quad \ell(y_1) \leq \ell(y_2). \tag{2}$$

Theorem 1. *Let Y be a linear subspace of B that contains a function y_0 greater than some positive constant, say 1:*

$$1 \leq y_0(s) \quad \text{for all } s \text{ in } S. \tag{3}$$

Let ℓ be a positive linear functional defined on Y.

Assertion. *ℓ can be extended to all of B as a positive linear functional.*

Proof. We define the function p on B as follows: for any x in B,

$$p(x) = \inf \ell(y), \quad x \leq y; \ y \text{ in } Y. \tag{4}$$

This function p is well defined; for it follows from (1) and (3) that

$$-cy_0 \leq x \leq cy_0. \tag{5}$$

which shows that the inf in (4) is over a nonempty set, and that $p(x) \leq c\ell(y_0)$ where c is any constant satisfying (1). The smallest such constant is $c = \sup_{s \text{ in } S} |x(s)|$. It follows from (5) that any $y \geq x$ satisfies $-cy_0 \leq x \leq y$. Since ℓ is linear and positive, for such y it follows from (2) that $-c\ell(y_0) \leq \ell(y)$, and so by (4)

$$-c\ell(y_0) \leq p(x). \tag{6}$$

Lemma 2. *The function p defined by (4) is*

  (i) *positive homogeneous.*
 (ii) *subadditive.*
(iii) *negative:* $p(x) \leq 0$ *for* $x \leq 0$.
(iv) $p(x) = \ell(x)$ *for x in Y.*

Proof.

  (i) It follows from the definition that $x \leq y$ implies $ax \leq ay, a > 0$. Positive homogeneity follows from definition (4).
 (ii) Let x_1 and x_2 be any two functions in B, y_1 and y_2 any two functions in Y satisfying

$$x_1 \leq y_1, \quad x_2 \leq y_2.$$

Adding the two we obtain $x_1 + x_2 \leq y_1 + y_2$; so by definition (4) of p,

$$\begin{aligned} p(x_1 + x_2) &= \inf_{x_1 + x_2 \leq y} \ell(y) \leq \inf_{\substack{x_1 \leq y_1 \\ x_2 \leq y_2}} \ell(y_1 + y_2) \\ &= \inf_{x_1 \leq y_1} \ell(y_1) + \inf_{x_2 \leq y_2} \ell(y_2) = p(x_1) + p(x_2) \end{aligned} \tag{7}$$

This proves subadditivity.
(iii) Suppose that $x \leq 0$; then $y = 0$ is admissible in the inf on the right in (4), giving $p(x) \leq \ell(0) = 0$, as asserted in (iii).
(iv) Suppose that x belongs to Y; then by (2), $x \leq y$ implies $\ell(x) \leq \ell(y)$, equality holding for $y = x$. Setting this into (4) gives $p(x) = \ell(x)$, as asserted in (iv).

$\square$

It follows from lemma 2 that we can apply the Hahn-Banach theorem to extend ℓ from Y to all of B so that ℓ remains dominated by p:

$$\ell(x) \leq p(x). \tag{8}$$

Suppose that x is nonpositive. Then by (iii), $p(x) \leq 0$, so by (8),

$$\ell(x) \leq 0 \qquad \text{for } x \leq 0. \tag{9}$$

This shows that ℓ is positive, as asserted in theorem 1.

$\square$

Theorem 1 is a special case of a very general theorem of Mark Krein; see p. 20 of Kelley and Namioka.

4.2 BANACH LIMITS

B denotes the space of *bounded infinite sequences* x of real numbers,

$$x = (a_1, a_2, \ldots). \tag{10}$$

B is a linear space over the reals when vector addition and multiplication by a scalar are defined componentwise. We define the function p on B as follows:

$$p(x) = \limsup_{n \to \infty} a_n, \tag{11}$$

where x is given by (10). It follows from this definition that p is a *positive homogeneous* function of x; we leave it as an textitexercise to the reader to prove that p is *subadditive*.

Define $\mathbf{A}$ as *left translation*, that is,

$$\mathbf{A}x = (a_2, a_3, \ldots). \tag{12}$$

It is an immediate consequence of definition (11) that p is *translation invariant*, namely that

$$p(\mathbf{A}x) = p(x). \tag{13}$$

We define Y as the space of *convergent* sequences of real numbers. Clearly, Y is a linear subspace of B. On Y, we define the linear functional ℓ by

$$\ell(y) = \lim_{n \to \infty} b_n, \tag{14}$$

where

$$y = (b_1, b_2, \ldots). \tag{14'}$$

Clearly, ℓ is linear. Comparing definitions (11) and (14), we conclude that

$$\ell(y) = p(y) \qquad \text{for } y \text{ in } Y. \tag{15}$$

Clearly, Y is mapped into itself by translation; equally clearly, ℓ is invariant on Y under translation:

$$\ell(\mathbf{A}y) = \ell(y) \qquad \text{for } y \text{ in } Y. \tag{16}$$

We apply now theorem 7 in chapter 3 to conclude that ℓ can be extended to all bounded sequences x in B so that

   (i) ℓ is linear

  (ii) ℓ is invariant under translation

 (iii) ℓ is dominated by p.

Theorem 3. *To each bounded sequence (10) we can assign a generalized limit (or Banach limit), denoted as*

$$\mathop{\mathrm{LIM}}_{n\to\infty} a_n$$

so that

  (i) For convergent sequences the generalized limit agrees with the usual limit.

  (ii)

$$\mathop{\mathrm{LIM}}_{n\to\infty} (a_n + b_n) = \mathop{\mathrm{LIM}}_{n\to\infty} a_n + \mathop{\mathrm{LIM}}_{n\to\infty} b_n.$$

(iii) For any k

$$\mathop{\mathrm{LIM}}_{n\to\infty} a_{n+k} = \mathop{\mathrm{LIM}}_{n\to\infty} a_n.$$

(iv)

$$\liminf_{n\to\infty} a_n \leq \mathop{\mathrm{LIM}}_{n\to\infty} a_n \leq \limsup_{n\to\infty} a_n.$$

Proof. We set, in the notation of (10),

$$\mathop{\mathrm{LIM}}_{n\to\infty} a_n = \ell(x).$$

Part (i) follows from (14), (14′); part (ii) expresses the linearity of ℓ; part (iii) is the translation invariance of ℓ. Part (iv) expresses the domination of ℓ by p, as defined by (11), and applied to $\ell(x)$ and $\ell(-x)$:

$$-p(-x) \leq \ell(x) \leq p(x). \qquad \square$$

Exercise 1. Show that if in section 4.1 we take $S = \{$positive integers$\}$, Y the space of convergent sequences, ℓ defined by (14), the function p given by (4) is the same as defined by (11).

Exercise 2. Show that a Banach limit can be so chosen that for any bounded sequence $(c_1, c_2, \ldots)$ that is Cesaro summable; namely the arithmetic means of the partial sums converge to c,

$$\mathop{\mathrm{LIM}}_{n\to\infty} c_n = c.$$

Exercise 3. Show that a generalized limit as $t \to \infty$ can be assigned to all bounded functions $x(t)$ defined on $t \geq 0$ that has properties (i) to (iv) in theorem 3.

4.3 FINITELY ADDITIVE INVARIANT SET FUNCTIONS

The Lebesgue measure on the unit circle is invariant under rotation. This measure can be extended to a considerably larger σ-algebra than the Lebesgue measurable sets on the unit circle so that rotational invariance is retained. However it is well known, and easy to show, that if we accept the *axiom of choice*, then there is *no* rotationally invariant *countably* additive measure defined for *all* subsets of the circle. We show now

Theorem 4. *One can define a nonnegative finitely additive set function $m(P)$, for all subsets P of the circle, that is invariant under rotation.*

Proof. We take S to be the unit circle, and B the set of all bounded real-valued functions on S. We take Y to be the space of bounded, Lebesgue measurable functions on S, and take $\ell(y)$ to be the Lebesgue integral of y:

$$\ell(y) = \int_S y(\theta)\,d\theta. \tag{17}$$

The space Y contains the function $y_0 \equiv 1$, so condition (3) of theorem 1 of section 4.1 is fulfilled. Therefore the function p described there by equation (4) is well defined.

We denote by $\{A_\rho\}$ the action on function of rotations ρ of the circle. As remarked above, ℓ is invariant under rotation:

$$(A_\rho y)(\theta) = y(\theta + \rho), \quad \ell(A_\rho y) = \ell(y). \tag{18}$$

Since the relation $x \leq y$ also is invariant under rotation, it follows that p as defined by (4) is rotation invariant:

$$p(A_\rho x) = p(x). \tag{18'}$$

Rotations of the circle commute, and so the linear maps $\{A_\rho\}$ form a commuting group of maps. We apply now theorem 7 of chapter 3 to conclude that ℓ can be extended to all of B so that ℓ is

(i) linear.
(ii) invariant under rotation.
(iii) dominated by p.

Let P be any set of points of the circle S; denote by c_P its characteristic function:

$$c_P(\theta) = \begin{cases} 1 & \text{if} \quad \theta \text{ is in } P \\ 0 & \text{otherwise.} \end{cases} \tag{19}$$

We define the set function m by setting,

$$m(P) = \ell(c_P). \tag{19'}$$

As shown in theorem 1, it follows from $\ell(x) \leq p(x)$ that ℓ is positive. Since c_P is a nonnegative function, it follows from definition $(19')$ of m that m is nonnegative:

$$m(P) \geq 0.$$

Let ρ be any rotation; denote the set P rotated by ρ as $P + \rho$. It follows from the definition (19) of c_P that

$$c_{P+\rho} = \mathbf{A}_\rho c_P. \tag{20}$$

Since ℓ is rotation invariant, it follows from the definition $(19')$ of m that

$$m(P + \rho) = m(P),$$

meaning that m is rotationally invariant.

Let P_1 and P_2 be disjoint subsets. Then, by definition (19),

$$c_{P_1 \cup P_2} = c_{P_1} + c_{P_2}.$$

Setting this into the definition (19) of m, and using the linearity of ℓ, we deduce that

$$m(P_1 \cup P_2) = m(P_1) + m(P_2).$$

This proves that m is finitely additive. $\square$

NOTE. Rotations of the circle commute with each other, and so the operators $\mathbf{A}_\rho$ commute; this was needed in invoking theorem 7 of chapter 3. Rotations of the three-dimensional sphere do *not* commute, and neither do the corresponding operators $\mathbf{A}_\rho$. Therefore the above proof cannot be used to extend theorem 4 to three dimensions. In fact Hausdorff has shown that the three-dimensional analogue of theorem 4 is *false*; there is *no* rotational invariant, finitely additive set function on the 2-sphere. The proof is based on a finite decomposition of the 2-sphere, sometimes called the Banach-Tarski paradox.

In conclusion, we point out that the duality theory of Banach spaces constitues the richest applications of the Hahn-Banach theorem. These are described in chapters 8 and 9.

HISTORICAL NOTE. His name is etched into the foundations of modern analysis: Hausdorff space, Hausdorff maximality principle, and Hausdorff measures are household concepts. He was a German mathematician, born in 1868; as a young man he published several volumes of poetry and aphorisms. He spent most of his professional life as professor in Bonn. Because he was Jewish, in 1942 he was ordered deported, part of the "Final Solution" to kill all the Jews in Europe. Knowing what awaited them, Hausdorff, his wife, and sister-in-law committed suicide.

BIBLIOGRAPHY

Hausdorff, F. *Grundzüge der Mengenlehre*. Verlag von Veit, Leipzig, 1914. Reprinted by Chelsea Publishing, New York.

Kelley, J. L. and Namioka, I. *Linear Topological Spaces*. Van Nostrand, Princeton, NJ, 1963.

5

NORMED LINEAR SPACES

5.1 NORMS

Let X denote a linear space over $\mathbb{R}$ or $\mathbb{C}$. A *norm* in X is a real-valued function: $X \to \mathbb{R}$, denoted as $|x|$, with the following properties:

(i) *Positivity*,

$$|x| > 0 \qquad \text{for } x \neq 0; |0| = 0. \tag{1}$$

(ii) *Subadditivity*,

$$|x + y| \leq |x| + |y|. \tag{2}$$

(iii) *Homogeneity*. For all scalars a,

$$|ax| = |a|\,|x|. \tag{3}$$

With the aid of a norm we can introduce a *metric* in X, by defining the distance of two points to be

$$d(x, y) = |x - y|. \tag{4}$$

It is easy to verify that this has all properties of a metric. Conversely, it is easy to show that every metric in a linear space that is translation invariant and homogeneous:

$$d(x + z, y + z) = d(x, y), \quad d(ax, ay) = |a|\,d(x, y) \tag{4'}$$

comes from a norm via (4).

With a metric (4) we can employ topological notions such as *convergent series, open sets, closed sets,* and *compact sets*. Those notions turn out to be crucial.

36

Definition. Two *different* norms, $|x|_1$ and $|x|_2$, defined on the same space X are called *equivalent* if there is a constant c such that

$$c|x|_1 \leq |x|_2 \leq c^{-1}|x|_1 \tag{5}$$

for all x in X.

The significance of this notion is that equivalent norms induce the same topology.

In chapter 1 we looked at various ways of building new linear spaces; the same constructions can be used to build new *normed* linear spaces. Specifically we observed the following:

(i) A *subspace* Y of a normed linear space X is again a normed linear space.

(ii) Given two linear spaces Z and U, their Cartesian product, denoted as a direct sum $Z \oplus U$, consists of all ordered pairs (z, u), $z \in Z$, $u \in U$. When Z and U are normed, $Z \oplus U$ can be normed, such as by setting

$$|(z, u)| = |z|+|u|, \quad |(z, u)|' = \max\{|z|, |u|\}, \quad \text{or} \quad |(z, u)|'' = (|z|^2+|u|^2)^{1/2}. \tag{6}$$

Exercise 1.

(a) Show that (6) are norms.

(b) Show that they are equivalent norms in the sense of (5).

Let X be a normed linear space, Y a subspace. We saw in chapter 1 that we can define their quotient X/Y as a linear space. We raise now the question: is there a natural way to introduce a *norm* in the quotient space? The answer is yes, provided that Y is *closed*:

Theorem 1. *Let Y be a closed subspace of a normed linear space X. Let $\{x_j\}$ be an equivalence class of elements of X mod Y. We define*

$$\|\{x_j\}\| = \inf_{x_j \in \{\}} |x_j|. \tag{7}$$

Assertion. *(7) has all properties of a norm in the quotient space X/Y.*

Proof. Property (3), homogeneity, holds trivially. To verify subadditivity, let $\{x_j\}$ and $\{z_j\}$ denote two equivalence classes. For any $\epsilon > 0$ we can, by definition (7), choose representatives so that

$$|x_j| < \|\{x_j\}\| + \epsilon, \quad |z_j| < \|\{z_j\}\| + \epsilon. \tag{8}$$

By definition of addition in X/Y, $x_j + z_j$ belongs to $\{x_j\} + \{z_j\}$; therefore, by definition (7),

$$\|\{x_j\} + \{z_j\}\| \leq |x_j + z_j|,$$

which by subadditivity in X, and (8), is

$$\leq |x_j| + |z_j| < |\{x_j\}| + |\{z_j\}| + 2\epsilon.$$

Since this is true for all $\epsilon > 0$, subadditivity of the norm (7) follows.

Clearly, (7) is nonnegative. To show positivity, suppose that $|\{x_j\}| = 0$. By definition (7), there is a sequence of elements x_n in $\{x_j\}$ such that

$$\lim_{n \to \infty} |x_n| = 0. \tag{9}$$

By definition of equivalence, the equivalent elements x_n differ from each other by elements that belong to Y. In particular, we can write

$$x_n = x_1 - y_n, \qquad n = 2, 3, \ldots, \qquad y_n \text{ in } Y.$$

Setting this into (9), we see that

$$\lim_{n \to \infty} |x_1 - y_n| = 0,$$

which by (4) means in the language of metric spaces that

$$\lim_{n \to \infty} y_n = x_1. \tag{9'}$$

In a metric space, the limit of a sequence of elements in a subset Y belongs to the *closure* of Y. Now, since Y is assumed to be closed, (9') implies that x_1 belongs to Y. But then the whole equivalence class $\{x_j\}$ consists of elements of Y, which is the zero element in X/Y. $\square$

Theorem 2. *Let X be a normed linear space, Y a subspace of X. The closure of Y is a linear subspace of X.*

Exercise 2. Prove theorem 2.

For purposes of analysis, in the construction of objects with desirable properties through limiting processes, we need metric spaces that are *complete* in the sense that every Cauchy sequence has a limit. So it is with normed linear spaces:

Definition. A *Banach space* is a normed linear space that is *complete*.

We recall the process of *completion* of a metric space whereby any metric space S is embedded in a *complete* metric space denoted as $\overline{S}$, consisting of equivalence classes of Cauchy sequences. S is a *dense* subset of $\overline{S}$, i.e. the closure of S is $\overline{S}$.

Theorem 3. *The completion $\overline{X}$ of a normed linear space X under the metric (4) has a natural linear structure that makes $\overline{X}$ a complete normed linear space.*

Proof. Recall that the points of the completion of a metric space are equivalence classes of Cauchy sequences. The term-by-term sum of two Cauchy sequences is again a Cauchy sequence, and sums of equivalent Cauchy sequences are equivalent.

$\square$

Exercise 3. Show that if X is Banach space, Y a closed subspace of X, the quotient space X/Y is complete. (Hint: Use a Cauchy sequence $\{q_n\}$ in X/Y that satisfies $|q_n - q_{n+1}| < 1/n^2$.)

The process of *completion* of a normed linear space is one of the royal roads to obtaining complete normed linear spaces. This is extremely important for the success of functional analysis. We describe now a number of the most important normed linear spaces. These are the household items of modern analysis.

(a) The space of all vectors with infinite number of components

$$x = \{a_1, a_2, \ldots\}, \qquad a_j \text{ complex},$$

where the $|a_j|$ are bounded. The norm is

$$|x|_\infty = \sup_j |a_j|. \tag{10}$$

This space is denoted as ℓ^∞; it is complete.

(b) The space of all vectors with infinitely many components such that $\sum |a_j|^p < \infty$, p some fixed number ≥ 1. The norm is

$$|x|_p = \left(\sum |a_j|^p\right)^{1/p}. \tag{11}$$

This space is denoted as ℓ^p; it is complete.

(c) S an abstract set, X the space of all complex-valued functions f that are bounded. The norm is

$$|f|_\infty = \sup_S |f(s)|. \tag{12}$$

This space is complete.

(d) Q a topological space, X the space of all complex valued, continuous, bounded functions f on Q. The norm is

$$|f| = \sup_Q |f(q)|. \tag{13}$$

This space is complete.

(e) Q a topological space, X the space of all complex-valued, continuous functions f with compact support. The norm is

$$|f|_{\max} = \max_Q |f(q)|. \tag{13'}$$

This space is not complete unless Q is compact.

(f) D some domain in $\mathbb{R}^n$, X the space of continuous functions f with compact support. The norm is

$$|f|_p = \left(\int_D |f(x)|^p \, dx \right)^{1/p}, \qquad 1 \leq p. \tag{14}$$

This space is not complete; its completion is denoted by L^p.

(g) D some domain in $\mathbb{R}^n$, the space of all C^∞ functions f in D with the following property: for some integer k and $p \geq 1$,

$$\int_D |\partial^\alpha f|^p \, dx < \infty \qquad \text{for all } |\alpha| \leq k,$$

where ∂^α is any partial derivative:

$$\partial^\alpha = \partial_1^{\alpha_1} \ldots \partial_n^{\alpha_n}, \quad \partial_j = \frac{\partial}{\partial x^j}, \quad |\alpha| = \alpha_1 + \cdots + \alpha_n.$$

The norm is

$$|f|_{k,p} = \left(\sum_{|\alpha| \leq k} \int |\partial^\alpha f|^p \, dx \right)^{1/p}. \tag{15}$$

This space is not complete; its completion is denoted as $W^{k,p}$, and is called a *Sobolev space*.

Theorem 4. *The norms defined in examples (a) through (g) have properties (1) through (3) imposed on a norm.*

Proof. Properties (1) and (3)—positivity and homogeneity—are obviously satisfied. We turn now to property (2), subadditivity. For the sake of brevity we consider only examples (a) and (b). Note that (a) can be regarded as a limiting case of (b), with $p = \infty$.

Define x and y as

$$x = \{a_1, a_2, \ldots\}, \quad y = \{b_1, b_2, \ldots\}.$$

Then

$$x + y = \{a_1 + b_1, \ldots\}$$

We take first $p = \infty$. By (10),

$$|x + y|_\infty = \sup_j |a_j + b_j| \leq \sup_j |a_j| + |b_j|$$
$$\leq \sup_j |a_j| + \sup_j |b_j| = |x|_\infty + |y|_\infty.$$

Next we turn to $p = 1$. By (11),

$$|x + y|_1 = \sum |a_j + b_j| \leq \sum |a_j| + |b_j| = |x|_1 + |y|_1.$$

For $1 < p < \infty$ we need Hölder's inequality. To state it, we introduce vectors u with finite q-norm:

$$u = \{c_1, c_2, \ldots\}, \quad \left(\sum |c_j|^q\right)^{1/q} = |u|_q < \infty, \tag{16}$$

where q is conjugate to p, in the sense

$$\frac{1}{p} + \frac{1}{q} = 1. \tag{17}$$

We define now a *scalar product* between vectors in ℓ^p and ℓ^q as follows:

$$(x, u) = \sum a_j c_j. \tag{18}$$

Hölder's Inequality. For x in ℓ^p, u in ℓ^q the series defining the scalar product (18) converges, and

$$|(x, u)| \leq |x|_p |u|_q, \tag{19}$$

provided that p and q are conjugate in the sense of (17).

For a proof we refer to Courant's *Calculus*, Vol 2. The sign of equality holds in (19) iff

$$\arg a_j c_j \quad \text{and} \quad |a_j|^p / |c_j|^q \quad \text{are independent of } j. \tag{20}$$

Since for given x in ℓ^p we can always choose u in ℓ^q so that (20) is satisfied, and so that $|u|_q = 1$, we can restate Hölder's inequality thus:

Theorem 5. *For any x in ℓ^p,*

$$|x|_p = \max_{|u|_q = 1} |(x, u)|. \tag{21}$$

Note that the scalar product (18) is *bilinear* as function of x and u. Applying (21) to $x + y$ in place of x and using the linear dependence, we get

$$|x + y|_p = \max_{|u|_q = 1} |(x + y, u)| \leq \max_{|u|_q = 1} |(x, u)| + |(y, u)|. \tag{22}$$

By Hölder's inequality (19), for $|u|_q = 1$,

$$|(x, u)| \leq |x|_p, \quad |(y, u)| \leq |y|_p.$$

Setting this into (22) gives

$$|x + y|_p \leq |x|_p + |y|_p,$$

as asserted in theorem 4.   □

The self-conjugate case $p = q = 2$ is an instance of a supremely important class of norms, to be discussed in the next chapter.

The norms defined in examples (f) and (g) satisfy important inequalities due to Sobolev: If

$$mp \leq n \quad \text{and} \quad p \leq q \leq \frac{np}{n - kp} \tag{23}$$

and if Q is a cube, then

$$|f|_q \leq \text{const.} \, |f|_{k, p}, \tag{23'}$$

where the constant depends only on p, q, k, n. These inequalities hold of course for all Q that are the images of cubes under a smooth mapping. Even more generally, they hold for all domains Q that satisfy a cone condition. For a proof, see Adams or Mazya.

Since the spaces L^q and $W^{k, p}$ are constructed by completing the space of smooth functions in the appropriate norms, it follows that if condition (23) is fulfilled, $W^{m, p}$ is contained in L^q.

The normed linear spaces studied and used in analysis are infinite-dimensional. According to Cantor's theory of sets, there is a gradation among infinites; the least of them are the countable sets.

Definition. A normed linear space is called *separable* if it contains a countable set of points that is *dense*, namely, whose closure is the whole space.

Most, but not all, spaces that are used in analysis are separable. Here is an important example that is not:

(h) The space of all signed measures m on, say, the interval $[0, 1]$, of finite total mass. We define the norm to be the total mass:

$$|m| = \int_0^1 |dm|.$$

Denote by m_y the unit mass located at the point y. Clearly, for $y \neq z$, $|m_y - m_z| = 2$. Since there are nondenumerably many points y in the interval $[0, 1]$, this shows that the space of measures is not separable.

5.2 NONCOMPACTNESS OF THE UNIT BALL

Many existence theorems in *finite-dimensional spaces* rest on the fact that the *closed unit ball*, meaning that the set of points

$$B_1 = \{x; \; |x| \leq 1\}, \tag{24}$$

is compact, that is to say, that any sequence of points in B_1 has a convergent subsequence. F. Riesz has shown that this property characterizes finite-dimensional spaces:

Theorem 6. *Let X be an infinite-dimensional normed linear space; then the unit ball B_1 defined by (24) is not compact.*

Proof. We require first a lemma:

Lemma 7. *Let Y be a closed, proper subspace of the normed linear space X. Then there is a vector z in X of length 1,*

$$|z| = 1, \tag{25}$$

and that satisfies

$$|z - y| > \tfrac{1}{2} \quad \text{for all } y \text{ in } Y. \tag{25'}$$

Proof. Since Y is a proper subspace of X, some point x of X does not belong to Y. Since Y is closed, x has a positive distance to Y:

$$\inf_{y \text{ in } Y} |x - y| = d > 0. \tag{26}$$

There is then a y_0 in Y such that

$$|x - y_0| < 2d. \tag{27}$$

Denote $z' = x - y_0$; we can then write (27) as

$$|z'| < 2d. \tag{27'}$$

It follows from (26) that

$$|z' - y| \geq d \quad \text{for all } y \text{ in } Y. \tag{28}$$

We set

$$z = \frac{z'}{|z'|}.$$

Clearly, (25) holds, and (25′) follows from combining (27) and (28).   □

REMARK 1. Clearly, the number $\frac{1}{2}$ on the right of (25′) can be replaced by any number < 1.

We turn now to the proof of theorem 6. We construct a sequence $\{y_n\}$ of unit vectors recursively as follows: y_1 is chosen arbitrarily. Suppose that $y_1, \ldots, y_{n-1}$ have been chosen; denote by Y_n the linear space spanned by them. Since Y_n is finite-dimensional, it is closed; since X is infinite-dimensional, Y_n is a proper subspace of X. So lemma 7 is applicable and a z with properties (25), (25′) exists. We set

$$y_n = z.$$

Since y_j, $j < n$ belongs to Y_n

$$|y_n - y_j| > \tfrac{1}{2}, \qquad j < n.$$

This shows that the distance of any two distinct y_j exceeds $\frac{1}{2}$. Therefore *no* subsequence can form a Cauchy sequence. Since all y_j belong to the unit ball B_1, it follows that B_1 is *not* compact.   □

Exercise 4. Prove that every finite-dimensional subspace of a normed linear space is closed. (Hint: Use the fact that all norms are equivalent on finite-dimensional spaces to show that every finite-dimensional subspace is complete.)

Next we describe a kind of a substitute for the compactness that is lacking in the unit ball.

Definition. A norm is called *strictly subadditive* if in (2) strict inequality holds except when x or y is a nonnegative multiple of the other.

Exercise 5. Show that the sup norms of examples (a), (c), (d), and (e) are *not* strictly subadditive.

Exercise 6. Show that the norms in examples (b) and (f) are *not* strictly subadditive for $p = 1$.

All the norms in examples (b) and (f) are strictly subadditive when $1 < p < \infty$. Furthermore for each of these norms the condition holds *uniformly*, in the following sense:

For any pair of unit vectors x, y, the norm of $(x + y)/2$ is strictly less than 1 by an amount that depends only on $|x - y|$. More explicitly, there is an increasing function $\epsilon(r)$ defined for positive r,

$$\epsilon(r) > 0, \quad \lim_{r \to 0} \epsilon(r) = 0, \tag{29}$$

such that for all x, y in the unit ball $|x| \leq 1$, $|y| \leq 1$, the inequality

$$\left| \frac{x + y}{2} \right| \leq 1 - \epsilon(|x - y|). \tag{30}$$

holds.

Definition. A normed linear space whose norm satisfies (30) for all vectors x, y of unit length, where $\epsilon(r)$ is some function satisfying (29), is called *uniformly convex*.

Theorem 8. *Let X be a uniformly convex Banach space. Let K be a closed, convex subset of X, z any point of X. Then there is a unique point y of K which is closer to z than any other point of K.*

Proof. We may take $z = 0$, provided that we assume that 0 does not lie in K. Denote by s the distance of 0 to K, that is,

$$s = \inf |y|, \quad y \text{ in } K. \tag{31}$$

Since 0 does not lie in K, and since K is closed, $s > 0$. Let $\{y_n\}$ be a minimizing sequence for (31), that is,

$$y_n \text{ in } K, \quad |y_n| = s_n \to s. \tag{31'}$$

Define the unit vectors x_n as

$$x_n = \frac{y_n}{s_n}; \tag{31''}$$

we can write

$$\frac{x_n + x_m}{2} = \frac{1}{2s_n} y_n + \frac{1}{2s_m} y_m$$

$$= \left(\frac{1}{2s_n} + \frac{1}{2s_m} \right) (c_n y_n + c_m y_m). \tag{32}$$

Clearly, c_n and c_m are positive, and $c_n + c_m = 1$. Since K is convex, it follows that $c_n y_n + c_m y_m$ belongs to K. Therefore, by (31),

$$|c_n y_n + c_m y_m| \geq s.$$

Setting this into (32), we get that

$$\left| \frac{x_n + x_m}{2} \right| \geq \frac{s}{2s_n} + \frac{s}{2s_m}.$$ (33)

Since $\{y_n\}$ is a minimizing sequence for (31), $s_n \to s$; therefore the right side of (33) tends to 1. So it follows from (33), (30), and (29) that $\lim_{n,m \to \infty} |x_n - x_m| = 0$. It follows then from (31') that also $\lim_{n,m \to \infty} |y_n - y_m| = 0$, meaning that the minimizing sequence $\{y_n\}$ is a Cauchy sequence. Since X is complete and K is closed, the sequence $\{y_n\}$ converges to an element y of K. Clearly, $|y| = s$.   □

The power of theorem 8 lies in the fact that it asserts the existence of a minimum, when the set K over which we wish to minimimze is not compact; according to theorem 6, a Banach space has many closed, bounded sets K that are not compact.

The notion of uniform convexity is due to Clarkson, as is the result that the L^p spaces are uniformly convex for $1 < p < \infty$.

We give now an example that shows that in the space C, which is not uniformly convex—indeed, the maximum norm is not even strictly subadditive—the conclusion of theorem 8 fails.

We take X to be $C[-1, 1]$, the space of continuous real-valued functions defined on the closed interval $-1 \leq t \leq 1$. We take K to consist of all functions $k(t)$ that satisfy

$$\int_{-1}^{0} k \, dt = 0, \qquad \int_{0}^{1} k \, dt = 0.$$ (34)

K is a linear subspace and therefore convex, and clearly closed.

We take for $z(t)$ any function in C for which

$$\int_{-1}^{0} z \, dt = 1, \qquad \int_{0}^{1} z \, dt = -1.$$

It follows from (34) that for any k in K

$$\int_{-1}^{0} (z - k) \, dt = 1, \qquad \int_{0}^{1} (z - k) \, dt = -1.$$

From this it follows that

$$\max_{-1 \leq t \leq 0} [z(t) - k(t)] \geq 1,$$ (35)

equality holding iff

$$z(t) - k(t) \equiv 1 \qquad \text{for } -1 \leq t \leq 0,$$ (35')

and similarly that

$$\min_{0 \le t \le 1} [z(t) - k(t)] \le -1, \tag{36}$$

equality holding iff

$$z(t) - k(t) \equiv -1 \quad \text{for } 0 \le t \le 1. \tag{36'}$$

Conditions (35') and (36') cannot both hold at $t = 0$, so it follows that in at least one of (35) or (36) inequality holds. This proves that

$$|z - k|_{\max} > 1 \tag{37}$$

for any k in K. On the other hand, one could choose k in K so that the max and min in (35) and (36) are as close to 1 and -1, respectively, as one wishes. So

$$\inf_{k \text{ in } K} |z - k|_{\max} = 1. \tag{37'}$$

This combined with (37) shows that there is *no* closest point to z in K. $\qquad \square$

5.3 ISOMETRIES

We turn now to *isometries* of a Banach space X onto itself, meaning mappings $\mathbf{M}$ of X onto X which preserve the distance of any pair of points:

$$|\mathbf{M}(x) - \mathbf{M}(y)| = |x - y| \quad \text{for all } x, y \text{ in } X. \tag{38}$$

Clearly, translations $\mathbf{M}(x) = x + u$, u fixed, are isometries. Also the isometries of X form a *group*. We want to investigate those isometries that map 0 into 0; all others can be obtained by composing these with a translation.

Theorem 9. *Let X be a linear space over the reals with a strictly subadditive norm. Let $\mathbf{M}$ be an isometric mapping of X into itself that maps the origin into itself. Then $\mathbf{M}$ is linear.*

Proof. Denote for simplicity $\mathbf{M}(x)$ by x'. Take any pair of points x and y and define

$$z = \frac{x + y}{2}. \tag{39}$$

Using isometry as stated in (38), and the definition of z, we have

$$|x' - z'| = |x - z| = \frac{|x - y|}{2},$$
$$|z' - y'| = |z - y| = \frac{|x - y|}{2}, \tag{40}$$

and

$$|x' - y'| = |x - y|. \tag{40'}$$

These imply that

$$|x' - y'| = |x' - z' + z' - y'| = |x' - z'| + |z' - y'|.$$

Since the norm is strictly subadditive, $x' - z'$ and $z' - y'$ must be positive multiples of each other. Since by (40) they have the same norm, they must be equal: $x' - z' = z' - y'$. Hence

$$2z' = x' + y'. \tag{41}$$

$\square$

Exercise 7. Deduce from (41) that **M** is linear.

It is a fact of life that some Banach spaces are very rich in isometries; others are very poor. Among the rich ones are the Hilbert spaces discussed in chapter 6; among the poor ones are function spaces with the max norm. Here is an example due to Schur:

Denote by X the space of null sequences of complex numbers

$$x = \{a_n\}, \quad \lim_{n \to \infty} a_n = 0 \tag{42}$$

normed by

$$|x| = \max_n |a_n|. \tag{42'}$$

Exercise 8. Show that X is complete.

Let $\{b_n\}$ be an arbitrary sequence of complex numbers of absolute value 1: $|b_n| = 1$. Define the mapping **U** by

$$\mathbf{U}x = \{b_n a_n\}. \tag{43}$$

Clearly, **U** is a linear map of X onto X, and satisfies $|\mathbf{U}x| = |x|$; thus **U** is an isometry.

Let p be a *permutation* of the positive integers. Define the map **P** by

$$\mathbf{P}x = \{a'_n\}, \quad a'_n = a_{p(n)}. \tag{44}$$

Clearly, **P** is a linear map of X onto X, and an isometry.

Theorem 10. *Every linear isometry of the Banach space X defined by (42), (42') is the composite of an isometry of type (43) and (44).*

Proof. Let u_j be a jth *unit vector*, i.e. a vector whose jth component has absolute value 1, all others are zero. Denote by T_j the linear subspace of X consisting of all vectors whose jth component is zero. Clearly,

$$T_j \text{ is closed and codim } T_j = 1, \tag{45}$$
$$|u_j + t| = 1 \quad \text{for all } t \text{ in } T_j, |t| \le 1. \tag{46}$$

Conversely, we have

Lemma 11. *Let u be a vector in X, $|u| = 1$, and T a subspace of X of codimension 1 so that (45) and (46) hold. Then u is a unit vector and T the corresponding subspace T_j.*

Proof. By definition $(42')$ of the norm

$$1 = |u| = |u_m| \quad \text{for some index } m.$$

It follows from (46) that no vector t in T can have an mth component $\ne 0$. Since T is assumed to have codimension 1, it follows that T consists of all null sequences whose mth component is zero. From this it follows, by (46), that all components of u other than the mth must be zero.   □

Let $\mathbf{M}$ be a linear isometry of X onto X; let u_j be any unit vector and T_j the corresponding subspace. Since $\mathbf{M}$ is linear, isometric, and onto, it follows that u'_j and T'_j, the image of u_j and T_j under $\mathbf{M}$, satisfy (45) and (46). Then by Lemma 11, u' is a unit vector; from this and the linearity of isometry theorem 10 follows readily.

We conclude this chapter with the following result due to Mazur and Ulam:

Theorem 12. *Let X and X' be two normed linear spaces over ther reals, $\mathbf{M}$ an isometric mapping of X onto X' that carries 0 into 0. Then $\mathbf{M}$ is linear.*

Proof. The case where the norms are strictly subadditive is covered in theorem 9. In the general case we take, as before, x and y to be any pair of points, and z their midpoint:

$$z = \frac{x+y}{2}.$$

As before, in (40), z is halfway between x and y, but when the norm in X is not strictly subadditive, this no longer characterizes the midpoint z. There may be other points u also halfway between x and y:

$$|x - u| = |y - u| = \frac{|x - y|}{2}. \tag{47}$$

We denote the set of all such u by A. We claim that this set A is *symmetric* with respect to the midpoint z. That is, that if u belongs to A, then so does

$$v = 2z - u. \tag{48}$$

To see this, we note that $2z = x + y$, and so

$$v - x = y - u, \quad \text{and} \quad v - y = x - u.$$

It follows from (47) that v is halfway between x and y.

We define the *diameter* d_A of A as the greatest distance between pairs of points of A:

$$d_A = \sup_{u,w \text{ in } A} |u - w|. \tag{49}$$

Since A is symmetric with respect to z, for all u in A,

$$|u - z| \le \tfrac{1}{2}d_A.$$

Of course, there may be other points p in A with this property:

$$|u - p| \le \tfrac{1}{2}d_A \qquad \text{for all } u \text{ in } A. \tag{50}$$

We denote the set of all such p by A_1. We claim that A_1 is symmetric with respect to the midpoint z. That is, if p belongs to A_1, so does

$$q = 2z - p. \tag{51}$$

For using (48), we can write for any u in A,

$$q - u = 2z - u - p = v - p. \tag{51'}$$

We conclude from (51') that $|q - u| = |v - p|$. Since v belongs to A when u does, it follows from (50) that $|u - q| \le \tfrac{1}{2}d_A$.

It follows from (50) that the diameter of A_1 does not exceed half the diameter of A:

$$d_{A_1} \le \tfrac{1}{2}d_A. \tag{52}$$

We now repeat this construction, obtaining a nested sequence of sets $A \supset A_1 \supset A_2 \cdots$, each containing the midpoint z, each symmetric with respect to z, and their diameters satisfying

$$d_{A_{n+1}} \le \tfrac{1}{2}d_{A_n}.$$

Clearly, d_{A_n} tends to zero; it follows that the intersection of all the sets A_n consist of the single point z. This characterizes the midpoint z of x, y purely in terms of the metric structure of X.

Let $\mathbf{M}$ be an isometric mapping of X onto X'. Then the inverse of $\mathbf{M}$ maps X' isometrically onto X. Denote by x' and y' the images of x, y under $\mathbf{M}$, and denote by A', A'_1, ..., A'_n the sets analogous to the ones defined in X. Recall that the set A was defined by (47), and that the set A_1 was defined by (49) and (50). Since these inequalities refer only to distances, and since $\mathbf{M}$ is an isometry, it follows that $\mathbf{M}$ maps every point of A_n into A'_n. Since the inverse of $\mathbf{M}$ is isometric, it maps every point of A'_n into A_n. Thus $\mathbf{M}$ maps A_n onto A'_n, and thus the intersection of the A_n onto the intersection of the A'_n. Since these intersections are, respectively, $(x+y)/2$ and $(x'+y')/2$, it follows that

$$\mathbf{M}\left(\frac{x+y}{2}\right) = \frac{x'+y'}{2}. \tag{53}$$

Setting $y = 0$, and using the assumption that $\mathbf{M}(0) = y' = 0$, we get that $\mathbf{M}(x/2) = x'/2$. Applying this to equation (53), we get

$$\mathbf{M}(x+y) = x'+y' = \mathbf{M}(x)+\mathbf{M}(y).$$

This is the first property of linearity, see equation (1) in chapter 2. From this we deduce that $\mathbf{M}(kx) = k\mathbf{M}(x)$ for all rational k. Since $\mathbf{M}$ is an isometry, it is continuous, and so the relation holds for all real k. $\quad\square$

BIBLIOGRAPHY

Adams, R. A. *Sobolev Spaces*. Academic Press, New York, 1975.

Clarkson, J. A. Uniformly convex spaces. *Trans. AMS*, **40** (1936): 396–414.

Day, M. M. *Normed Linear Spaces*. Springer Verlag, 1958.

Mazur, S. and Ulam, S. Sur les transformation isométriques d'espace vectoriel normés, *C.R. Acad. Sci. Paris*, **194** (1932): 946–948.

Mazya, V. *Sobolev Spaces*. Springer Verlag, 1985.

6

HILBERT SPACE

6.1 SCALAR PRODUCT

A *scalar product* in a linear space X over $\mathbb{R}$ is a real valued function of two points x and y in X, denoted as (x, y), having the following properties:

(i) *Bilinearity.* For fixed y, (x, y) is a linear function of x, for fixed x a linear function of y.

(ii) *Symmetry*, $(y, x) = (x, y)$.

(iii) *Positivity*, $(x, x) > 0$ for $x \neq 0$.

When the field of scalars is $\mathbb{C}$, (x, y) is complex valued, and properties (i) and (ii) are altered as follows:

(i) *Sesquilinearity.* For fixed y, (1) is a linear function of x and for x fixed (1) is a *skewlinear* function of y, that is,

$$(ax, y) = a(x, y), \quad (x, ay) = \bar{a}(x, y). \tag{1}$$

(ii) *Skew symmetry*,

$$(y, x) = \overline{(x, y)}. \tag{2}$$

Given a scalar product, we can define a norm, denoted by $\| \ \|$, as follows:

$$\|x\| = (x, x)^{1/2}. \tag{3}$$

We claim that $\| \ \|$ has the obvious properties of a norm:
Positivity follows from (iii), and homogeneity from (1). To show subadditivity we need

Theorem 1 (Schwarz Inequality). *A scalar product satisfying (i), (ii), and (iii) satisfies*

$$|(x, y)| \leq \|x\| \, \|y\|, \tag{4}$$

where the norm is defined by (3). Equality holds for $x = ay$ or $y = 0$.

Proof. Let t be a real scalar and $y \neq 0$. Using bilinearity and skew symmetry, we can write

$$\|x + ty\|^2 = \|x\|^2 + 2t\mathrm{Re}(x, y) + t^2\|y\|^2. \tag{5}$$

By (iii), this is nonnegative. Set $t = -\mathrm{Re}(x, y)/\|y\|^2$ and multiply by $\|y\|^2$. We get

$$(\mathrm{Re}(x, y))^2 \leq \|x\|^2\|y\|^2.$$

Replacing x by ax, $|a| = 1$ so chosen that $a(x, y)$ is real, we deduce (4). Note that *equality* holds in (4) iff x and y are scalar multiples of one another. $\qquad\square$

Corollary 1′. *For every vector x in a scalar product space*

$$\|x\| = \max_{\|y\|=1} |(x, y)|.$$

Now we are ready to prove that the norm is subadditive. Set $t = 1$ in (5) and estimate the middle term by (4). We get

$$\|x + y\|^2 \leq (\|x\| + \|y\|)^2,$$

which is subadditivity of the norm. $\qquad\square$

We set $t = \pm1$ in (5) and add to obtain the *parallelogram identity:*

$$\|x + y\|^2 + \|x - y\|^2 = 2\|x\|^2 + 2\|y\|^2. \tag{6}$$

Exercise 1. Show that a norm that satisfies (6) comes from a scalar product, an observation due to von Neumann.

Exercise 2. Show that the scalar product depends continuously on its factors; that is, if $x_n \to x$, $y_n \to y$ in the sense of $\|x_n - x\| \to 0$, $\|y_n - y\| \to 0$, then (x_n, y_n) tends to (x, y). (Use the Schwarz inequality.)

Definition. Two vectors x and y are called *orthogonal* if $(x, y) = 0$.

Definition. A linear space with a scalar product that is *complete* with respect to the induced norm is called a *Hilbert space.*

Given a linear space with a scalar product, it can be completed with respect to the norm derived from the scalar product. It follows from the Schwarz inequality that the scalar product is a continuous function of its factors; therefore it can be extended to the completed space. Thus the completion is a Hilbert space.

We give some examples of linear spaces with inner product:

Example 1. The space of continuous functions $x(t)$ on the interval $[0, 1]$, with

$$(x, y) = \int_0^1 x(t)\,\overline{y}(t)\,dt.$$

This space is incomplete.

Example 2. The space ℓ^2 of vectors with infinitely many components:

$$x = (a_1, a_2, \ldots), \quad y = (b_1, b_2, \ldots)$$

subject to the restriction

$$\sum |a_j|^2 < \infty, \quad \sum |b_j|^2 < \infty.$$

We define the scalar product as

$$(x, y) = \sum a_j \overline{b}_j.$$

Exercise 3. Show that ℓ^2 is complete.

Example 3. The space L^2 of all functions square integrable according to Lebesgue on some domain in $\mathbb{R}^n$. This space is complete.

Many other examples will come up in the applications presented in subsequent chapters.

6.2 CLOSEST POINT IN A CLOSED CONVEX SUBSET

Theorem 2. *Given a nonempty closed, convex subset K of a Hilbert space H, and a point x in H, there is a unique point y in K that is closer to x than any other point of K.*

Proof. Define

$$\inf_{z \text{ in } K} \|x - z\| = d. \tag{7}$$

Let y_n in K be a minimizing sequence:

$$\lim d_n = d, \quad d_n = \|x - y_n\|. \tag{8}$$

We apply the parallelogram identity (6) to $x = (x - y_n)/2$, $y = (x - y_m)/2$:

$$\left\| x - \frac{y_n + y_m}{2} \right\|^2 + \frac{1}{4}\| y_n - y_m \|^2 = \frac{1}{2}(d_n^2 + d_m^2). \tag{9}$$

Since K is convex, $(y_n + y_m)/2$ belongs to K, and so by (7), $\|x - (y_n + y_m)/2\| \geq d$. Using this and (8) in (9), we deduce that y_n is a Cauchy sequence. Since H is complete and K closed, $y = \lim y_n$ belongs to K. Since $\|x - y\| = \lim \|x - y_n\| = d$, y minimizes the distance from x. That there is only one minimizer follows from (6): suppose that y' is another minimum, and apply (6) to $x - y$, $x - y'$. □

Theorem 2 is a special case of theorem 8 in chapter 5.

Definition. Let Y be a linear subspace; its *orthogonal complement* consists of all vectors v orthogonal to Y, that is, satisfying $(v, y) = 0$. It is denoted as $Y^\perp$.

Theorem 3. *Let H be a Hilbert space, Y a closed subspace of H, $Y^\perp$ the orthogonal complement of Y. We claim that*

(i) $Y^\perp$ is a closed linear subspace of H;

(ii) Y and $Y^\perp$ are complementary subspaces, meaning that every x can be decomposed uniquely as a sum of a vector in Y and in $Y^\perp$;

(iii) $(Y^\perp)^\perp = Y$.

Proof. It follows from the bilinearity of scalar product that the set of vectors v orthogonal to all vectors of any set Y form a linear space. This shows that $Y^\perp$ is a linear space. Let $\{v_j\}$ be a convergent sequence of elements of $Y^\perp$:

$$\lim v_j = v. \tag{10}$$

We claim that v belongs to $Y^\perp$, namely, that

$$(v, z) = 0 \quad \text{when } z \text{ is in } Y.$$

Since v_j belongs to $Y^\perp$,

$$(v, z) = (v - v_j, z) + (v_j, z) = (v - v_j, z).$$

By the Schwarz inequality applied on the right,

$$|(v, z)| \leq \| v - v_j \| \, \| z \|; \tag{11}$$

by (10), $\| v - v_j \|$ tends to zero. So (11) shows that $(v, z) = 0$, meaning that $Y^\perp$ is closed, as asserted in (i).

We turn now to (ii). Given any x in H, there is, according to theorem 2, a vector y in Y closest to x. Set

$$v = x - y. \tag{12}$$

The minimum property of y means that for any z in Y and any real t,

$$\|v\|^2 \le \|v + tz\|^2.$$

Using (5), we can rewrite the right side as $\|v\|^2 + 2t\,\mathrm{Re}(v, z) + t^2\|z\|^2$ and conclude that

$$\mathrm{Re}(v, z) = 0 \qquad \text{for all } z \text{ in } Y. \tag{13}$$

This shows that v belongs to $Y^\perp$; (12) gives the decomposition of x as $y + v$, the sum of a vector from Y and one from $Y^\perp$.

This decomposition is unique, for if $x = y + v = y' + v'$, then $y - y' = v' - v$ would belong to both Y and $Y^\perp$ and thus would be orthogonal to itself. But then, by positivity, $y - y' = v' - v' = 0$. Thus (ii) is proved. Part (iii) is an immediate consequence of (ii). $\qquad\square$

COMMENT. It follows from theorem 3 that every closed linear subspace of a Hilbert space has a closed complement. This is not true for all Banach spaces; examples will be given later.

6.3 LINEAR FUNCTIONALS

A Hilbert space comes equipped with a whole set of built-in linear functionals. For y fixed, $(x, y) = \ell(x)$ is a *linear functional* of x, that is, a *linear mapping of H into* $\mathbb{C}$. Furthermore, according to the Schwarz inequality (4), $\ell(x)$ is *bounded* by a constant multiple of $\|x\|$. It turns out that conversely:

Theorem 4. *Let $\ell(x)$ be a linear functional on a Hilbert space H that is bounded:*

$$|\ell(x)| \le \text{const.}\,\|x\|. \tag{14}$$

Then ℓ is of the form

$$\ell(x) = (x, y), \quad y \text{ in } H. \tag{15}$$

The point y is uniquely determined.

Proof. We will use the following facts:

Lemma 5.

(i) *The nullspace of a linear functional that is not $\equiv 0$ is a linear subspace of codimension 1.*

(ii) *If two linear functionals ℓ and m have the same nullspace, they are constant multiples of each other:*

$$\ell = cm. \tag{16}$$

(iii) *The nullspace of a linear functional that is bounded in the sense of (14) is a closed subspace.*

Exercise 4. Prove lemma 5.

Note that lemma 5 holds in any Banach space, and parts (i) and (ii) in any linear space.

Suppose now that ℓ is $\not\equiv 0$. Then its nullspace is a closed subspace Y of H of codimension 1. Its orthogonal complement $Y^{\perp}$ (see theorem 3) is one dimensional. Let p be any nonzero vector in $Y^{\perp}$, and define the linear functional m by

$$m(x) = (x, p).$$

Clearly, the nullspace of m is Y. So by (16) of part (ii) of lemma 5,

$$\ell(x) = cm(x) = (x, \bar{c}p). \qquad \Box$$

Theorem 4 is called the *Riesz-Frechet representation theorem.*

The following useful generalization has been given by Milgram and Lax:

Theorem 6 (Lax-Milgram Lemma). *Let H be a Hilbert space, and $B(x, y)$ a function of two vectors with the following properties:*

(i) *$B(x, y)$ is for fixed y a linear function of x, for fixed x a skewlinear function of y.*

(ii) *B is bounded: there is a constant c so that for all x and y in H*

$$|B(x, y)| \le c\|x\| \, \|y\|. \tag{17}$$

(iii) *There is a positive constant b such that*

$$|B(y, y)| \ge b\|y\|^2 \tag{18}$$

for all y in H.

Assertion. *Every linear functional ℓ on H that is bounded in the sense of (14) is of the form*

$$\ell(x) = B(x, y), \qquad y \text{ a uniquely determined vector in } H. \tag{19}$$

Proof. By (i) and (ii), for y fixed $B(x, y)$ is a bounded linear functional of x. Therefore by theorem 4 it can be written as

$$B(x, y) = (x, z), \qquad z \text{ in } H. \tag{20}$$

Since z is uniquely determined by y, it is a function of y. It follows from (20) that the relation of z to y is *linear*; it follows from this that the set of z appearing in (20) as y takes on all values in H is a *linear subspace* of H. We claim that it is a *closed* linear subspace. To see this, set $x = y$ in (20):

$$B(y, y) = (y, z). \tag{20'}$$

Using (18) on the left and the Schwarz inequality on the right, we get after dividing by $\|y\|$ that

$$b\|y\| \le \|z\|. \tag{21}$$

Let $\{z_n\}$ be a sequence of vectors appearing in (20), with corresponding y_n:

$$B(x, y_n) = (x, z_n). \tag{22}$$

Subtraction and skew linearity gives $B(x, y_n - y_m) = (x, z_n - z_m)$. By (21), $b\|y_n - y_m\| \le \|z_n - z_m\|$. From this it follows that if z_n converges to z, the corresponding y_n form a Cauchy sequence. Since H is complete, the sequence $\{y_n\}$ converges to a limit y. It follows from (17) that the left side of (22) converges to $B(x, y)$, and it follows from (4) that the right side converges to (x, z). So

$$B(x, y) = (x, z),$$

which proves that the set of z appearing in (20) form a closed subspace of H.

We claim that this closed subspace is all of H; for if not, then according to theorem 3 there would be a nonzero vector x orthogonal to all z. It follows from (20) that such an x satisfies $B(x, y) = 0$ for all y. Setting $y = x$ gives $B(x, x) = 0$; using (18), we get $\|x\| = 0$, contrary to $x \ne 0$.

According to theorem 4, all linear functionals $\ell(x)$ can be represented as (x, z), z in H. Combined with (20), this establishes (19). It follows from (18) that y is uniquely determined. $\qquad\square$

6.4 LINEAR SPAN

We recall from chapter 1 that the *linear span* of a collection of points $\{y_j\} = S$ is the smallest linear subspace containing them. The *closed linear span* of a collection of points S in a Hilbert space H is defined to be the *smallest closed linear subspace* containing S, that is, the *intersection* of all such subspaces.

Exercise 5. Show that the closed linear span of a set is the closure of its linear span.

Theorem 7. *The point y of a Hilbert space H belongs to the closed linear span Y of the set $\{y_j\}$ iff every vector z that is orthogonal to all y_j is orthogonal also to y:*

$$(y, z) = 0 \quad \text{for all } z \text{ that satisfy} \quad (y_j, z) = 0 \quad \text{for all } j. \tag{23}$$

Proof. We claim that the set Z of vectors z orthogonal to all y_j form the orthogonal complement of Y. Since every vector z orthogonal to all y_j is orthogonal to all linear combinations of y_j, and by continuity to limits of linear combinations, $Z \subset Y^{\perp}$. Conversely, every vector in $Y^{\perp}$ is orthogonal to all the y_i, and so belongs to Z. This shows that $Z = Y^{\perp}$. We appeal to part (iii) of theorem 3 to conclude that $Y = (Y^{\perp})^{\perp} = Z^{\perp}$, as asserted in theorem 7. $\qquad\qquad\square$

We stated in chapter 5 that every isometry of a Banach space onto itself that maps 0 into 0 is linear. We give now a new proof of this in Hilbert space:

Denote by $x \to x'$ an isometry of a Hilbert space that maps $0 \to 0$. Let x, y be any pair of vectors, x' and y' their images. Since distances are preserved, $d(0, x) = d(0, x')$, $d(0, y) = d(0, y')$, and $d(x, y) = d(x', y)$, which can be expressed as

$$\|x\| = \|x'\|, \quad \|y\| = \|y'\|, \tag{24}$$

$$\|x - y\|^2 = \|x' - y'\|^2. \tag{24'}$$

Expanding both sides in (24$'$) and using (24), we get

$$(x, y) = (x', y'). \tag{25}$$

Now denote $x + y$ by z, and let u be any vector in H. Using (25) we have

$$(z', u') = (z, u) = (x + y, u) = (x, u) + (y, u) = (x', u') + (y', u') = (x' + y', u').$$

Thus

$$(z' - x' - y', u') = 0$$

for all u'. This can be only if $z' = x' + y'$. $\qquad\qquad\square$

The virtue of this proof is that it applies even when the scalar product is not positive, as long as it is nondegenerate, meaning that no u is orthogonal to all points.

We turn now to orthonormal sets:

Definition. A collection of vectors in an inner product space, $\{x_j\}$, is called *orthonormal* if

$$(x_j, x_k) = 0 \quad \text{for } j \notin k, \quad \text{and} \quad \|x_j\| = 1 \quad \text{for all } j. \tag{26}$$

Definition. A collection of vectors $\{x_j\}$ is called an *orthonormal base* if the vectors are orthonormal, and if the closed linear span of $\{x_j\}$ is the whole space.

Lemma 8. *Let H denote a Hilbert space, $\{x_j\}$ an orthonormal set in H. The closed linear span of $\{x_j\}$ consists of all vectors of the form*

$$x = \sum a_j x_j, \tag{27}$$

where the a_j are complex numbers so chosen that

$$\sum |a_j|^2 < \infty. \tag{27'}$$

The sum (27) converges in the sense of the Hilbert space norm. Furthermore

$$\|x\|^2 = \sum |a_j|^2, \tag{28}$$

and

$$a_j = (x, x_j). \tag{28'}$$

Exercise 6. Prove lemma 8.

Theorem 9. *Every Hilbert space contains an orthonormal basis.*

Proof. Consider all orthonormal sets, partially ordered by inclusion. Given a totally ordered collection, the union of all vectors contained in the sets in the collection includes all of them. Therefore, by Zorn's lemma, there is an orthonormal set that is maximal. We claim that the closed linear span X of a maximal orthonormal set $\{x_j\}$ is the whole space. We argue indirectly: suppose that there is a y that does not belong to the closed linear span X. Define a_j by

$$a_j = (y, x_j). \tag{29}$$

We claim that *Bessel's inequality* holds:

$$\sum |a_j|^2 \leq \|y\|^2; \tag{30}$$

for consider

$$\|y - \sum_F a_j x_j\|^2, \tag{31}$$

where F means a finite collection of j. Using the orthonormality of $\{x_j\}$, we find that (31) equals

$$\|y\|^2 - \sum_F \bar{a}_j (y, x_j) - \sum_F a_j (x_j, y) + \sum_F |a_j|^2,$$

which by (29) equals

$$\|y\|^2 - \sum_F |a_j|^2.$$

Since (31) is nonnegative, (30) follows for every finite collection F, and therefore for the infinite sum.

It follows then from lemma 8 that we can define a vector x by (27), and that x belongs to X. Now using (29) and (28'), we have

$$(y - x, x_j) = (y, x_j) - (x, x_j) = a_j - a_j = 0,$$

meaning that the $y - x$ is orthogonal to all x_j. The difference $y - x$ is not zero, since by assumption y does not belong to X, while x does; so

$$\frac{y - x}{\|y - x\|}$$

could be joined to the orthonormal set $\{x_j\}$, enlarging it, contradicting maximality.

□

Suppose that H is a separable Hilbert space; that is, it contains a denumerable set of points that is dense. In this case every orthogonal basis is denumerable, and the basis elements can be constructed without appealing to transcendental arguments such as Zorn's lemma:

Theorem 9'. *Let $\{y_j\}$ be a sequence of vectors in a Hilbert space whose closed linear span is all of H. Then there exists an orthonormal basis $\{x_j\}$ such that the linear span of $\{x_1, \ldots, x_n\}$ contains $y_1, \ldots, y_n$.*

Exercise 7. Prove theorem 9'.

The construction of the orthonormal basis $\{x_j\}$ in theorem 9' is called the *Gram-Schmidt process*.

Exercise 8. Let H be a Hilbert space; prove that any two orthonormal bases in H have the same cardinality.

Theorem 10. *Let H denote a Hilbert space, $\{x_j\}$ and $\{y_j\}$ two orthonormal bases. According to theorem 8, every x can be written as*

$$x = \sum a_j x_j, \quad a_j = (x, x_j).$$

Then the mapping

$$x \to y = \sum a_j y_j$$

is an isometry of H onto H, mapping $0 \to 0$. Furthermore every isometry of H onto H mapping $0 \to 0$ can be obtained in this fashion.

Exercise 9. Prove theorem 10.

Exercise 10. Show that every infinite-dimensional separable Hilbert space is isomorphic with the space ℓ^2 consisting of all vectors with infinitely many components: $x = (a_1, a_2, \ldots)$, subject to the restriction $\|x\|^2 = \sum |a_j|^2 < \infty$.

NOTES. The abstract notion of Hilbert space described in this chapter is due to von Neumann in 1929. Earlier, Hilbert and his school had used the concrete spaces described in Examples 2 and 3; hence the name.

Theorem 2 is essentially due to Beppo Levi, in a concrete context.

BIBLIOGRAPHY

Frechet, M. Sur lés opérations linéaires, III, *Trans. AMS,* **8** (1907): 433–446.

Hilbert, D. *Grundzüge einer allgemeinen Theorie der linearen Integralgleichungen.* Teubner, Leipzig, 1912.

Lax, P. D. and Milgram, A. *Parabolic Equations. Contributions to the Theory of Partial Differential Equations.* Annals of Math. Studies **33**. Princeton University Press, Princeton, 1954.

Levi, B. Sul Principio di Dirichlet. *Rend. del Circolo Mat. di Palermo,* **22** (1906): 293–300.

Riesz, F. Sur un espiece de géométrie analytiques des systemes de fonctions sommables. *C.R. Acad. Sci. Paris,* **144** (1907): 1409–1411.

von Neumann, J. Allgemeine Eigenwert-theorie Hermitescher Funktionaloperatoren. *Math. An.,* **102** (1929): 49–131.

7

APPLICATIONS OF HILBERT SPACE RESULTS

7.1 RADON-NIKODYM THEOREM

Let ν and μ be finite nonnegative measures on the same σ-algebra. ν is said to be *absolutely continuous* with respect to μ if every set that has μ-measure zero has ν-measure zero. The Radon-Nikodym theorem asserts that such a ν-measure can be expressed as

$$\nu(E) = \int_E g \, d\mu, \tag{1}$$

where g is a nonnegative integrable function with respect to μ.

Von Neumann showed how to derive this from the Riesz representation theorem for linear functionals in Hilbert space:

Let H be the real Hilbert space $L^2(\mu + \nu)$, with the norm

$$\|x\|^2 = \int x^2 \, d(\mu + \nu). \tag{2}$$

Assume, for simplicity, that the μ and ν measure of the whole space is finite; then it follows, via the Schwarz inequality, that every square integrable function is integrable. The linear functional

$$\ell(x) = \int x \, d\mu \tag{3}$$

is bounded with respect to the $L^2(\mu)$-norm, so even more with respect to the $L^2(\mu + \nu)$-norm. Then, by theorem 4 of chapter 6, $\ell(x)$ can be represented as a scalar product (x, y) for some y in $L^2(\mu + \nu)$:

$$\int x \, d\mu = \int xy \, d(\mu + \nu);$$

y depends only on the measures μ and ν. We rewrite this as

$$\int x(1-y)\,d\mu = \int xy\,d\nu. \tag{4}$$

We claim that

$$0 < y \leq 1 \tag{5}$$

except for a set of μ-measure zero. To show this, we denote by F the set on which $y \leq 0$, and claim that

$$\mu(F) = 0. \tag{6}$$

Set $x = 1$ on F, $x = 0$ off F; with this choice, (4) becomes

$$\int_F (1-y)\,d\mu = \int_F y\,d\nu. \tag{7}$$

Since $y \leq 0$ on F, the right side of (7) is ≤ 0, while the left side is $\geq \mu(F)$; this proves (6).

Denote by G the set where $y > 1$; suppose that $\mu(G) > 0$. Set $x = 1$ on G, $x = 0$ off G; with this choice (4) becomes

$$\int_G (1-y)\,d\mu = \int_G y\,d\nu. \tag{8}$$

Since $y > 1$ on G, the left side of (8) is negative, and the right side is positive, a contradiction. This completes the proof of (5).

We modify, if necessary, the function y on a set of μ-measure 0 so that (5) holds everywhere. Since ν is absolutely continuous with respect to μ, this does not affect (4).

We claim that the function g in (1) is given by $g = (1-y)/y$. To see this, denote $u = xy$, and rewrite (4) as

$$\int u\,g\,d\mu = \int u\,d\nu. \tag{9}$$

Let E be any measurable set; we choose x so that u is 1 on E, 0 off E. Then (9) gives

$$\int_E g\,d\mu = \nu(E). \tag{10}$$

This is relation (1). $\square$

Exercise 1. Prove the Radon-Nikodym theorem for measures that are only σ-finite.

7.2 DIRICHLET'S PROBLEM

First, let D be a bounded domain in $\mathbb{R}^n$. Denote by $C_0^\infty(D)$ the space of real-valued infinitely differentiable functions f whose support is contained in a compact subset of D. On the space $C_0^\infty(D)$ we introduce two scalar products:

$$(f, g)_0 = \int_D fg \, dx \quad \text{and} \quad (f, g)_1 = \int_D \sum f_j g_j \, dx, \qquad (11)$$

where $f_j = \partial f / \partial x_j$, $j = 1, \ldots, n$.

Exercise 2. Verify that $C_0^\infty(D)$ is an inner product space under each of these scalar products.

The following inequality connecting these two norms is due to Zaremba:

Lemma 1. *For all f in $C_0^\infty(D)$,*

$$\|f\|_0 \le d \|f\|_1, \qquad (12)$$

where d is the width of D.

Proof. Since f is zero on the boundary of D, at any point x in D,

$$f(x) = \int_{x^b}^x f_1 \, dx_1$$

where x^b is a boundary point of D with the same $x_2, \ldots, x_n$ coordinate as x. Applying the Schwarz inequality above gives

$$f^2(x) \le d \int |f_1|^2 \, dx_1.$$

Integrating this over D gives (12).   □

Denote by H_1^0 the completion of $C_0^\infty(D)$ with respect to the norm $\|\ \|_1$, by H_0 its completion with respect to the norm $\|\ \|_0$.

Lemma 2. *Every element v of H_1^0 belongs to H_0, and has partial derivatives v_j of first order that belong to H_0; these partial derivatives satisfy*

$$(z, v_j)_0 = -(\partial z / \partial x_j, v)_0 \qquad (13)$$

for any C_0^∞ function z. Furthermore formula (11) holds for f, g in H_1^0.

Proof. Let $\{v^{(n)}\}$ be a sequence of C_0^∞ functions that tends to v in the $\| \ \|_1$ norm. That means that the first derivatives $v_j^{(n)}$ converge in the $\| \ \|_0$ norm; we call these limits v_j. By lemma 1, $\{v^{(n)}\}$ converges in the $\| \ \|_0$ to a limit in H_0, which we identify with the limit v in H_1^0. Integration by parts gives relation (13) for $v^{(n)}$ in place of v; letting $n \to \infty$ gives (13); relation (11) follows similarly. $\qquad\Box$

We claim that the identification of elements v in H_1^0 with elements of H_0 is one-to-one, that is, an embedding of H_1^0 in H_0. We have to show that if v is zero in H_0, then it is zero in H_1^0. Clearly, it follows from (13) that if $v = 0$ in H_0, then $v_j = 0$ in H_0 for all j. This makes $v = 0$ in H_1^0.

Relation (13) asserts that v_j are the first partial derivatives of v in the sense of distributions (see Appendix B).

Let f be any element of H_0; define the linear functional ℓ by

$$\ell(u) = (u, f)_0. \tag{14}$$

By the Schwarz inequality, and inequality (12) of lemma 1,

$$|\ell(u)| \le \|f\|_0 \|u\|_0 \le d \|f\|_0 \|u\|_1 \tag{15}$$

for all u in H_1^0. According to the Riesz-Frechet representation theorem, theorem 4 of chapter 6, the functional (14) can be represented as an inner product. That is, there exists v in H_1^0 such that

$$(u, f)_0 = (u, v)_1 \tag{16}$$

for all u in H_1^0. By definition (11) and lemma 2, with $u_j = \partial u / \partial x_j$,

$$(u, v)_1 = \sum (u_j, v_j)_0. \tag{17}$$

Take now u to be C_0^∞. We can, using the theory of distributions, rewrite the right side of (17) as

$$-\sum (u, v_{jj})_0 = -(u, \Delta v)_0, \tag{17'}$$

where v_{jj} are the second partial derivatives of v, and Δ the Laplace operator, acting in the sense of distributions. Combining (17), (17'), and (16), we deduce that

$$(u, f)_0 = -(u, \Delta v)_0$$

for all u in C_0^∞. From this it follows that in the sense of distribution theory,

$$f = -\Delta v. \tag{18}$$

Thus v is a *distribution solution* of the inhomogeneous equation (18).

Next we show that by virtue of belonging to H_1^0, $v(x)$ tends to zero in an average sense as x tends to the boundary of D. The precise statement is lemma 3:

Lemma 3. *Suppose that D is a domain in $\mathbb{R}^2$ whose boundary ∂D is a C^1 curve. For any point p on the boundary of D, choose a coordinate system x_1, x_2, with p as the origin, and the positive x_1 axis perpendicular to the boundary of D and pointing inward. Denote by $R(p, d)$ all points of D where $x_1 < d$ and $|x_2| < d$. Let v be any function in H_1^0. We claim that the mean value of $|v|$ over $R(p, d)$ tends to zero as d tends to zero.*

Proof. Since the area of $R(p, d)$ is proportional to d^2, the claim is that

$$\int_R |v|\, dx \le o(d^2). \tag{19}$$

To deduce this, we need to estimate

$$\int_R |f|\, dx$$

for functions f in $C_0^\infty(D)$. Integrate by parts with respect to x_1:

$$\int_R |f|\, dx = \int_R |f|_1 (d - x_1)\, dx \le \int_R |f_1||d - x_1|\, dx$$

$$\le \left(\int_R (d - x_1)^2\, dx \int f_1^2\, dx \right)^{1/2} \le d^2 \left(\int_R f_1^2\, dx \right)^{1/2}; \tag{20}$$

in the second step we have used the Schwarz inequality. Now approximate v in the 1-norm by a sequence $v^{(n)}$ of $C_0^\infty(D)$ functions. The limit of (20) for $f = v^{(n)}$ is

$$\int_R |v|\, dx \le d^2 \left(\int_R v_1^2\, dx \right)^{1/2}. \tag{21}$$

Since the integral of v_1^2 over $R(d)$ tends to zero as d tends to zero, (19) follows from (21). $\qquad\square$

We have thus succeeded in constructing a generalized function v that solves the differential equation (18) in the sense of distribution theory, and that vanishes on the boundary in a mean value sense.

The argument above can be extended to solve the Dirichlet problem in this generalized sense for any second-order partial differential equation that is self-adjoint and positive. We show now how the Lax-Milgram lemma can be used to extend the argument above to non-self-adjoint partial differential operators. For example, consider for any pair u, v in H_1^0 the functional B defined by

$$B(u, v) = \int_D \left(\sum u_j v_j + \sum u v_j + uv \right) dx\, dy. \tag{22}$$

Clearly, B is *bilinear*; it follows from lemma 1 that it is *bounded*. Estimating the middle term by the Schwarz inequality, we see that B is positive in the sense of theorem 6 of chapter 6. It follows then from that theorem that the linear functional (14) can be represented in terms of B, meaning that there exists a v in H_1^0 such that

$$(u, f)_0 = B(u, v) \qquad (23)$$

for all u in H_1^0. Integrating by parts on the right in (23), we deduce that

$$f = -\Delta v + \sum v_j + v \qquad (24)$$

in the sense of distributions.

We present now another method for solving the Dirichlet problem for the homogeneous Laplace equation

$$\Delta v = 0 \quad \text{in} \quad D \qquad (25)$$

whose value on the boundary is prescribed. This method exploits some special properties of harmonic functions and yields genuine solutions that satisfy the boundary condition in the usual sense. We assume that ∂D is once differentiable, and that the boundary values of v are also once differentiable. We can then construct a C^1 function f on $D \cup \partial D$ that has the prescribed value on ∂D, and we state the boundary condition thus:

$$v = f \quad \text{on} \quad \partial D. \qquad (26)$$

We reformulate the boundary value problem (25), (26): decompose f as

$$f = v + z, \qquad (27)$$

where v is harmonic and z vanishes on ∂D. When v and z have continuous first partial derivatives up to the boundary, we can apply Green's formula:

$$(v, z)_1 = \int_D \sum v_j z_j \, dx \, dy = -\int_D (\Delta v) z \, dx \, dy + \int_{\partial D} \frac{dv}{dn} z \, ds.$$

Since by (25), $\Delta v = 0$ in D, and by (26), $z = 0$ on ∂D, the right side $= 0$. In words, *the space of harmonic functions and the space of functions that vanish on the boundary are orthogonal to each other in the* $(\ ,\ )_1$ *scalar product.*

Performing the decomposition (27) thus appears as the task of splitting f into the sum of two functions from two orthogonal function spaces. We show how this can be accomplished by appealing to theorem 3 of chapter 6. The scalar product defined by (11) for all f in $C^\infty(D)$ is not positive because $(f, f)_1 = 0$, not only for $f \equiv 0$ but for all constant functions f. We overcome this slight blemish by considering two functions as equivalent if they differ by a constant.

Denote by H_1 the completion in the $\|\ \|_1$ norm of all C^1 functions on $D \cup \partial D$; the space H_1^0 is a closed subspace of H_1. We apply now the orthogonal decomposition theorem, theorem 3 of chapter 6, to conclude that every f in H_1 can be decomposed uniquely as in (27), where z in H_1^0, $v \perp H_1^0$. The condition $v \perp H_1^0$ asserts that

$$(u, v)_1 = \sum (u_j, v_j)_0 = 0$$

for all u in H_1^0. Taking u to be C_0^∞, we can integrate by parts in the sense of distributions to get

$$0 = \sum (u_j, v_j)_0 = -\sum (u, v_{jj})_0 = -(u, \Delta v)_0,$$

which implies that

$$\Delta v = 0$$

in the sense of distributions. It is a well-known result of Hermann Weyl that a function harmonic in the distribution sense is harmonic in the classical sense; a proof is provided in section 4 of Appendix B.

We claim that when f is continuous up to the boundary, $z(q)$ tends to zero as q approaches the boundary. Denote by d the distance of q to ∂D, and let C be the circular disc with center q and radius $r = d/2$ and denote mean values over C by bars: taking the mean value of (27), we obtain

$$\bar{f}(q) = \bar{z}(q) + \bar{v}(q). \tag{28}$$

Since f is continuous up to the boundary, its mean value on C differs from its value at the center q of C by an amount w that tends to zero as d tends to zero. According to the elementary theory of partial differential equations, the mean value of the harmonic function v on a circular disc C is equal to its value at the center of C. So (28) can be rewritten as

$$\bar{f}(q) = \bar{z}(q) + v(q).$$

Subtracting (27) from this, we conclude that

$$w = \bar{z}(q) - z(q). \tag{29}$$

We appeal now to lemma 3, according to which the mean value over $R(p, d)$ of the absolute value of a function z in H_1^0 tends to zero as the distance d of q to ∂D tends to zero. It follows that the mean value of $|z|$ over the disc C tends to zero. Combined with (29), we conclude that $z(q)$ itself tends to zero as q approaches the boundary. Using (27), we see that the harmonic function v is continuous up to the boundary, and its boundary value equals f. Thus we succeeded in constructing not a generalized, but a genuine solution of the Dirichlet boundary value problem.

Here is another way of looking at (27). We saw that the boundary value problem (25), (26) for the Laplace equation amounts to decomposing f as a sum of a harmonic function and of one vanishing on the boundary. We showed that these spaces are orthogonal to each other in the $(\ ,\ )_1$ scalar product. The decomposition was accomplished by appealing to the orthogonal decomposition theorem, theorem 3 of chapter 6. Here we show how to perform an orthogonal decomposition with respect to the *subspace V consisting of harmonic functions whose first derivatives are square integrable in D*. It is an easy fact in the theory of harmonic functions that V is *complete* in the $\| \ \|_1$-norm. So, according to theorem 3 of chapter 6, we can decompose any f in H_1 as

$$f = v + z, \tag{30}$$

where v is in the space V of harmonic functions with square integrable first derivatives, and z is orthogonal to V. Our aim is to show that z vanishes on the boundary.

We take D to lie in $\mathbb{R}^2$, and assume that the boundary of D is twice differentiable; we assume f to be twice differentiable. For any pair of points p and q in the plane, we define $k(p, q)$ as the fundamental singular solution of the Laplace equation:

$$k(p, q) = -\frac{1}{2\pi} \log |p - q|. \tag{31}$$

Suppose that q lies in D. If we knew that the function z vanished on the boundary of D, then by Green's formula, see section 4 of Appendix B, we would deduce that

$$z(q) = \int_D (z_x k_x + z_y k_y) \, dx \, dy,$$

where (x, y) are the coordinates of p. However, we do not at this point know that z vanishes on ∂D, and so we denote the function defined by the integral above as u:

$$u(q) = \frac{1}{2\pi} \int_D \left(z_x \frac{x' - x}{|p - q|^2} + z_y \frac{y' - y}{|p - q|^2} \right) dx \, dy, \tag{32}$$

where x', y' denote the coordinates of q.

Lemma 4. *If ∂D is twice differentiable, $u(q)$ defined by (32) is continuous up to the boundary and vanishes there.*

Proof. It is easy to show that u is continuous inside D. Let q be a point of D near the boundary; denote the nearest boundary point by b. Since ∂D is assumed to be twice differentiable, there are two circular disks S and $\overline{S}$ with the same radius, d, tangent to ∂D at b, S contained in D and $\overline{S}$ exterior to D. Denote by $\bar{q} = (\bar{x}, \bar{y})$ the image of q under *inversion* across the circle bounding S. For q near enough to ∂D, $\bar{q}$ lies in $\overline{S}$.

Since $\bar{q}$ lies outside D, $k(p, \bar{q})$ is a regular harmonic function in D. In particular, $k(p, \bar{q})$ belongs to V, and thus is orthogonal to z in the $(\ , \ )_1$ scalar product. Therefore $u(\bar{q}) = 0$, and we can write

$$u(q) = u(q) - u(\bar{q}) = \frac{1}{2\pi} \int_D z_x \left(\frac{x' - x}{|p - q|^2} - \frac{\bar{x} - x}{|p - \bar{q}|^2} \right)$$

$$+ z_y \left(\frac{y' - y}{|p - q|^2} - \frac{\bar{y} - y}{|p - \bar{q}|^2} \right) dx \, dy. \tag{33}$$

As q approaches the boundary point b, so does the point $\bar{q}$. Therefore the integrand on the right in (33) tends to zero uniformly at all points of D whose distance from b exceeds any positive quantity r. It can be shown (see Lax for details) that also the integral over the remaining portion of D tends to zero. This completes the proof of lemma 4. $\qquad\square$

Lemma 5. *The function u defined in D by (32) is twice differentiable in D, and*

$$\Delta u = \Delta z. \tag{34}$$

Proof. Since f was assumed twice differentiable, and since by (30), z differs from f by a harmonic function v, it follows that z is twice differentiable in D and that $\Delta z = \Delta f$. Let D' be a subdomain of D that contains q, and whose closure is contained in D. We split the integral on the right of (32) and integrate by parts over D':

$$u(q) = z(q) + \int_{\partial D'} z \frac{\partial}{\partial n} k \, ds + \int_{D-D'} \left(z_x k_x + z_y k_y \right) dx \, dy. \tag{35}$$

The two integrals on the right are harmonic functions in D'; therefore, since D' is arbitrary,

$$u = z + h, \tag{36}$$

h harmonic. This proves lemma 5. $\qquad\square$

Express z from (36) as $u - h$ and set into (30):

$$f = v - h + u.$$

Since $v - h$ is harmonic, and $u = 0$ on ∂D, $v - h$ *solves the boundary value problem* (25), (26).

The preceding proof is a reworking of an argument of Garabedian and Schiffer.

BIBLIOGRAPHY

Garabedian, P. and Schiffer, M. On existence theorems of potential theory and conformal mapping. *An. Math.*, **73** (1950): 107–121.

Lax, P. D. A remark on the method of orthogonal projection. *CPAM*, **4** (1951): 457–464.

von Neumann, J. On rings of operators III. *An. Math.*, **4** (1940): 94–161; see p. 127.

Nikodym, O. M. Sur une généralisation des integrals de M. J. Radon. *Fund. Math.*, **15** (1930): 131–179.

Radon, J. Theorie und Anwendung der absolut additiven Mengenfunktionen. *S. B. Akad. Wiss. Wien*, **122** (1913): 1295–1438.

Weyl, H. The method of orthogonal projection in potential theory. *Duke Math. J.*, **7** (1940): 411–444.

Zaremba, S. *Sur le Principe de minimum.* Krakauer Akademieberichte, 1909.

8

DUALS OF NORMED LINEAR SPACES

8.1 BOUNDED LINEAR FUNCTIONALS

In this chapter we deal with normed linear spaces X over the *real* or the *complex* numbers. We will study *linear functionals*, namely mappings ℓ of X into $\mathbb{R}$ or $\mathbb{C}$ satisfying

$$\ell(ax) = a\ell(x), \quad \ell(x + y) = \ell(x) + \ell(y), \tag{1}$$

that are in addition *continuous*. They are continuous in that they satisfy

$$\lim_{n \to \infty} \ell(x_n) = \ell(x) \quad \text{when} \quad \lim_{n \to \infty} |x_n - x| = 0. \tag{2}$$

Definition. The collection of all continuous linear functionals is called the *dual* of X. It is denoted by X'.

Clearly, the sum and constant multiple of continuous linear functionals is continuous and linear; thus X' is a linear space.

Definition. A linear functional ℓ on X is called *bounded* if there is a positive number c such that

$$|\ell(x)| \leq c|x| \qquad \text{for all } x \text{ in } X, \tag{3}$$

where $|\ |$ on the left denotes the absolute value.

Theorem 1. *A linear functional ℓ on X is continuous if and only if it is bounded.*

Proof. Set $x_n - x = y_n$ in (2); using (1) and (3), we get

$$|\ell(x_n) - \ell(x)| = |\ell(y_n)| \leq c|y_n|;$$

this shows that boundedness implies continuity.

Suppose that ℓ is not bounded; then for any choice of $c = n$, (2) is violated by some x_n:

$$\ell(x_n) > n|x_n|.$$

Clearly, x_n can be replaced by any multiple of x_n; if we normalize x_n so that

$$|x_n| = \frac{1}{\sqrt{n}},$$

then $x_n \to 0$ but $\ell(x_n) \to \infty$. This shows that lack of boundedness implies lack of continuity. $\qquad\square$

Theorem 2. *The nullspace of a bounded linear functional ℓ on a normed linear space is a closed linear subspace. For ℓ nontrivial, meaning $\not\equiv 0$, the nullspace has codimension 1.*

Proof. The nullspace of any linear map is a linear subspace. Since a bounded linear functional is continuous, it follows that the inverse image of 0 is closed. That for $\ell \not\equiv 0$ the nullspace has codimension 1 is immediate. $\qquad\square$

Definition. The *norm* of a bounded linear functional is the smallest c for which (3) holds; it is denoted as $|\ell|$:

$$|\ell| = \sup_{x \neq 0} \frac{|\ell(x)|}{|x|}; \tag{4}$$

by homogeneity, we may take x to have norm equal to 1.

Theorem 3. *The dual X' of any normed linear space is a complete normed linear space under the norm defined by (4).*

Proof. Homogeneity and positivity are obvious. For subadditivity, consider two bounded linear functionals ℓ and m:

$$
\begin{aligned}
|\ell + m| &= \sup_{|x|=1} |(\ell + m)(x)| \leq \sup_{|x|=1} (|\ell(x)| + |m(x)|) \\
&\leq \sup_{|x|=1} |\ell(x)| + \sup_{|x|=1} |m(x)| = |\ell| + |m|.
\end{aligned}
$$

We show now completeness: let $\{\ell_m\}$ be a Cauchy sequence in X':

$$|\ell_n - \ell_m| \to 0 \qquad \text{as } n, m \to \infty. \tag{5}$$

According to definition (4) of the norm for functionals and (5),

$$|(\ell_n - \ell_m)(x)| = |\ell_n(x) - \ell_m(x)| \leq |\ell_n - \ell_m||x| \to 0 \qquad \text{as } n, m \to \infty,$$

for every x in X. Since the field of scalars, $\mathbb{R}$ or $\mathbb{C}$, is complete,

$$\lim_{n\to\infty} \ell_n(x) = \ell(x)$$

exists. It is easy to show that $\ell(x)$ is linear and bounded, and it is not hard to deduce from (5) that if $|\ell_n - \ell_m| \le \epsilon$ for $m > n$, then also $|\ell_n - \ell| \le \epsilon$. Therefore

$$\lim_{n\to\infty} |\ell_n - \ell| = 0. \qquad \square$$

8.2 EXTENSION OF BOUNDED LINEAR FUNCTIONALS

So far we have not shown the existence of a single linear functional except $\ell \equiv 0$. Certainly there are lots of them in a Hilbert space; we show now that there are just as many in a Banach space. The tool needed is the Hahn-Banach theorem, specialized to the case

$$p(x) = c|x|.$$

Theorem 4. *Let X be a normed linear space over the real or complex numbers, Y a subspace, and ℓ a linear functional defined on Y and bounded there:*

$$|\ell(y)| \le c|y|, \qquad y \text{ in } Y.$$

Then ℓ can be extended as a bounded linear functional to all of X so that its bound on X equals its bound on Y.

This theorem is a special case of theorem 8 of chapter 3. We give now some applications.

Theorem 5. *Say that $y_1, \ldots, y_N$ are N linearly independent vectors in a normed linear space X, $a_1, \ldots, a_N$ arbitrary complex numbers. Then there exists a bounded linear functional ℓ such that*

$$\ell(y_j) = a_j, \qquad j = 1, \ldots, N. \tag{6}$$

Proof. Denote by Y the linear space spanned by $y_1, \ldots, y_m$; it consists of vectors of the form

$$y = \sum b_j y_j.$$

Since the y_j are linearly independent, this representation of y is unique. Now define ℓ on Y by

$$\ell(y) = \sum b_j a_j.$$

Clearly, ℓ is linear and bounded on Y, and satisfies (6); by theorem 4, it can be boundedly extended to all of X. $\qquad \square$

Corollary 4′. *Every finite-dimensional subspace Y of a normed linear space X has a closed complement.*

Proof. Choose a basis $y_1, \ldots, y_N$ in Y. According to theorem 5, there exist N bounded linear functionals ℓ_j, $j = 1, \ldots, N$, such that

$$\ell_j(y_k) = \delta_{jk};$$

according to theorem 2, the nullspace Z_j of ℓ_j is closed. So then is their intersection

$$Z = Z_1 \cap \ldots \cap Z_N.$$

It is easy to check that Z and Y are complementary, namely that $X = Y \oplus Z$. $\qquad\square$

Theorem 6. *For every y in a normed linear space X over the real or complex field,*

$$|y| = \max_{|\ell|=1} |\ell(y)|. \tag{7}$$

Proof. By definition (4) of $|\ell|$, $|\ell(y)| \leq |\ell|\,|y|$. It follows that the right side of (7) is $\leq$ the left side. Therefore to prove the result, we have to exhibit for every y in X an ℓ in X' such that

$$\ell(y) = |y|, \quad |\ell| = 1.$$

To accomplish this, we note that this defines ℓ on all scalar multiples Y of y as $\ell(ay) = a|y|$. Clearly, on this one dimensional space Y, ℓ has norm $= 1$. By theorem 4, ℓ can be extended to all of X so that $|\ell| = 1$.

$\qquad\square$

Corollary 5′. *When the field of scalars is $\mathbb{R}$, for every x in X*

$$|x| = \max_{|\ell|\leq 1} \ell(x). \tag{8}$$

The following is a far-reaching generalization of theorem 6.

Theorem 7. *X is a normed linear space over $\mathbb{C}$, Y a linear subspace of X. For any z in X, denote by $m(z)$ its distance from Y:*

$$m(z) = \inf_{y \text{ in } Y} |z - y| \tag{9}$$

We claim that for every z in X

$$m(z) = M(z), \tag{10}$$

where

$$M(z) = \max_{|\ell|\leq 1, \ell=0 \text{ on } Y} |\ell(z)|. \tag{11}$$

Proof. Since the functionals ℓ entering the maximum problem (11) vanish on Y, and since $|\ell| \leq 1$, $|\ell(z)| = |\ell(z - y)| \leq |z - y|$ holds for all y in Y; therefore

$$|\ell(z)| \leq \inf_{y \text{ in } Y} |z - y| = m(z).$$

It follows from this and the definition (11) of $M(z)$ that

$$M(z) \leq m(z). \tag{12}$$

To show equality, we look at the linear space Y_0 consisting of all vectors of the form $y + az$, y in Y, a complex, and define on Y_0 the linear functional ℓ_0:

$$\ell_0(y + az) = am(z). \tag{13}$$

By definition (9) of m, it follows that ℓ_0 is bounded on Y_0 by 1; so by theorem 4, it can be extended to all of X so that $|\ell_0| = 1$. Set $y = 0$, $a = 1$ in (13):

$$\ell_0(z) = m(z).$$

Combined with (12) this shows that ℓ_0 solves the maximum problem (11), and that (10) holds.   □

REMARK 1. In case Y is the trivial subspace consisting of $\{0\}$, theorem 7 reduces to theorem 6.

Theorem 7 is an example of *dual variational problems*: a pair of a minimum and a maximum problem whose extreme values are *equal*.

Definition. The set of linear functionals ℓ that vanish on a subspace Y of X is called the *annihilator* of Y, and is denoted by $Y^\perp$.

Exercise 1. Show that $Y^\perp$ is a closed linear subspace of X'.

Exercise 2. Let Y be a closed subspace of a normed linear space X. Show that the dual of (X/Y) is isometrically isomorphic with $Y^\perp$.

Theorem 7'. *X is a normed linear space over $\mathbb{C}$, Y a subspace of X. For any ℓ in X', define*

$$|\ell|_Y = \sup_{\substack{y \text{ in } Y \\ |y|=1}} |\ell(y)|. \tag{14}$$

We claim that

$$|\ell|_Y = \min_{m \text{ in } Y^\perp} |\ell - m|. \tag{15}$$

Proof. For any m in $Y^\perp$, and any y in Y with $|y| = 1$,

$$|\ell(y)| = |(\ell - m)(y)| \le |\ell - m|. \qquad (16)$$

It follows that $|\ell|_Y$ is $\le$ the right side of (15).

According to theorem 4, the restriction of ℓ to Y has an extension to X, call it ℓ_0, whose norm on X equals its norm on Y:

$$|\ell_0| = |\ell|_Y. \qquad (17)$$

Since ℓ_0 and ℓ are equal on Y, $\ell - \ell_0 = m$ belongs to $Y^\perp$; furthermore by (17),

$$|\ell - m| = |\ell_0| = |\ell|_Y. \qquad (17')$$

This combined with (16) proves that equality holds in (15). $\qquad\qquad\qquad\square$

Theorem $7'$ is another example of dual variational problems.

Exercise 3. Show that Y' is isometrically isomorphic with $X'/Y^\perp$.

Definition. The *closed linear span* of a subset $\{y_j\}$ of a normed linear space is the *smallest* closed linear space containing all y_j, that is, the *intersection* of all closed linear spaces containing all y_j.

Exercise 4. Show that the closed linear span of $\{y_j\}$ is the closure of the *linear span* Y of $\{y_j\}$, consisting of all finite linear combinations of the y_j:

$$y = \sum_F a_j y_j. \qquad (18)$$

The following result, called the *spanning criterion*, is one of the workhorses of functional analysis.

Theorem 8. *A point z of a normed linear space X belongs to the closed linear span Y of a subset $\{y_j\}$ of X iff every bounded linear functional ℓ that vanishes on the subset vanishes at z; that is,*

$$\ell(y_j) = 0 \qquad \text{for all } y_j \qquad (19)$$

implies that $\ell(z) = 0$.

Proof. Since ℓ is linear, (19) implies that $\ell(y) = 0$ for y of form (18); since ℓ is continuous, it vanishes on all limits of points of form (18). Conversely, suppose that z does not belong to the closed linear span Y of $\{y_j\}$; then

$$\inf_{y \text{ in } Y} |z - y| = d > 0. \qquad (20)$$

Define the subspace Z to consist of all points of the form

$$y + az, \qquad y \text{ in } Y, \tag{21}$$

and define on Z the linear functional ℓ_0 by

$$\ell_0(y + az) = a.$$

It follows from (20) that

$$|y + az| \geq d|a|.$$

Combining this with the definition of ℓ_0, we deduce that on Z, ℓ_0 is bounded by d^{-1}. So by theorem 4, ℓ_0 can be extended boundedly to all of X. By definition,

$$\ell_0(y_j) = 0 \qquad \text{for all } y_j, \ell_0(z) = 1. \qquad\qquad \square$$

NOTE. Theorem 8 is a generalization to Banach spaces of theorem 7 in chapter 6 on Hilbert spaces.

8.3 REFLEXIVE SPACES

The dual X' of a normed linear space has its own dual, denoted as X''. Since $\ell(x)$ is a bilinear function of ℓ and x, and is bounded, definition (4), it follows that, for fixed x, $\ell(x)$ is a bounded linear functional of ℓ. It follows from theorem 6 that the norm of this linear functional is $|x|$. Thus the space X is, in this natural way, isometrically embedded in X''. It is a basic result of the theory of finite dimensional vector spaces that $X'' \equiv X$. This is no longer true for all Banach spaces.

Definition. A Banach space is called reflexive if $X'' = X$, that is, if X is all of X''.

Theorem 9. *Every Hilbert space is reflexive.*

Proof. This is an immediate consequence of theorem 4 in chapter 6. $\qquad\qquad \square$

The following result is due to Milman:

Theorem 10. *A uniformly convex Banach space is reflexive.*

For *proof* we refer to Milman.

We stated in chapter 5 that the L^p spaces, $1 < p < \infty$, are uniformly convex. Combining this result of Clarkson's with the preceding result of Milman, we conclude that L^p, $1 < p < \infty$, are reflexive.

Theorem 11. *The dual of L^p is L^q,*

$$\frac{1}{p} + \frac{1}{q} = 1.$$

Proof. We saw in chapter 5 that for any u in L^q we can define a bounded linear functional ℓ on L^p by

$$\ell(f) = (f, u) = \int f(s) u(s) dm.$$

Furthermore we showed, as theorem 5 of chapter 5, that the norm of this linear functional is $|u|_q$. Thus L^q is isometrically embedded in $(L^p)'$. We claim that L^q is all of $(L^p)'$; for if not, there would be some z in $(L^p)'$ not in L^q. Since L^q is closed, it would follow from the spanning criterion, theorem 8, that there is an $\ell \neq 0$ in $(L^p)''$ such that $\ell(u) = 0$ for all u in L^q. Since L^p is reflexive, ℓ lies in L^p, and it follows that $(\ell, u) = 0$ for all u in L^q. By theorem 5 of chapter 5, this implies that $\ell = 0$, a contradiction. $\qquad\qquad\qquad\qquad\qquad\qquad\qquad\qquad\qquad\qquad\qquad\qquad\qquad \square$

We give now a second proof that for $p < 2$ the dual of L^p is L^q, without appealing to uniform convexity.

We assume for simplicity that the total measure with respect to which we form the L^p norms equals 1. Then, by Hölder's inequality with $p' = 2/p, q' = 2/(2-p)$,

$$\|f\|_p^p = \int |f|^p dm \leq \|1\|_{2/(2-p)} \|f^p\|_{2/p} = \|f\|_2^p.$$

So $\|f\|_p \leq \|f\|_2$ for all f in L^2.

Let ℓ be a linear functional defined on all L^2 functions that is bounded in the L^p norm:

$$|\ell(f)| \leq \text{ const. } \|f\|_p.$$

Since the p-norm is less than the 2-norm, ℓ is also bounded in the L^2-norm. According to the representation theorem, theorem 4 in chapter 6, we can express ℓ as

$$\ell(f) = \int f u \, dm, \qquad u \text{ in } L^2. \tag{22}$$

We claim that in fact u lies in L^q. To see this, we choose $f = f_k$ as follows:

$$f_k(x) = |u_k|^{q-1}(x) \text{ sgn } u(x),$$

where

$$|u_k|(x) = \min\{|u(x)|, k\},$$

k a constant. Setting $f = f_k$ in (22) gives

$$\ell(f_k) = \int f_k u \, dm = \int |u_k^{q-1}| \, |u| \, dm \geq \int |u_k|^q \, dm.$$

On the other hand,

$$\| f_k \|_p^p = \int |u_k|^{(q-1)p} \, dm = \int |u_k|^q \, dm.$$

Since, by assumption, $|\ell(f_k)| \leq c |f_k|_p$, the last two inequalities imply that

$$\int |u_k|^q \, dm \leq c \left(\int |u_k|^q \, dm \right)^{1/p}.$$

Dividing both sides by the right side gives

$$\| u_k \|_q \leq c.$$

Letting $k \to \infty$ we deduce that u is in L_q. This completes the proof. $\qquad\square$

Exercise 5. Show that if the total measure equals 1, then $\| f \|_p$ is an increasing function of p.

Theorem 12. $C[-1, 1]$, *normed by the maximum norm, is not reflexive.*

Proof. If it were, C would be the dual of C'. According to theorem 6 applied to $X = C'$, for every ℓ in C' there is an f in $C'' = C$ such that

$$|\ell| = \ell(f), \quad |f|_{\max} = 1. \tag{23}$$

Now define

$$\ell(g) = \int_{-1}^{0} g(t) \, dt - \int_{0}^{1} g(t) \, dt.$$

Clearly, for every g in $C[-1, 1]$,

$$|\ell(g)| < 2 |g|_{\max}, \tag{23'}$$

but given any $\epsilon > 0$, we can choose g so that

$$|\ell(g)| > (2 - \epsilon) |g|_{\max}.$$

This result shows that $|\ell| = 2$. Along with (23'), it contradicts (23) for $g = f$. $\quad\square$

Theorem 13. *Let Z be a normed linear space over $\mathbb{C}$. If Z' is separable, so is Z.*

Proof. Separability means that Z' contains a dense denumerable set $\{\ell_n\}$. By definition of the norm in Z', there is a z_n in Z such that

$$|z_n| = 1, \quad \ell_n(z_n) > \tfrac{1}{2}|\ell_n|. \tag{24}$$

We claim that the denumerable set $\{z_n\}$ has Z as its closed linear span. According to theorem 8 this means that a linear functional ℓ that vanishes on every z_n vanishes everywhere. Suppose, on the contrary, that there is an ℓ such that

$$\ell(z_n) = 0 \quad \text{for all } n, \text{ and } |\ell| = 1. \tag{25}$$

Since $\{\ell_n\}$ are dense in Z', we can find an ℓ_n such that

$$|\ell - \ell_n| < \tfrac{1}{3}. \tag{26}$$

Since $|\ell| = 1$, it follows that

$$|\ell_n| > \tfrac{2}{3}. \tag{26'}$$

Since $\ell(z_n) = 0$, it follows from (26), (24) that

$$\tfrac{1}{3} > |(\ell - \ell_n)(z_n)| = |\ell_n(z_n)| > \tfrac{1}{2}|\ell_n|.$$

This contradicts (26'), and shows that no ℓ satisfying (25) exists. It proves, by theorem 8, that finite linear combinations of the z_n are dense in Z. But then finite linear combinations of the z_n with rational coefficients also are dense in Z; since these are denumerable, Z is separable. $\qquad\square$

Theorem 13 furnishes another proof of theorem 12. For $C[-1, 1]$ is separable: every continuous function can be approximated by piecewise linear functions with rational nodes and rational ordinates. On the other hand, C' is *not* separable; the linear functionals ℓ_s defined by

$$\ell_s(f) = f(s), \quad -1 \le s \le 1,$$

are clearly each bounded by 1, and equally clearly

$$|\ell_s - \ell_t| = 2 \quad \text{for } s \ne t.$$

Since the $\{\ell_s\}$ form a nondenumerable collection, C' cannot contain a dense denumerable subset. It follows now that $C'' \ne C$. If it were, we could apply theorem 13 to $Z = C'$ and conclude that since $C'' \equiv C$ is separable, so is C', but C' is not separable. $\qquad\square$

The conclusion of theorem 12 is applicable to the space $C(Q)$, Q any Hausdorff space containing more than a discrete set of points. This is the precise state of affairs:

Theorem 14. *Let Q be a compact Hausdorff space, C(Q) the space of continuous real-valued functions on Q, normed by the max norm.*

(i) *C′ consists of all signed measures m of finite total mass, defined over all Borel sets. That is, every bounded linear functional ℓ on C(Q) can be written as*

$$\ell(f) = \int_Q f \, dm. \tag{27}$$

The norm of ℓ is

$$|\ell| = \int_Q |dm|. \tag{28}$$

The measure m is uniquely determined by ℓ.

(ii) *C″ is $L^\infty(Q)$, the space of all bounded, Borel-measurable functions on Q.*

The prototype of this basic result is due to F. Riesz; the general result is due to Kakutani. A functional analytic proof for Q metric is supplied in Appendix A.

NOTE. Theorem 14 is emphatically *false* when Q is *not compact* and C(Q) is the space of all *bounded* continuous functions on Q, normed by the sup norm. Here is what happens:

Take Q to be the real line $\mathbb{R}$, $\{t_k\}$ a sequence of points $\to \infty$. We take Y to be the subspace of $C(\mathbb{R})$ consisting of all functions f for which

$$\lim f(t_k) = f_\infty$$

exists. For f in Y we define the functional ℓ by

$$\ell(f) = f_\infty.$$

Clearly, ℓ is linear and ℓ is bounded on Y: $|\ell|_Y \leq 1$. By the Hahn-Banach theorem, ℓ can be extended to all of $C(\mathbb{R})$ as a bounded linear functional.

We claim that this ℓ cannot be of the form (27). If it were, the value of $\ell(f)$ would depend on values of f on any compact interval I on which

$$\int_I |dm| \neq 0.$$

But clearly, we can alter the values of f in Y on I without changing the value of $\ell(f)$; so there cannot be such a dependence.

The following result is of some interest:

Theorem 15. *A closed linear subspace Y of a reflexive Banach space X is reflexive.*

Proof. Every bounded linear functional ℓ on X, when restricted to Y, becomes a bounded linear functional on Y; we denote this functional by ℓ_0. Since by Hahn-Banach every bounded linear functional on Y can be extended to X, this *restriction* map $\ell \to \ell_0$,

$$X' \longrightarrow Y'$$

maps X' *onto* Y'. The restriction map induces the following mapping from Y'' to X'':
For any η in Y'' we define ζ in X'' by setting, for any ℓ in X',

$$\zeta(\ell) = \eta(\ell_0), \qquad (29)$$

where ℓ_0 is the restriction of ℓ to Y. Since X is reflexive, ζ can be identified with an element z of X:

$$\zeta(\ell) = \ell(z);$$

setting this into (29) gives

$$\ell(z) = \eta(\ell_0). \qquad (29')$$

We claim that z belongs to Y. To show this, we note that if ℓ belongs to $Y^\perp$, meaning it vanishes on Y, then $\ell_0 = 0$, and so by (29'), $\ell(z) = 0$. We appeal now to theorem 8 to conclude that z belongs to the closure of Y. But since Y is closed, z belongs to Y. So we can rewrite (29') as

$$\ell_0(z) = \eta(\ell_0). \qquad (30)$$

Since every functional in Y' occurs as ℓ_0, (30) shows that every η in Y'' can be identified with some z in Y. $\qquad \square$

8.4 SUPPORT FUNCTION OF A SET

We recall from chapter 1 the notion of the *convex hull* of a pointset M in a linear space X over the reals as the *smallest* convex set in X containing M, that is, the *intersection* of all convex sets that contain M. The convex hull of M is denoted by $\tilde{M}$.

As remarked in theorem 6 of chapter 1, $\tilde{M}$ consists of all convex combinations of points of M. These are points of the form

$$x = \sum_F a_j x_j, \qquad x_j \text{ in } M \qquad (31)$$

$$a_j \geq 0, \qquad \sum_F a_j = 1. \qquad (31')$$

Definition. The *closed convex hull* of a subset M of a normed linear space X is the *smallest closed convex set* containing M, that is the *intersection* of all closed convex sets containing M. We denote this set as $\check{M}$.

Exercise 6. Show that the closed convex hull of M is the closure of the convex hull of M.

Definition. For any bounded subset M of a normed linear space X over $\mathbb{R}$, we define the *support function* S_M as the following function on X':

$$S_M(\ell) = \sup_{y \text{ in } M} \ell(y). \tag{32}$$

Theorem 16. *Support functions have the following properties:*

  (i) Subadditivity, for all ℓ, m in X', $S_M(\ell + m) \leq S_M(\ell) + S_M(m)$.

 (ii) $S_M(0) = 0$.

(iii) Positive homogeneity, $S_M(a\ell) = a S_M(\ell)$ for $a > 0$.

(iv) Monotonicity, for $M \subset N$, $S_M(\ell) \leq S_N(\ell)$.

 (v) Additivity, $S_{M+N} = S_M + S_N$.

 (vi) $S_{-M}(\ell) = S_M(-\ell)$.

(vii) $S_{\overline{M}} = S_M$.

(viii) $S_{\check{M}} = S_M$.

Exercise 7. Prove theorem 16.

We give now some examples.

  (a) M consists of a single point x_0,

$$S_{\{x_0\}}(\ell) = \ell(x_0).$$

  (b) M is the ball B_R of radius R around 0: $\{|x| \leq R\}$,

$$S_{B_R}(\ell) = R|\ell|.$$

  (c) M is the ball $B_R(x_0)$: $\{|x - x_0| \leq R\}$, using examples (a) and (b), and part (v) of theorem 16, we get

$$S_{B_R(x_0)}(\ell) = \ell(x_0) + R|\ell|. \tag{33}$$

Theorem 17. *X is a normed linear space over $\mathbb{R}$, M a bounded subset of X. A point z of X belongs to the closed, convex hull $\check{M}$ of M iff for all ℓ in X',*

$$\ell(z) \leq S_M(\ell). \tag{34}$$

Proof. By definition (32) of support function, for all ℓ in X' and any z in $\check{M}$, $\ell(z) \leq S_{\check{M}}(\ell)$. By parts (vii) and (viii) of theorem 16, $S_{\check{M}} \equiv S_M$, so that (34) is satisfied for all z in $\check{M}$.

Conversely, suppose that z does not belong to $\check{M}$. Since $\check{M}$ is closed, some open ball $B_R(z)$ centered at z does not intersect $\check{M}$. By the *extended hyperplane separation theorem*, theorem 6 of chapter 3, there is a nonzero linear functional ℓ_0 and a real number c such that

$$\ell_0(u) \leq c \leq \ell_0(v) \tag{35}$$

for all u in $\check{M}$, all v in $B_R(z)$. It follows from the right half of inequality (35) that ℓ_0 is a *bounded* linear functional.

The points v of $B_R(z)$ are of the form $v = z + Rx$, $|x| < 1$. By the right half of inequality (35),

$$c \leq \ell_0(z) + R\ell_0(x).$$

It follows from the definition of the norm of a linear functional that

$$\inf_{|x|<1} \ell_0(x) = -|\ell_0|;$$

it follows from the inequality above that

$$c \leq \ell_0(z) - R|\ell_0|. \tag{36}$$

From the left half of inequality (35) and the definition (32) of S_M, we conclude that

$$S_M(\ell_0) \leq c. \tag{36'}$$

Combining (36) and (36') gives

$$S_M(\ell_0) + R|\ell_0| \leq \ell_0(z). \tag{37}$$

Since $\ell_0 \neq 0$, $|\ell_0| > 0$; thus (37) shows that if z does not belong to $\check{M}$, (34) fails for some ℓ_0, as asserted in theorem 17. $\qquad\square$

Theorem 18. *K denotes a closed, convex subset of a real linear space X, z a point of X not in K. Then*

$$\inf_{u \ in \ K} |z - u| = \sup_{|\ell|=1} [\ell(z) - S_K(\ell)]. \tag{38}$$

Proof. By definition (32) of support function,

$$S_K(\ell) \geq \ell(u) \qquad \text{for all } \ell, \text{ all } u \in K.$$

So for $|\ell| = 1$,

$$S_K(\ell) \geq \ell(u) = \ell(z) + \ell(u - z) \geq \ell(z) - |u - z|,$$

which is the same as $|u - z| \geq \ell(z) - S_K(\ell)$. It follows from this that

$$\inf_{u \text{ in } K} |u - z| \geq \sup_{|\ell|=1} [\ell(z) - S_K(\ell)]. \qquad (39)$$

To show the opposite inequality, let R be any positive number *less* than the inf on the left in (38). Denote by B_R the ball of radius R around the origin; then the set $K + B_R$ has a positive distance from z. So it follows from theorem 17, with $K + B_R$ in place of M, that for some ℓ_0 in X',

$$S_{K+B_R}(\ell_0) < \ell_0(z). \qquad (40)$$

We use now additivity and example (b):

$$S_{K+B_R}(\ell_0) = S_K(\ell_0) + R|\ell_0|.$$

Since we may choose ℓ_0 to have norm $= 1$, it follows from (40) and (40') that

$$R < \ell_0(z) - S_K(\ell_0)$$

for some ℓ_0 with $|\ell_0| = 1$; it follows from this that the sup on the right of (38) is $\geq R$. Since R can be any number less than the inf on the left of (38), it follows that

$$\inf_{u \text{ in } K} |u - z| \leq \sup_{|\ell|=1} [\ell(z) - S_K(\ell)]. \qquad (39')$$

Combined with (39) this proves (38).   □

Theorem 18 presents another example of dual variational problems. Theorem 7 is a special case of theorem 18, if we extend the definition of support function to a linear space Y:

$$S_Y(\ell) = \begin{cases} 0 & \text{if } \ell \text{ in } Y^\perp \\ \infty & \text{if } \ell \text{ not in } Y^\perp. \end{cases}$$

BIBLIOGRAPHY

Day, M. M. Normed linear spaces. *Ergebnisse der Math. und ihrer Grenzgebiete*, **21**, 1962.

Kakutani, S. Concrete representation of abstract *(M)-s* spaces. (A characterization of the space of continuous functions). *An. Math.*, **42** (1941): 994–1024.

Milman, D. P. On some criteria for the regularity of spaces of type (B). *Dokl. Akad. Nauk SSSR (N.S.)*, **20** (1938): 234.

Riesz, F. Sur les opérations fonctionnelles linéairès. *C.R. Acad. Sci. Paris*, **149** (1909): 615–619.

9

APPLICATIONS OF DUALITY

9.1 COMPLETENESS OF WEIGHTED POWERS

Let $w(t)$ be a given positive function defined on $\mathbb{R}$ that decays exponentially as $|t| \to \infty$:

$$0 < w(t) < ae^{-c|t|}, \qquad c > 0. \tag{1}$$

Denote by C the set of continuous functions on $\mathbb{R}$ that vanish at ∞:

$$\lim_{|t| \to \infty} x(t) = 0. \tag{2}$$

C is a Banach space under the maximum norm.

Theorem 1. *The functions $t^n w(t)$ belong to C; their closed linear span is all of C. That is, every function in C can be approximated uniformly on $\mathbb{R}$ by weighted polynomials.*

Proof. We will use theorem 8 of chapter 8. Let ℓ be any bounded linear functional over C that vanishes on the functions $t^n w$:

$$\ell(t^n w) = 0, \qquad n = 0, 1 \ldots. \tag{3}$$

Let ζ be a complex variable, $|\operatorname{Im} \zeta| < c$. Then $w(t)e^{i\zeta t}$ belongs to C, and so

$$f(\zeta) = \ell(we^{i\zeta t}) \tag{4}$$

is defined in the strip $|\operatorname{Im} \zeta| < c$. We claim that $f(z)$ is analytic there. For the complex difference quotients of $we^{i\zeta t}$ tend to $iwt\, e^{i\zeta t}$ in the norm of C, and so

$$f'(\zeta) = \lim_{\delta \to 0} \frac{f(\zeta + \delta) - f(\zeta)}{\delta} = \lim_{\delta \to 0} \ell\left(w \frac{e^{i(\zeta + \delta)t} - e^{i\zeta t}}{\delta} \right) = \ell(iwt\, e^{i\zeta t}).$$

Similarly for the higher derivatives; in particular, using (3),

$$\frac{d^n f}{d\zeta^n}\bigg|_{\zeta=0} = i^n \ell(wt^n) = 0, \qquad n = 0, 1, \ldots.$$

Since f is analytic, the vanishing of all its derivatives at $\zeta = 0$ means that $f(\zeta) \equiv 0$ in the strip, in particular,

$$f(\zeta) = \ell(w\, e^{i\zeta t}) = 0 \qquad \text{for all } \zeta \text{ real.}$$

By theorem 8, chapter 8, it follows that all functions $w\, e^{i\zeta t}$ belong to the closed linear span of $t^n w$.

According to the Weierstrass approximation theorem, every continuous *periodic* function $h(t)$ is the uniform limit of trigonometric polynomials. It follows that wh belongs to the closed linear span of the functions $w\, e^{i\zeta t}$, ζ real, hence of the functions $t^n w$. Let y be any continuous function of compact support; define x by

$$x = \frac{y}{w}. \tag{5}$$

Denote by h a $2p$ periodic function such that

$$x(t) \equiv h(t) \qquad \text{for } |t| < p, \tag{5'}$$

p chosen so large that the support of x is contained in the interval $|t| < p$. Then

$$|x - h|_{\max} \le |x|_{\max},$$

and so, by (5), (5'), and (1),

$$|y - wh| \le a\, e^{-cp} |x|_{\max}.$$

This show that as $p \to \infty$, $wh \to y$. Since wh belongs to the closed linear span of the functions $t^n w$, so does y. The functions y of compact support are dense in C, and the proof is complete.   □

9.2 THE MÜNTZ APPROXIMATION THEOREM

According to the Weierstrass approximation theorem, any continuous function $x(t)$ on the interval $[0, 1]$ can be approximated uniformly by polynomials in t. Let n be any integer. Clearly, if $x(t)$ is continuous on $[0, 1]$, so is

$$y(s) = x(s^{1/n}).$$

Now $y(s)$ can be approximated arbitrarily closely in the maximum norm by polynomials $p(s)$. Setting $s = t^n$, we conclude that $x(t)$ can be approximated arbitrarily closely by linear combination of t^{jn}, $j = 0, 1, \ldots$. Thus in the Weierstrass approximation theorem not all powers of t are needed.

Serge Bernstein posed the following question: What sequences of positive numbers $\{\lambda_j\}$ tending to ∞ have the property that the closed linear span of the functions

$$1, \quad \{t^{\lambda_j}\}, \qquad j = 1, 2, \ldots \tag{6}$$

is the space C of all continuous functions $[0, 1]$? After some preliminary results were obtained by Bernstein, Müntz proved the following:

Theorem 2. *Let λ_j be a sequence of positive numbers tending to ∞. The functions (6) span the space of all continuous functions C on $[0, 1]$ that vanish at $t = 0$ iff*

$$\sum \frac{1}{\lambda_j} = \infty. \tag{7}$$

Proof. We will use the spanning criterion, theorem 8, chapter 8. Let ℓ be a bounded linear functional on C that vanishes on all the functions (6):

$$\ell(t^{\lambda_j}) = 0, \qquad j = 1, 2, \ldots. \tag{8}$$

Let ζ be a complex variable, Re $\zeta > 0$. For such ζ, t^ζ belongs to C and depends analytically on ζ, in the sense that

$$\lim_{\delta \to 0} \frac{t^{\zeta+\delta} - t^\zeta}{\delta} = (\log t)t^\zeta$$

exists in the sense of the norm in C, the maximum norm. Define

$$f(\zeta) = \ell(t^\zeta). \tag{9}$$

It follows that f is an analytic function of ζ. Furthermore, since $|t^\zeta| \leq 1$ when $0 \leq t < 1$ and Re $\zeta > 0$, and since ℓ is bounded, say $|\ell| \leq 1$, it follows from (9) that

$$|f(\zeta)| \leq 1 \qquad \text{for Re } \zeta > 0. \tag{10}$$

Relation (8) can be expressed as

$$f(\lambda_j) = 0. \tag{11}$$

We define a *Blaschke product* $B_N(\zeta)$ as follows:

$$B_N(\zeta) = \prod_1^N \frac{\zeta - \lambda_j}{\zeta + \lambda_j}. \tag{12}$$

It has the following properties:

$$B_N(\lambda_j) = 0, \qquad j = 1, \ldots, N. \tag{13a}$$

$$B_N(\zeta) \neq 0 \qquad \text{for } \zeta \neq \lambda_j. \tag{13b}$$

$$|B_N(\zeta)| \to 1 \qquad \text{as Re } \zeta \to 0. \tag{13c}$$

$$|B_N(\zeta)| \to 1 \qquad \text{as } |\zeta| \to \infty. \tag{13d}$$

Since the zeros of $B_N(\zeta)$ are shared by $f(\zeta)$

$$g_N(\zeta) = \frac{f(\zeta)}{B_N(\zeta)} \tag{14}$$

is regular analytic in Re $\zeta > 0$. We claim that

$$|g_N(\zeta)| \leq 1 \qquad \text{for Re } \zeta > 0. \tag{15}$$

For combining (10) and (13c), (13d) we conclude that for any $\epsilon > 0$, $|g_N(\zeta)| \leq 1+\epsilon$ for Re $\zeta = \delta$, and for $|\zeta| = \delta^{-1}$, δ small enough. By the maximum principle for the analytic function g_N on the domain Re $\zeta \geq \delta$, $|\zeta| \leq \delta^{-1}$, $|g_N(\zeta)| \leq 1 + \epsilon$ there. Letting $\delta, \epsilon \to 0$, we obtain (15). Let k be a positive number such that $f(k) \neq 0$; from (14) and (15) we conclude that

$$\prod_1^N \left| \frac{\lambda_j + k}{\lambda_j - k} \right| \leq \frac{1}{|f(k)|}. \tag{16}$$

We can write the factors on the left in (16) as

$$1 + \frac{2k}{\lambda_j - k}.$$

Since $\lambda_j \to \infty$, all but a finite number of the factors above are > 1. So from the uniform boundedness of the product (16) for all N, we conclude the uniform boundedness, for all N, of the sum

$$\sum_1^N \frac{1}{\lambda_j - k}.$$

This contradicts (7); so we conclude that $f(k) = 0$ for all k. In view of the definition (9) of f, and property (8) of ℓ, this says that any linear functional that vanishes on the functions t^{λ_j} vanishes on t^k, k positive. So by theorem 8, chapter 8, the spanning criterion, we conclude that all functions t^k can be approximated uniformly by linear combination of the functions $\{t^{\lambda_j}\}$. Taking in particular $k = 1, 2, 3, \ldots$ and appealing to the Weierstrass approximation theorem, we conclude that the functions (6) span C.

We omit the proof of the necessity of condition (7). $\qquad\qquad\qquad\qquad\Box$

NOTE. Szász has extended Müntz's theorem to complex λ_j.

Exercise 1. Formulate and prove the extension of theorem 2 to complex exponents.

9.3 RUNGE'S THEOREM

Theorem 3. *Let D be a bounded simply connected domain in $\mathbb{C}$. Every analytic function $f(\zeta)$ in D can be approximated, uniformly on compact subsets K, by polynomials in ζ.*

Proof. Since D is simply connected, every compact subset of D is contained in a simply connected compact subset K of D. Choose a closed smooth curve in $D - K$ that winds once around every point of K, and express $f(\zeta)$, ζ in K, by Cauchy's integral formula. This integral can be approximated, uniformly for all points ζ of K, by a sum. This sum is a linear combination of functions of the form $(\chi - \zeta)^{-1}$, χ on the curve. Therefore to prove the theorem, it suffices to show that all functions of ζ of form $(\chi - \zeta)^{-1}$, χ not in K, can be approximated on K by polynomials in ζ. This is clear when $|\chi| > R$, $R = \max |\zeta|$, ζ in K. Then the geometric series

$$(\chi - \zeta)^{-1} = \sum_0^\infty \frac{\zeta^n}{\chi^{n+1}}$$

converges uniformly on K. To show it for all ζ, we will use the spanning criterion. Let ℓ be any bounded linear functional on $C(K)$ that vanishes on all polynomials:

$$\ell(p) = 0.$$

We claim that ℓ vanishes on all functions of the form $(\chi - \zeta)^{-1}$, χ not in K. Define

$$g(\chi) = \ell((\chi - \zeta)^{-1}).$$

Since $(\chi - \zeta)^{-1}$, as element of $C(K)$, depends analytically on χ, it follows that $g(\chi)$ is an analytic function of χ in the exterior of K. Since for $|\chi| > R$, $(\chi - \zeta)^{-1}$ belongs to the closure of polynomials p, and since $\ell(p) = 0$, it follows by continuity that $\ell((\chi - \zeta)^{-1}) = 0$ for such χ, and so

$$g(\chi) = 0 \qquad \text{for } |R| < |\chi|.$$

Since g is analytic in the exterior, and since the exterior of a simply connected set K is connected, it follows that $g(\chi) = 0$ for all χ not in K. Then, by the spanning criterion, theorem 8 of chapter 8, for all χ outside K, $(\chi - \zeta)^{-1}$ is in the closure of the space of polynomials. □

This beautiful proof is due to Lars Hörmander.

9.4 DUAL VARIATIONAL PROBLEMS IN FUNCTION THEORY

Theorem 4. *Let D be a bounded domain in $\mathbb{C}$, whose boundary consists of a finite number of C^1 arcs. Denote by A the space of functions analytic in D and continuous* *up to the boundary, and by ζ_0 any point in D. Define*

$$M = \sup |f'(\zeta_0)|, \qquad f \text{ in } A, |f|_{\max} \leq 1. \tag{17}$$

Denote by u_0 the function

$$u_0(\zeta) = \frac{1}{2\pi i} \frac{1}{(\zeta - \zeta_0)^2} \tag{18}$$

on ∂D, and define

$$m = \inf_{g \text{ in } A} \int_{\partial D} |u_0(\zeta) - g(\zeta)| \, |d\zeta|. \tag{19}$$

We claim that $m = M$, and that the supremum M in (17) is attained.

Proof. We show first that $M \leq m$. Using the Cauchy integral formula, and the Cauchy integral theorem, we can represent $f'(\zeta_0)$ as follows:

$$f'(\zeta_0) = \int_{\partial D} f u_0 \, d\zeta = \int_{\partial D} f(u_0 - g) \, d\zeta, \tag{20}$$

where g is any function in A. Since f in (17) is ≤ 1 in absolute value, we deduce from (20) that

$$|f'(\zeta_0)| \leq \int_{\partial D} |u_0 - g| \, |d\zeta|.$$

Choose g so that it nearly minimizes (19); we get

$$|f'(\zeta_0)| < m + \epsilon.$$

In view of (17) this implies $M \leq m + \epsilon$, and since $\epsilon > 0$ is arbitrary, $M \leq m$.

To prove the converse inequality, we look at the space C of continuous functions on ∂D; u_0 belongs to C, and A is a linear subspace of C. We impose the L^1-norm $|\ \ |_1$ on C, with respect to arclength $|d\zeta|$ along ∂D. The infimum (19) can be written as

$$m = \inf_{g \text{ in } A} |u_0 - g|_1.$$

According to theorem 7 of chapter 8,

$$m = \max_{|\ell| \leq 1} |\ell(u_0)|, \qquad \ell = 0 \text{ on } A. \tag{21}$$

We denote by ℓ_0 an ℓ that maximizes (21). We define the function f_0 for χ not on ∂D by

$$f_0(\chi) = \frac{1}{2\pi i} \ell_0 \left(\frac{1}{\zeta - \chi} \right). \tag{22}$$

We claim that f_0 has the following properties:

(i) $f_0(\chi)$ is analytic for χ not on ∂D.

(ii) $f_0(\chi) = 0$ for χ not in D.

(iii) $|f_0(\chi)| \leq 1$.

(iv) $f_0'(\zeta_0) = m$.

Property (i) follows from the fact that $1/(\zeta - \chi)$ as an element of C depends analytically on χ. Property (ii) follows since for χ not in D, $1/(\zeta - \chi)$ lies in A, and ℓ_0 vanishes on A. To prove (iii), choose χ near ∂D, and take χ' to be the reflection of χ across D. χ' lies outside D. We will use (ii) to write

$$f_0(\chi) = f_0(\chi) - f_0(\chi') = \frac{1}{2\pi i}\ell_0\left(\frac{1}{\zeta - \chi} - \frac{1}{\zeta - \chi'}\right).$$

Since $|\ell_0| \leq 1$,

$$|f_0(\chi)| \leq \frac{1}{2\pi}\left|\frac{1}{\zeta - \chi} - \frac{1}{\zeta - \chi'}\right|_1 = \frac{1}{2\pi}\int_\partial \frac{|\chi - \chi'|}{|\zeta - \chi||\zeta - \chi'|}|d\zeta|.$$

A simple estimate shows that the integral on the right is $\leq 1 + \epsilon$ for χ near the boundary. So $|f_0(\chi)| \leq 1 + \epsilon$ for χ near the boundary; from this (iii) follows by the maximum principle.

Differentiate (22) and set $\zeta = \zeta_0$; using (18) and (21), we get

$$f_0'(\zeta_0) = \ell_0(u_0) = m. \tag{23}$$

Since we have already shown that $|f'(\zeta_0)| \leq m$ for $|f| \leq 1$ in D, it follows that f_0 solves the maximum problem (17), and that $M = m$. $\qquad\square$

Suppose that D is simply connected. We claim that f_0, the function maximizing (17), maps D conformally onto the unit disc. Here is a sketch of a proof:

It can be shown that the minimum problem (19) has a solution, call it g_0. Setting $f = f_0$ and $g = g_0$ into (20), and using (23), we get

$$m = \int_{\partial D} f_0(u_0 - g_0)\frac{d\zeta}{ds}\,ds,$$

where s is arclength. Using the fact that $|f_0(\zeta)| \leq 1$, and that, by (19),

$$m = \int_{\partial D} |u_0 - g_0|\left|\frac{d\zeta}{ds}\right|\,ds,$$

we conclude that $|f_0(\zeta)| = 1$ on ∂D and that

$$f_0(u_0 - g_0)\frac{d\zeta}{ds} > 0 \qquad \text{on } \partial D. \tag{24}$$

Denote by $2\pi[h]$ the *change of argument* of a nonzero complex valued function h around ∂D. From (24) we deduce that

$$[f_0] + [u_0 - g_0] + \left[\frac{d\zeta}{ds}\right] = 0. \tag{25}$$

According to the argument principle, for the boundary values h of functions meromorphic in D,

$$[h] = \# \text{ zeros } - \# \text{ poles in } D.$$

But f_0 and g_0 have no poles, and u_0 has a single pole of order 2. For D simply connected, $[d\zeta/ds] = 1$; so we deduce from (25) that

$$\# \text{ zeros of } f_0 + \# \text{ zeros of } (u_0 - g_0) - 2 + 1 = 0. \tag{26}$$

It follows from (26) that f_0 has at most one zero in D. We claim that it has at least one zero; otherwise, f_0^{-1} would be analytic in D. Since $|f_0| = 1$ on ∂D, it would follow from the maximum principle that $|f_0(\zeta)| \equiv 1$ in D, which implies that $f_0 \equiv \text{const.}$, contrary to $f'(\zeta_0) = m$. Combining the two statements above, we conclude that f_0 has exactly one zero in D.

According to the argument principle, $[f_0]$ equals 1. Since $|f_0(\zeta)| = 1$ for ζ on ∂D, $[f_0 - w] = 1$ for all w inside the unit disc; it follows that $f(\zeta)$ takes on every value w exactly once in D. This shows that f_0 maps D one-to-one onto the open unit disc.

The arguments used in this section combine methods introduced by Rogosinski and Shapiro with results of Garabedian and Schiffer.

9.5 EXISTENCE OF GREEN'S FUNCTION

Definition. Let D be a plane domain whose boundary B is once continuously differentiable. Green's function $G(p, q)$ of the domain D is defined for p, q in D by the requirements that

(i) $\Delta_p G = \delta(p - q)$, where Δ_p is the Laplace operator with respect to p, and δ is the Dirac distribution, see Appendix B.
(ii) $G(p, q) = 0$ for p on B.

In this definition, the variable p and q play unsymmetric roles, q appearing merely as a parameter in a boundary value problem.

The significance of Green's function is that it can be used to represent every harmonic function h in D in terms of the boundary values of h by using Green's formula:

$$h(q) = \int_{\partial D} h(p)\, G_n(p, q)\, dp,$$

where $G_n(p, q)$ is the derivative of G with respect to p in the direction normal to the boundary B, and dp is arclength. Furthermore, for D simply connected, G is the logarithm of the absolute value of the analytic function f mapping D onto the unit disk, carrying q into the origin.

Green's function can be split into its singular and regular part:

$$G(p, q) = -\frac{1}{2\pi} \log |p - q| + g_0(p, q). \tag{27}$$

The function g_0 is called the regular part of Green's function. Referring to (i) and (ii) above, we can characterize g_0 as the solution of the following boundary value problem:

$$\Delta_p g_0 = 0 \qquad \text{in } D, \tag{28}$$

$$g_0(p, q) = \overline{\log}|p - q| \qquad \text{for } p \text{ on } B, \tag{29}$$

where $\overline{\log} r$ is an abbreviation for $(1/2\pi) \log r$.

Clearly, the definition of Green's function rests on the fact that the boundary value problem (28), (29) can be solved. Classically this is deduced from the solvability of the Dirichlet boundary value problem for Δ with arbitrarily prescribed boundary values. Here we show how to solve (28), (29) without appealing to the general theory. We denote by C the space of continuous functions on the boundary B normed by the max norm. We denote by H the subspace consisting of the boundary values of functions h that are *harmonic* in D and continuous up to the boundary. In what follows the point q is fixed once and for all in D. We define the functional ℓ_q for functions h in H by setting

$$\ell_q(h) = h(q), \tag{30}$$

where $h(q)$ denotes the value at q of the harmonic function whose value on B is denoted by h. It is well known in the theory of harmonic functions that h at q is uniquely determined by h on the boundary B, and that the *maximum principle* holds:

$$h(q) \leq \max_{z \in B} |h(z)| = |h|.$$

This inequality can be expressed so: as defined on H, the norm of ℓ_q is ≤ 1. It follows then from the Hahn-Banach theorem that ℓ_q can be extended from H to all of C so that

$$|\ell_q| \leq 1. \tag{31}$$

We denote by w any point of the plane not on the boundary B of D, and define the element $k(w)$ of C by

$$k(p, w) = \overline{\log}|p - w|, \qquad p \in B. \tag{32}$$

Two observations on the manner of dependence of k on the parameter w:

(i) $k(w)$ *is a differentiable function of* w, *and satisfies*

$$\Delta_w k = 0 \qquad (33)$$

in each component of the complement of B.

(ii) *For* w *in the exterior of* D, $k(w)$ *belongs to* H.

We define now the function $g(w, q)$ by

$$g(w, q) = \ell_q(k(w)). \qquad (34)$$

where k is defined by (32).

Lemma 5.

(i) $g(w, q)$ *is a harmonic function of* w *in each component of the complement of* B.

(ii) *For* w' *in the exterior of* D,

$$g(w', q) = \overline{\log}|q - w'|. \qquad (35)$$

Proof. Since ℓ_q is linear, by (34),

$$\ell_q\left(\frac{k(w + du) - k(w)}{d}\right) = \frac{g(w + du, q) - g(w, q)}{d}.$$

We let d tend to 0. Since ℓ_q is bounded, we deduce that

$$\ell_q(\partial_w k) = \partial_w g(w, q).$$

Applying this to second derivatives and using (33), we get

$$\Delta_w g = \ell_q(\Delta_w k) = 0,$$

as asserted in part (i).

For w' in the exterior of D, $k(w')$ belongs to H. Applying the original defintion (30) of ℓ_q in (34) yields (35). □

Lemma 6. $g(w, q)$ *depends continuously on* w *as* w *crosses the boundary.*

To see this, let w be a point in D close to the boundary, and w' the *reflection* of w across the boundary. Reflection w' is obtained by drawing a straight line from w to the nearest boundary point p_0, and choosing w' so that

$$\frac{w + w'}{2} = p_0.$$

By definition (34) and since ℓ_q is linear,

$$g(w, q) - g(w', q) = \ell_q(k(w) - k(w')) = \ell_q\left(\overline{\log}\frac{|p - w|}{|p - w'|}\right). \qquad (36)$$

It is easy to verify that since the boundary B has a continuously turning tangent

$$\frac{|p - w|}{|p - w'|} \to 1 \qquad (37)$$

as w tends to the boundary B, *uniformly* for all points p in B. It follows that

$$\overline{\log}\frac{|p - w|}{|p - w'|} \to 0 \qquad (38)$$

uniformly for all p on B, and thus in the maximum norm. Since ℓ_q is a bounded linear functional, it follows that the right side of (36) $\to 0$ as w approaches the boundary.

For w in D and near the boundary, w' lies outside D. According to (35), $g(w', q)$ tends to $\overline{\log}|q - p|$ as w' approaches p on the boundary. This proves lemma 6, and shows that as w in D tends to p in B,

$$\lim_{w \to p} g(w, q) = \overline{\log}|q - p|. \qquad (39)$$

$\square$

According to part (i) of lemma 5 we see that $g(w, q)$ is a harmonic function in D. According to (39) its boundary values are $\overline{\log}|q - p|$. These two facts characterize the regular part g_0 of Green's function and therefore

$$g(w, q) = g_0(w, q).$$

This concludes the proof of the *existence of Green's function.* $\square$

REMARK 1. The argument above shows that $k(w)$ defined by (32) is the boundary value of the harmonic function $g(p, w)$, and therefore belongs to H even when w lies in D. Therefore the original definition (30) of ℓ_q applies:

$$\ell_q(k(w)) = g(q, w).$$

By (34), it follows that

$$g(w, q) = g(q, w). \qquad (40)$$

This shows that the regular part of Green's function depends symmetrically on its two arguments. But then, by (27), so does Green's function itself.

REMARK 2. In case of a boundary curve B that is *twice* differentiable, relation (37) can be sharpened to

$$\left| \frac{p - w}{p - w'} \right| = 1 + O(d),$$

where $d = |w - p|$ is the distance of w to the boundary. Using this, we can sharpen (38):

$$\overline{\log} \left| \frac{p - w}{p - w'} \right| = O(d).$$

We set this into (36). Since by (35), $g(w', q) = \overline{\log}|q - w|$, it differs from $g(p, q) = \overline{\log}|q - p|$ by $O(d)$. So we conclude that as w in D tends to the nearest boundary point p,

$$|g(w, q) - g(p, q)| \leq O(d). \tag{41}$$

We claim that the first derivatives of g are uniformly bounded in D up to the boundary. This is because we can express the first derivatives of the harmonic function g at w as integrals over the circle of radius d centered at w:

$$2\pi \ \text{grad} \ g(w) = \frac{1}{d} \int \left[g(w + de(\theta)) - g(p) \right] e(\theta) \, d\theta, \tag{42}$$

where $e(\theta) = (\cos\theta, \ \sin\theta)$. It follows from (41) that the integrand in (42) is $O(d)$; from this the uniform boundedness of grad g follows.

We know from the Cauchy-Riemann equations that the conjugate harmonic function to g also has uniformly bounded first derivatives in D. This shows that for D simply connected, the analytic function mapping D onto a disk is uniformly Lipschitz continuous.

Green's functions of more than two variables were constructed in Lax.

BIBLIOGRAPHY

Garabedian, P. R. and Schiffer, M. On existence theorems of potential theory and conformal mapping. *An. Math.*, **52** (1950): 164–187.

Garabedian, P. R. and Shiffman, M. On solutions of partial differential equations by the Hahn-Banach theorem. *Trans. AMS*, **76** (1954): 288–299.

Lax, P. D. On the existence of Green's function. *Proc. AMS*, **3** (1952): 526–531.

Lax, P. D. Reciprocal extremal problems in function theory. *CPAM*, **8** (1955): 437–454.

Müntz, Ch. H. Über die Approximationssatz von Weierstrass. *Math. Abhandlungen H. A. Schwarz gewidmet*, Berlin (1914): 303–312.

Rogosinski, W. W. and Shapiro, H. S. On certain extremum problems for analytic functions. *Acta Math.*, **84** (1953): 287–318.

Szász, O. Über die Approximation stetiger Funktionen durch lineare Aggregate von Potenzen. *Math. An.*, **77** (1915–1916): 482–496.

10

WEAK CONVERGENCE

Definition. A sequence $\{x_n\}$ in a normed linear space X is said to *converge weakly* to x if

$$\lim_{n \to \infty} \ell(x_n) = \ell(x) \tag{1}$$

for every ℓ in X'. This relation is indicated by a half arrow:

$$x_n \rightharpoonup x. \tag{1'}$$

Another notation is

$$w - \lim_{n \to \infty} x_n = x. \tag{1''}$$

The notion is to be contrasted with converge in sense of the norm:

$$\lim_{n \to \infty} |y_n - y| = 0. \tag{2}$$

In this case we say that $\{y_n\}$ tends *strongly* to y, and denote it as

$$y_n \longrightarrow y. \tag{2'}$$

Another suggestive notation for strong convergence is

$$s - \lim_{n \to \infty} y_n = y. \tag{2''}$$

Clearly, a sequence that converges to x strongly also converges weakly to x, but in general not vice versa. Here are some examples:

Example 1. $X = \ell^2$; its points are vectors

$$x = (a_1, a_2, \ldots) \tag{3}$$

with denumerably many components, such that

$$\|x\|^2 = \sum |a_j|^2 < \infty. \tag{3'}$$

Since ℓ^2 is a Hilbert space, according to theorem 4 of chapter 6 all bounded linear functionals on ℓ^2 are of the form

$$\ell(x) = (x, y) = \sum a_j b_j, \quad \sum |b_j|^2 < \infty. \tag{4}$$

Define x_n as the nth unit vector, that is, the vector whose nth component is 1, all others zero: $x_n = (0, \ldots, 0, 1, 0, \ldots)$. It is easy to show, and left as an *exercise*, that the sequence $\{x_n\}$ tends weakly to zero, but not strongly.

Example 2. H any Hilbert space, $\{x_n\}$ an orthonormal sequence; such a sequence tends weakly, but not strongly, to zero.

Proof. It follows from Bessel's inequality (31) of chapter 6, that for any y in H

$$\sum |(x_n, y)|^2 \leq \|y\|^2, \tag{5}$$

from which it follows that

$$\ell(x_n) = (x_n, y) \to 0. \tag{5'}$$

Since according to the Riesz representation theorem, theorem 4 in chapter 6, all linear functionals are of the form (5′), weak convergence to zero follows. Since $\|x_n\| = 1$ for all n, $\{x_n\}$ does not tend to zero strongly. $\qquad\square$

Example 3. $X = C[0, 1]$.

$$x_n(t) = \begin{cases} nt & \text{for} \quad 0 \leq t \leq \frac{1}{n} \\ 2 - nt & \text{for} \quad \frac{1}{n} \leq t \leq \frac{2}{n} \\ 0 & \text{for} \quad \frac{2}{n} \leq t \leq 1. \end{cases}$$

CLAIM. x_n tends to zero weakly but not strongly.

Proof. Let ℓ be a bounded linear functional; we claim that $\lim \ell(x_n) = 0$.

Now suppose not; then there would be infinitely many n such that $|\ell(x_n)| > \delta > 0$, say

$$\ell(x_n) > \delta. \tag{6}$$

Choose a subsequence $\{n_k\}$, $n_{k+1} > 2n_k$, for which (6) holds. It is not hard to show that for all t in $[0,1]$

$$y_K = \sum_1^K x_{n_k}(t) < 4, \tag{7}$$

which implies that $|y_K| < 4$ for all K. From (6) it follows that

$$\ell(y_K) = \sum_1^K \ell(x_{n_k}) > K\delta.$$

Since this holds for all K, and since $|y_K| < 4$ for all K, the boundedness of ℓ is contradicted. Since $|x_n| = \max x_n(t) = 1$, $\{x_n\}$ does not tend to zero strongly.   □

Exercise 1. Prove inequality (7). (Draw a picture of the graph of $x_n(t)$).

Example 4. $X = \ell^1$, consisting of vectors $x = (a_1, a_2, \ldots)$, $|x| = \sum |a_k| < \infty$. As we saw in chapter 8 (exercise after theorem 11), the dual of ℓ^1 is ℓ^∞. The following observation is due to *Schur*:

If a sequence $\{x_n\}$ in ℓ^1 converges weakly, it converges strongly.

Exercise 2. Prove the preceding statement.

10.1 UNIFORM BOUNDEDNESS OF WEAKLY CONVERGENT SEQUENCES

The following result is useful in proving weak convergence:

Theorem 1. *Suppose that a sequence $\{x_n\}$ of points in a normed linear space satisfies*

(i) *$\{|x_n|\}$ are uniformly bounded:*

$$|x_n| \le c.$$

(ii) *$\lim \ell(x_n) = \ell(x)$ for a set of ℓ dense in X'. Then*

$$w - \lim x_n = x.$$

Exercise 3. Prove theorem 1.

Surprisingly, the converse of theorem 1 holds in Banach spaces. To see this, we appeal to the *principle of uniform boundedness* for a complete metric space S: if a collection $\{f_\nu\}$ of continuous real-valued functions f_ν on S is bounded at each point x of S,

$$|f_\nu(x)| \le M(x) \qquad \text{for all } \nu, \tag{8}$$

then the functions f_ν are uniformly bounded,

$$|f_\nu(u)| \le M, \tag{8'}$$

for all u in some nonempty open set O. We specialize this to the case where S is a Banach space X and each f_ν is *subadditive* and *absolutely homogeneous*:

$$f(x + y) \le f(x) + f(y), \qquad f(ax) = |a| f(x). \qquad (9)$$

Theorem 2. *Let X be a Banach space, $\{f_\nu\}$ a collection of real-valued continuous subadditive and absolutely homogeneous functions on X, bounded at each point x of X as in (8). Then the $\{f_\nu\}$ are uniformly bounded, that is, there is a number c such that*

$$|f_\nu(x)| \le c|x| \qquad (10)$$

for all f_ν and all x in X.

Proof. By the principle of uniform boundedness for metric spaces, $|f_\nu(u)| \le M$ for all f_ν and all u in some open ball $u = z + y$, $|y| < r$. Using subadditivity, we have that

$$|f_\nu(y)| = |f_\nu(u - z)| \le |f_\nu(u)| + |f_\nu(z)| \le 2M \qquad (11)$$

holds for all y with $|y| = r/2$. For any x in X define y by $y = rx/2|x|$. By construction, $|y| = r/2$, so (11) holds. Using absolute homogeneity we get from (11) that

$$|f_\nu(x)| = \left| f_\nu \left(\frac{2|x|}{r} y \right) \right| = \frac{2|x|}{r} f(y) \le \frac{4M}{r} |x|;$$

this proves (10), with $c = 4M/r$.   □

An immediate consequence of theorem 2 is

Theorem 3. *X is a Banach space, $\{\ell_\nu\}$ a collection of bounded linear functionals such that at every point x of X*

$$|\ell_\nu(x)| \le M(x) \qquad \text{for all } \ell_\nu. \qquad (12)$$

Then there is a constant c such that

$$|\ell_\nu| \le c \qquad \text{for all } \ell_\nu. \qquad (13)$$

Proof. $|\ell(x)|$ is a continuous subadditive and absolutely homogeneous function of x. Therefore theorem 2 is applicable; its conclusion (10) yields (13).   □

Another immediate consequence is

Theorem 4. *X is a normed linear space, $\{x_\nu\}$ a collection of points in X such that for every bounded linear functional ℓ*

$$|\ell(x_\nu)| \le M(\ell) \qquad \text{for all } x_\nu. \tag{12'}$$

Then there is a constant c such that

$$|x_\nu| \le c \qquad \text{for all } x_\nu. \tag{13'}$$

Proof. Theorem 4 follows from theorem 3 applied to the Banach space X', on which the elements x_ν of X act as bounded linear functionals. $\qquad\square$

An immediate consequence of theorem 4 is

Theorem 4'. *A weakly convergent sequence $\{x_n\}$ in a normed linear space X is uniformly bounded in norm.*

Proof. Weak convergence means that $\ell(x_n)$ is convergent for every ℓ in X'. Since a convergent sequence of numbers is bounded, hypothesis $(12')$ of theorem 4 is satisfied; therefore $(13')$ holds. $\qquad\square$

Theorems 2, 3, and 4 are called the *principle of uniform boundedness.*

Theorem 5. *Let $\{x_n\}$ be a sequence in a normed linear space converging weakly to x. Then*

$$|x| \le \lim \inf |x_n|. \tag{14}$$

Proof. According to theorem 6 of chapter 8, there is an ℓ in X', such that

$$|x| = |\ell(x)|, \quad |\ell| = 1.$$

Since weak convergence means that

$$\ell(x) = \lim \ell(x_n),$$

and since

$$|\ell(x_n)| \le |\ell| \, |x_n| = |x_n|,$$

(14) follows. $\qquad\square$

The following far reaching generalization of theorem 5 is due to Mazur:

Theorem 6. *Let K be a closed, convex subset of a normed linear space X, $\{x_n\}$ a sequence of points in K, converging weakly to a point x. Then x belongs to K.*

Proof. Let S_K be the support function of K, defined by equation (32) of chapter 8 as $\sup_{x \text{ in } K} \ell(x)$. It follows from that definition that for any ℓ in X'

$$\ell(x_n) \leq S_K(\ell). \tag{15}$$

Since $\ell(x_n)$ tends to $\ell(x)$, it follows that also

$$\ell(x) \leq S_K(\ell).$$

But according to theorem 17 of chapter 8, this guarantees that x belongs to K. □

Exercise 4. Deduce theorem 5 from theorem 6 applied to balls centered at the origin $K = B_R : \{x | |x| \leq R\}$.

10.2 WEAK SEQUENTIAL COMPACTNESS

Definition. A subset C of a Banach space X is called *weakly sequentially compact* if any sequence of points in C has a subsequence weakly convergent to a point of C.

Exercise 5. Show that a weakly sequentially compact set is bounded.

The importance of weak sequential compactness is the same as that of compactness in the sense of strong convergence. Weak compactness is a valuable tool in constructing, as weak limits, mathematical objects of interest. To wield this tool, we need simple, easily verifiable criteria for weak compactness; the following is such a criterion:

Theorem 7. *In a reflexive Banach space X the closed unit ball is weakly sequentially compact.*

Proof. Let $\{y_n\}$ be a sequence of points in the unit ball, that is, $|y_n| \leq 1$. Denote by Y the closed linear subspace spanned by the set $\{y_n\}$; Y is separable. Since X is assumed reflexive, it follows from theorem 15 of chapter 8 that Y is reflexive. Since $Y = Y''$ is separable as well, it follows from theorem 13 of chapter 8 that Y' also is separable, meaning that it contains a dense, denumerable subset $\{m_j\}$. Using the classical diagonal process, we can select a subsequence $\{z_n\}$ of $\{y_n\}$ such that

$$\lim_{n \to \infty} m_j(z_n) \tag{16}$$

exists for every m_j. Since all z_n satisfy $|z_n| \leq 1$, and since the $\{m_j\}$ are dense, it follows from (16) and theorem 1 that for all m in Y', $m(z_n)$ tends to a limit as $n \to \infty$. This limit is a linear functional of m:

$$\lim_{n \to \infty} m(z_n) = y(m). \tag{16'}$$

Since $|m(z_n)| \leq |m| |z_n| \leq |m|$, it follows from (16') that the linear functional $y(m)$ has norm ≤ 1. Since Y is reflexive, there is a y in Y such that $y(m) = m(y)$, $|y| \leq 1$, and so (16) says that for all m in Y', $m(z_n)$ tends to $m(y)$ as $n \to \infty$. Since the restriction of any ℓ in X' to Y is an m in Y', this proves that z_n converges weakly to a point y in the unit ball. $\qquad\qquad\qquad\qquad\qquad\qquad\qquad\qquad\qquad\qquad\qquad\qquad$ □

Note the sharp contrast between theorem 7 and theorem 6 of chapter 5, according to which the unit ball is never compact in the norm topology. Compactness is gained by replacing strong with weak convergence.
Eberlein has proved the converse of theorem 7:

Theorem 8. *The closed unit ball in a Banach space X is weakly sequentially compact only if X is reflexive.*

Combining theorems 6 and 7 gives the following useful result:

Theorem 9. *In a reflexive Banach space every bounded, closed, convex set is weakly sequentially compact.*

Here is a useful application of theorem 9.

Theorem 10. *Let X be a reflexive Banach space, K a closed, convex subset of X, z any point of X. Then there is a point y of K which is as close to z as any other point of K.*

Proof. We may take $z = 0$, and assume that $0 \notin K$. Denote by s the distance of 0 to K, that is,

$$s = \inf |y|, \qquad y \text{ in } K. \tag{17}$$

Let $\{y_n\}$ be a minimizing sequence for (17). We may assume that each y_n lies in the intersection of K and that the ball of radius is $2s$ around the origin. This is a bounded, closed, convex set, therefore, by theorem 9, a subsequence $\{z_n\}$ of $\{y_n\}$ converges weakly to some point z of K. According to theorem 5,

$$|z| \leq \liminf |z_n|. \tag{18}$$

Since $\{z_n\}$ is the subsequence of a minimizing sequence, $\lim |z_n| = s$. Combining this with (17) and (18) gives $|z| = s$; that is, z is a point of K closest to 0. $\qquad$ □

Theorem 10 is a generalization of theorem 8 of chapter 5. There we assumed that X is uniformly convex; here we assume only that X is reflexive.

10.3 WEAK* CONVERGENCE

In a Banach space U that is the dual X' of another Banach space X, there is a subclass of linear functionals associated with elements x of X:

$$x(u) = u(x).$$ (19)

One can define sequential convergence in U with respect to this subclass of linear functionals:

Definition. A sequence $\{u_n\}$ in a Banach space U that is the dual of another Banach space X is said to be weak* convergent to u if

$$\lim u_n(x) = u(x)$$ (20)

for all x in X. We denote this relation as

$$w^* - \lim u_n = u.$$

REMARK 1. Of course, if X is reflexive, weak* convergence is no different than weak convergence.

Example 5. U is the space of all signed Borel measures m on $[-1, 1]$, of finite total mass. According to theorem 14 of chapter 8, U is the dual of $C[-1, 1]$.

Consider the sequence $\{m_n\}$:

$$m_n(h) = \int h\,dm_n = \frac{n}{2}\int_{-1/n}^{1/n} h(t)\,dt.$$ (21)

Clearly, for any continuous h,

$$\lim_{n\to\infty} m_n(h) = h(0).$$ (22)

This shows that m_n is weak* convergent to the unit mass at the origin. The dual of U is $L^\infty[-1, 1]$, the space of all bounded measurable functions. Since (22) is not true for some h discontinuous at 0, m_n does not converge weakly.

Theorem 11. *A weak* convergent sequence $\{u_n\}$ of points in a Banach space $U = X'$ is uniformly bounded.*

   Proof. Weak* convergence implies that the boundedness condition (12) holds at every point x of X. So theorem 3 implies (13), uniform boundedness.   □

Exercise 6. Show that if the sequence $\{u_n\}$ is weak* convergent to u,

$$|u| \leq \liminf |u_n|.$$

Definition. A subset C of a Banach space U that is the dual of another Banach space X is called *weak* sequentially compact* if every sequence of points in C has a subsequence that is weak* convergent to a point of C.

The following important result is due to Helly:

Theorem 12. *Let X be a separable Banach space, $U = X'$. The closed unit ball in U is weak* sequentially compact.*

Proof. Given a sequence $\{u_n\}$ in U,

$$|u_n| \leq 1, \tag{23}$$

and a denumerable set $\{x_k\}$ in X, we can, by the diagonal process, select a subsequence $\{v_n\}$ of $\{u_n\}$ such that

$$\lim_{n \to \infty} v_n(x_k) \tag{24}$$

exists for all x_k. It follows from (23) and (24) that $v_n(x)$ tends to a limit for all x that lie in the closure of the set $\{x_k\}$. So, if we take for $\{x_k\}$ a set dense in X, $v_n(x)$ tends to a limit for all x in X. It is easy to see that this limit is a linear function of x, and, using (23), that it is bounded by 1. $\qquad\square$

BIBLIOGRAPHY

Eberlein, W. F. Weak compactness in Banach spaces. *I. Proc. Nat. Acad. Sci. USA*, **33** (1947): 51–53.

Helly, E. Über lineare Funkionaloperationen. *S.-B. K. Akad. Wiss. Wien Math.–Naturwiss. Kl.*, **121**, (1912): 265–297.

Mazur, S. Über konvexe Mengen in linearen normierten Räumen. *Studia Math.*, **4** (1933): 70–84.

11

APPLICATIONS OF
WEAK CONVERGENCE

11.1 APPROXIMATION OF THE δ FUNCTION
BY CONTINUOUS FUNCTIONS

Definition. A sequence $\{k_n\}$ of continuous functions on $[-1, 1]$ tends to the δ function if

$$\lim_{n \to \infty} \int_{-1}^{1} f(t) \, k_n(t) \, dt = f(0) \tag{1}$$

for all continuous functions f on $[-1, 1]$.

Theorem 1 (Toeplitz). *The sequence $\{k_n\}$ of continuous functions on $[-1, 1]$ tends to the δ function in the sense of (1) if and only if it satisfies the following conditions:*

(i)

$$\lim_{n \to \infty} \int_{-1}^{1} k_n(t) \, dt = 1. \tag{2}$$

(ii) For every C^∞ function g whose support does not contain 0,

$$\lim_{n \to \infty} \int_{-1}^{1} g(t) \, k_n(t) \, dt = 0. \tag{3}$$

(iii) There is a constant c for which

$$\int_{-1}^{1} |k_n(t)| \, dt \leq c \tag{4}$$

holds for all n.

Proof. Suppose that $f(0) = 0$; let g be a C^∞ function that differs from f by less than ϵ at all points t in $[-1, 1]$, and that is zero in some interval around $t = 0$. Then by (4),

$$\left| \int_{-1}^{1} (f - g) k_n \, dt \right| \leq \epsilon \int |k_n| \, dt \leq c\epsilon. \tag{5}$$

By assumption (3), $\int g k_n \, dt$ tends to zero, so it follows from (5) that

$$\limsup \left| \int f k_n \, dt \right| \leq \epsilon. \tag{6}$$

Since ϵ is arbitrary, (1) is verified in case $f(0) = 0$. Every function can be decomposed as $b + f$, b some constant and $f(0) = 0$, so (1) follows from (2) for every continuous f.

Now to the converse: condition (2) is clearly necessary, for it is a special case of (1) when $f(t) \equiv 1$; the same goes for (3).

We can regard $\{k_n\}$ as a sequence in $C'[-1, 1]$, and state (1) as

$$w^* - \lim k_n = \delta.$$

According to theorem 3 of chapter 10, the norms

$$|k_n| = \int |k_n(t)| \, dt$$

must be uniformly bounded. This proves the necessity of (4), and even more:

Corollary 1'. *If (4) is violated, there exists a continuous function f for which the left side of (1) tends to infinity.* □

11.2 DIVERGENCE OF FOURIER SERIES

Theorem 2. *There exists a periodic continuous function $f(\theta)$ whose Fourier series diverges at one point.*

Proof. The Fourier series of a continuous function f on the unit circle S^1 is

$$f(\theta) \approx \sum_{-\infty}^{\infty} a_n e^{in\theta}, \tag{7}$$

where

$$a_n = \int_{-\pi}^{\pi} f(\theta) e^{-in\theta} \, d\theta \qquad \text{where} \quad d\theta = \frac{d\theta}{2\pi}. \tag{7'}$$

The convergence of the series at, say, $\theta = 0$ means that

$$f(0) = \lim_{N \to \infty} \sum_{-N}^{N} a_n. \qquad (8)$$

Using (7'), we can write

$$\sum_{-N}^{N} a_n = \int_{-\pi}^{\pi} f(\theta) \, k_N(\theta) \, d\theta, \qquad (9)$$

where

$$2\pi k_N(\theta) = \sum_{-N}^{N} e^{-in\theta}. \qquad (10)$$

Using the formula for the sum of a finite geometric series, we readily get, for $\theta \neq 0$,

$$2\pi k_N(\theta) = \frac{\sin(N + 1/2)\,\theta}{\sin \theta/2}. \qquad (10')$$

Thus the convergence of the Fourier series of every continuous function is equivalent to the sequence $\{k_n\}$ defined by (10) approximating the δ function. According to theorem 1 this is the case iff conditions (2), (3), and (4) are satisfied. We show now that condition (4) fails! To see this, we use the inequality $|\sin \phi| \leq |\phi|$, which implies that $|1/(\sin \theta/2)| \geq 2/|\theta|$. Using (10') and a little calculus, we get

$$\int_{-\pi}^{\pi} |k_N(\theta)| \, d\theta \geq \frac{1}{\pi} \int_{-\pi}^{\pi} \left| \sin\left(N + \frac{1}{2}\right) \theta \right| \frac{d\theta}{|\theta|} = \frac{2}{\pi} \int_{0}^{(N+1/2)\pi} |\sin \phi| \frac{d\phi}{\phi}.$$

This last integral is, it is easy to show, $\geq$ const. $\log N$. Thus condition (4) fails, and so, by Corollary 1', for some function f, the Fourier series of f diverges at $\theta = 0$ to infinity. $\qquad \Box$

Exercise 1. Show that there exists a continuous periodic function whose Fourier series diverges at n aribtrarily given points.

11.3 APPROXIMATE QUADRATURE

An approximate quadrature formula is an approximation to the integral of a continuous function f on, say, $[-1, 1]$. Take N points t_j in $[-1, 1]$, called *nodes*, and N numbers w_j called *weights*; define $q(f)$ by

$$q(f) = \sum_{1}^{N} w_j f(t_j). \qquad (11)$$

We regard q as an approximation to

$$\int_{-1}^{1} f(t)\, dt. \tag{11'}$$

Theorem 3. *Let q_N be a sequence of quadrature formulas of form* (11), *satisfying the following conditions:*

(i) *For every nonnegative integer k,*

$$\lim_{N\to\infty} q_N(t^k) = \int_{-1}^{1} t^k\, dt. \tag{12}$$

(ii) *For all N,*

$$\sum_{1}^{N} |w_j(N)| \le c, \tag{13}$$

c *a constant. Then*

$$\lim_{N\to\infty} q_N(f) = \int_{-1}^{1} f(t)\, dt \tag{14}$$

for all continuous f. Conversely, if (14) *holds for all continuous f,* (12) *and* (13) *must be satisfied.*

Proof. It follows from (12) that (14) holds for all polynomials f. Inequality (13) asserts that the linear functionals q_N on $C[-1, 1]$ have uniformly bounded norms. Since the polynomials are dense in $C[-1, 1]$, (14) follows for all continuous f. The converse follows from theorem 3 of chapter 10. □

Exercise 2. Prove that if the weights w_j are positive, (13) follows from (12).

11.4 WEAK AND STRONG ANALYTICITY OF VECTOR-VALUED FUNCTIONS

Let f be a function defined in some domain G of the complex ζ plane, whose values lie in some complex Banach space X.

Definition. $f(\zeta)$ is *strongly analytic* in G if the limit

$$\lim_{h\to 0} \frac{f(\zeta + h) - f(\zeta)}{h}$$

exists in the norm topology at every point of G.

Definition. $f(\zeta)$ is *weakly analytic* in G if for every bounded linear functional ℓ, $\ell(f(\zeta))$ is an analytic function of ζ in the classical sense.

N. Dunford has proved the following surprising result:

Theorem 4. *A weakly analytic function is strongly analytic.*

Proof. If $\ell(f(\zeta))$ is analytic in G, we can represent it by the Cauchy integral formula

$$\ell(f(\zeta)) = \int_C \frac{\ell(f(\chi))}{\chi - \zeta} \, d\chi, \qquad \text{where} \quad d\chi = \frac{d\chi}{2\pi i}, \tag{15}$$

C some rectifiable curve winding around ζ. Similar formulas hold when ζ is replaced by $\zeta + k$ and $\zeta + h$, k and h small enough. Assume that $k \neq 0$, $h \neq 0$, $h \neq k$; then we can express the difference quotient of difference quotients thus:

$$\frac{1}{h-k} \left\{ \frac{\ell(f(\zeta+h)) - \ell(f(\zeta))}{h} - \frac{\ell(f(\zeta+k)) - \ell(f(\zeta))}{k} \right\}$$
$$= \int_C \ell(f(\chi)) \frac{d\chi}{(\chi - \zeta - h)(\chi - \zeta - k)(\chi - \zeta)}. \tag{16}$$

For fixed ℓ, and $|k|$, $|h|$ small enough, the right side of (16) is bounded by a constant M independent of h and k. We can rewrite the left side as $\ell(x_{h,k})$ where

$$x_{h,k} = \frac{1}{h-k} \left\{ \frac{f(\zeta+h) - f(\zeta)}{h} - \frac{f(\zeta+k) - f(\zeta)}{k} \right\}. \tag{17}$$

Weak analyticity thus implies that for each ℓ and all h and k sufficiently small, $|\ell(x_{h,k})| \leq M(\ell)$. We appeal now to the principle of uniform boundedness, theorem 4, chapter 10, and conclude that $|x_{h,k}| \leq c$ for all h, k sufficiently small. By definition (17) of $x_{h,k}$ this implies the norm inequality

$$\left| \frac{f(\zeta+h) - f(\zeta)}{h} - \frac{f(\zeta+k) - f(\zeta)}{k} \right| \leq c|h - k|. \tag{18}$$

Since X is complete, it follows that the difference quotients of $f(\zeta)$ tend to a limit in the strong sense. $\qquad\square$

11.5 EXISTENCE OF SOLUTIONS OF PARTIAL DIFFERENTIAL EQUATIONS

We denote by L a first-order partial differential operator of the following form, acting on vector-valued functions:

$$L = \sum_{j=1}^{m} A_j \, \partial_j + B. \tag{19}$$

Here A_j and B are square matrix valued functions of the independent variables s_j, $A_j(s)$ being once differentiable, $B(s)$ continuous, and

$$\partial_j = \frac{\partial}{\partial s_j}.$$

We assume, for simplicity, that A_j and B, as well as the functions on which L acts, are periodic in all variables s and that they are real valued. We denote, as customary, the *formal adjoint* of L by L^*:

$$L^* = -\sum \partial_j A_j^T + B^T, \tag{19'}$$

where A^T, B^T denote transposes. Integration by parts shows that for any pair of C^1 vector-valued periodic functions u and v,

$$(v, Lu) = (L^* v, u). \tag{20}$$

Here the bracket denotes the L^2 scalar product over a period cube:

$$(v, w) = \int_F v(s) \cdot w(s)\, ds;$$

the dot denotes the dot product between vectors, and F a period cube.

Suppose that every A_j is symmetric: $A_j^T = A_j$. Then comparing (19) and (19'), we deduce that

$$L^* = -L - \sum A_{j,j} + B + B^T.$$

Here $A_{j,j}$ denotes the partial derivative of A_j with respect to s_j. Setting this into (20), and choosing $v = u$, we get

$$2(u, Lu) = ((L + L^*) u, u) = \left(\left[B + B^T - \sum A_{j,j} \right] u, u \right). \tag{20'}$$

Using the language of distributions, see Appendix B, we state

Theorem 5. *Suppose the matrix on the right in (20') is positive definite:*

$$B + B^T - \sum A_{j,j} > kI, \qquad k > 0. \tag{21}$$

Then for every periodic, square integrable f the equation

$$Ly = f \tag{22}$$

has a solution y in the sense of distributions that is periodic and square integrable.

Proof. It follows from (20') and (21) that every periodic C^1 function satisfies the inequality

$$(u, Lu) \geq \frac{k}{2} \|u\|^2, \tag{21'}$$

where $\|u\|$ denotes the $L^2(F)$-norm. Denote the Hilbert space $L^2(F)$ by H. Let Y be any finite-dimensional subspace of H consisting of periodic C^1 functions; denote the orthogonal complement of Y in H by $Y^\perp$. Consider the equation

$$Ly - f \in Y^\perp \qquad (22_N)$$

for y in Y. These are N linear equations for y in Y, $N = \dim Y$. According to linear algebra, such a system of linear equations has a solution for every f iff the homogeneous equation

$$Lz \in Y^\perp, \qquad z \text{ in } Y, \qquad (23)$$

is satisfied only by $z = 0$. Take the scalar product of (23) with z; using (21′), we get

$$0 = (z, Lz) \geq \frac{k}{2} \|z\|^2,$$

which implies that $z = 0$. So it follows that (22_N) has a unique solution y. Take the scalar product of (22_N) with y; using (21′) and the Schwarz inequality, we get

$$\frac{k}{2} \|y\|^2 \leq (y, Ly) = (y, f) \leq \|y\| \, \|f\|.$$

This implies that

$$\|y\| \leq \frac{2}{k} \|f\|. \qquad (24)$$

Now let Y_N be an increasing sequence of subspaces of C^1 functions whose union is dense in H. Denote by y_N the solution of (22_N). It follows from (24) that $\|y_N\|$ is a uniformly bounded sequence. Therefore, since H is reflexive, we appeal to theorem 7 of chapter 10 to conclude that a subsequence of $\{y_N\}$, also denoted as $\{y_N\}$, converges weakly:

$$w - \lim y_N = y.$$

Let v belong to $\cup \, Y_N$; since each Y_N consists of differentiable functions, v is differentiable, for it belongs to some Y_M. For y_N in Y_N,

$$Ly_N - f \in Y_N^\perp.$$

Take the scalar product of this with v; for $N > M$ we get

$$(v, Ly_N) - (v, f) = 0.$$

Since v is differentiable, this can be rewritten by (20), as

$$(L^* v, y_N) - (v, f) = 0.$$

Since the sequence y_N converges weakly to y, we conclude that for every v in $\cup\, Y_N$

$$(L^* v, y) - (v, f) = 0. \tag{25}$$

We can choose the spaces Y_N so that their union is dense not only in H but also in H_1, the space of all periodic L^2 functions whose first derivatives belong to L^2. That means that given any periodic C^1 function v, there is a sequence $\{v_k\}$ of functions in $\cup\, Y_N$ such that v_k converges in the L^2-norm to v, and the first derivatives of v_k converge to the first derivatives of v in the L^2-norm. Since L^* is a first-order operator, it follows that $L^* v_k$ tends to $L^* v$ in the L^2-norm. Setting $v = v_k$ in (25), we can pass to the limit and conclude that (25) holds for all v in C^1.

A function y that satisfies (25) for all C^1 functions v is said to satisfy the differential equation (22) in the *weak sense*. Clearly, such a v is a solution of (22) in the sense of distributions, which requires (25) to hold for all C_0^∞ functions v. □

Friedrichs has shown that a weak solution y of (22) is a *strong solution* in the following sense: there is a sequence of C^1 functions z_n that converge to y in the L^2 sense, and at the same time Lz_n converges to f in the L^2 sense. It is easy to show, using (21′), that equation (22) has only one strong solution. It follows that not only a subsequence, but the whole sequence y_N converges.

The method described in this section to obtain the solution y of equation (22) as the weak limit of the solutions y_N of equation (22_N) is called *Galerkin's method*. It is more than a theoretical device for proving the existence of a solution of (22); it is also a practical method for constructing it.

11.6 THE REPRESENTATION OF ANALYTIC FUNCTIONS WITH POSITIVE REAL PART

Let $f(\zeta)$ be an analytic function in the unit disk $|\zeta| < 1$ whose real part is *positive*:

$$h(\zeta) = \mathrm{Re}\, f(\zeta) \geq 0, \qquad |\zeta| < 1.$$

Every analytic function defined in a disk and continuous up to the boundary can be expressed—up to an imaginary constant—in terms of its real part on the boundary by the Poisson integral. On the disk of radius $R < 1$ we have for $|\zeta| < R$

$$f(\zeta) = \int_0^{2\pi} \frac{R + \zeta\, e^{-i\theta}}{R - \zeta\, e^{-i\theta}}\, h\left(Re^{i\theta}\right) d\theta + ic. \tag{26}$$

Setting $\zeta = 0$, we see that

$$h(0) = \int_0^{2\pi} h\left(R\, e^{i\theta}\right) d\theta. \tag{26′}$$

Let $R \to 1$ through a sequence $R_n \to 1$. The functions $h(R_n\, e^{i\theta})$ are nonnegative functions of θ whose integrals over the whole circle are, by (26′), all equal to $h(0)$.

We associate with each R_n a linear functional

$$\ell_n(u) = \int_0^{2\pi} h\left(R_n \, e^{i\theta}\right) u(\theta) \, d\theta \tag{27}$$

acting on the space C of continuous functions u on the circle S^1. It follows from $h \geq 0$ and (26') that

$$|\ell_n| = h(0).$$

Since $C(S^1)$ is separable, we can appeal to Helly's theorem, theorem 12 of chapter 10, and conclude that a subsequence of $\{\ell_n\}$ is weak* convergent to some limit ℓ:

$$\lim_{n \to \infty} \ell_n(u) = \ell(u) \tag{28}$$

for all continuous functions u. It follows from (28) and the uniform boundedness of $|\ell_n|$ that for any sequence u_n strongly convergent to u

$$\lim_{n \to \infty} \ell_n(u_n) = \ell(u). \tag{28'}$$

We apply this now to

$$u_n = \frac{R_n + \zeta \, e^{-i\theta}}{R_n - \zeta \, e^{-i\theta}}, \qquad u = \frac{1 + \zeta \, e^{-i\theta}}{1 - \zeta \, e^{-i\theta}},$$

ζ any complex number with $|\zeta| < 1$; using (26), (27) and (28'), we get

$$f(\zeta) = \ell\left(\frac{1 + \zeta \, e^{-i\theta}}{1 - \zeta \, e^{-i\theta}}\right).$$

The functionals ℓ_n defined by (27) are clearly nonnegative; therefore so is their weak* limit ℓ. According to corollary 14' of the Riesz representation theorem, chapter 8, such a nonnegative functional acting on $C(S^1)$ can be represented as an integral with respect to a positive measure m. Thus we have proved the first part of

Theorem 6 (Herglotz-Riesz). *Every analytic function f in the unit disk $|\zeta| < 1$ whose real part is positive there can be expressed as*

$$f(\zeta) = \int \frac{1 + \zeta \, e^{-i\theta}}{1 - \zeta \, e^{-i\theta}} \, dm + ic, \qquad m \text{ a positive measure, } c \text{ real.} \tag{29}$$

Conversely every function f so represented is analytic in the unit disk and has positive real part there. The representation (29) is unique.

Proof. That (29) represents an analytic function with positive real part in the unit disk for any positive measure m is evident from formula (30) below. To see that the representation is unique, we note that the real part of (29) is

$$h(\zeta) = \int \frac{1 - r^2}{1 - 2r\cos(\phi - \theta) + r^2}\, dm, \qquad \zeta = r\, e^{i\phi}. \tag{30}$$

Take any continuous function $u(\phi)$, multiply (30) by $u(\phi)$, and integrate with respect to ϕ over S^1. We get, after interchanging the order of integration on the right,

$$\int h\left(r\, e^{i\phi}\right) u(\phi)\, d\phi = \int u_r(\theta)\, dm, \tag{31}$$

where

$$u_r(\theta) = \int \frac{1 - r^2}{1 - 2r\cos(\phi - \theta) + r^2}\, u(\phi)\, d\phi.$$

Suppose that $h(\zeta)$ can be represented in the form (30) by two different measures m and m'. Let $r \to 1$ in (31); by theorem 1 of this chapter, $u_r \to u$ in the maximum norm. Since the left side of (31) does not depend on the representing measure, it follows that

$$\int u(\theta)\, dm = \int u(\theta)\, dm'$$

for all continuous functions u. We appeal now to the uniqueness of measure in the Riesz representation theorem to conclude that $m \equiv m'$. This completes the proof of theorem 6. $\qquad\square$

From the uniqueness of the measure m it follows that the limit (28) exists not only for a subsequence but every sequence of R.

BIBLIOGRAPHY

Friedrichs, K. O. The identity of the weak and strong extension of differential operators. *Trans. AMS*, **55** (1944): 132–151.

Herglotz, G. Über Potenzreihen mit positivem reellem Teil in Einheitskreis. *S.B. Sächs. Akad. Wiss.*, **63** (1911): 501–511.

Riesz, F. Sur certains systémes singuliers d'equations intégrales. *An. l'École Normale Sup.* (3), **28** (1911): 33–62.

Toeplitz. O. Über allgemeine lineare Mittelbildungen. *Prace Math.–Fiz.*, **22** (1911): 113–119.

12

THE WEAK AND
WEAK* TOPOLOGIES

Definition. The *weak topology* in a Banach space is the weakest topology in which *all* bounded linear functionals are continuous. Since bounded linear functionals are continuous in the norm (strong) topology it follows that the weak topology is coarser than the strong topology. We show now that—except in finite-dimensional spaces— the weak topology is genuinely coarser than the strong topology:

The open sets in the weak topology are unions of *finite* intersections of sets of the form

$$\{x : a < \ell(x) < b\}. \tag{1}$$

Clearly, in an infinite-dimensional space the intersection of a finite number of sets of form (1) is unbounded. This shows that *every set that is open in the weak topology is unbounded.* In particular, the balls

$$\{x : |x| < R\}, \tag{2}$$

open in the strong topology, are *not* open in the weak topology.

Next we show that the weak topology is coarser than weak sequential convergence, in the following sense: define the *weak sequential closure* of a set S in a Banach space X as the weak limit of all weakly convergent sequences in S.

Theorem 1.

(i) *The weak sequential closure of any set S belongs to the closure of S in the weak topology.*

(ii) *In every infinite-dimensional Banach space there are sets weakly sequentially closed, but not closed in the weak topology.*

Proof. Part (i) is an immediate consequence of the definitions of weak convergence and weak topology. Part (ii) is exemplified by the following set S:

$$S = S_2 \cup S_3 \cup \dots . \tag{3}$$

Each set S_k is finite, constructed as follows: Choose any sequence of subspaces X_k of X, $\dim X_k = k$. S_k consists of points $x_{k,j}$, finite in number, so chosen that for every point x in X_k of norm k, x in X_k, $|x| = k$, there is a point $x_{k,j}$ in S_k such that

$$|x - x_{k,j}| < \frac{1}{k}, \ |x_{k,j}| = k. \tag{4}$$

We claim that the origin belongs to the closure of S in the weak topology. Any open set containing the origin contains a subset of the form

$$\{x : |\ell_i(x)| < \epsilon, \quad i = 1, \dots, n\}, \tag{5}$$

where the ℓ_i are linear functionals of norm 1. Since X_k is k-dimensional, for $k > n$ it contains a nonzero vector x_k such that

$$\ell_i(x_k) = 0, \qquad i = 1, \dots, n,$$
$$|x_k| = k. \tag{6}$$

By construction, there is an $x_{k,j}$ in S_k satisfying condition (4) for $x = x_k$. Using (4), (6), and $|\ell_i| = 1$, we get for any $\ell = \ell_i, i = 1, \dots, n$, that

$$\ell(x_{k,j}) = \ell(x_{k,j} - x_k) \le |x_{kj} - x_k| \le \frac{1}{k}. \tag{7}$$

Therefore for $k > 1/\epsilon$, the point $x_{k,j}$ belongs to the subset (5). This proves that the origin belongs to the closure of S in the weak topology.

On the other hand, S contains only a finite number of points in any ball of radius R. So, by the principle of uniform boundedness (see theorem 4 of chapter 10), S contains *no* weakly convergent sequences other than the trivial ones. $\square$

Despite the coarseness of the weak topology compared to the strong topology, the following is true:

Theorem 2. *Every convex subset K of a Banach space X that is closed in the strong topology is closed in the weak topology.*

Proof. We will show that if z in X does not belong to K, then z is not in the weak closure of K. Since K is closed in the strong topology, there is an open ball $B_R(z)$ centered at z that is disjoint from K. According to the hyperplane separation theorem, theorem 6 of chapter 3, there is a nonzero functional ℓ and a constant c such that

$$\ell(u) \le c \le \ell(v) \tag{8}$$

for all u in K and all v in $B_R(z)$. As explained in the proof of theorem 17 of chapter 8, the norm of ℓ is bounded by $1/R$. The points of $B_R(z)$ are of the form $v = z + x$,

$|x| < R$. Since $\ell(v) = \ell(z) + \ell(x)$, it follows that

$$\inf_{v \text{ in } B_R(z)} \ell(v) = \ell(z) + \inf_{|x| < R} \ell(x) = \ell(z) - |\ell|R.$$

Setting this into (8), we get that $\ell(z) > c$; so we conclude, again by (8), that the hyperplane

$$\{x : \ell(x) > c\} \tag{8'}$$

contains z but contains no point of K. Since (8)' is an open set in the weak topology, it follows that z does not belong to the closure of K in the weak topology.   □

Theorem 2 bears an analogy to theorem 6, chapter 10.

Suppose that U is a Banach space that is the dual of another Banach space X:

$$U = X'. \tag{9}$$

Then there is in U a natural class of linear functionals, those associated with elements of X:

$$x(u) = u(x). \tag{10}$$

Definition. The weak* topology in a Banach space U that is the dual of another Banach space X is the crudest topology in which all linear functionals (10) are continuous.

For U of form (9) and nonreflexive, the weak* topology is genuinely coarser than the weak topology, as will be clear from the following theorems. The first is due to Alaoglu.

Theorem 3. *The closed unit ball B in a Banach space U that is the dual of another Banach space X is compact in the weak* topology.*

Proof. To u in B we assign the array of numbers $\{u(x)\}$, x in X. Since $|u| \leq 1$, $|u(x)| \leq |x|$. Regard each array of numbers as a point in the product space P:

$$P = \prod_{x \text{ in } X} I_x, \qquad I_x = [-|x|, |x|], \tag{11}$$

and regard

$$u \to \{u(x)\} \tag{12}$$

as a mapping of B into P. The elements of P are vectors whose components are points in the factors I_x. The natural mappings in P are the projections of vectors onto their components, and the natural topology is the weakest one in which all these mappings are continuous. The mapping (12) is one-to-one, so it embeds B in P. It follows from the definitions that the weak* topology of B is the same as the inherited topology from P under the embedding. Each I_x is a compact interval of $\mathbb{R}$; it follows from Tychonov's theorem that P is compact. Since a closed subset of a compact set is compact, it suffices to prove that (12) maps B into a *closed* subset of P.

Let p be a point in the closure of the image of B under (12); we will show that then p is the image of some u in B, that is,

$$p_x = u(x) \qquad \text{for all } x \text{ in } X, \tag{13}$$

where $\{p_x\}$ denote the components of p. Equation (13) defines a function $u(x)$ on X. We have to show that it is bounded by 1 and linear. Boundedness follows from the fact that p_x belongs to $I_x = [-|x|, |x|]$. Linearity means that

$$p_{x+y} = p_x + p_y, \qquad p_{ax} = a p_x. \tag{14}$$

For every q that is the image of B in P under the mapping (12),

$$q_{x+y} = q_x + q_y, \qquad q_{ax} = a q_x. \tag{15}$$

Since p lies in the weak* closure of $\{q\}$, and since these relations involve only 3 (resp. 2) components of q, (14) follows from (15). $\qquad\qquad\square$

Theorem 3′. *A subset S of $U = X'$ that is closed in the w^* topology is w^* compact iff it is bounded in norm.*

Proof. If S is bounded in norm, it belongs to some closed ball B_R. According to theorem 3, B_R is w^*−compact, but then so is its w^*−closed subset S.

Conversely, suppose that S is w^*−compact; then so is its image $\{u(x)\}$ under the continuous mapping $u \to u(x)$ for every x. A compact set in $\mathbb{R}$ (or $\mathbb{C}$) is bounded, so for every x in X $\{|u(x)|\} \leq b(x)$ for every u in S. But then by theorem 3 of chapter 10, the principle of uniform boundedness, $|u| \leq b$ for all u in S. $\qquad\square$

Theorem 4. *The closed unit ball in a Banach space Z is compact in the weak topology iff Z is reflexive.*

Proof. The "if" part follows from theorem 3. The "only if" is due to Eberlein and Smulyan; see Dunford and Schwartz.

BIBLIOGRAPHY

Alaoglu, L. Weak topologies of normed linear spaces. *An. Math.*, **41** (1940): 252–267.

Chernoff, P. R. A simple proof of Tychonoff's theorem via nets. *Am. Math. Monthly*, **99** (1992): 932–934.

Dunford, N. and Schwartz, J. *Linear Operators: Part 1: General Theory.* Wiley–Interscience, 1957, pp. 423–425.

Eberlein, W. F. Weak compactness in Banach spaces. *Proc. Nat. Acad. Sci. USA*, **33** (1947): 51–53.

Smulyan, V. I. Über lineare topologische Räume. *Math. Sbornik, N.S.*, **7** (1940): 425–448.

Tychonoff, A. Über die topologische Erweiterung von Räumen. *Math. An.*, **102** (1929–30): 544–561.

13

LOCALLY CONVEX TOPOLOGIES AND THE KREIN-MILMAN THEOREM

The weak and weak* topologies are the weakest in which certain linear functionals are continuous. If one demands the continuity of even fewer functionals, one gets even weaker topologies. All these topologies have the property that openness can be defined in terms of convex sets. In this chapter we develop, and apply, the theory of such topologies.

Definition. A locally convex topological (LCT) linear space is a linear space over the reals with a Hausdorff topology that has the following properties:

  (i) Addition is continuous; that is, $(x, y) \rightarrow x + y$ is a continuous mapping of $X \times X$ into X.

 (ii) Multiplication by scalars is continuous; that is, $(k, x) \rightarrow kx$ is a continuous mapping of $\mathbb{R} \times X$ into X.

(iii) There is a basis for the open sets at the origin consisting of convex sets; that is, every open set containing the origin contains a convex open set containing the origin.

Note that the norm topology of a Banach space is a locally convex topology; the convex, open sets containing the origin that form a basis for the topology are the open balls centered at the origin.

Exercise 1. Show that the weak and weak* topologies are locally convex.

Exercise 2. Let $\{\ell_\alpha\}$ be a collection of linear functions in a linear space X over $\mathbb{R}$ that separates points; that is, for any two distinct points x and y of X there is an ℓ_α such that $\ell_\alpha(x) \neq \ell_\alpha(y)$.

(a) Show that the weakest topology in which all the ℓ_α are continuous is locally convex.

(b) Show that a linear functional ℓ is continuous in the topology above iff it is a *finite* linear combination of the ℓ_α.

Exercise 3. Show that a continuous function $f(a, b)$ on a product of two topological spaces whose values lie in a topological space is a continuous function of a when b is fixed.

Theorem 1.

(i) *In a LCT linear space X the collection of open sets is translation invariant; that is, if T is an open set, so $T - x$, for any x in X.*

(ii) *If T is an open set, so is kT, for $k \neq 0$; in particular, $-T$ is open.*

(iii) *Every point of an open set T is interior to T.*

Proof. According to exercise 3, $x + y$ is a continuous function of y for x fixed. The inverse image of the open set T under this mapping is $T - x$; this proves part (i). Part (ii) follows similarly.

(iii) By exercise 3, kx is a continuous function of k for x fixed. So the set of k for which kx lies in some open set T is an open subset of $\mathbb{R}$. Suppose that T contains the origin; then $k = 0$ belongs to this set, and by the above observation, so does an open interval containing $k = 0$. That means that for k small enough, kx belongs to T, but that is what it means for the origin to be an interior point of T. This proves (iii), for by (i) any point of T can be shifted to the origin. □

13.1 SEPARATION OF POINTS BY LINEAR FUNCTIONALS

Theorem 2. *The continuous linear functionals in a LCT linear space X separate points. That is, if y and z are distinct points of X, there is a continuous linear functional ℓ such that*

$$\ell(y) \neq \ell(z). \tag{1}$$

Proof. We construct a linear functional separating y and z. Without loss of generality we take $y = 0$. Since the topology is Hausdorff, there is an open set T containing $y = 0$ but not z; by (iii) of the definition of LCT space, we may take T to be convex. By theorem 1, 0 is an interior point of T, so the gauge function p_T of T is finite, and

$$p_T(u) < 1 \qquad \text{for all } u \text{ in } T. \tag{2}$$

According to the hyperplane separation theorem, theorem 5 of chapter 3, there exists a linear functional ℓ satisfying

$$\ell(z) = 1, \tag{3}$$

$$\ell(x) \le p_T(x) \qquad \text{for all } x. \tag{4}$$

Clearly, since $\ell(y) = \ell(0) = 0$, ℓ separates y and z.

To complete the proof, we have to show that ℓ is continuous. We first show that every halfspace $w : \ell(w) < c$ is open. We claim that if w belongs to the halfspace, so does the open set

$$w + rT, \quad r = c - \ell(w). \tag{5}$$

Using (4), (5), and (2), we get that for u in T,

$$\ell(w + ru) = \ell(w) + r\ell(u) \le \ell(w) + rp_T(u) < \ell(w) + c - \ell(w) = c, \tag{6}$$

showing that every point of (5) lies in the halfspace; so the halfspace is open. We can show similarly that every halfspace of the form

$$w : \ell(w) > d$$

is open. For this argument p_T needs to be an even function; this will be the case if T is symmetric around this origin. This can be accomplished by replacing T, if necessary, by $T \cap (-T)$. $\square$

Theorem 2 can be sharpened as follows:

Theorem 2′. *Denote by K a closed, convex set in a LCT linear space X, z a point in X not in K. Then there is a continuous linear functional ℓ such that*

$$\ell(y) \le c \qquad \text{for all } y \text{ in } K, \ \ell(z) > c. \tag{7}$$

Proof. The proof is similar to that of theorem 2, except that in place of the hyperplane separation theorem we use the extended version, theorem 6 in chapter 3. $\square$

Exercise 4. Let K denote a convex subset of a LCT linear space X. Show that the closure $\overline{K}$ of K also is convex.

13.2 THE KREIN-MILMAN THEOREM

We recall from the end of chapter 1 the notion of an *extreme subset* of a convex set K and, in particular, the notion of an *extreme point*. A subset E of a convex set K is called an *extreme subset* of K if:

  (i) E is convex and nonempty.
  (ii) Whenever a point x of E is expressed as a convex combination of y and z in K, then both y and z belong to E.

An extreme set consisting of a single point is called an *extreme point*.

Exercise 5. Show that the nonempty intersection of extreme sets is extreme.

The elementary properties of extreme set are contained in theorems 7 and 8 of chapter 1. The basic result concerning convex sets in finite dimensional spaces is the following theorem of Carathéodory:

Every compact convex subset K in $\mathbb{R}^N$ has extreme points, and every point of K can be written as a convex combination of $N + 1$ extreme points.

Exercise 6. Furnish a proof of Carathéodory's theorem by induction on N.

M. G. Krein and D. P. Milman have given the following beautiful—and useful—generalization of this result:

Theorem 3. *Let X be a LCT linear space, K a nonempty, compact, convex subset of X.*

(i) *K has at least one extreme point.*

(ii) *K is the closure of the convex hull of its extreme points.*

Proof. Consider the collection $\{E_j\}$ of all nonempty closed extreme subsets of K. This collection is nonempty, for it contains K itself. Partially order this collection by inclusion. We claim that every totally ordered subcollection $\{E_j\}$ has a lower bound. That lower bound is the intersection $\cap E_j$. To see this, we have to show that $\cap E_j$ is nonempty, closed, and extreme.

We claim that every *finite* subset of the totally ordered collection $\{E_j\}$ has a nonempty intersection. This is because in being totally ordered by inclusion, the intersection of a finite subset of the collection $\{E_j\}$ is the smallest member of that subset. To conclude that $\cap E_j$ is nonempty, we argue indirectly: suppose that the intersection is empty. Then the union of the complements of the E_j cover K. Since K is compact, a finite collection of these already cover K, but then the intersection of these finite number of E_j is empty, contrary to what we have already shown. Being the intersection of closed sets, $\cap E_j$ is closed. By exercise 5 the nonempty intersection of extreme subsets of a convex set K is itself an extreme subset of K.

We conclude from Zorn's lemma that K has a closed extreme subset E that is *minimal* with respect to inclusion. We claim that such an E consists of a single point. To see this, suppose, on the contrary, that E contains two distinct points. According to theorem 2, there exists a continuous linear functional ℓ that separates these points. Since E is compact, and ℓ continuous and not constant on E, ℓ achieves its maximum on some proper subset M of E. Since ℓ is continuous and E is closed, M is closed. Since the inverse image of an extreme subset is extreme (see corollary 8′ in chapter 1), the set M where a linear functional ℓ assumes its maximum on a convex set

E is an extreme subset of E. It is easy to show further (see theorem 7 of chapter 1) that if E is an extreme subset of K, and M an extreme subset of E, then M is an extreme subset of K. Since E is a minimal extreme subset of K, and M an extreme subset smaller than E, we have a contradiction, into which we got by assuming that E contains more than one point. We conclude therefore that a minimal E consists of a single point. This single point is an extreme point of K. This completes the proof of part (i), and gives a little more:

(i′) Every closed, extreme subset of K contains an extreme point.

We turn now to the proof of part (ii). Denote by K_e the set of extreme points of K, and by $\widehat{K}_e$ the convex hull of K_e. To show that every point of K belongs to the closure of $\widehat{K}_e$ is the same as showing that a point z that does not belong to the closure of $\widehat{K}_e$ does not belong to K. According to exercise 4, the closure of $\widehat{K}_e$ is convex. So, if z does not belong to the closure, then according to theorem 2′ there is a continuous linear functional ℓ such that

$$\ell(y) \le c \qquad \text{for all } y \text{ in } \widehat{K}_e, \ \ell(z) > c. \tag{8}$$

Since K is compact and ℓ continuous, ℓ achieves its maximum over K on some closed subset E of K. According to corollary 8′, chapter 1, E is an extreme subset of K. According to part (i′) of theorem 3 noted above, E contains some extreme point p of K. Since p belongs to K_e, and so to $\widehat{K}_e$, it follows from (8) that $\ell(p) \le c$. Since by construction $\ell(p) = \max_K \ell(x)$, $\ell(x) \le \ell(p) \le c$ for all x in K. Since by (8), $\ell(z) > c$, this proves that z does not belong to K. $\qquad\square$

13.3 THE STONE-WEIERSTRASS THEOREM

Theorem 4. *Let S be a compact Hausdorff space, $C(S)$ the set of all real-valued continuous functions on S. Let E be a subalgebra of $C(S)$, that is,*

(i) *E is a linear subspace of $C(S)$.*
(ii) *The product of two functions in E belongs to E.*

In addition we impose the following conditions on E:

(iii) *E separates points of S, that is, given any pair of points p and q, $p \ne q$, there is a function f in E such that $f(p) \ne f(q)$.*
(iv) *All constant functions belong to E.*

Conclusion: E is dense in $C(S)$ in the maximum norm.

The classical Weierstrass theorem is a special case of this proposition, with S an interval of the x axis, and E the set of all polynomials in x. We present Louis de Branges's elegant proof, based on the Krein-Milman theorem, of Stone's generalization of the Weierstrass theorem.

Proof. According to the spanning criterion, theorem 8 of chapter 8, E is dense in $C(S)$ if the only bounded linear functional ℓ on $C(S)$ that is zero on E is the zero functional. According to the Riesz-Kakutani representation theorem, theorem 14 of chapter 8, the bounded linear functionals on $C(S)$ are of the form

$$\ell(f) = \int_S f \, dv,$$

v a signed measure of finite total variation $\|v\| = \int |dv|$. So what we have to show is that if $\int_S f \, dv = 0$ for all f in E, $v = 0$.

Suppose not; denote by U the set of signed measures of finite total mass is ≤ 1 that annihilate all functions in E. This is a convex set, and according to Alaoglu's theorem, theorem 3 in chapter 12, compact in the weak* topology. So according to the Krein-Milman theorem, if U contained a nonzero measure, it would contain a nonzero extreme point; call it μ. Since μ is extreme, $\|\mu\| = 1$. Since E is an algebra, if f and g belong to E, so does gf. Since μ annihilates every function in E,

$$\int (fg) d\mu = 0.$$

It follows that the measure $gd\mu$ also annihilates every function in E.

Let g be a function in E whose values lie between 0 and 1:

$$0 < g(p) < 1 \qquad \text{for all } p \text{ in } S.$$

Denote

$$a = \|g\mu\| = \int g|d\mu|, \quad b = \|(1-g)\mu\| = \int (1-g)|d\mu|.$$

Clearly a and b are positive. Add them:

$$a + b = \int |d\mu| = 1.$$

The identity

$$\mu = a\frac{g\mu}{a} + b\frac{(1-g)\mu}{b}$$

represents μ as a nontrivial convex combination of $g\mu/a$ and $(1-g)\mu/b$, both points in U. Since μ is an extreme point, μ must be equal to $g\mu/a$.

Define the *support* of the measure μ to be the set of points p that have the property that $\int_N |d\mu| > 0$ for any open set N containing p. If $\mu = g\mu/a$, it follows that g has the same value at all points of the support of μ.

We claim that the support of μ consists of a single point. To see this, suppose that both p and q, $p \neq q$, belong to the support μ. Since the functions in E separate points of S, there is a function h in E, $h(p) \neq h(q)$. Adding a large enough constant to h and dividing it by another large constant, we obtain a function g whose values lie between 0 and 1, and $g(p) \neq g(q)$. This contradicts our previous conclusion.

A measure μ whose support consists of a single point p, and $\|\mu\| = 1$, is a unit point mass at p. Therefore

$$\int f d\mu = f(p) \text{ or } -f(p).$$

Since, by hypothesis, the constant 1 belong to E, $\int f d\mu \neq 0$ for $f \equiv 1$ in E, a contradiction. $\square$

13.4 CHOQUET'S THEOREM

The following further extension of Caratheodory's theorem holds on locally convex linear spaces:

Theorem 5. *Let X be a LCT linear space, K a nonempty compact, convex subset of X, K_e the set of extreme points of K. For any point u of K there is a probability measure m_u on $\overline{K}_e$, the closure of K_e, tht is a measure satisfying:*

$$m_u \geq 0, \quad \int_{\overline{K}_e} dm_u = 1, \tag{9}$$

such that in the weak sense

$$u = \int_{\overline{K}_e} e \, dm_u. \tag{10}$$

The weak sense of the integral representation above is that for every continuous linear functional ℓ over X,

$$\ell(u) = \int_{\overline{K}_e} \ell(e) \, dm_u(e). \tag{10'}$$

Proof. For any continuous ℓ and K compact, ℓ achieves its minimum and maximum on K. According to corollary $8'$ of chapter 1, the sets where ℓ achieves its minimum and maximum are extreme subsets of K. According to part (i') of theorem 3, these extreme subsets contain extreme points. Therefore for any u in K and for every continuous linear functional ℓ,

$$\min_{p \text{ in } K_e} \ell(p) \leq \ell(u) \leq \max_{p \text{ in } K_e} \ell(p). \tag{11}$$

It follows from (11) applied to $\ell_1 - \ell_2$ that if ℓ_1 and ℓ_2 are equal on K_e, they are equal on K; therefore ℓ on $\overline{K}_e$ uniquely determines the value of $\ell(u)$ for every u in K. Denote the restriction of ℓ to $\overline{K}_e$ by f:

$$f(q) = \ell(q), \quad q \text{ in } \overline{K}_e. \tag{12}$$

Since $\ell(u)$ is determined by f, we can write

$$\ell(u) = u(f), \tag{12'}$$

clearly a linear functional of f. We rewrite (11) as

$$\min_{q \text{ in } \overline{K}_e} f(q) \le u(f) \le \max_{q \text{ in } \overline{K}_e} f(q). \tag{13}$$

The set L of functions f defined in (12) form a linear subspace of the space $C(\overline{K}_e)$ of continuous functions on $\overline{K}_e$. We claim that we can extend the linear functional $u(f)$ defined in (12') from L to all of $C(\overline{K}_e)$ so that property (13) is preserved. To see this, we adjoin the function $f_0 \equiv 1$ to L and define $u(f_0) = 1$. Then we appeal to theorem 1 of chapter 4, according to which a positive linear functional can be extended positively to the space of all functions. Since (13) implies that $u(f)$ is positive, such an extension is possible; imagine it done.

$\overline{K}_e$ is a closed subset of the compact set K and is therefore compact. We appeal now to the Riesz-Kakutani representation theorem (see chapter 8, theorem 14) according to which a bounded linear functional u on $C(\overline{K}_e)$ can be represented as

$$u(f) = \int_{\overline{K}_e} f \, dm. \tag{14}$$

Since the functional is positive, so is the representing measure m; it follows from $u(f_0) = 1$ that $m(\overline{K}_e) = 1$. Setting (12) and (12') into (14), we obtain (10'). $\quad\square$

Theorem 5 asserts that every point u of the compact convex set K can be represented as a continuous convex combination of points of the *closure* of the set of extreme points. This proviso is needed because the set of extreme points may not be closed, not even in a finite-dimensional space. Say we take in $\mathbb{R}^3$ the convex hull of the circle: $x^2 + y^2 = 1, z = 0$, and the interval:

$$x = 1, \quad y = 0, \quad -1 \le z \le 1.$$

The extreme points of the convex hull are $x = 1, y = 0, z = \pm 1$, and all points of the circle $x^2 + y^2 = 1, z = 0$ *except* $x = 1, y = 0, z = 0$.

Exercise 7. Let v be a point in $\overline{K}_e$ that does not belong to K_e; show that v can then be represented as

$$v = \int_{\overline{K}_e} e \, dm,$$

where m is a probability measure on $\overline{K}_e$ such that $m(v) = 0$.

Exercise 8. Deduce part (ii) of theorem 3 from theorem 5.

Choquet gave the following sharpening of theorem 5:

Theorem 6 (Choquet). *Let K be a nonempty compact, convex subset of a LCT linear space, and assume in addition that K is metrizable. Then every point u of K can be represented in the weak sense as*

$$u = \int_{K_e} e\, dm_u, \tag{10''}$$

where m_u is a probability measure on the set of extreme points.

Proof. For proof, see Phelps.   □

We call $(10'')$ a Choquet-type representation. In the next chapter we give many examples of Choquet-type representations of convex sets.

We present now a useful result that extends to LCT spaces, the following intuitively clear property of convex hulls of compact sets S in finite-dimensional spaces: the points of $\hat{S}$ that are added to S to make it convex contain no extreme points.

Theorem 7. *Let X be a LCT linear space, S a compact subset of X. Suppose K, the closure of the convex hull of S, is compact. Then every extreme point of K belongs to S.*

Proof. Let N be any open convex set that contains the origin. The open sets $y + N$, y in S form an open cover of S; since S is compact, a finite number of them cover S:

$$\cup(y_i + N) \supset S. \tag{15}$$

Denote by S_i the intersection $(y_i + N) \cap S$; it follows from (15) that

$$\cup S_i = S. \tag{16}$$

Denote the closure of the convex hull of S_i by K_i. Since $S_i \subset S$, it follows that $K_i \subset K$; since K_i is closed and K assumed compact, it follows that each K_i is compact. Next we need

Lemma 8. *Let K_1 and K_2 be a pair of compact convex sets in a LCT linear space. Then the convex hull of their union is compact.*

Proof. Since K_1 and K_2 themselves are convex, it is easy to see that the convex hull of their union consists of all points of the form

$$ay_1 + (1-a)y_2, \quad y_1 \in K_1, \quad y_2 \in K_2, \quad 0 \le a \le 1. \tag{17}$$

These points are images of the triple product

$$K_1 \times K_2 \times I, \quad I = [0, 1] \tag{18}$$

under the mapping (17). The triple product (18) is compact, and according to the definition of a LCT space, the mapping (17) is continuous. It follows that the image of the compact set (18) is compact, as claimed in lemma 8. □

From lemma 8 we deduce inductively that the convex hull of the union of a finite number of compact sets is compact. We turn to the compact sets K_i defined above and claim that the convex hull of their union, denoted as $\mathrm{CH}[K_1 \cup \ldots \cup K_n]$, contains K:

$$K \subset \mathrm{CH}[K_1 \cup \ldots \cup K_n]. \tag{19}$$

We note that K_i contains S_i. Therefore, by (16), $K_1 \cup \ldots \cup K_n$ contains $S_1 \cup \ldots \cup S_n = S$. By lemma 8, the right side of (19) is compact, and therefore closed. Thus it is a closed, convex set that contains S. But K is defined to be the smallest such set, and it therefore is contained in $\mathrm{CH}[K_1 \cup \ldots \cup K_n]$. This proves (19). In words, every point of K is a convex combination of points of K_j. Since each K_j is contained in K, it follows from the definition of extreme point that each extreme point p of K belongs to a K_i.

By definition, S_i is contained in $y_i + N$. Since N is convex, the convex hull of S_i belongs to $y_i + N$: $\widehat{S}_i \subset y_i + N$. For any set R, $R + N$ contains the closure of R. Since K_i is the closure of $\widehat{S}_i$, it follows that $K_i \subset y_i + N + N = y_i + 2N$. Therefore, since each y_i belongs to S,

$$\cup K_i \subset S + 2N.$$

We have shown that each extreme point of K belongs to some K_i, so it follows from the result above that every extreme point p of K belongs to $S + 2N$. Now N is arbitrary; since S is closed, the intersection of all sets $S + 2N$ is S itself. Thus every extreme point p of K is contained in S. □

Theorem 7 is useful in identifying extreme points.

Exercise 9. Show that if S is a compact set in a Banach space, its closed convex hull is compact. Is this true in every LCT space?

NOTE. The prime examples of LCT linear spaces are Banach spaces in the weak and weak* topologies. Other important examples are spaces of distributions. In view of the enourmous success of the theory of distributions in the theory of partial differential equations and in harmonic analysis, it was thought that other locally convex topologies might play a similarly fruitful role; that hope has not yet been realized.

Other applications of the Krein-Milman theorem and its generalizations are described in Diestal and Uhl.

BIBLIOGRAPHY

Choquet, G. Existence des représentation intégrales au moyen des points extremaux, dans les cones convexes. *C. R. Acad. Sci. Paris*, **243** (1956): 699–702.

de Branges, L. The Stone-Weierstrass theorem. *Proc. AMS*, **10** (1959): 822–824.

Diestel J. and Uhl, J. J. *Vector Measures*. American Mathematical Society, Providence, RI, 1970.

Kelley, J. L. *General Topology*. Van Nostrand, Princeton, NJ, 1955.

Kelley, J. L. and Namioka, I. *Linear Topological Spaces*. Van Nostrand, Princeton, NJ, 1963.

Krein, M. G. and Milman, D. On extreme points of regularly convex sets. *Studia Math.*, **9** (1940): 133–138.

Phelps, R. R. *Lectures on Choquet's Theorem*. Van Nostrand, Princeton, NJ, 1966.

14

EXAMPLES OF CONVEX SETS AND THEIR EXTREME POINTS

In this chapter we present a great variety of examples of convex sets, their extreme points, and Choquet-type integral representations of points of the set in terms of the extreme points. In some examples the extreme points are determined by a direct argument. Then a locally convex topology is introduced so that the convex set in question is compact and the Choquet-type representation is then derived via Choquet's theorem. In other examples the Choquet-type representation is derived directly by an analytic argument. The representation is then used to identify the extreme points of the set. In most of these examples the representation is unique.

14.1 POSITIVE FUNCTIONALS

Let Q denote a compact Hausdorff space, and $C(Q)$ the space of continuous functions on Q whose values are real. We denote a linear functional ℓ defined on $C(Q)$ as *positive* if $\ell(f) \geq 0$ for all nonnegative f in $C(Q)$. Recall from chapter 8 that a positive linear functional is bounded. Denote by P the collection of all positive linear functionals ℓ that satisfy

$$\ell(1) = 1. \tag{1}$$

Theorem 1. *P is a convex set whose extreme points are the point evaluations* e_r, *defined as*

$$e_r(f) = f(r), \tag{2}$$

r any point of Q.

Proof. The convexity of P is obvious. To see that every e_r, defined by (2), is an extreme point, imagine e_r represented as

$$e_r = am + (1-a)\ell, \qquad m \text{ and } \ell \text{ in } P, 0 < a < 1. \tag{3}$$

Let f be any function in $C(Q)$ that is nonnegative and satisfies

$$f(r) = 0. \tag{4}$$

Set f into (3); using (2) and (4) obtains

$$e_r(f) = f(r) = 0 = am(f) + (1-a)\ell(f).$$

We claim that $m(f)$ and $\ell(f)$ are both zero; otherwise, one of them would have to be positive, the other negative, in contradiction to $f \geq 0$ and ℓ, m both positive functionals.

Every continuous f can be decomposed into its positive and negative part:

$$f = f_+ - f_-, \qquad f_+ = \max(f, 0).$$

Both f_+ and f_- are nonnegative, and if $f(r) = 0$, $f_+(r) = f(r) = 0$. It follows from this and the foregoing that if $f(r) = 0$, $m(f) = \ell(f) = 0$. In other words, the nullspaces of m and of ℓ contain the nullspace of e_r. Since the nullspace of a nontrivial functional has codimension 1, it follows that ℓ and m are constant multiples of e_r; since all functionals in P satisfy (1), it follows that the constant multiplier is 1. This proves that $\ell = m = e_r$, so e_r is extreme.

We show now that the e_r are the only extreme points of P. Let ℓ be any positive linear functional on $C(Q)$, normalized by (1). According to the Riesz-Kakutani representation theorem, there exists a nonnegative measure m on Q such that for every continuous function f on Q,

$$\ell(f) = \int f \, dm. \tag{5}$$

Because of the normalization (1), $m(Q) = 1$; the measure m is uniquely determined by the functional ℓ. We claim that the only extreme points of the set of positive linear functionals normalized by (1) are those where the measure m is concentrated at a single point. Otherwise, m can be split as $am_1 + (1-a)m_2$, where both m_1 and m_2 are nonnegative measures of total mass 1, and $m_1 \neq m_2$. Setting

$$\ell_j(f) = \int f \, dm_j, \qquad j = 1, 2,$$

we get $\ell = a\ell_1 + (1-a)\ell_2$. If ℓ were an extreme point, $\ell_1 = \ell_2 = \ell$, and so ℓ can be represented in form (5) by the distinct measures m_1 and m_2. This contradicts uniqueness of the representing measure. This completes the proof of theorem 1. $\square$

We can use formula (5) to rewrite the formula 2 symbolically as

$$\ell = \int e_r \, dm(r),\tag{6}$$

a Choquet-type representation.

14.2 CONVEX FUNCTIONS

In this section we make use of the notions and results of the theory of distributions as explained in Appendix B.

Definition. A real-valued function f in $\mathbb{R}^n$ is *convex* if it satisfies

$$f\left(\sum a_j x_j\right) \leq \sum a_j f(x_j)\tag{7}$$

for all choices of $x_1, \ldots, x_N$ in $\mathbb{R}^n$ and all a_j satisfying $a_j \geq 0$, $\sum a_j = 1$.

Here we consider convex functions f of a single variable. It suffices to assume (7) to hold for $N = 2$:

$$f(ax + (1-a)z) \leq af(x) + (1-a)f(z), \qquad 0 \leq a \leq 1,\tag{8}$$

for all x, z. Setting $ax + (1-a)z = y$, condition (7) is easily seen to be equivalent with the following: for $x < y < z$,

$$\frac{f(y) - f(x)}{y - x} \leq \frac{f(z) - f(y)}{z - y}.\tag{9}$$

It follows from (9) that every convex function is continuous and has right and left derivatives.

The second difference quotients

$$\frac{f(x+h) - 2f(x) + f(x-h)}{h^2}\tag{10}$$

converge in the sense of distributions to f'' as h tends to zero. It follows from (9) that the difference quotients (10) are nonnegative; since the limit in the sense of distributions of a nonnegative distribution is nonnegative, it follows that for convex f,

$$0 \leq f''\tag{11}$$

in the sense of distributions.

Convexity in an interval is defined in the same way; note that a function convex in an interval need not be continuous at the endpoints.

We denote by C the set of a functions f convex in the interval $[0, 1]$ and satisfying

$$f(0) = 0, \quad f(1) = 1, \quad f(x) \geq 0 \qquad \text{for } 0 \leq x \leq 1. \tag{12}$$

Theorem 2. *C is a convex set, whose extreme points are the functions*

$$e_r(x) = \begin{cases} 0 & \text{for } x \leq r \\ \frac{x-r}{1-r} & \text{for } r \leq x, \end{cases} \tag{13}$$

where $0 \leq r < 1$, and

$$e_1(x) = \begin{cases} 0 & \text{for } x < 1 \\ 1 & \text{for } x = 1. \end{cases} \tag{13'}$$

Proof. First we show that all the functions e_r defined above are extreme points of C. Suppose that e_r is represented as

$$e_r = af + (1 - a)g. \tag{14}$$

We claim that both f and g are zero on $[0, r]$; for e_r is, and so otherwise by (14) one of the functions f or g would be negative. Since this is contrary to (12),

$$f(x) = g(x) = 0, \qquad 0 \leq x \leq r,$$

follows. Similarly we claim that $f(x)$ and $g(x)$ are both equal to e_r on $[r, 1]$. For if not, one of them would be $> e_r$ at some point $y > r$; a short calculation shows that this contradicts (9) with $x = r, z = 1$. This shows that e_r is extreme.

Let f be any convex function in $[0, 1]$ satisfying (12). We set $f(r) = 0$ for $r < 0$. Clearly, f thus extended remains convex. Let ϕ be any C_0^∞ test function that is zero for $r \geq 0$. Then according to the theory of distributions

$$\int f\phi'' \, dr = \int f''\phi \, dr \tag{15}$$

where $'$ denotes differentiation with respect to r. We choose now x in $0 < x < 1$, and define the function $\phi_x(y)$ by

$$\phi_x(r) = \begin{cases} x - r & \text{for } r \leq x, \\ 0 & \text{for } r \geq x. \end{cases} \tag{16}$$

The function ϕ_x is piecewise linear and $\phi_x'' = \delta(r - x)$. If we could substitute $\phi = \phi_x$ in (15), we would obtain

$$f(x) = \int \phi_x(r) f''(r) \, dr. \tag{17}$$

Since ϕ_x is not C^∞, this is not legitimate, so we approximate ϕ_x by a sequence ϕ_x^ϵ of C^∞ functions. Since $f(r)$ is continuous for $r < 1$, the left side of (15) tends to the left side of (17). On the other hand, the nonnegative distribution f'' is a nonnegative measure; therefore the right side of (15) tends to the right side of (17). This proves (17) for $x < 1$. We can rewrite it using notation (13) as follows:

$$f(x) = \int e_r(x)(1-r) \, f''(r) \, dr, \qquad x < 1. \tag{18}$$

We let $x \to 1$ and obtain

$$m_1 = \lim_{x \to 1} f(x) = \int (1-r) \, f''(r) \, dr. \tag{18'}$$

Since f is an increasing function, and $f(1) = 1, m_1 \leq 1$. We can combine formulas (18) and (18') into one and write

$$f(x) = \int_0^1 e_r(x) \, dm(r), \tag{19}$$

where

$$m(r) = \int_0^r (1-s) \, f''(s) \, ds \qquad \text{for } r < 1, \text{ and } m(1) = 1 - m_1. \tag{19'}$$

The measure m is uniquely determined by the convex function f; this follows readily from formula (19). It follows, as in section 14.1, that the only extreme points of the set C of convex functions satisfying conditions (12) are those where the measure m is concentrated at a single point of the interval [0,1]. This completes the proof of theorem 2. □

We can rewrite (19) symbolically as

$$f = \int_0^1 e_r \, dm, \tag{20}$$

a Choquet-type representation of convex functions.

Exercise 1. Find a version of theorem 2 for convex functions of n variables.

Exercise 2. Prove theorem 2 without the theory of distributions, using the theorem of Krein-Milman.

14.3 COMPLETELY MONOTONE FUNCTIONS

The difference operator D_a, acting on functions of a single variable, is defined by

$$(D_a f)(t) = f(t + a) - f(t). \tag{21}$$

For $a > 0$, D_a maps functions f defined on $\mathbb{R}_+$ into functions defined on $\mathbb{R}_+$.

Definition. A real-valued function f defined on $\mathbb{R}_+$ is called *completely monotone* (c.m.) if

$$(-1)^n \left(\prod_1^n D_{a_j} \right) f \geq 0 \qquad \text{on } \mathbb{R}_+ \tag{22}$$

for all $a_j > 0$ and all $n = 0, 1, \ldots$.

The following result is due to S. Bernstein:

Theorem 3. *Every completely monotone function f on $\mathbb{R}_+$ can be represented as*

$$f(t) = \int_0^\infty e^{-\lambda t} dm(\lambda), \tag{23}$$

m some nonnegative measure, $m(\mathbb{R}_+) < \infty$. Conversely, every function of form (23) is completely monotone.

Proof. To show that f of form (23) is completely monotone we write

$$D_a f = \int_0^\infty D_a e^{-\lambda t} dm(\lambda) = \int (e^{-a\lambda} - 1) e^{-\lambda t} dm(\lambda);$$

then

$$(-1)^n \left(\prod_1^n D_{a_i} \right) f = \int \prod (1 - e^{-a_i \lambda}) e^{-\lambda t} \, dm$$

is clearly nonnegative.

Turning to the direct part, we will make use of the following properties of c.m. functions.

Lemma 4.

  (i) *The sum of two c.m. functions is c.m.*

  Suppose that f is a c.m. function; then

 (ii) *f is nonnegative.*
(iii) *af is c.m. for $a > 0$.*
(iv) *$-D_a f$ is c.m. for $a > 0$.*
  (v) *$T_a f = f(t + a)$ is c.m. for $a > 0$.*
 (vi) *$H_b f = f(bt)$ is c.m. for $b > 0$.*
(vii) *f is nonincreasing.*
(viii) *f is convex.*

Proof. Parts (i) and (iii) follow from the fact that the operators D_a appearing in condition (22) characterizing c.m. functions are linear. Part (ii) is (22), parts (iv) and (v) can be deduced by applying the operators D_a, respectively T_a, to (22), and noting that these operators commute with D_a. Part (vi) follows by applying H_b to (22), and noting that

$$H_b D_a = D_{ab^{-1}} H_b.$$

Part (vii) follows from (22) for $n = 1$, and (viii) from (22) for $n = 2$. □

We define X to be the space of all real-valued functions on $\mathbb{R}_+$, and take K to be the subset of all c.m. functions, in the sense of (22), normalized by

$$f(0) = 1. \tag{24}$$

It follows from parts (i) and (iii) of lemma 4 that K is a convex set.

Lemma 5. *The extreme points of K are of the form*

$$e_\lambda(t) = e^{-\lambda t}, \qquad 0 \le \lambda < \infty, \tag{25a}$$

and

$$e_\infty(t) = \begin{cases} 0 & for\ t > 0 \\ 1 & for\ t = 0. \end{cases} \tag{25b}$$

Proof. By parts (ii) and (vii) of lemma 4, every c.m. function is nonnegative and nonincreasing. It follows from (24) that every f in K satisfies

$$0 \le f(t) \le 1. \tag{26}$$

Let e be an extreme point of K; then in particular,

$$0 \le e(t) \le 1. \tag{26'}$$

Suppose that for all $a > 0$, strict inequality holds:

$$0 < e(a) < 1. \tag{26''}$$

We define two auxiliary functions as follows:

$$f(t) = \frac{e(t) - e(t + a)}{1 - e(a)},$$

$$g(t) = \frac{e(t + a)}{e(a)}. \tag{27}$$

It follows from part (iii), (iv), and (v) of lemma 4 and (26'') that f and g belong to K.

Clearly,

$$e = (1 - e(a))f + e(a)g. \tag{27'}$$

By definition of extreme point given in the previous chapter, it follows from (27')
that $f \equiv g \equiv e$. In particular, by (27), this implies that for all t and a,

$$e(t)e(a) = e(t + a). \tag{28}$$

All continuous solutions of this equation are exponential functions. It follows from
part (viii) of lemma 4 that every f in K is convex, and so is continuous for $t > 0$.
We can conclude that

$$e(t) = e^{-\lambda t}.$$

That λ is ≥ 0 follows from part (vii) of lemma 4.

The cases where (26'') fails to hold for some $a > 0$ can be easily handled. When
$e(a) = 1$ for some $a > 0$, $e = e_0$; when $e(a) = 0$ for some $a > 0$, $e = e_\infty$.   □

We introduce now the topology for functions that is the coarsest in which all the
linear functionals ℓ_t:

$$\ell_t(f) = f(t), \qquad 0 \leq t, \tag{29}$$

are continuous. The topology is the product topology

$$\prod_{0 \leq t} f(t).$$

According to (26), the values of f in K lie between 0 and 1. So K is a subset of

$$\prod_{0 \leq t} [0, 1],$$

which by Tychonov's theorem is compact. So to show K compact, it suffices to show
that K is closed. But this is easy: for fixed a_j and t, the set of f that satisfy (22) is
clearly closed. K, being the intersection of these sets for all a_j and all $t \geq 0$, is
closed.

We showed in lemma 5 that the extreme points of K are contained in the set $\{e_\lambda\}$,
$0 \leq \lambda \leq \infty$ defined by (25a), (25b). The set $\{e_\lambda\}$, it is easy enough to show, is closed
and therefore contains the closure of the set of extreme points.

We appeal now to formula (10) in theorem 5 of chapter 13. That formula with e
given by (25) and ℓ by (29), is precisely the desired representation formula (23).

□

Some corollaries and addenda:

Theorem 6.

(i) *Every completely monotone function is C^∞.*

(ii) *The representation (23) is unique.*

(iii) *Every e_λ defined by (25a), $0 < \lambda < \infty$ is an extreme point.*

Proof. Since the measure m is ≥ 0 and $m(\mathbb{R}_+) < \infty$, we can differentiate (23) with respect to t under the integral sign; this proves (i).

For (ii) suppose that some f in K had two distinct representations. Subtracting them, we get

$$\int_0^\infty e^{-\lambda t} d\nu(\lambda) \equiv 0,$$

ν some signed measure of finite total mass over $\mathbb{R}$. The function

$$F(\zeta) = \int_0^\infty e^{-\lambda \zeta} d\nu(\lambda)$$

is then an analytic function continuous in the right half-plane $\text{Re}\,\zeta \geq 0$ that vanishes on the real axis $\zeta = t$. It follows that $F(\zeta) \equiv 0$, in particular, that

$$F(i\tau) = 0 \qquad \text{for all real } \tau.$$

$F(i\tau)$ is the Fourier transformation of the measure $d\nu$. By uniqueness of the Fourier transform, we conclude that $d\nu = 0$, meaning no f can have two different representations of form (23).

For (iii), since K is compact, by Krein-Milman, it has at least one extreme point. By lemma 5, the extreme points must be of the form e_λ given by (25). They must include more than e_0 and e_∞ since the convex combinations of e_1 and e_∞ do not include all of K. So some e_λ, $\lambda \neq 0, \infty$ is an extreme point of K. According to lemma 4 (vi), the operator H_b maps K into K; being a one-to-one linear map, H_b carries every extreme point of K into an extreme point of K. So, if e_λ is extreme, so is

$$H_b e_\lambda = e_{\lambda b}, \qquad b > 0.$$

This completes part (iii) of theorem 6. $\qquad\square$

We remark that an analogue of Bernstein's theorem holds in n-dimensional space, namely for functions defined on $\mathbb{R}_+^n$. It also holds for functions on $\mathbb{Z}_+^n$.

14.4 THEOREMS OF CARATHÉODORY AND BOCHNER

Definition. A *skew-symmetric* doubly infinite sequence $\{a_n\}$ of complex numbers:

$$a_{-n} = \bar{a}_n, \tag{30}$$

is called *positive definite* if

$$\sum_{n,k} a_{n-k}\phi_n\bar{\phi}_k \geq 0 \tag{31}$$

for all finite sets of complex numbers ϕ_n, $-N \leq n \leq N$.

The following result is due to Toeplitz, Carathéodory, and Herglotz:

Theorem 7. *All positive definite sequences can be represented uniquely as*

$$a_n = \int_{-\pi}^{\pi} e^{in\theta} dm(\theta), \tag{32}$$

where m is a nonnegative measure on S^1. Conversely, every sequence of form (32) is positive definite.

Proof. First we show that every sequence of form (32) is positive definite. Substitute (32) for a_{n-k} into the left side of (31):

$$\sum_{n,k} \int e^{i(n-k)\theta} dm\, \phi_n\bar{\phi}_k = \int \sum_{n,k} e^{in\theta} e^{-ik\theta} \phi_n\bar{\phi}_k\, dm$$

$$= \int \left(\sum_n e^{in\theta}\phi_n\right)\left(\sum_k e^{-ik\theta}\bar{\phi}_k\right) dm$$

$$= \int \left|\sum_n e^{in\theta}\phi_n\right|^2 dm$$

is clearly nonnegative.

We turn now to the direct part of the theorem. We claim that if $\{a_n\}$ is a positive definite sequence, then

$$|a_m| \leq a_0 \qquad \text{for all integers } m. \tag{33}$$

To see this, set $\phi_0 = 1$, $\phi_m = \phi$, and all other $\phi_n = 0$. Substituting this into (31), we get, using (30), that

$$a_0 + a_m\phi + \bar{a}_m\bar{\phi} + a_0|\phi|^2 \geq 0$$

for all complex ϕ; this implies (33).

According to the theory of distributions it follows from (33) that there exists a distribution a whose Fourier coefficients are a_n:

$$a_n = \int_{S^1} e^{in\theta} a\, d\theta. \tag{34}$$

For any C^∞ function ψ,

$$\int \overline{\psi} a \, d\theta = \sum_n \overline{\psi}_n a_n, \tag{34'}$$

where ψ_k are the Fourier coefficients of ψ. It follows from (33) that the right side converges. We claim that a is nonnegative; to see this, take any trigonometric polynomial q_N of degree N,

$$q_N(\theta) = \sum_{-N}^{N} \phi_n \, e^{in\theta}.$$

Then $|q_N(\theta)|^2 = \sum_{n,k} \phi_n \overline{\phi}_k \, e^{i(n-k)\theta}$. Set $\psi = |q_N|^2$ in (34'). We get an expression on the right that, by (31), is nonnegative:

$$\int a|q_N(\theta)|^2 d\theta = \sum a_{n-k}\phi_n\overline{\phi}_k \geq 0. \tag{35}$$

Let $q(\theta)$ be any C^∞ function on S^1; it is easy and classical to show that q can be approximated by a sequence $\{q_N\}$ of trigonometric polynomials in the C^∞ topology. By definition of distribution, as N tends to ∞, and (35) tends to

$$\int |q(\theta)|^2 a \, d\theta \geq 0, \tag{36}$$

for all C^∞ functions q. Let $p(\theta)$ be any C^∞ function that is *positive* on S^1. Then

$$q(\theta) = \sqrt{p(\theta)}$$

is a C^∞ function; therefore (36) implies that

$$\int p(\theta) a \, d\theta \geq 0 \tag{37}$$

for any positive C^∞ function p. A distribution a with this property is called *nonnegative*. It is a classical result of the theory of distributions, see Appendix B, that every nonnegative distribution is a nonnegative *measure*. Thus $a \, d\theta = dm$ and formula (34) is the desired formula (32). $\qquad\square$

Theorem 7 can be extended to functions a defined on $\mathbb{Z}^k$, k any positive integer; the proof is the same.

NOTE. Carathéodory's own proof made use of his theorem on convex sets in finite-dimensional spaces. In section 14.6 we will give yet another proof, using the theory of positive harmonic functions.

Exercise 3. Denote by P the set of all positive-definite sequences normalized by

$$a_0 = 1. \tag{38}$$

(a) Show that P is a convex subset of the space ℓ^∞ of all bounded sequences. Show that P is a compact subset of ℓ^∞ in the product topology.

(b) Show that the extreme points of P are of the form

$$a_n = e^{in\theta}, \tag{39}$$

and deduce the representation (32) using theorem 4 of chapter 13.

An important extension of Carathéodory's theorem is due to Bochner:

Definition. A skew-symmetric complex-valued continuous function $a(s)$ on $\mathbb{R}$:

$$a(-s) = \overline{a(s)}, \tag{40}$$

is called *positive-definite* if

$$\sum a(s_j - s_k)\phi_j\overline{\phi}_k \geq 0 \tag{41}$$

for all choices of $s_1, \ldots, s_N$ on $\mathbb{R}$, and for all complex numbers $\phi_1, \ldots, \phi_N$.

Exercise 4. Show that condition (41) is equivalent to the requirement that

$$\int\int a(s - t)\phi(s)\overline{\phi}(t)\, ds\, dt \geq 0 \tag{41'}$$

for all continuous, complex-valued functions ϕ with compact support.

Theorem 8. *Every continuous positive-definite function a can be represented uniquely as*

$$a(s) = \int e^{i\sigma s} dm(\sigma), \tag{42}$$

m a nonnegative measure on $\mathbb{R}$, $m(\mathbb{R}) < \infty$. Conversely, every function of form (12) is positive-definite.

Proof. We show first that every function of form (42) is positive definite. Setting (42) into the left side of (41') yields

$$\int\int\int e^{i\sigma(s-t)}\phi(s)\,\overline{\phi}(t)\, ds\, dt\, dm(\sigma) = \int |\tilde{\phi}(\sigma)|^2\, dm(\sigma),$$

where $\tilde{\phi}$ is the Fourier transform of ϕ. Clearly, the right side is nonnegative.

To construct the measure m for a given positive-definite function, we proceed as in the discrete case. We deduce from (41), analogously to (33), that

$$|a(s)| \leq a(0). \tag{43}$$

We recall from section B.5 of Appendix B the Schwartz class of functions S, consisting of functions $f(s)$ all of whose derivatives $\partial_s^n f(s)$, $n = 0, 1, \ldots$, tend to zero faster than $|s|^{-k}$ as $|s| \to \infty$, for any k. S' is the dual of S; its elements are called *tempered distributions*. The function a, is, according to (43), bounded; therefore it belongs to S'. Therefore a has a Fourier inverse b that also belongs to S'. The Parseval relation

$$\int b\tilde{f} \, d\sigma = \int af \, ds \tag{44}$$

holds for all f in S, where $\tilde{f}$ denotes the Fourier transform of f.

According to exercise 4, (41′) holds for all C^∞ functions ϕ with compact support. Introduce in (41′) $s - t = r$ and s as new variables:

$$\int \int a(r)\phi(s)\overline{\phi}(s - r) \, ds \, dr \geq 0. \tag{45}$$

Denote by f the convolution

$$\int \phi(s)\overline{\phi}(s - r) \, ds = f(r); \tag{46}$$

f belongs to C^∞ and has compact support, and (45) can be written as

$$\int a(r)f \, dr \geq 0. \tag{45′}$$

Denote by ψ the Fourier transform of ϕ; taking the Fourier transform of (46) gives

$$|\psi(\sigma)|^2 = \tilde{f}(\sigma). \tag{46′}$$

Formula (44) expresses the left side of (45′) in terms of b and $\tilde{f}$; using formula (46′) for $\tilde{f}$, we get

$$\int b(\sigma)|\psi(\sigma)|^2 d\sigma \geq 0. \tag{45″}$$

Let $p(\sigma)$ be any nonnegative C^∞ function with compact support on $\mathbb{R}$. Then $\psi = p^{1/2}$ too is C^∞ with compact support, and so belongs to S. Setting $\psi^2 = p$ into (45″), we get

$$\int b(\sigma) \, p(\sigma) \, d\sigma \geq 0$$

for every nonnegative C^∞ function p with compact support. Such a distribution b is called nonnegative. According to theorem 13 of Appendix B, b is a nonnegative measure: $b(\sigma)\,d\sigma = dm$.

We claim that the total mass of m is finite:

$$\int dm = \int_{-\infty}^{\infty} b(\sigma)\,d\sigma < \infty. \tag{47}$$

For let g be any nonnegative C^∞ function of compact support that is equal 1 on $[-1, 1]$. Define $g_n(\sigma) = g(\sigma/n)$. Denote by f the Fourier inverse of g. Then the Fourier inverse of g_n is $f_n(s) = nf(ns)$. Set f_n into (44):

$$\int b(\sigma) g\left(\frac{\sigma}{n}\right) d\sigma = \int a(s)n\,f(ns)\,ds. \tag{48}$$

The measure b and the function g are nonnegative, and $g(\sigma) = 1$ for $|\sigma| \le 1$. Therefore the left side of (48) is greater than

$$\int_{-n}^{n} b(\sigma)\,d\sigma. \tag{48'}$$

On the other hand, according to (43), $|a(s)| \le a(0)$. Therefore the right side of (48) is less than

$$a(0) \int n\,|f(ns)|\,ds = a(0) \int |f(s)|\,ds,$$

a quantity independent of n. This shows that the integrals (48') are bounded independently of n, proving (47).

It follows from (47) that $a(s)$ can be represented *pointwise* as the Fourier transform of b:

$$a(s) = \int e^{i\sigma s}\,dm$$

for every s, as claimed in (42). The uniqueness of a representation of form (42) follows from the uniqueness of the Fourier transform. □

Denote by P the set of positive definite functions a normalized by $a(0) = 1$. It follows from theorem 8 that the extreme points of P are the exponentials $e^{i\sigma s}$, σ real. Thus (42) is seen as a Choquet-type representation of positive definite functions.

Theorem 8 is easily extended to n dimensions; see Rudin's book, *Fourier Analysis on Groups*.

Laurent Schwartz has given the following extension of Bochner's theorem.

Definition. A skew-symmetric complex-valued *tempered distribution* $a(s)$ on $\mathbb{R}$ is called *positive definite* if

$$\int\int a(s-t)\phi(s)\bar{\phi}(t)\,ds\,dt \geq 0$$

for all C_0^∞ functions ϕ.

Theorem 8'. *Every positive definite tempered distribution is the Fourier transform of a nonnegative measure of class S'.*

Schwartz has extended his theorem to $\mathbb{R}^n$.

14.5 A THEOREM OF KREIN

Definition. Let p be a continuous real-valued even function defined on $\mathbb{R}$:

$$p(-t) = p(t).$$

p is called *evenly positive definite* if

$$\int\int p(s-t)\phi(s)\phi(t)\,ds\,dt \geq 0 \tag{49}$$

for all real-valued, continuous, *even* functions of compact support:

$$\phi(-s) = \phi(s).$$

Clearly, every even function of form (42) has this property. These functions can be written as

$$p(s) = \int_0^\infty \cos\sigma s\,dm(\sigma), \qquad dm \geq 0.$$

These are, however, not all. For all real λ, and all even, real-valued continuous functions ϕ with compact support

$$\int\int \cosh\lambda(s-t)\phi(s)\phi(t)\,ds\,dt = \int e^{\lambda s}\phi(s)\,ds \int e^{\lambda t}\phi(t)\,dt.$$

This shows that $\cosh\lambda s$ is evenly positive definite. But then so is

$$p(s) = \int \cosh\lambda s\,dn(\lambda),$$

n any nonnegative measure for which the integral converges for all s.

Similarly an even, real-valued function on $\mathbb{R}$ is called *oddly positive definite* if (49) holds for all real-valued, continuous *odd* functions ϕ. Examples of such a function are $-\cosh\lambda s$ and superpositions.

M. G. Krein has proved the following result:

Theorem 9. *Every real, even, continuous function p on $\mathbb{R}$ that is evenly positive-definite can be represented uniquely as*

$$p(s) = \int_0^\infty \cos \sigma s \, dm(\sigma) + \int_0^\infty \cosh \lambda s \, dn(\lambda),$$

m and n nonnegative measures. Similarly, every oddly positive-definite function can be represented as

$$q(s) = \int_0^\infty \cos \sigma s \, dm(\sigma) - \int_0^\infty \cosh \lambda s \, dn(\lambda).$$

An easy consequence is

Theorem 9′. *Denote by P the set of evenly positive-definite functions, normalized by $p(0) = 1$. P is a convex set, and its extreme points are*

$$\cos \sigma s, \quad \sigma \geq 0, \quad and \quad \cosh \lambda s, \quad \lambda \geq 0.$$

For a proof we refer to Krein.

14.6 POSITIVE HARMONIC FUNCTIONS

In chapter 11, section 11.6, it was shown that every harmonic function h defined in the unit disk and positive there can be represented uniquely by Poisson's formula

$$h(re^{i\chi}) = \int \frac{1 - r^2}{1 - 2r \cos(\chi - \theta) + r^2} \, dm(\theta), \tag{50}$$

m a nonnegative measure.

It is easy to verify that the Poisson kernel has for $r < 1$ the following Fourier expansion:

$$\frac{1 - r^2}{1 - 2r \cos \lambda + r^2} = \sum_{-\infty}^\infty r^{|\ell|} e^{i\ell\lambda}. \tag{51}$$

Setting this into (50) gives the Fourier expansion of h:

$$h(re^{i\chi}) = \sum b_\ell r^{|\ell|} e^{i\ell\chi}, \tag{52}$$

where

$$b_\ell = \int e^{-i\ell\theta} \, dm(\theta). \tag{52'}$$

We show now how to deduce Carathéodory's theorem, theorem 7, from (50); this proof is due to Herglotz. Let $\{a_n\}$ be a positive definite sequence in the sense of (31). It follows then from (33) that the sequence a_n is bounded; therefore the series

$$k(re^{i\chi}) = \sum_{-\infty}^{\infty} a_\ell\, r^{|\ell|}\, e^{i\ell\chi} \tag{53}$$

converges for $r < 1$, uniformly for $r < 1 - \delta$. Clearly, k is a harmonic function of x, y, where $x + iy = re^{i\chi}$ in the unit disk. We claim that k is positive. To see this, we rewrite the left side of (31) by introducing $n - k = \ell$ as a new variable. We get

$$\sum_\ell a_\ell \sum_n \phi_n \bar\phi_{n-\ell} \geq 0. \tag{53'}$$

We want to choose $\{\phi_n\}$ so that for all ℓ and $r < 1$, χ given

$$\sum_n \phi_n \bar\phi_{n-\ell} = r^{|\ell|}\, e^{i\ell\chi}. \tag{54}$$

To satisfy (54) we multiply it by $e^{-i\ell\theta}$ and sum over ℓ. We get

$$\sum_\ell \sum_n \phi_n \bar\phi_{n-\ell} e^{-i\ell\theta} = \sum_\ell r^{|\ell|}\, e^{i\ell(\chi-\theta)}. \tag{54'}$$

The left side can be written as

$$\sum \phi_n e^{-in\theta} \bar\phi_{n-\ell} e^{i(n-\ell)\theta} = \Big| \sum \phi_n e^{-in\theta} \Big|^2;$$

the right side, by (51), is the Poisson kernel, which is positive. Therefore we can set

$$\sum \phi_n e^{-in\theta} = \left(\frac{1 - r^2}{1 - 2r\cos(\chi - \theta) + r^2} \right)^{1/2}. \tag{55}$$

This choice of $\{\phi_n\}$ satisfies (54'), from which (54) follows. Setting (54) into (53') shows that $k(re^{i\chi})$ is positive for $r < 1$.

Once k has been shown to be positive, we can appeal to the Herglotz-Riesz theorem, theorem 6 in chapter 11, and obtain a representation of form (50) for k. As was shown above, this in turn gives formula (52') for the coefficients a_ℓ in the Fourier expansion of k. This is the desired formula (32). $\qquad\square$

Denote by H the linear space of real-valued harmonic functions in the open unit disk. Denote by P the subset of positive harmonic functions h, normalized by

$$h(0) = 1. \tag{56}$$

Clearly, P is a convex subset of H. From the uniqueness of the measure in representation (50), we deduce, as in earlier sections of this chapter, that all extreme points of P are of the form

$$e_\theta = \frac{1 - r^2}{1 - 2r(\chi - \theta) + r^2}. \tag{57}$$

This shows that the Herglotz-Riesz representation of positive harmonic functions is a Choquet-type representation.

Exercise 5. We impose on H the weakest topology that makes continuous all linear functionals

$$h \to h(z), \quad |z| < 1. \tag{58}$$

Show that the convex set P defined above is compact in this topology. (Hint: Use Harnack's theorem on positive harmonic functions.)

14.7 THE HAMBURGER MOMENT PROBLEM

A sequence of real numbers $a_0, a_1, \ldots$ is called *positive* in the sense of *Hankel* (*H* positive) if

$$\sum_{n,k} a_{n+k} \xi_n \xi_k \geq 0 \tag{59}$$

for all finite collection of real numbers $\xi_n, n = 0, 1, \ldots, N$.

Let m be a nonnegative measure on $\mathbb{R}$ all of whose moments are finite.

$$\int_{\mathbb{R}} t^{2n} dm(t) < \infty, \qquad n = 0, 1, \ldots. \tag{60}$$

Define

$$a_\ell = \int_{\mathbb{R}} t^\ell dm(t), \qquad \ell = 0, 1, \ldots. \tag{61}$$

We claim that this sequence is H positive, for

$$\sum_{n,k} a_{n+k} \xi_n \xi_k = \int \sum_{n,k} t^{n+k} \xi_n \xi_k \, dm(t) = \int \left(\sum t^n \xi_n \right)^2 dm(t) \geq 0. \tag{62}$$

Conversely, Hamburger has proved:

Theorem 10. *Every H positive sequence $\{a_n\}$ can be represented in the form (61).*

For proof, we refer to chapter 33. An interesting fact is that there are H positive sequences that can be represented in form (61) in only one way, as well as others that have several distinct representations. Why this is so will be explained in chapter 33 on self-adjoint operators.

Denote by H_0 the set of all H positive sequences normalized by $a_0 = 1$. It follows from theorem 10 that every extreme point e of H_0 is of the form

$$e_k(t) = t^k, \qquad k \text{ in } \mathbb{Z}_+; t \text{ real.} \qquad (63)$$

It is not hard to show that conversely, every sequence $e(t)$ of form (63) is an extreme point of H_0. Thus (61) is a Choquet-type representation of the set H_0.

Examples of Hankel positive sequences:

Take in (61) as

$$\frac{dm(t)}{dt} = \begin{cases} t^{\delta-1} & \text{for } 0 \le t \le 1, \delta > 0 \\ 0 & \text{otherwise.} \end{cases} \qquad (64)$$

Then

$$a_\ell = \int_0^1 t^\ell t^{\delta-1}\, dt = \frac{1}{\ell + \delta}.$$

Thus

$$0 \le \sum \frac{\xi_n \xi_k}{n + k + \delta} \qquad (65)$$

for all real ξ_n.

In conclusion, we note that theorem 10 *fails* to hold in more than one variable.

14.8 G. BIRKHOFF'S CONJECTURE

Definition. An $n \times n$ matrix $S = (s_{ij})$ is called *doubly stochastic* if

(i) all entries are nonnegative,

$$s_{ij} \ge 0. \qquad (66)$$

(ii) all row sums and column sums are equal to 1,

$$\sum_j s_{ij} = 1, \qquad \sum_i s_{ij} = 1 \qquad \text{for all } i, \text{resp. } j. \qquad (66')$$

It is obvious that the set D of all doubly stochastic matrices forms a convex set in $\mathbb{R}^{n^2}$.

A *permutation* p of n objects is a one-to-one map of the indices $1, \dots, n$ onto themselves. The associated *permutation matrix* P is defined by

$$p_{ij} = \begin{cases} 1 & \text{if } j = p(i) \\ 0 & \text{if } j \neq p(i). \end{cases}$$

Clearly, each row, and each column of a permutation matrix P, contains exactly one entry equal to 1 and all others are zero. This shows that each P is doubly stochastic, that is, $P \in D$. We claim that each P is an extreme point of the set D. To see this, suppose that P were the midpoint of an interval, whose endpoints

$$P \pm Q$$

both belong to D. Clearly, it follows from (66) that if $p_{ij} = 0$, then $q_{ij} = 0$, and from (66') that if $p_{ij} = 1$, then $q_{ij} = 0$. Since all entries of P are either 0 or 1, it follows that $Q = 0$. This proves that P is extreme.

Conversely, D. König and G. Birkhoff have shown that *all extreme points of D are permutation matrices P*.

Exercise 6. Prove the König-Birkhoff theorem.

By Carathéodory's theorem, it follows that all doubly stochastic matrices are convex combinations of permutation matrices. This representation, however, is not unique in general.

Definition. An $n \times n$ matrix $S = (s_{ij})$ is called *doubly substochastic* if

(i) all entries are nonnegative,

$$s_{ij} \geq 0. \tag{67}$$

(ii) all row sums and column sums are ≤ 1,

$$\sum_j s_{ij} \leq 1, \quad \sum_i s_{ij} \leq 1 \tag{67'}$$

for all i, respectively j.

We denote the set of all doubly substochastic matrices by D_0. Clearly, D_0 is a convex set that contains the set D. We call a matrix P_0 a *subpermutation matrix* if its entries are either 0 or 1, and if each row and column contains *at most* a single entry 1. Every P_0 belongs to D_0. The argument used above to show that every P is an extreme point of D can be used to prove that every P_0 is an extreme point of D_0. Conversely,

Exercise 7. Show that *all extreme points of D_0 are subpermutation matrices P_0*.

We turn now to infinite matrices $S = (s_{ij})$, i, j in $\mathbb{Z}_+$. The notions of *doubly stochastic, doubly substochastic, permutation,* and *subpermutation matrices* are defined exactly as in the $n \times n$ case. We denote by X the linear space of all matrices S with real entries whose rows and columns have uniformly bounded ℓ^1 norms:

$$\sup_i \sum_j |s_{ij}| < \infty, \qquad \sup_j \sum_i |s_{ij}| < \infty. \tag{68}$$

We deal first with doubly substochastic matrices. As topology we use the coarsest topology in which all the linear functionals that map the matrix S into its ijth component are continuous:

$$\ell_{ij}(S) = s_{ij}. \tag{69}$$

We recall the following result from chapter 13, exercise 3: the only linear functionals in X that are continuous are finite linear combinations of the ℓ_{ij}.

Theorem 11.

(i) *The extreme points of the convex set D_0 consisting of all doubly substochastic matrices is the set $\{P_0\}$ of subpermutation matrices.*

(ii) *D_0 is the closure of the convex hull of $\{P_0\}$ in the topology induced by the functionals (69).*

Proof. We first prove part (ii). Suppose, on the contrary, that some doubly substochastic matrix Z does not belong to the closure of the convex hull of $\{P_0\}$. According to theorem $2'$ of chapter 13, there would exist a continuous linear functional ℓ such that

$$\ell(Z) > c, \tag{70}$$

but for all T in the closure of the convex hull of $\{P_0\}$,

$$\ell(T) \leq c. \tag{71}$$

We may, in particular, set $T =$ any subpermutation matrix P_0:

$$\ell(P_0) \leq c. \tag{71'}$$

The only linear functionals ℓ that are continuous in the coarsest topology that makes the functionals ℓ_{ij} in (69) continuous are the finite linear combinations of ℓ_{ij},

$$\ell = \sum_{i,j \leq n} b_{ij} \ell_{ij}. \tag{72}$$

Denote by S_n the projection of any substochastic matrix S onto the $n \times n$ matrix formed by the intersection of the first n rows and columns of S. Clearly, S_n is a doubly substochastic $(n \times n)$ matrix. Then it follows from (69) and (72) that for any S,

$$\ell(S) = \ell(S_n). \tag{73}$$

Denote the projection of Z by Z_n. As remarked earlier, the extreme points of the set of doubly substochastic $n \times n$ matrices are the $n \times n$ subpermutation matrices. It follows from Carathéodory's theorem that on a compact convex set, a continuous linear functional takes its maximum one of the extreme points. We have that

$$\ell(Z_n) \leq \sup_{P_n} \ell(P_n), \tag{74}$$

where the P_n are $n \times n$ subpermutation matrices. Such a P_n is the projection of a subpermutation matrix P_0 of infinite order whose elements not in the first n rows and columns are set $= 0$. By (73),

$$\ell(P_n) = \ell(P_0) \quad \text{and} \quad \ell(Z_n) = \ell(Z). \tag{75}$$

Combining (74) with (75), we obtain

$$\ell(Z) \leq \sup_{P_0} \ell(P_0).$$

This combined with $(71')$ shows that (70) cannot hold. Therefore every Z in D_0 belongs to the closure of the convex hull of $\{P_0\}$.

To show that, conversely, all points of the closure of the convex hull of $\{P_0\}$ belong to D_0, we rewrite the criterion (67) and $(67')$ for belonging to D_0 as follows:

$$\ell_{ij}(S) \geq 0, \tag{76}$$

$$\sum_{j<n} \ell_{ij}(S) \leq 1, \qquad \sum_{i<n} \ell_{ij}(S) \leq 1, \tag{76'}$$

for all positive integers n. Since by definition of the topology, the functionals (76), $(76')$ are continuous, and since these inequalities hold on the convex hull of $\{P_0\}$, it follows that they hold on its closure. This completes the proof of part (ii).

The proof of part (i) is based on theorem 6 of chapter 13, which says that the extreme points of the closure of the convex hull of a set S belong to S, provided both sets are compact. In order to apply that theorem to $S = \{P_0\}$, we have to verify that both D_0 and $\{P_0\}$ are compact sets. To see this, we note that the topology we have imposed is the weak product topology

$$\prod s_{ij}.$$

The entries of S in D_0 lie in $[0, 1]$, so D_0 is a subset of

$$\prod [0, 1].$$

According to Tychonov's theorem, this is a compact set. We have shown already in part (ii) that D_0 is a closed set; being a closed subset of a compact set makes D_0 compact.

Similarly, in order to show that $\{P_0\}$ is compact, it suffices to show that this set is closed. The matrices P_0 are characterized by the inequalities (76') and

$$\ell_{ij}(S) \in \{0, 1\}.$$

Each of these sets is closed; therefore so is their intersection $\{P_0\}$.

We now appeal to theorem 6 of chapter 13. It states that given a compact set such as $\{P_0\}$ whose closed convex hull—which by part (ii) of theorem 11 is D_0—is also compact, then all extreme points of the closed convex hull belong to the original compact set. This completes the proof of part (i) of theorem 11.   □

Theorem 12.

(i) *Every extreme point of the set D of doubly stochastic infinite matrices is a permutation matrix P.*

(ii) *D is the closure of the convex hull of the set $\{P\}$ of permutation matrices in the coarsest topology that makes continuous the linear functionals ℓ_{ij}, ℓ_i, and ℓ^j, where ℓ_{ij} are defined by (69) and*

$$\ell_i(S) = \sum_j s_{ij}, \quad \ell^j(S) = \sum_i s_{ij}.$$

Proof. (i) We start by showing that D is an extreme subset of D_0. Suppose that S in D lies on an interval

$$S = aT + bR, \quad T \text{ and } R \text{ in } D_0, \quad a + b = 1, \quad 0 < a, 0 < b.$$

Form the row and column sum of both sides:

$$\sum_j s_{ij} = a \sum_j t_{ij} + b \sum_j r_{ij}, \quad \sum_i s_{ij} = a \sum_i t_{ij} + b \sum_i r_{ij}.$$

Since S belongs to D, the sums on the left are $= 1$; since T and R belong to D_0, the sums on the right are ≤ 1. Since the two sides are equal, the sums on the right must all be $= 1$. But this means that T and R belong to D and this proves that D is an extreme subset of D_0.

We noted in theorem 7 of chapter 1 that being an extreme subset is a transitive relation among convex sets. Thus an extreme point E of D is an extreme point of D_0. According to part (ii) of theorem 11, all extreme points of D_0 are subpermutation matrices P_0, so $E = P_0$. Since E belongs to D, it follows that E is not a sub- but a genuine permutation matrix P.

This completes the proof of part (i) of theorem 12; part (ii) can be proved along the lines of the argument presented for part (ii) of theorem 11. Note, however, that part (ii) cannot be proved by appealing to the Krein-Milman theorem, since the set D is not closed, and therefore not compact.

Theorem 12 was conjectured by Garrett Birkhoff. The preceding theorems and proofs are due to Kiefer and Kendall; see D. G. Kendall.   □

14.9 De FINETTI'S THEOREM

The setting of probability theory is a space Ω, in which a σ-algebra Σ is specified. The sets in Σ represent all possible events; a probability measure on Σ represents the probability of their occurrence. An infinite sequence of occurrences is modeled by the direct product $\mathbb{Z} \times \Omega$. The events in $\mathbb{Z} \times \Omega$ form the smallest σ-algebra that contains all the *cylinder sets*, formed by the product sets

$$\prod E_j,$$

where E_j belongs to the σ-algebra Σ in Ω, and all but a finite number of the sets E_j are the whole space Ω.

A probability measure m on the cylinder sets of $\mathbb{Z} \times \Omega$ is called *invariant under permutations* if for all cylinder sets

$$m\left(\prod E_j\right) = m\left(\prod E_{p(j)}\right);$$

where $j \to p(j)$ is a permutation of the indices, such that $p(j) = j$ for all but a finite number of j.

The set of probability measures on $\mathbb{Z} \times \Omega$ that are invariant under permutations is clearly a *convex set*, where convex combinations of measure is defined in the obvious way.

A probability measure m on the σ-algebra Σ in Ω induces the *product measure* on the cylinder sets of $\mathbb{Z} \times \Omega$ by the formula

$$m\left(\prod E_j\right) = \prod m(E_j).$$

Since all but a finite number of the E_j are equal to Ω, all but a finite number of the factors on the right are equal to 1. Clearly, the product measure on $\mathbb{Z} \times \Omega$ induced by a measure on Ω is invariant under permutation.

De Finetti proved the following important result:

Theorem 13.

(i) *The extreme points of the set of permutation invariant probability measures on $\mathbb{Z} \times \Omega$ are the product measures.*

(ii) *Each measure on $\mathbb{Z} \times \Omega$ invariant under permutation can be expressed in a unique fashion as an integral over the product measures.*

Clearly this is a Choquet type of result. For a proof, see de Finetti or any advanced text on probability.

14.10 MEASURE-PRESERVING MAPPINGS

In this section Ω denotes a compact metric space, T a homeomorphism of Ω onto Ω. It can be shown that there exists at least one probability measure on the Borel subsets of Ω that is invariant under T. There may be many. The collection of all invariant probability measures form a convex set. The following result of John Oxtoby sheds light on the structure of this collection:

Theorem 14.

(i) *The extreme points of the convex set of probability measures invariant under T are those measures with respect to which T is ergodic.*

(ii) *Every invariant measure can be represented as an integral over the ergodic measures. This representation is unique.*

Proof. We will sketch the proof of part (i). We recall the definition of ergodicity: a mapping T is ergodic with respect to a measure m on Ω if it is not possible to decompose Ω into two parts,

$$\Omega = \Omega_1 \cup \Omega_2,$$

both of positive measure,

$$m(\Omega_1) > 0, \quad m(\Omega_2) > 0,$$

so that both Ω_1 and Ω_2 are invariant under T.

Suppose now that m is invariant under T but that T is not ergodic with respect to m. Then there is a decomposition of Ω as above. We define two new measures m_1 and m_2 as the restrictions of m to Ω_1, Ω_2, respectively. That is, for any Borel set S,

$$m_1(S) = \frac{m(S \cap \Omega_1)}{m(\Omega_1)},$$

$$m_2(S) = \frac{m(S \cap \Omega_2)}{m(\Omega_2)}.$$

Clearly, m_1 and m_2 are probability measures, and they are invariant under T. The measure m is a convex combination of them:

$$m = m(\Omega_1)m_1 + m(\Omega_2)m_2.$$

Since $m_1 \neq m_2$, this shows that if m is not ergodic, it is not an extreme point.

Conversely, we show that if m is not an extreme point, it is not ergodic. Suppose that

$$m = am_1 + (1 - a)m_2, \qquad 0 < a < 1, m_1 \neq m_2.$$

We first take the case that m_1 is absolutely continuous with respect to m_2. By the Radon-Nikodym theorem,

$$m_1 = fm_2, \qquad f \text{ nonnegative and in } L^1(m_2).$$

Since both m_1 and m_2 are invariant under T, so is f. Since $m_1 \neq m_2$, $f \not\equiv 1$; therefore there exists a positive number c such that the sets $\Omega_1 = \{\omega | f(\omega) > c\}$ and $\Omega_2 = \{\omega | f(\omega) \leq c\}$, both having positive m_2 measure. Since f is invariant under T, T maps Ω_1 and Ω_2 onto themselves.

Substituting $m_1 = fm_2$ into the expression for m as convex combination of m_1 and m_2 gives

$$m = [af + 1 - a]m_2.$$

It follows that Ω_1 and Ω_2 have positive m measures; this shows that T is not ergodic with respect to m.

The case when m_1 is not absolutely continuous with respect to m_2 is just as simple. Then there exist sets E whose m_2 measure is zero, but $m_1(E) > 0$. Denote by s the quantity

$$s = \sup m_1(E), \qquad m_2(E) = 0.$$

Let E_n be a maximizing sequence:

$$\lim m_1(E_n) = s, \qquad m_2(E_n) = 0.$$

Denote by F the union of the E_n. Clearly, $m_1(F) = s$, $m_2(F) = 0$. It follows that F is invariant under T; for if not, the set $(F \cup TF)$ would have m_1 measure greater than $m(F) = s$ but m_2 measure 0 contrary to the definition of s.

We claim that in the decomposition $\Omega = F \cup F^c$, both pieces have positive m-measure. For, using the expression of m as linear combination of m_1 and m_2, we get

$$m(F) \geq am_1(F) = as$$

and

$$m(F^c) \geq (1 - a)m_2(F^c) = (1 - a).$$

This shows that T is not ergodic with respect to m.

For the proof of part (ii), see the article by Oxtoby. $\square$

NOTE. In their work on X-ray crystallography, honored by the Nobel Prize in physics in 1986, H. Hauptman and J. Karle made crucial use of Toeplitz's characterization (32) of the Fourier coefficients of positive measures.

HISTORICAL NOTE. Dénes König (1884–1944), professor at the Technical University in Budapest, was the founding father of graph theory. He developed many of the basic concepts, and wrote the first book on the subject in 1936. His proof of the Birkhoff-König theorem is graph theoretical. The brilliant Hungarian school in graph theory is his legacy.

König supervised the Eötvös mathematical competitions for high school students. He was extremely kind and encouraging to budding young mathematicians, including the writer of these pages.

When the German army occupied Hungary in 1944, putting Hungarian Nazis in power, König saw what was coming and threw himself out the window of his apartment.

BIBLIOGRAPHY

Akhiezer, N. I. *The Classical Moment Problem and Some Related Questions in Analysis* (English trans.). Oliver and Boyd, Edinburgh, 1965.

Bernstein, S. Sur les fonctions absolument monotone. *Acta Math.*, **52** (1929): 1–66.

Birkhoff, G. Three observations on linear algebra. *Rev. Univ. Nac. Tucuman* (A), **5** (1946): 147–151.

Bochner, S. *Vorlesungen über Fouriersche Integrale*. Akademische Verlagsgesellschaft, Leipzig, 1932.

Carathéodory, C. Über den Variabilitätsbereich der Koefficienten von Potenzreihen die gegebene Werte nicht annemen. *Math. An.*, **54** (1907): 95–115.

de Finetti, B. Funzione caratteristica di un fenomeno aleatorio. *Atti Accad. Nax. Lincei Rend. Cl. Sci. Fiz. Mat. Nat.*, (1930): 86–133.

Gelfand, I. M. and Do-Shing, S. On positive definite distributions. *Usp. Mat. Nauk.*, **15** (1960): 185–190.

Hamburger, H. Über eine Erweiterung des Stieltjesschen Moment Problems. *Math. An.*, **81** (1920): 235–319; **82** (1921): 120–164, 168–187.

Herglotz, A. Über Potenzreihen mit Positiven Reellen Teil in Einheitskreise. *Berichte Verh Sächs, Akad. Wiss. Leipzig, Math.-phys. Kl.*, **63** (1911): 501–511. See also *Collected Works*, Vandenhoeck and Ruprecht, Göttingen, 1979.

Kendall, D. G. On infinite doubly stochastic matrices and Birkhoff's problem III. *London Math. Soc. J.*, **35** (1960): 81–84.

König, D. *Theory of Finite and Infinite Graphs*. Täubner 1936; Birkhäuser, Boston, 1990, p. 327.

Krein, M. G. On a general method of decomposing Hermite-positive nuclei into elementary products. *Dokl. Akad. Nauk SSSR*, **53** (1946): 3–6.

Landau, H. J. Classical background of the moment problem. *Proc. Symp. Appl. Math, AMS*, **37** (1987): 1–16.

Oxtoby, J. Ergodic sets. *Bull. AMS*, **58** (1952): 116–136.

Rudin, W. *Fourier Analysis on Groups*. Interscience, New York, 1962.

Shohat, J. A. and Tamarkin, J. D. *The Problem of Moments*. American Mathemtical Society, New York, 1943.

Schwartz, L. *Théorie des Distribution's*. 2 vols. Hermann, Paris, 1959.

Toeplitz, O. Über die Fouriersche Entwickelung positive Funktionen. *Rend. di Circ. Mat. di Palermo*, **32** (1911): 191–192.

15

BOUNDED LINEAR MAPS

In chapter 2 we studied some rudimentary properties of linear maps **M** of one linear space into another. Here we impose topological structures on the linear spaces and on the mappings themselves. Alternative names for maps, and synonymous with it, are *operator* and *transformation*.

15.1 BOUNDEDNESS AND CONTINUITY

Definition. X and U are a pair of Banach spaces. A linear map (actually any map)

$$\mathbf{M} : X \to U$$

is called *continuous* if it maps convergent sequences into convergent ones, that is, if

$$x_n \longrightarrow x \quad \text{implies} \quad \mathbf{M}x_n \longrightarrow \mathbf{M}x. \tag{1}$$

Here convergence is reckoned in the sense of the norm in X and U, respectively.

Definition. A linear map $\mathbf{M} : X \to U$ of one Banach space X into another U is called *bounded* if there is a constant c such that for all x in X

$$|\mathbf{M}x| \le c|x|. \tag{2}$$

Theorem 1. *A linear map* $\mathbf{M} : X \to U$ *of one Banach space* X *into another* U *is continuous if and only if it is bounded.*

Proof. It is easy to show that a bounded linear map is continuous, even Lipschitz continuous.

Conversely, if **M** were not bounded, (2) fails for any c, say n, for some x, say x_n:

$$|\mathbf{M}x_n| > n|x_n|.$$

Normalize x_n so that $|x_n| = 1/\sqrt{n}$; x_n tends to zero but Mx_n does not. Clearly, (1) is violated, so **M** is not continuous. $\qquad \square$

Suppose that the spaces X and U on which and into which $\mathbf{M}$ acts are merely normed linear spaces, not complete, and suppose that $\mathbf{M}$ is bounded in the sense of (2). Then $\mathbf{M}$ can be extended *by continuity* to a bounded mapping of the *completion* of X into the completion of U. This observation is as trivial as it is important, since most maps of interest are constructed in the fashion described above, being first defined in an incomplete space and then extended by a flick of the wrist to the completed space. The incomplete space usually consists of smooth functions, the complete space of functions less smooth, or not at all smooth.

Definition. Let $\mathbf{M} : X \to U$ be a bounded linear map of one Banach space into another. Its *norm*, denoted as $|\mathbf{M}|$, is defined by

$$|\mathbf{M}| = \sup_{x \neq 0} \frac{|\mathbf{M}x|}{|x|}. \tag{3}$$

Clearly, for any x in X, (2) holds with $c = |\mathbf{M}|$:

$$|\mathbf{M}x| \leq |\mathbf{M}||x|. \tag{2'}$$

Equally clearly, $|\mathbf{M}|$ is the smallest value of c for which (2) holds for all x in X. A useful reformulation of (3) is

$$|\mathbf{M}| = \sup_{|x|=1} |\mathbf{M}x|. \tag{3'}$$

Theorem 2. *Norm of bounded maps has the following properties:*

(ii) *Homogeneity, for any scalar, real, or complex,* $|a\mathbf{M}| = |a||\mathbf{M}|$.

(ii) *Positivity,* $|\mathbf{M}| \geq 0$, $|\mathbf{M}| = 0$ *if and only if* $\mathbf{M} \equiv 0$.

(iii) *Subadditivity,* $|\mathbf{M} + \mathbf{K}| \leq |\mathbf{M}| + |\mathbf{K}|$.

Proof. Properties (i) and (ii) are obvious. $\qquad\qquad\qquad\qquad\qquad\qquad\qquad$ $\square$

Exercise 1. Prove property (iii).

Definition. The set of all bounded maps of one Banach space X into another U is denoted by

$$\mathcal{L}(X, U).$$

Theorem 3. $\mathcal{L}(X, U)$ *is a Banach space under the norm (3).*

Proof. Properties (i), (ii), and (iii) in theorem 2 show that $\mathcal{L}(X, U)$ forms a normed linear space under the norm (3). What remains to be shown is the completeness of $\mathcal{L}(X, U)$.

Let $\{\mathbf{M}_n\}$ be a Cauchy sequence in $\mathcal{L}(X, U)$:

$$\lim_{n,k \to \infty} |\mathbf{M}_n - \mathbf{M}_k| = 0. \tag{4}$$

It follows from (4) that for any x in X

$$\lim_{n,k \to \infty} |\mathbf{M}_n x - \mathbf{M}_k x| = 0. \tag{4'}$$

This shows that $\{\mathbf{M}_n x\}$ is a Cauchy sequence in U; since U is complete, the limit $\lim \mathbf{M}_n x = u$ exists. We define the mapping $\mathbf{M}$ to be $\mathbf{M}x = u$; clearly, $\mathbf{M}$ is linear. By definition of a norm,

$$|\mathbf{M}_n - \mathbf{M}| = \sup_{|x|=1} |\mathbf{M}_n x - \mathbf{M}x| = \sup_{|x|=1} \lim_{k \to \infty} |\mathbf{M}_n x - \mathbf{M}_k x| \le \sup_{n < k} |\mathbf{M}_n - \mathbf{M}_k|.$$

Using (4) it follows that $|\mathbf{M}_n - \mathbf{M}| \to 0$.   □

In the special case is where the target space U of the linear mappings is *one-dimensional*, that is, isomorphic with $\mathbb{R}$ or $\mathbb{C}$, the bounded linear maps are bounded linear functionals, and $\mathcal{L}(X, U)$ is just the dual space X' of X.

We recall from chapter 2 the notion of the *nullspace* $N_\mathbf{M}$ of a linear map $\mathbf{M}$: $X \to U$; it consists of all points x of X mapped into 0 by $\mathbf{M}$:

$$\mathbf{M}x = 0. \tag{5}$$

Theorem 4. *Let X and U denote normed linear spaces, $\mathbf{M} : X \to U$ a bounded, linear map.*

(i) $N_\mathbf{M}$, *the nullspace of* $\mathbf{M}$, *is a closed linear subspace of X.*

(ii) $\mathbf{M}$, *when regarded as a map*

$$\mathbf{M}_0 : \left(\frac{X}{N_\mathbf{M}} \right) \longrightarrow U$$

is one-to-one, bounded, with $|\mathbf{M}_0| = |\mathbf{M}|$. The range of $\mathbf{M}_0$ is the same as the range of $\mathbf{M}$.

Proof. (i) $N_\mathbf{M}$ is the inverse image in X of $\{0\}$ in U. Since $\{0\}$ is a closed set, and $\mathbf{M}$ is continuous, $N_\mathbf{M}$ is closed.

(ii) x_1 and x_2 belong to the same equivalence class mod $N_\mathbf{M}$ if $x_1 - x_2 \in N_\mathbf{M}$. By (5), and linearity, $\mathbf{M}x_1 = \mathbf{M}x_2$; therefore the mapping $\mathbf{M}_0$ is unequivocally defined. We recall from chapter 5 that the norm in the quotient space X/N is defined as

$$|\{x\}| = \inf_{y \equiv x \bmod N} |y|. \tag{6}$$

We have shown there, in theorem 1, that if N is closed, as N_M is, then the quantity $|\{x\}|$ is a norm. Using the definition (3) of the norm of a map, and some obvious manipulations, we have

$$|\mathbf{M}| = \sup_{x \neq 0} \frac{|\mathbf{M}x|}{|x|} = \sup_{\{x\} \neq 0} \sup_{y \equiv x} \frac{|\mathbf{M}x|}{|y|} = \sup_{\{x\} \neq 0} \frac{|\mathbf{M}x|}{\inf_{y \equiv x} |y|} = \sup_{\{x\} \neq 0} \frac{|\mathbf{M}\{x\}|}{|\{x\}|} = |\mathbf{M}_0|. \quad \square$$

We turn to defining the *transpose* of a bounded linear map $\mathbf{M} : X \to U$, X and U normed linear spaces. Let ℓ be a point of U', the dual of U; that is, ℓ is a bounded linear functional on U. The composite $\ell(\mathbf{M}x)$ is a linear and bounded functional of x:

$$\ell(\mathbf{M}x) = \xi(x). \tag{7}$$

The linear functional $\xi \in X'$ clearly depends linearly on ℓ:

$$\xi = \mathbf{M}'\ell. \tag{7'}$$

$\mathbf{M}' : U' \to X'$ is called the *transpose* of $\mathbf{M}$.

The transpose of a bounded linear map is the infinite-dimensional generalization of the transpose of a matrix; it is an enormously useful concept. In studying and using the transpose, it is convenient to denote the action of the linear functionals by parentheses as follows:

$$\ell(u) = (u, \ell), \quad \xi(x) = (x, \xi),$$

where $u \in U, \ell \in U', x \in X, \xi \in X'$. In this notation the relations (7), (7') defining the transpose can be rewritten as

$$(\mathbf{M}x, \ell) = (x, \mathbf{M}'\ell). \tag{8}$$

We recall from chapter 8, (see theorem 7') the definition of the *annihilator* $R^\perp$ of a subspace R of a normed linear space U as the subspace $R^\perp$ of U' consisting of all bounded linear functionals ℓ that vanish on R. Similarly, for any subset S of X', we define $S^\perp$ as the subset of those vectors in X that are annihilated by every vector ξ in S. Clearly, $S^\perp$ is a closed linear subspace of X. The basic properties of transposition are summarized in

Theorem 5.

(i) *The transpose $\mathbf{M}'$ of a bounded linear map $\mathbf{M}$ is bounded, and*

$$|\mathbf{M}'| = |\mathbf{M}|. \tag{9}$$

(ii) *The nullspace of $\mathbf{M}'$ is the annihilator of the range of $\mathbf{M}$,*

$$N_{\mathbf{M}'} = R_{\mathbf{M}}^\perp. \tag{10}$$

(iii) The nullspace of $\mathbf{M}$ *is the annihilator of the range of* $\mathbf{M}'$,

$$N_{\mathbf{M}} = R_{\mathbf{M}'}^{\perp}. \tag{11}$$

(iv) $(\mathbf{M} + \mathbf{N})' = \mathbf{M}' + \mathbf{N}'$.

Proof. By (3'), applied to $\mathbf{M}'$,

$$|\mathbf{M}'| = \sup_{|\ell|=1} |\mathbf{M}'\ell|. \tag{12}$$

By definition, the norm of ξ in X' is

$$|\xi| = \sup_{|x|=1} |(x, \xi)|. \tag{13}$$

Setting $\xi = \mathbf{M}'\ell$ into (13) and combining this with (12), we get, using (8), that

$$|\mathbf{M}'| = \sup_{|\ell|=1} \sup_{|x|=1} |(x, \mathbf{M}'\ell)| = \sup_{|\ell|=1=|x|} |(\mathbf{M}x, \ell)|. \tag{12'}$$

According to theorem 6 of chapter 8, for every u in U,

$$|u| = \max_{|\ell|=1} |(u, \ell)|. \tag{14}$$

On the right side of (12') we maximize first with respect to ℓ. Using (14), with $u = \mathbf{M}x$, we get

$$|\mathbf{M}'| = \sup_{|x|=1} |\mathbf{M}x|,$$

which by (3') equals $|\mathbf{M}|$; this proves (9).

To prove (10), we note that for any x in X and any ℓ in $N_{\mathbf{M}'}$, the right side of (8) is zero. Therefore so is the left side; this shows that $N_{\mathbf{M}'} \subset R_{\mathbf{M}}^{\perp}$. Conversely, if ℓ annihilates the range of $\mathbf{M}$, the left side of (8) is zero for every x. Therefore so is the right side, which can only be if $\mathbf{M}'\ell = 0$. This shows that $N_{\mathbf{M}'} \supset R_{\mathbf{M}}^{\perp}$; these two relations taken together prove (10).

To prove (11), we note that the left side of (8) is zero when x belongs to the nullspace of $\mathbf{M}$. This shows that every x in $N_{\mathbf{M}}$ belongs to the nullspace of every $\xi = \mathbf{M}'\ell$, ℓ in U'. Conversely, suppose that x belongs to the nullspace of all such ξ; then the right side of (8) is zero for all ℓ in U'. But then so is the left side; this can be only if $\mathbf{M}x = u = 0$, that is, if x belongs to $N_{\mathbf{M}}$. This proves (11).

Part (iv) is obvious. $\qquad\qquad\qquad\qquad\qquad\qquad\qquad\qquad\qquad\qquad\qquad\qquad\square$

Exercise 2. Let X and U be Banach spaces, U reflexive. Let $\mathbf{M}$ be a bounded linear map: $X \longrightarrow U$. Let x_n be a sequence in X weakly convergent to x. Then $\mathbf{M}x_n$ converges weakly to $\mathbf{M}x$.

Exercise 3. Denote by **I** the identity map $X \to X$. Show that $\mathbf{I}'$ is the identity map: $X' \to X'$.

In a complex Hilbert space the notion of transpose is replaced by *adjoint*, defined by the analogue of (8) and denoted by an asterisk:

$$(\mathbf{M}x, y) = (x, \mathbf{M}^* y).$$

For matrices, the adjoint is the conjugate transpose.

Exercise 4. Show that theorem 5 is valid for the adjoint operation.

15.2 STRONG AND WEAK TOPOLOGIES

The norm of linear maps $X \to U$ defines a metric topology in $\mathcal{L}(X, U)$ that is sometimes called the *uniform topology*, in deference to two other topologies that are also useful and therefore important:

Definition. The *strong topology* in $\mathcal{L}(X, U)$ is the weakest topology in which all functions $\mathcal{L} \to U$ of the form $\mathbf{M} \longrightarrow \mathbf{M}x$ are continuous, x being any point of X.

Definition. The *weak topology* in $\mathcal{L}(X, U)$ is the weakest topology in which all linear functionals of the form $\mathbf{M} \longrightarrow (\mathbf{M}x, \ell)$ are continuous, x being any point of X and ℓ any point in U'.

Exercise 5. Define the weak* topology in $\mathcal{L}(X, U')$, X, U Banach spaces. Show that there is a natural one-to-one correspondence between $\mathcal{L}(X, U')$ and $\mathcal{L}(U, X')$, and that this correspondence is continuous in the weak* topology.

Of equal importance are the corresponding notions of sequential convergence:

Definition. A *sequence* $\{\mathbf{M}_n\}$ of bounded linear maps: $X \to U$, X, U Banach spaces, is called *strongly convergent* if

$$s - \lim_{n \to \infty} \mathbf{M}_n x \tag{15}$$

exists for every x in X.

$\{\mathbf{M}_n\}$ is called *weakly* convergent if

$$w - \lim_{n \to \infty} \mathbf{M}_n x \tag{15'}$$

exists for all x in X.

It is easy to show, and is left as an exercise, that a strongly or weakly convergent sequence of maps has a limit $\mathbf{M}$, in the sense that (15), (15′) are equal to $\mathbf{M}x$. We will denote these relations as $s - \lim \mathbf{M}_n = \mathbf{M}$ and $w - \lim \mathbf{M}_n = \mathbf{M}$.

Exercise 6. Prove that if $w - \lim \mathbf{M}_n = \mathbf{M}$, then $w - \lim \mathbf{M}'_n = \mathbf{M}'$ provided that X is reflexive. (Hint: Use the definition of weak convergence; see (18) below.)

No such result holds for strong convergence; take X and U both to be the Hilbert space ℓ^2 (see chapter 6) consisting of vectors

$$x = (a_1, a_2, \ldots), \quad \|x\|^2 = \sum |a_j|^2.$$

Define $\mathbf{M}_n$ to be

$$\mathbf{M}_n x = (a_n, 0, 0, \ldots).$$

It is easy to see that $s - \lim \mathbf{M}_n = 0$. Since ℓ^2 is a Hilbert space, it is self-dual; take $\ell = (b_1, b_2, \ldots)$; the relation $(\mathbf{M}_n x, \ell) = a_n b_1 = (x, \mathbf{M}'_n \ell)$ shows that $\mathbf{M}'_n \ell = (0, \ldots, b_1, \ldots)$. Clearly, $s - \lim \mathbf{M}'_n \ell$ does not exist unless $b_1 = 0$.

The significance of these notions is that maps of interest are often—one is tempted to say usually—constructed as limits, uniform, strong, or weak of sequences of approximate maps. The following result, as important as it is trivial, is used all the time:

Theorem 6. *Let X, U be Banach spaces, $\mathbf{M}_n$ a sequence of linear maps: $X \to U$, uniformly bounded in norm:*

$$|\mathbf{M}_n| \leq c \quad \text{for all } n. \tag{16}$$

Suppose further that

$$s - \lim \mathbf{M}_n x$$

exists for a dense set of x in X. Then $\{\mathbf{M}_n\}$ converges strongly, i.e. the s-limit exists for all x in X.

Exercise 7.

(a) Prove theorem 6.

(b) Formulate and prove an analogous theorem for weak convergence.

15.3 PRINCIPLE OF UNIFORM BOUNDEDNESS

Uniform boundedness turns out to be not only convenient for proving strong or weak convergence, but it is necessary as well.

Theorem 7. *Let X and U be Banach spaces, $\{M_\nu\}$ a collection of bounded maps $X \to U$, such that for each x in X and each ℓ in U', $(M_\nu x, \ell)$ is bounded by a constant that only depends on x and ℓ:*

$$|(M_\nu x, \ell)| \leq c(x, \ell) \qquad \text{for all } M_\nu. \tag{17}$$

Conclusion: $\{M_\nu\}$ is uniformly bounded, meaning that (16) holds.

Proof. We appeal to the principle of uniform boundedness, theorem 4 of chapter 10: If $\{u_\nu\}$ is a collection of points in a normed linear space U such that for every linear function ℓ in U', $|\ell(u_\nu)| \leq c(\ell)$ for all u_ν, then there is a constant c such that $|u_\nu| \leq c$. We apply this result to $u_\nu = M_\nu x$, and conclude that for all ν,

$$|M_\nu x| \leq c(x). \tag{17'}$$

Next we appeal to theorem 2 of chapter 10: If $\{f_\nu\}$ is a collection of real-valued, continuous, subadditive, and positive homogeneous functions defined on a Banach space X, and if at each point x of X, $f_\nu(x) \leq c(x)$ for all ν, then there is a number c such that

$$f_\nu(x) \leq c|x| \qquad \text{for all } x, \text{ all } \nu.$$

We identify the functions f_ν with $f_\nu(x) = |M_\nu x|$. Clearly the f_ν are homogeneous, subadditive, and continuous. According to (17') above, the $f_\nu(x)$ are bounded at every point. Therefore the f_ν are uniformly bounded, which in our case means that $|M_n x| \leq c|x|$ for all x in X, as asserted in (16). $\square$

The relation $w - \lim M_n = M$ means that (15') holds for every x in X, which in turn means (see definition 1 of chapter 10) that

$$\lim_{n \to \infty} (M_n x, \ell) = (Mx, \ell) \tag{18}$$

for all ℓ in U' and all x in X. Since a convergent sequence is bounded, it follows that condition (17) of theorem 7 is satisfied; therefore by theorem 7, the sequence M_n is uniformly bounded.

Corollary 7'. *A weakly convergent sequence of maps of one Banach space into another is uniformly bounded.*

15.4 COMPOSITION OF BOUNDED MAPS

We turn now to the discussion of composition, called the *product*, of a map $M : X \to U$ with another map $N : U \to W$. This operation was studied in chapter 2 from the point of view of linear algebra. Here we study some further properties of it in case X, U and W are Banach spaces, and M and N are bounded linear maps.

Theorem 8. *Let X, U, W denote Banach spaces, $\mathbf{M}$ and $\mathbf{N}$ bounded linear maps,*

$$\mathbf{M}: X \to U, \quad \mathbf{N}: U \to W.$$

Then the composite $\mathbf{NM}$ is a bounded linear map: $X \to W$ with the following properties:

(i) Submultiplicativity, $|\mathbf{NM}| \leq |\mathbf{N}|\,|\mathbf{M}|$.
(ii) $(\mathbf{NM})' = \mathbf{M}'\,\mathbf{N}'$.

Proof. Applying inequality (4) twice, we get

$$|\mathbf{NM}x| \leq |\mathbf{N}|\,|\mathbf{M}x| \leq |\mathbf{N}|\,|\mathbf{M}|\,|x|.$$

Applying definition (3), we get

$$|\mathbf{NM}| = \sup \frac{|\mathbf{NM}x|}{|x|} \leq |\mathbf{N}|\,|\mathbf{M}|. \tag{19}$$

We turn to (ii): applying (8) twice, we get

$$(\mathbf{NM}x, m) = (\mathbf{M}x, \mathbf{N}'m) = (x, \mathbf{M}'\mathbf{N}'m). \tag{20}$$

$\square$

Exercise 8. Prove that multiplication of maps is a continuous operation in the strong topology on the unit balls of $\mathcal{L}(X, U)$ and $\mathcal{L}(U, W)$.

Definition. Two maps $\mathbf{A}$ and $\mathbf{M}$ of a linear space X into itself are said to *commute* if $\mathbf{AM} = \mathbf{MA}$.

Exercise 9. Let X denote a Banach space, $\mathbf{A}$ a bounded map: $X \to X$ that commutes with each of a collection $\{\mathbf{M}_\nu\}$ of bounded maps $X \to X$. Show that then $\mathbf{A}$ commutes with every map $\mathbf{M}$ that lies in the closed linear span of the set of maps $\{\mathbf{M}_\nu\}$ in the weak topology.

Exercise 10. Show that in a complex Hilbert space $(\mathbf{NM})^* = \mathbf{M}^*\mathbf{N}^*$.

15.5 THE OPEN MAPPING PRINCIPLE

The next group of results, the open mapping principle, and the closed graph theorem, goes considerably deeper than the foregoing material. These ideas are due to Stefan Banach; their validity is far from being intuitively clear at first glance, or even a second one.

Theorem 9. *X and U are Banach spaces, and $\mathbf{M}: X \to U$ a bounded linear mapping of X onto all of U. Then there is a $d > 0$ such that the image of the open* *unit ball in X under* **M** *contains the ball of radius d in U:*

$$\mathbf{M}B_1(0) \supset B_d(0). \tag{21}$$

Proof. Denote by B_n the open ball of radius n around the origin in either the space X or U. Since **M** is assumed to map X *onto* U, and since the union of all the B_n is all of X, it follows that $\cup \mathbf{M}B_n = U$. Since the Banach space U is complete, it follows from the Baire category principle that at least one of the sets $\mathbf{M}B_n$ is *dense* in some open set. Some translate of this set is dense in some ball around the origin; since the range of **M** is all of U, by linearity of **M** we take that translate to be of the form $\mathbf{M}(B_n - x_0)$. The set $B_n - x_0$ is contained in the ball of radius $n + |x_0|$ around the origin. So by homogeneity of **M**, we conclude that $\mathbf{M}B_1(0)$ is dense in $B_r(0)$ for some $r > 0$. Consequently for any $c > 0$,

$$\mathbf{M}B_c(0) \quad \text{is dense in} \quad B_{cr}(0). \tag{22}$$

We want to show now that any point u in $B_r(0)$ is the image of some point x in $B_2(0)$:

$$\mathbf{M}x = u. \tag{23}$$

This point x in $B_2(0)$ is constructed as an infinite series

$$x = \sum_1^\infty x_j. \tag{23'}$$

The terms x_j are constructed recursively: x_1 is taken as a point satisfying

$$|u - \mathbf{M}x_1| < \frac{r}{2}, \qquad |x_1| < 1; \tag{24a}$$

by (22), with $c = 1$, there is such an x_1. We choose x_2 as a point satisfying

$$|u - \mathbf{M}x_1 - \mathbf{M}x_2| < \frac{r}{4}, \qquad |x_2| < \frac{1}{2}; \tag{24b}$$

it follows from (22), with $c = \frac{1}{2}$, and (24a) that such an x_2 exists. Generally, we choose x_m to satisfy

$$\left| u - \sum_1^m \mathbf{M}x_j \right| < \frac{r}{2^m}, \qquad |x_m| < \frac{1}{2^{m-1}}; \tag{24c}$$

It follows from (22), with $c = 1/2^{m-1}$ and (24c) that there is such an x_m.

We noted in chapter 5 on the geometry of normed spaces that if the sum of the norms $\sum |x_j|$ of a series in a complete normed linear space X converges, the series $\sum x_j$ converges strongly. Since by (24c), $|x_j| < 1/2^{j-1}$, it follows that $\sum_1^\infty x_j$

converges to a point x in X, and

$$|x| \leq \sum_1 |x_j| < \sum_1^\infty \frac{1}{2^{j-1}} = 2. \tag{25}$$

Since $\mathbf{M}$ is a bounded map, letting $m \to \infty$ in (24c), we conclude that $\mathbf{M}x = \sum_1^\infty \mathbf{M}x_j = u$. $\quad\square$

Theorem 9 has a number of interesting and important consequences; the first one is the *open mapping principle*:

Theorem 10. *X and U are Banach spaces, $\mathbf{M} : X \to U$ a bounded linear map onto all of U. Then $\mathbf{M}$ maps open sets onto open sets.*

This is an immediate corollary of theorem 9. $\quad\square$

Theorem 11. *X and U are Banach spaces, $\mathbf{M} : X \to U$ a bounded linear map that carries X one-to-one onto U. Then the algebraic inverse of $\mathbf{M}$ is a bounded linear map of $U \to X$.*

Proof. It follows from (21) of theorem 9 that for every u in U of norm $d/2$, there is an x in the unit ball of X such that $\mathbf{M}x = u$; note that $|x| \leq 1 = 2|u|/d$. Since $\mathbf{M}$ is homogeneous, it follows that for every u in U there is an x in X, such that

$$\mathbf{M}x = u, \quad |x| \leq 2|u|/d. \tag{26}$$

Since $\mathbf{M}$ is assumed one-to-one, $x = \mathbf{M}^{-1}u$. Clearly, from (26), $|\mathbf{M}^{-1}| \leq 2/d$. $\quad\square$

Definition. A map $\mathbf{M} : X \to U$ from one Banach space into another is called *closed* if whenever $\{x_n\}$ is a sequence in X such that

$$x_n \longrightarrow x \quad \text{and} \quad \mathbf{M}x_n \to u \tag{27}$$

then

$$\mathbf{M}x = u. \tag{27'}$$

If $\mathbf{M}$ is continuous, it is obviously closed. It is surprising but true that conversely, a closed linear map of a Banach space into another is continuous:

Theorem 12. *X and U are Banach spaces, $\mathbf{M} : X \to U$ is a closed linear map.*

Assertion. *$\mathbf{M}$ is continuous.*

Proof. Define the linear space G to consist of all pairs g of form

$$g = \{x, \mathbf{M}x\}, \quad x \text{ in } X. \tag{28}$$

We define the following norm for g in G:

$$|g| = |x| + |\mathbf{M}x| \tag{28'}$$

Clearly, this is a norm. It follows from (27), (27') and the completeness of X and U that G is complete under this norm. Define the mapping $\mathbf{P} : G \rightarrow X$ to be the projection onto the first component, that is,

$$g = \{x, \mathbf{M}x\}, \quad \mathbf{P}g = x. \tag{29}$$

By definition (28') of $|g|$, $|\mathbf{P}g| \leq |g|$, meaning that $\mathbf{P}$ is a bounded operator, $|\mathbf{P}| \leq 1$. Clearly, $\mathbf{P}$ is linear and maps G one-to-one onto X. Therefore, by theorem 11, the inverse of $\mathbf{P}$ is bounded; that is, there is a constant c such that $c|\mathbf{P}g| \geq |g|$. In view of the definition (29) of $\mathbf{P}$ and (28') of $|g|$, it follows that $(c - 1)|x| \geq |\mathbf{M}x|$, meaning that $\mathbf{M}$ is bounded. $\qquad\qquad\qquad\qquad\qquad\qquad\qquad\qquad\qquad\qquad\qquad\qquad\qquad\square$

The space G defined by (28) is called the *graph* of the mapping $\mathbf{M}$. Requiring $\mathbf{M}$ to be closed is the same as requiring its graph to be closed. Theorem 12 is known as the *closed graph theorem*. The closed graph theorem has many surprising applications.

Theorem 13. *X is a linear space equipped with two norms $|x|_1$ and $|x|_2$ that are compatible in the following sense: If a sequence $\{x_n\}$ converges in both norms, the two limits are equal.*

Suppose that X is complete with respect to both norms; then the two norms are equivalent. That is to say, there is a constant c such that for all x in X, $|x|_1 \leq c|x|_2$, $|x|_2 \leq c|x|_1$.

Proof. Denote by X_1, resp. X_2 the space X under the 1-, resp. 2-norm. By hypothesis, both X_1 and X_2 are complete. Compatibility clearly means that the identity map between X_1 and X_2 is closed. Therefore, by the closed graph theorem, it is bounded in both directions. $\qquad\qquad\qquad\qquad\qquad\qquad\qquad\qquad\qquad\qquad\qquad\qquad\qquad\qquad\square$

Theorem 14. *X and U are Banach spaces, $\mathbf{M} : X \rightarrow U$ a bounded linear map. Assume that the range $R_\mathbf{M}$ is a finite-codimensional subspace of U; then $R_\mathbf{M}$ is closed.*

Exercise 11. Prove theorem 14. (Hint: Extend $\mathbf{M}$ to $X \oplus Z$ so that its range is all of U.)

Exercise 12. Show that for every infinite-dimensional Banach space there are linear subspaces of finite codimension that are not closed. (Hint: Use Zorn's lemma.)

Theorem 15. *X is a Banach space, Y and Z closed subspaces of X that complement each other: $X = Y \oplus Z$, in the sense that every x in X can be decomposed uniquely* *as* $x = y + z$, y *in* Y, z *in* Z. *Denote the two components* y *and* z *of* x *by*

$$y = \mathbf{P}_Y x, \quad z = \mathbf{P}_Z x.$$

(i) $\mathbf{P}_Y$ *and* $\mathbf{P}_Z$ *are linear maps of* X *on* Y *and* Z, *respectively.*
(ii) $\mathbf{P}_Y^2 = \mathbf{P}_Y$, $\mathbf{P}_Z^2 = \mathbf{P}_Z$, $\mathbf{P}_Y \mathbf{P}_Z = 0$.
(iii) $\mathbf{P}_Y$ *and* $\mathbf{P}_Z$ *are continuous.*

Proof. Parts (i) and (ii) are obvious. To prove part (iii) we observe that since Y and Z are closed, and the decomposition is unique, it follows that the graphs of $\mathbf{P}_Y$ and $\mathbf{P}_Z$ are closed. The closed graph theorem does the rest.   □

A map satisfying $\mathbf{P}^2 = \mathbf{P}$ is called a *projection.*
We conclude this chapter by observing that complete metric spaces have proper subsets, called *sets of second category*, that are not unions of a denumerable number of nowhere dense sets. This allows a sharpening of the open mapping principle:

Theorem 16. X *and* U *are Banach spaces,* $\mathbf{M}: X \to U$ *a bounded linear map whose range* $R_\mathbf{M}$ *is a subset of* U *of second category. Then the range of* $\mathbf{M}$ *is all of* U.

Exercise 13. Prove theorem 16.

HISTORICAL NOTE. Stefan Banach (1892–1945), a Polish mathematician, was one of the founding fathers of functional analysis. Banach spaces are named in recognition of his numerous and deep contributions, and for having written the first monograph on the subject (1932). He was the inspiration of the brilliant Polish school of functional analysis.

During the Second World War, Banach was one of a group of people whose bodies were used by the Nazi occupiers of Poland to breed lice, in an attempt to extract an anti-typhoid serum. He died shortly after the conclusion of the war.

The Nazi attitude toward Poles is epitomized by the following story. After the conquest of France in 1940, when Hitler ruled most of Europe, a leading German mathematician, a member of the Nazi party, called on Elie Cartan, the dean of French mathematicians, to discuss the organization of mathematical life in the new European order. Cartan wanted to know how the Polish mathematicians would fit in. "Oh," the German replied, "the Führer has declared the Poles to be subhuman."

BIBLIOGRAPHY

Banach, S. Sur les fonctionelles linéaires, I, II. *Studia Math.*, **1** (1929): 211–216, 223–339.

Schauder, J. Über die Umkehrung linearer, stetiger Funktionaloperationen. *Studia Math.*, **2** (1930): 1–6.

16

EXAMPLES OF BOUNDED LINEAR MAPS

An important class of linear maps is furnished by *integral operators*. The first part of this chapter is devoted to investigating their boundedness in various norms. Let T and S be Hausdorff spaces, equipped with measures n and m. $\mathbf{K}$ denotes an integral operator mapping complex-valued functions f on T into complex-valued functions g on S:

$$g(s) = (\mathbf{K}f)(s) = \int_T K(s, t)\, f(t)\, dn(t). \tag{1}$$

The complex-valued function $K(s, t)$ is called the *kernel* of $\mathbf{K}$; f, K are assumed to be measurable and restricted so that (1) defines a measurable function g. Each subsequent theorem reveals a natural class for f, K, and g. We recall from chapter 4 the L^p-norms:

$$|f|_{L^p} = \left(\int_T |f(t)|^p\, dn(t) \right)^{1/p}, \qquad 1 \le p \le \infty.$$

The space $L^p(T, n)$ is the completion of the space $C_0(T)$ in the L^p-norm. The space $L^p(S, m)$ is defined analogously. The space L^∞ is the space of essentially bounded, measurable functions.

16.1 BOUNDEDNESS OF INTEGRAL OPERATORS

We start with conditions that guarantee that (1) is a bounded map from $L^1(T, n)$ or $L^\infty(T, n)$ to $L^1(S, m)$ or $L^\infty(S, m)$.

Theorem 1.

(i) *The map* $\mathbf{K}$ *defined by (1) is bounded as a mapping* $L^1 \to L^\infty$, *and*

$$|\mathbf{K}| \le \sup_{s,t} |K(s,t)|, \tag{2_i}$$

provided that the quantity on the right is $< \infty$.

(ii) **K** is bounded as a mapping $L^\infty \to L^1$, and

$$|\mathbf{K}| \le \int \int |K(s,t)| \, dm(s) \, dn(t), \tag{2_{ii}}$$

provided that the quantity on the right is $< \infty$.

(iii) **K** is bounded as a mapping $L^\infty \to L^\infty$,

$$|\mathbf{K}| \le \sup_{s} \int |K(s,t)| \, dn(t) \tag{2_{iii}}$$

if the quantity on the right is $< \infty$.

(iv) K is bounded as a mapping $L^1 \to L^1$, and

$$|\mathbf{K}| \le \sup_{t} \int |K(s,t)| \, dm(s), \tag{2_{iv}}$$

if the quantity on the right is $< \infty$.

Proof. By (1), for any s in S,

$$|g(s)| \le \int_T |K(s,t)| \, |f(t)| \, dn(t). \tag{3}$$

The right side is $\le \sup_t |K(s,t)| \, |f|_{L^1}$, so

$$|g|_\infty = \sup_s |g(s)| \le \sup_{s,t} |K(s,t)| \, |f|_{L^1}.$$

This proves (2_i).

Integrate (3) with respect to dm over S:

$$|g|_{L^1} = \int_S |g(s)| \, dm(s) \le \int \int |K(s,t)| \, |f(t)| \, dn(t) \, dm(s)$$
$$= \int \Big[\int |K(s,t)| \, dm(s) \Big] |f(t)| \, dn(t). \tag{4}$$

The right side is

$$\le \int \int |K(s,t)| \, dm(s) \, dn(t) \, |f|_\infty;$$

this proves (2_{ii}).

The right side of (4) is also less than

$$\sup_t \int |K(s,t)|\, dm(s)\, |f|_{L^1};$$

this proves (2_{iv}).

The right side of (3) is less than

$$\int_T |K(s,t)|\, dn(t)\, |f|_{L^\infty};$$

this combined with (3) proves (2_{iii}).   □

Note that when $K(s,t)$ and $f(t)$ are both positive, the sign of equality holds in both (3) and (4). From this it is not hard to deduce

Corollary 2′. *When the kernel $K(s,t)$ in (1) is nonnegative, the sign of equality holds in (2_{iii}) and (2_{iv}).*

The transpose of **K** is easily written down; denote by $(\ ,\ )_S$ and $(\ ,\ )_T$ the standard L^2 scalar product on S and T with respect to dm and dn, respectively:

$$(g,h)_S = \int g(s)\, h(s)\, dm(s), \tag{5}$$

$$(k,f)_T = \int_T k(t)\, f(t)\, dn(t). \tag{5′}$$

Multiplying (1) by $h(s)$ and integrating gives

$$(\mathbf{K}f, h)_S = \iint\limits_{ST} K(s,t)\, f(t)\, h(s)\, dn(t)\, dm(s) = (f, \mathbf{K}'h)_T, \tag{6}$$

where

$$(\mathbf{K}'h)(t) = \int_S K(s,t)\, h(s)\, dm(s). \tag{6′}$$

In words, the kernel of the transpose **K**′ is the same as the kernel of **K**, with the roles of the variables s and t interchanged.

We recall now theorem 5 of chapter 15, according to which the norm of **K**′ equals the norm of **K**. We verify this in the case when the kernel K is nonnegative, and **K** is regarded as mapping $L^1(T)$ into $L^1(S)$. According to corollary 1′, $|\mathbf{K}|$ is given by formula (2_{iv}). **K**′, on the other hand, maps $L^\infty(S)$ into $L^\infty(T)$, and its norm is given by formula (2_{iii}), with the roles of s and t reversed. Clearly, $|\mathbf{K}'| = |\mathbf{K}|$, as it should be.

We turn now to the L^2-norms, which we denote as $\| \ \|$; the corresponding norm of $\mathbf{K}$ is denoted as $\|\mathbf{K}\|$.

Theorem 2. *The map $\mathbf{K}$ defined by (1) is bounded as a map: $L^2 \to L^2$, and*

$$\|\mathbf{K}\|^2 \leq \iint\limits_{ST} |K^2(s,t)| \, dm \, dn, \tag{7}$$

provided that the quantity on the right is $< \infty$.

Proof. Applying the Schwarz inequality (see chapter 6) to the integral on the right in (1), we get

$$|g(s)|^2 \leq \int_T |K^2(s,t)| \, dn \int_T |f(t)|^2 \, dn.$$

Integrating both sides dm gives

$$\|g\|^2 \leq \iint\limits_{ST} |K(s,t)|^2 \, dn \, dm \|f\|^2,$$

as asserted in (7). □

Inequality (7) is due to Hilbert and E. Schmidt. Another criterion has been given by Holmgren:

Theorem 3. $\mathbf{K}$ *as defined by (1) is bounded: $L^2 \to L^2$, and*

$$\|\mathbf{K}\| \leq \left(\sup_s \int |K(s,t)| dn\right)^{1/2} \left(\sup_t \int |K(s,t)| dm\right)^{1/2}, \tag{8}$$

provided that the quantity on the right is $< \infty$.

Proof. According to theorem 1 in chapter 6,

$$\|g\| = \max_{\|h\|=1} (g,h)_S. \tag{9}$$

We are going to use (9) to estimate $g = \mathbf{K}f$. By (6),

$$(g,h)_S = \int \int K(s,t) f(t) h(s) \, dn \, dm. \tag{10}$$

For any three positive numbers f, h and c, $fh \leq cf^2/2 + h^2/2c$, so the right side of (10) is

$$\leq \int \int |K(s,t)| \left\{ \frac{c}{2}|f(t)|^2 + \frac{1}{2c}|h(s)|^2 \right\} dm \, dn.$$

In the first term we integrate first with respect to s, in the second with respect to t. We get the estimate

$$\frac{c}{2} \sup_t \int |K(s,t)| \, dm \, \|f\|^2 + \frac{1}{2c} \sup_s \int |K(s,t)| \, dn \|h\|^2. \qquad (10')$$

We take now $\|f\| = 1 = \|h\|$ and choose c so that $(10')$ is as small as possible. This minimum is

$$\left(\sup_t \int \right)^{1/2} \left(\sup_s \int \right)^{1/2}. \qquad (10'')$$

Combining (9) and (10) with $(10'')$, we conclude that for $\|f\| = 1$, $\|\mathbf{K}f\|$ is $\leq$ the quantity in $(10'')$. In view of the definition $\|\mathbf{K}\| = \sup \|\mathbf{K}f\|$, $\|f\| = 1$, this proves (8). $\qquad\qquad\qquad \square$

16.2 THE CONVEXITY THEOREM OF MARCEL RIESZ

The two factors appearing on the right in (8) are the square roots of the quantities appearing on the right in (2_{iii}) and (2_{iv}). We noted in corollary $1'$ that for a positive kernel these quantities are not merely upper bounds for the norm of $\mathbf{K} : L^\infty \to L^\infty$ and $L^1 \to L^1$ but equal to these norms.

Definition. Denote by $M(p,q)$ the norm of

$$\mathbf{K} : L^p(T,n) \longrightarrow L^q(S,m). \qquad (11)$$

For integral operators whose kernel is ≥ 0, we can restate inequality (8) as follows:

$$M(2,2) \leq M^{1/2}(1,1) \, M^{1/2}(\infty,\infty).$$

This turns out to be a special case of a far more general theorem due to M. Riesz.

Theorem 4. *Let* $\mathbf{M}$ *be a linear map of complex-valued functions defined on* T *into complex-valued functions defined on* S. *Suppose that* $\mathbf{M}$ *carries functions measurable with respect to* n *into functions measurable with respect to* m. *Suppose further that* $\mathbf{M}$ *is bounded with respect to two pairs of norms:*

$$L^{p_0}(T,n) \longrightarrow L^{q_0}(S,m) \quad and \quad L^{p_1}(T,n) \longrightarrow L^{q_1}(S,m).$$

Conclusion: then $\mathbf{M}$ *is a bounded map of* $L^{p(a)}(T,n) \to L^{q(a)}(S,m)$ *where*

$$\left(\frac{1}{p(a)}, \frac{1}{q(a)} \right) = (1-a) \left(\frac{1}{p_0}, \frac{1}{q_0} \right) + a \left(\frac{1}{p_1}, \frac{1}{q_1} \right), \qquad 0 \leq a \leq 1. \quad (12)$$

Furthermore, $M(p, q)$ is a log-convex function of its arguments:

$$M(p(a), q(a)) \leq M^{1-a}(p_0, q_0) \, M^a(p_1, q_1); \tag{12'}$$

here $M(p, q)$ is the norm of the operator $\mathbf{K}$ defined in (11).

Proof. We sketch Thorin's beautiful proof of this theorem. The starting point is the following result due to Hadamard:

Three Lines Theorem. *Let $\phi(\zeta)$ be a bounded analytic function in the strip $0 \leq$ $\mathrm{Re}\ \zeta \leq 1$. Denote*

$$N(a) = \sup_{\eta} |\phi(a + i\eta)| \tag{13}$$

Then

$$N(a) \leq N^{1-a}(0) \, N^a(1). \tag{13'}$$

Proof. Set $c = \log N(0)/N(1)$; by (13), the function $\phi(\zeta)e^{c\zeta}$ is in absolute value $\leq N(0)$ for $\mathrm{Re}\ \zeta = 0$ and $\mathrm{Re}\ \zeta = 1$. So, by the maximum principle applied in the strip $0 \leq \mathrm{Re}\ \zeta \leq 1$,

$$|\phi(a + i\eta)|e^{ca} \leq N(0);$$

from this and the definition of c, (13') follows. $\square$

We turn now to the mapping $\mathbf{M}$; by definition of the norm,

$$M(p, q) = \sup_{|f|_{L^p} = 1} |\mathbf{M}f|_{L^q}.$$

Furthermore, according to theorem 5 of chapter 5—Hölder's inequality and equality—for any g in L^q, $|g|_{L^q} = \sup_{|h|_{L^{q'}} = 1} |(g, h)s|$, where q' is dual to q,

$$\frac{1}{q} + \frac{1}{q'} = 1.$$

Combining the last two, with $g = \mathbf{M}f$, we get

$$M(p, q) = \sup_{|f|_{L^p} = 1, |h|_{L^{q'}} = 1} |(\mathbf{M}f, h)|. \tag{14}$$

We take $p = p(a), q = q(a)$ as defined in (12). The complex-valued functions f and h can be factored as $f = |f|e^{i\mu}$, $h = |h|e^{i\nu}$. For any ζ in the strip $0 \leq \mathrm{Re}\ \zeta \leq 1$ we define

$$f(\zeta) = |f|^{p(a)/p(\zeta)}e^{i\mu}, \quad h(\zeta) = |h|^{q'(a)/q'(\zeta)}e^{i\nu}, \tag{15}$$

where $p(a)$, $p(\zeta)$, and so on, are defined by formula (12). Note that $f(a) = f$, $h(a) = h$. Since $1/p(\zeta)$ and $1/q(\zeta)$ are linear functions of ζ, so is $1/q'(\zeta)$. Therefore $f(\zeta)$ and $h(\zeta)$ are *analytic* functions of ζ, and so is

$$\phi(\zeta) = (\mathbf{M}f(\zeta), h(\zeta))_S = \int \mathbf{M}f(\zeta) h(\zeta) \, dm(s). \tag{15'}$$

Lemma 5. *Let f and h be functions of unit norm, $|f|_{p(a)} = 1$ $|h|_{q'}(a) = 1$, and $\varphi(\zeta)$ defined as above. Define $N(a)$ as the supremum of $|\phi(\zeta)|$ on the line $\mathrm{Re}\ \zeta = a$; we claim that*

$$N(0) \le M(p_0, q_0), \quad N(1) \le M(p_1, q_1). \tag{16}$$

Proof. Let's take $\mathrm{Re}\ \zeta = 0$. Then $\zeta = i\eta$, and so by formula (12),

$$\frac{p(a)}{p(\zeta)} = \frac{p(a)}{p_0} + \text{imag.}, \quad \frac{q'(a)}{q'(\zeta)} = \frac{q'(a)}{q'_0} + \text{imag.} \tag{17}$$

From (15),

$$|f(i\eta)|_{L^{p_0}}^{p_0} = |f|_{L^{p(a)}}^{p(a)}, \quad |h(i\eta)|_{L^{q'_0}}^{q_0} = |h|_{L^{q'(a)}}^{q'(a)}. \tag{18}$$

Since f and h were chosen so that $|f|_{L^{p(a)}} = 1$, $|h|_{L^{q'(a)}} = 1$, it follows from (18) that $|f(i\eta)|_{L^{p_0}} = 1$, $|h(i\eta)|_{L^{q'_0}} = 1$. Therefore,

$$|\mathbf{M}f(i\eta)|_{L^{q_0}} \le M(p_0, q_0). \tag{19}$$

Estimate ϕ defined in (15') by Hölder's inequality; using (19), and $|h(i\eta)|_{L^{q'_0}} = 1$, we get

$$|\phi(i\eta)| = |(\mathbf{M}f(i\eta), h(i\eta))| \le |(\mathbf{M})f(i\eta)|_{L^{q_0}} |h(ik\eta)|_{L^{q'_0}} \le M(p_0, q_0).$$

This proves the first part of (16); the second part follows in exactly the same fashion. $\square$

We apply now the three lines theorem, (13'), to φ defined by (15'); using (16), we get

$$|\phi(a)| \le N(a) \le M^{1-a}(p_0, q_0) M^a(p_1, q_1) \tag{20}$$

Since $f(a) = f$, $h(a) = h$, by (15')

$$\phi(a) = (\mathbf{M}f, h).$$

According to (14) the supremum of the right side over all f, h of unit norm is the norm of $\mathbf{M}$:

$$M(p(a), q(a)) = \sup |\phi(a)|.$$

Using the estimate (20) for the right side, we obtain the desired inequality (12′). □

16.3 EXAMPLES OF BOUNDED INTEGRAL OPERATORS

Theorems 2 and 3 both furnish criteria for an integral operator to be bounded is $L^2 \to L^2$. These criteria are *very far* from being necessary for boundedness, and are insufficient for proving the $L^2 \to L^2$ boundedness of the most important and most beloved mappings. We illustrate this on a number of examples.

16.3.1 The Fourier Transform

The Fourier transform is defined by

$$(\mathbf{F}f)(s) = \int_{\mathbb{R}} e^{-ist} f(t) \frac{dt}{\sqrt{2\pi}}, \tag{21}$$

whose kernel is

$$K(s, t) = \frac{1}{\sqrt{2\pi}} e^{-ist}, \tag{21'}$$

$T = S = \mathbb{R}$, m and n Lebesgue measure.

Clearly, for this K the right side of both (7) and (8) is ∞, so neither theorem 2 nor theorem 3 can be used to show the $L^2 \to L^2$ boundedness of the Fourier transform. Yet it is well known that it is bounded; see theorem 21 of Appendix B. On the other hand, we can use part (i) of theorem 1 to conclude that $\mathbf{F} : L^1 \to L^\infty$ is bounded by $1/\sqrt{2\pi}$. We can now appeal to the M. Riesz convexity theorem, theorem 4, with $(p_0, q_0) = (2, 2)$, $(p_1, q_1) = (1, \infty)$ to conclude, after a brief calculation,

Theorem 6. *For* $1 \le p \le 2$, $\mathbf{F}$ *is a bounded map of* $L^p \to L^{p/(p-1)}$ *and*

$$|\mathbf{F}| \le \left(\frac{1}{\sqrt{2\pi}} \right)^{(2-p)/p}. \tag{22}$$

This inequality is called the Hausdorff-Young inequality after its discoverers.

16.3.2 The Hilbert Transform

Let $h(t)$ be a real-valued function on $\mathbb{R}$, fairly smooth—C^1 will do—and tending to zero as $|t| \to \infty$ at a reasonable rate, say $O(t^{-2})$.

The Cauchy integral

$$\frac{1}{\pi i} \int_{\mathbb{R}} \frac{h(t)}{t - \zeta} dt = f(\zeta) \tag{23}$$

defines a function $f(\zeta)$, or rather two functions, one analytic in the upper half-plane, the other in the lower half-plane. We will restrict ζ to the upper half-plane.

Writing $\zeta = \xi + i\eta$, we can express the real and imaginary parts of f as follows:

$$f(\zeta) = \frac{1}{\pi i} \int_{\mathbb{R}} \frac{h(t)(t - \bar{\zeta})}{|t - \zeta|^2}\, dt = \frac{1}{\pi} \int_{\mathbb{R}} \frac{\eta}{(\xi - t)^2 + \eta^2} h(t)\, dt$$

$$+ \frac{i}{\pi} \int_{\mathbb{R}} \frac{(\xi - t)}{(\xi - t)^2 + \eta^2} h(t)\, dt. \tag{23'}$$

Using the properties imposed on h, it is not hard to show that

(i) As $|\zeta| \to \infty$,

$$|f(\zeta)| = o(|\zeta|^{-1}). \tag{24}$$

(ii) $f(\zeta)$ is continuous up to the real axis, and its real part there equals h:

$$f(\xi) = h(\xi) + i\,k(\xi), \tag{25}$$

where k is expressed in terms of h as the principal value integral

$$k(\xi) = \frac{1}{\pi} PV \int \frac{h(t)}{\xi - t}\, dt \equiv (\mathbf{H}h)(\xi). \tag{25'}$$

The map $\mathbf{H}$ defined in (25') is called the *Hilbert transform*; it relates the real to the imaginary part of the boundary value of analytic functions in the upper half-plane satisfying (24).

Theorem 7. *The Hilbert transform is an isometry of $L^2(\mathbb{R}) \to L^2(\mathbb{R})$.*

Proof. Since f^2 is analytic in Im $\zeta > 0$, by Cauchy's theorem,

$$\oint f^2 d\zeta = 0 \tag{26}$$

over every closed contour there. We take now the contour to consist of a line segment $\xi + i\epsilon$, $-R \le \xi \le R$, and a semicircle $\xi = R\cos\theta$, $\eta = R\sin\theta + \epsilon$. Now let $\epsilon \to 0$ and $R \to \infty$. It follows from (24) that the integral over the semicircle in (26) tends to zero as $R \to \infty$, while it follows from (24) and (25) that the integral over the segment tends to

$$\int_{\mathbb{R}} (h + ik)^2 d\xi = 0. \tag{26'}$$

Taking the real part of (26') gives

$$\int h^2 d\xi = \int k^2 d\xi,$$

as asserted in theorem 7.   □

Exercise 1. Show that

$$\mathbf{H}^2 = -\mathbf{I}, \qquad \text{where } \mathbf{I} = \text{identity}. \tag{27}$$

(Hint: Consider the relation of the real part of $-if(\xi)$ to its imaginary part).

Note that the kernel of **H**,

$$K(s,t) = \frac{1}{s-t}, \tag{28}$$

fails miserably the tests for boundedness given in theorems 2 and 3.
    The argument used above to show that **H** is an isometry of $L^2 \to L^2$ can be used
to prove:

Theorem 8. *The Hilbert transform* **H** *is a bounded map of* $L^p \to L^p$ *for all* p,
$1 < p < \infty$.

Proof. Take $p = 4$, and consider the analytic function f^4. By Cauchy's theorem,

$$\oint f^4 d\zeta = 0. \tag{29}$$

We choose the same contour as in (26) and let $\epsilon \to 0$, $R \to \infty$, to obtain

$$\int_{\mathbb{R}} (h(\xi) + ik(\xi))^4 d\xi = 0.$$

The real part of this relation is

$$\int_{\mathbb{R}} (h^4 - 6h^2k^2 + k^4) \, d\xi = 0. \tag{29'}$$

According to a well-known inequality, for a, b, c positive $ab \leq ca^2/2 + b^2/2c$;
applying this to $a = h^2$, $b = k^2$, $c = 6$, gives

$$6h^2k^2 \leq 18h^4 + \tfrac{1}{2}k^4.$$

Setting this into (29'), we get

$$\frac{1}{2} \int k^4 d\xi \leq 17 \int h^4 d\xi.$$

This shows that $H : L^4 \to L^4$ is bounded, and that $|H| \leq 34$.

The same argument works for p any even integer. Then, using M. Riesz's convexity theorem, theorem 4, we deduce the boundedness of H as a map of $L^p \to L^p$ for any p, $2 \le p < \infty$.

To complete the proof, we turn to theorem 5 of chapter 15, according to which the transpose H' of H has the same norm as H. According to formulas (1), (1'), the kernel of H' is obtained from the kernel of H by interchanging the roles of the variables. According to (28), interchanging the variables in the kernel of H merely changes the sign of the kernel. So

$$H' = -H. \qquad (30)$$

If H maps $L^p \to L^p$, H' maps $(L^p)' \to (L^p)'$. According to theorem 11 of chapter 8, the dual of L^p is $L^{p'}$, where

$$\frac{1}{p'} + \frac{1}{p} = 1. \qquad (31)$$

Note that if $p > 2$, $p' < 2$. Combining (30) and (31) with theorem 5 of chapter 8, we conclude that the norm of $H : L^{p'} \to L^{p'}$ equals the norm of $H : L^p \to L^p$. Since the latter were shown to be finite for $2 < p < \infty$, it follows that they are bounded for $1 < p' < 2$ as well. $\qquad \square$

Theorem 8, and the astonishing proof above, are due to M. Riesz.

Exercise 2. Show that H is *not* bounded as a map: $L^\infty \to L^\infty$. Deduce from this that H is not bounded as a map: $L^1 \to L^1$.

16.3.3 The Laplace Transform

Let $f(t)$ be a complex-valued function on $\mathbb{R}_+$: $t \ge 0$. Its Laplace transform Lf is the function on $\mathbb{R}_+$: $s > 0$ defined by

$$g(s) = (Lf)(s) = \int_0^\infty f(t) e^{-st} \, dt. \qquad (32)$$

Theorem 9. *The Laplace transform* L *is a bounded map of* $L^2(\mathbb{R}_+) \to L^2(\mathbb{R}_+)$, *and*

$$\|L\| = \sqrt{\pi}. \qquad (33)$$

Proof. We estimate $g(s)$ by the Schwarz inequality:

$$|g(s)|^2 = \left(\int_0^\infty f(t) e^{-st} \, dt \right)^2 = \left(\int_0^\infty (f(t) e^{-\frac{st}{2}} t^{1/4})(e^{-\frac{st}{2}} t^{-1/4}) \, dt \right)^2$$

$$\le \int_0^\infty |f(t)|^2 e^{-st} t^{1/2} \, dt \int_0^\infty e^{-st} t^{-1/2} \, dt. \qquad (34)$$

By a change of variable we can write the second integral as

$$\int_0^\infty e^{-st} t^{-1/2} dt = \int_0^\infty e^{-u} u^{-1/2} du \; s^{-1/2} = Cs^{-1/2}, \tag{35}$$

where

$$C = \int_0^\infty e^{-u} u^{-1/2} du = \int_0^\infty e^{-x^2} x^{-1} 2x dx = 2 \int_0^\infty e^{-x^2} dx = \sqrt{\pi}. \tag{36}$$

Setting (35) into (34) gives

$$|g(s)|^2 \leq Cs^{-1/2} \int_0^\infty |f(t)|^2 e^{-st} t^{1/2} dt. \tag{37}$$

Integrating (37) gives

$$\|g\|^2 = \int_0^\infty |g(s)|^2 ds \leq C \int_0^\infty \int |f(t)|^2 e^{-st} t^{1/2} s^{-1/2} dt ds. \tag{38}$$

Interchange the order of integration, and change variables in the s-integral:

$$\int_0^\infty e^{-st} t^{1/2} s^{-1/2} ds = \int_0^\infty e^{-u} u^{-1/2} du = C.$$

So we get from (38) that

$$\|g\|^2 \leq C^2 \|f\|^2.$$

Using the value of C given by (36), we conclude that $\|L\| \leq \sqrt{\pi}$. To show that equality holds, take $f(t) = 1/\sqrt{t}$ for $a < t < b$, zero outside this interval; for this choice $\|f\|^2 = \log b/a$. Set $g = Lf$; it is not hard to show that as a tends to zero and b to ∞, $\|g\|^2 \geq \pi(1 - \varepsilon) \log b/a$. Combined with $\|L\| \leq \sqrt{\pi}$ this proves (33). $\qquad \square$

Again note that the kernel of L, e^{-st}, utterly fails the criteria for L^2 boundedness contained in either theorem 2 or theorem 3.

Exercise 3. Prove that the Laplace transform L is **not** bounded as a map of $L^p(\mathbf{R}_+) \to L^p(\mathbf{R}_+)$, except for $p = 2$. (Hint: Try $f(t) = e^{-at}$.)

As remarked in chapter 15, theorem 8, if L is bounded, so is L^2; therefore it follows by submultiplicity from (33) that

$$\|L^2\| \leq \|L\|^2 = \pi. \tag{39}$$

We claim that in (39) the sign of equality holds. To see this, we note that the kernel of the integral operator L, e^{-st}, is a *real symmetric* function of s and t. It is easily verified (see formulas (6),(6′)) that an integral operator L with a symmetric kernel satisfies

$$(Lu, v) = (u, Lv). \tag{40}$$

That is, such an operator is its own adjoint; such operators are called *symmetric*.

Theorem 10. *Let L be a bounded, symmetric mapping of a real Hilbert space into itself. Then*

$$\|L^2\| = \|L\|^2.$$

Proof. By submultiplicity, $\|L^2\| \le \|L\|^2$ is valid for all mappings, symmetric or not. To show the opposite inequality, we set $v = Lu$ into (40); we get

$$(Lu, Lu) = (u, L^2u).$$

The left side equals $\|Lu\|^2$; estimating the right side by the Schwarz inequality gives

$$\|Lu\|^2 \le \|u\|\|L^2u\| \le \|u\|^2\|L^2\|.$$

Since this holds for all vectors u in H, $\|L\|^2 \le \|L^2\|$ follows.   □

Clearly, it follows from theorem 10 that the sign of equality holds in relation (39). The mapping L^2 is easily computed:

$$(L^2 f)(r) = \int_0^\infty (Lf)(s)\, e^{-rs}\, ds = \int_0^\infty \int_0^\infty f(t)\, e^{-st}\, dt\, e^{-rs}\, ds$$

$$= \int_0^\infty f(t) \int_0^\infty e^{-(t+r)s}\, ds\, dt = \int_0^\infty \frac{f(t)}{t+r}\, dt.$$

So we have proved

Theorem 11. *The integral operator $f \to g$:*

$$g(r) = \int_0^\infty \frac{f(t)}{t+r}\, dt \tag{41}$$

is bounded as a map of $L^2(\mathbb{R}_+) \to L^2(\mathbb{R}_+)$, and its norm is equal to π.

The map (41) is called the *Hilbert-Hankel* operator. Note that its kernel, $1/s + r$, utterly fails the test for L^2 boundedness contained in either theorem 2 or 3.

Exercise 4. Prove that the Hilbert-Hankel operator is a bounded map of $L^p \to L^p$ for $1 < p < \infty$.

For further information about integral operators, see Halmos and Sunder.

16.4 SOLUTION OPERATORS FOR HYPERBOLIC EQUATIONS

We recall from section 5 of chapter 11 the class of symmetric hyperbolic operators of first-order. These are first-order partial differential operators of the form

$$L = \sum_1^m A_j \partial_j + B, \qquad \partial_j = \frac{\partial}{\partial s_j}. \tag{42}$$

The A_j, B are $n \times n$ matrices with real-valued entries that are reasonably smooth functions of s. We take them to be periodic in s. L acts on vector-valued functions $u(s)$, whose components are real valued, and assumed reasonably smooth, periodic functions of s. As scalar product for such functions we take the L^2 scalar product over a period parallelogram F:

$$(u, v) = \int_F u \cdot v \, ds, \tag{43}$$

where the dot is the standard inner product for vectors. We assume the coefficient matrices A_j to be *symmetric:*

$$A_j^T = A_j. \tag{44}$$

In this case the formal adjoint $\mathbf{L}^*$ of $\mathbf{L}$ takes the form

$$\mathbf{L}^* = -\mathbf{L} + \mathbf{K}, \tag{44'}$$

where

$$\mathbf{K} = B + B^T - \sum A_{j,j}, \qquad A_{j,j} = \partial_j A_j. \tag{45}$$

Adjointness means that for smooth functions u and v,

$$(v, \mathbf{L}u) = (\mathbf{L}^* v, u);$$

from this and (44') we deduce, upon setting $v = u$, that

$$2(u, \mathbf{L}u) = (u, \mathbf{K}u), \tag{45'}$$

see equation (20'), chapter 11.

Theorem 12. *Let $u(s, t)$ be a solution of*

$$u_t + \mathbf{L}u = 0, \tag{46}$$

and suppose that u is periodic in s. Then

$$\|u(T)\| \le c\|u(0)\|, \tag{47}$$

with c a constant that may depend on T. Here the norm is the L^2 norm (43) over a period parallelogram F.

REMARK 1. It follows from (47) that the solution u is uniquely determined by its initial value $u(s, 0)$. Thus $u(T)$ is related to $u(0)$, and since equation (46) is linear, this relation is clearly linear; denote by $S(T)$ the map relating $u(0)$ to $u(T)$:

$$S(T) : u(0) = u(T);\qquad(48)$$

$S(T)$ is called the *solution operator.* Theorem 12 states that for each T the solution operator is bounded in the norm $L^2(F) \to L^2(F)$.

Proof. Assume first that the matrix K in (45) is ≥ 0 for all s. Take the scalar product of (46) with $2u$, and integrate over F. Using (43), we can write

$$2(u, u_t) + 2(u, \mathbf{L}u) = 0,$$

so by (45')

$$2(u, u_t) + (u, Ku) = 0.\qquad(49)$$

The first term can be written as $d(u, u)/dt$. Therefore, if the symmetric matrix $K \geq 0$, it follows from (49) that $\|u(t)\|$ is decreasing as function of T; from this (47) follows for all $T > 0, c = 1$.

If K is not positive, introduce v by $u = e^{kt}v$ as new dependent variable; set this into (46) to obtain $v_t + (k + \mathbf{L})v = 0$. For k large enough, $k + K \geq 0$, so v satisfies (47) with $c = 1$. Therefore u satisfies (47) with $c = e^{kT}, T > 0$. $\qquad\square$

Obviously the proof works also if u is not periodic in s but tends to zero as $|s| \to \infty$ so fast that $u \in L^2(\mathbb{R}^n)$.

We take now a special example of an equation of form (42):

$$A_1 = \begin{pmatrix} 1 & 0 \\ 0 & -1 \end{pmatrix}, \quad A_2 = \begin{pmatrix} 0 & 1 \\ 1 & 0 \end{pmatrix}, \quad B = 0.$$

Denoting $u = (v, w)'$, we can write (42) componentwise as

$$v_t + v_x + w_y = 0,$$
$$w_t - w_x + v_y = 0.$$

We can eliminate one of the two components, obtaining after a brief calculation

$$v_{tt} - v_{xx} - v_{yy} = 0,$$
$$w_{tt} - w_{xx} - w_{yy} = 0,$$

the classical wave equation. There is an explicit solution formula for the wave equation that puts the solution operator $S(T)$ in the form of an integral operator. The kernel of $S(T)$ is in this case is not even a function but a distribution; it utterly fails to satisfy the L^2 boundedness criteria stated in theorems 2 and 3.

16.5 SOLUTION OPERATOR FOR THE HEAT EQUATION

We consider solutions $u(x, t)$ of the heat equation

$$u_t = u_{xx} \tag{50}$$

that are defined for all x and all $t \geq 0$, and which $\rightarrow 0$ sufficiently rapidly as $|x| \rightarrow \infty$.

Theorem 13. *Let $u(x, t)$ be a solution of the heat equation, as above. Then for all $T > 0$*

   (i) $|u(T)|_{\max} \leq |u(0)|_{\max}$.
   (ii) $|u(T)|_{L^1} \leq |u(0)|_{L^1}$.
   (iii) $|u(T)|_{L^2} \leq |u(0)|_{L^2}$.

REMARK 2. Since (50) is linear, these estimates show that u is uniquely determined by its initial value and that the dependence of u on its initial data is linear. Therefore the solution operator

$$\mathbf{S}(t) : u(0) \rightarrow u(t) \tag{51}$$

is well defined. In terms of it, theorem 13 can be formulated so: $|\mathbf{S}(t)| \leq 1$ as an operator mapping $L^p \rightarrow L^p$, $p = \infty, 1, 2$.

Proof. Let k be an arbitrary positive number. Define $v(x, t)$ to be

$$v = u\, e^{-kt}. \tag{52}$$

Then v satisfies the equation

$$v_t + kv = v_{xx}. \tag{50'}$$

Since $u(x, t)$ was assumed to $\rightarrow 0$ as $|x| \rightarrow \infty$, the same is true of $v(x, t)$. It follows that in the strip $0 \leq t \leq T, -\infty < x < \infty$ the function $|v(x, t)|$ takes on its maximum. We claim that this is at a point where $t = 0$. Say that the maximum occurs where $t = T$. If $v(x, T) > 0$ at this point, then the first term on the left in (50') is ≥ 0, and the second term on the left is > 0, while the term v_{xx} on the right is ≤ 0. At a negative minimum we find an analogous contradiction. So it follows that

$$\max_{0 \leq t \leq T, x} |v(x, t)| = \max_x |v(x, 0)|.$$

This shows that v satisfies property (i) of theorem 13. Letting $k \rightarrow 0$ in the definition (52) of v shows that also u satisfies (i) :

$$|\mathbf{S}(T)| \leq 1, \qquad \mathbf{S} : L^\infty \rightarrow L^\infty.$$

(ii) Consider the space of all solutions $w(x, t)$ of the *backward heat* equation

$$w_t = -w_{xx},\tag{53}$$

defined for $0 \le t \le T$, and all x, which tends to zero sufficiently rapidly as $|x| \to \infty$. Multiply equation (50) by w, (53) by u, and add; the result can be written as

$$(uw)_t = wu_{xx} - uw_{xx}.$$

Integrate this with respect to x on $\mathbb{R}$; integrate by parts. The fact that v, w tend to zero as $|x| \to \infty$ shows that the integral of the right is zero. So we get

$$0 = \int (uw)_t dx = \frac{d}{dt} \int uw \, dx;$$

that is, $\int u\,w\,dx = (u(t), w(t))$ is independent of t, in particular,

$$(u(0), w(0)) = (u(T), w(T)).\tag{54}$$

Denote the inintial value $u(0)$ by f; in the notation (51), $u(T) = S(T)f$. Similarly denote the final value $w(T)$ by g. Analogously to what we have shown about solutions of (50), for $t < T$, $w(t)$ is completely determined by $w(T)$, and there is a linear relation between $w(T)$ and $w(0)$ that we denote by S':

$$w(0) = S'(T)g.$$

We rewrite (54) in this new notation as

$$(f, S'(T)g) = (S(T)f, g).\tag{55}$$

The bracket (u, w) is a bilinear function: for fixed w, it is a linear functional of u and for fixed u, a linear functional of w. Thus (55) says that S and S' are *transposes* of each other with respect to this bilinear pairing.

It is easy to verify that

$$|u|_{L^1} = \sup_{|w|_{max}=1} |(u, w)|.$$

According to part (i), $|S'(T)g|_{max} \le |g|_{max}$, so we deduce from (55) that $|S(T)f|_{L^1} \le |f|_{L^1}$, as asserted in part (ii).

Part (iii), the boundedness of $S : L^2 \to L^2$, follows from the Marcel Riesz convexity theorem, theorem 4 above. $\square$

REMARK 3. Here is another, direct proof for part (iii). Multiply equation (50) by $2u$ and integrate with respect to x over $\mathbb{R}$. Integrate by parts on the right. By the fact that $u(x, t) \to 0$ as $|x| \to \infty$, we get

$$\frac{d}{dt} \int u^2 dx = -\int u_x^2 dx.$$

This shows that $\int u^2(x, t)\, dx$ is a decreasing function of t, from which part (iii) follows.

A similar direct proof can be given for part (ii). Let $x_j(t)$ be the points where $u(x, t)$ changes sign:

$$u(x, t) \begin{cases} > 0 & \text{for } x_j < x < x_{j+1}, \ j \text{ even} \\ < 0 & \text{for } x_j < x < x_{j+1}, \ j \text{ odd.} \end{cases} \tag{56}$$

Then

$$|u(t)|_{L^1} = \sum (-1)^j \int_{x_j(t)}^{x_{j+1}(t)} u(x, t)\, dx. \tag{57}$$

Differentiate this with respect to t. Using calculus and equation (50), we get

$$\frac{d}{dt}|u(t)|_{L^1} = \sum (-1)^j \int_{x_j}^{x_{j+1}} u_t\, dx = \sum (-1)^j \int_{x_j}^{x_{j+1}} u_{xx}$$

$$= \sum (-1)^j (u_x(x_{j+1}) - u_x(x_j)). \tag{57'}$$

It follows from (56) that the first x-derivative of u alternates in sign at the points x_j:

$$u_x(x_j, t) \begin{cases} \geq 0 & \text{for } j \text{ even} \\ \leq 0 & \text{for } j \text{ odd;} \end{cases}$$

therefore the right side of (57') is ≤ 0. This shows that $|u(t)|_{L^1}$ is a decreasing function of t, as asserted in part (ii). ☐

We give now yet another proof of theorem 13; the initial value problem for (50) can be solved explicitly:

$$u(x, t) = \frac{1}{2\sqrt{\pi t}} \int f(y)\, e^{-(x-y)^2/4t}\, dy.$$

This shows that S is an integral operator whose kernel is $\exp\{(x - y)^2/4t\}/2\sqrt{\pi t}$. We appeal to parts (iii) and (iv) of theorem 1 to prove parts (i) and (ii) of theorem 13, and to theorem 3 to prove part (iii). ☐

Theorem 13 holds for second-order parabolic equations in any number of space variables; the proofs sketched above apply to the general case, except of course the last one based on the explicit formula for the solution.

16.6 SINGULAR INTEGRAL OPERATORS, PSEUDODIFFERENTIAL OPERATORS AND FOURIER INTEGRAL OPERATORS

The above-named classes of operators play a dominant role in modern analysis, in particular, the modern theory of partial differential equations. They extend enormously the class of traditional integral operators; they unify integral and differential operators. In particular, inverses of various differential operators can be expressed in terms of such operators.

For the theory of these operators we refer to the literature, in particular, to Hörmander and Taylor. We call attention to a particularly sharp result concerning the L^2 boundedness of pseudodifferential operators due to David and Journée.

BIBLIOGRAPHY

David, G. and Journé, J-L. A boundedness criterion for generalized Calderon-Zygmund operators. *An. Math.*, **20** (1984): 371–397.

Halmos, P. R. and Sunder, V. S. *Bounded Integral Operators on L^p Spaces*. Ergebnisse der Math. und ihrer Grenzgebiete, **XV**. Springer, Berlin, 1978.

Hardy, G. H., Littlewood, J. E., and Polya, G. *Inequalities*, Cambridge University Press, Cambridge, 1934.

Hausdorff, F. Eine Ausdehnung des Parsevalschen Satzes über Fourierreihen. *Math. Zeitschrift* **16** (1923): 163–169.

Hörmander, L. *The Analysis of Linear Partial Differential Operators*. Springer Verlag, 1983.

Riesz, M. Sur les maxima des formes bilinéaires et sur les fonctionelles linéaires. *Acto Math.*, **49** (1926): 465–497.

Riesz, M. Sur les functions conjugées. *Math. Zeit.*, **27** (1927): 218–244.

Schur, I. Bemerkungen zur Theorie der beschränkten Bilinearformen mit unendichen vielen Veränderlichen. *J. für Math.*, **140** (1911): 1–28.

Taylor, M. E. *Pseudodifferential Operators*. Princeton University Press, Princeton, NJ, 1981.

Thorin, G. O. Convexity theorems generalizing those of M. Riesz and Hadamard with some applications. *Seminar Math. Lund*, **9** (1948).

Young, W. H. On the determination of the summability of a function by means of its Fourier coefficients. *Proc. London Math. Soc.*, **12** (1913): 71–88.

Weyl, H. *Singuläre Integralgleichungen mit besonderer Berücksichtigung des Fourierschen Integraltheorems*. Collected Works. Springer, Berlin.

17

BANACH ALGEBRAS AND THEIR ELEMENTARY SPECTRAL THEORY

17.1 NORMED ALGEBRAS

We saw in chapter 15 how to multiply, by composition, two linear maps of Banach spaces into other Banach spaces, provided that the target space of the first map is the domain space of the second map. In this chapter we specialize to the study of bounded, linear maps of a Banach space X into itself. Any two such maps may be composed so that the set $\mathcal{L}(X, X)$ of such maps forms an algebra, with a unit. Each element of $\mathcal{L}(X, X)$ has a norm, with the properties embodied in theorem 2 and theorem 8; that is to say, the norm is subadditive and submultiplicative. Such an algebra is called a *normed algebra*.

There is a group of important results about bounded linear maps of a Banach space X into itself that depend only on the algebraic and analytic structure of $\mathcal{L}(X, X)$. In this chapter and in chapter 18, we derive these results in the context of normed algebras.

Definition. A *normed algebra* is an associative algebra over the complex numbers. Each element has a positive norm, $|\mathbf{M}| = 0$ only for $\mathbf{M} = 0$. The norm satisfies the usual conditions:

$$|\mathbf{M} + \mathbf{N}| \le |\mathbf{M}| + |\mathbf{N}|, \quad |c\mathbf{M}| = |c|\,|\mathbf{M}|, \quad \text{and} \quad |\mathbf{NM}| \le |\mathbf{N}|\,|\mathbf{M}|. \tag{1}$$

A normed algebra with a unit is just that, where the norm of the unit is 1:

$$|\mathbf{I}| = 1. \tag{2}$$

Definition. A *normed algebra* $\mathcal{L}$ that is *complete* with respect to the norm is called a *Banach algebra.*

The theorems contained in this section are valid for all Banach algebras $\mathcal{L}$ with a unit, not just for $\mathcal{L}(X, X)$.

Definition. An element **M** of a Banach algebra $\mathcal{L}$ with a unit is called *invertible* if it has an inverse $\mathbf{N} = \mathbf{M}^{-1}$ in $\mathcal{L}$:

$$\mathbf{NM} = \mathbf{MN} = \mathbf{I}. \tag{3}$$

M is said to have a *left inverse* **A**, respectively a *right inverse* **B**, if

$$\mathbf{AM} = \mathbf{I}, \quad \mathbf{I} = \mathbf{MB}. \tag{4}$$

It is an elementary fact of algebra that if **M** has both a left inverse **A** and a right inverse **B**, then these are equal. For multiply the first relation in (4) by **B** on the right:

$$\mathbf{AMB} = \mathbf{B}. \tag{5}$$

Using *associativity* and the second relation (4) we get $\mathbf{A} = \mathbf{B}$.

Theorem 1.

(i) *If* **M** *and* **K** *in* $\mathcal{L}$ *are invertible, so is their product* **MK**, *and*

$$(\mathbf{MK})^{-1} = \mathbf{K}^{-1}\mathbf{M}^{-1}. \tag{6}$$

(ii) *If* **M** *and* **K** *commute,*

$$\mathbf{MK} = \mathbf{KM}, \tag{7}$$

and if their product is invertible, so are **M** *and* **K** *separately.*

Proof. (i) is an obvious consequence of associativity. To show (ii), denote the inverse of **MK** by **N**:

$$(\mathbf{MK})\mathbf{N} = \mathbf{I} = \mathbf{N}(\mathbf{MK}).$$

By **N** and **K**'s associativity, we conclude that **KN** is a *right inverse* of **M**. By commutativity of **M** and **K** and associativity, we get

$$\mathbf{I} = \mathbf{N}(\mathbf{MK}) = \mathbf{N}(\mathbf{KM}) = (\mathbf{NK})\mathbf{M},$$

from which we conclude that **NK** is a *left inverse* of **M**. So **M** is invertible. $\square$

Theorem 1 is purely algebraic; not so

Theorem 2. *Suppose that* **K** *in* $\mathcal{L}$ *is invertible; then so are all elements of* $\mathcal{L}$ *close enough to* **K**. *Specifically all elements of form* $\mathbf{L} = \mathbf{K} - \mathbf{A}$ *are invertible, provided that*

$$|\mathbf{A}| < \frac{1}{|\mathbf{K}^{-1}|}. \tag{8}$$

Proof. We treat first the special case $\mathbf{K} = \mathbf{I}$; we claim that all elements of the form $\mathbf{I} - \mathbf{B}$ are invertible, provided that

$$|\mathbf{B}| < 1. \tag{9}$$

The inverse of $\mathbf{I} - \mathbf{B}$ is given by the geometric series

$$\sum_{0}^{\infty} \mathbf{B}^n = \mathbf{S}. \tag{9'}$$

Clearly, since $|\mathbf{B}| < 1$, the sequence of partial sums is a Cauchy sequence; since $\mathcal{L}$ is complete, the series converges. It follows from (1) that a convergent series can be multiplied termwise; multiplying $(9')$ on the left by $\mathbf{B}$, we get

$$\mathbf{BS} = \mathbf{B} \sum_{0}^{\infty} \mathbf{B}^n = \sum_{1}^{\infty} \mathbf{B}^k = \mathbf{S} - \mathbf{I},$$

from which it follows that $(\mathbf{I} - \mathbf{B})\mathbf{S} = \mathbf{I}$. Similarly, multiplying $(9')$ by $\mathbf{B}$ on the right shows that $\mathbf{S}(\mathbf{I} - \mathbf{B}) = \mathbf{I}$. This shows that $\mathbf{S}$ is the inverse of $\mathbf{I} - \mathbf{B}$.

We return now to (8); we factor

$$\mathbf{K} - \mathbf{A} = \mathbf{K}\left(\mathbf{I} - \mathbf{K}^{-1}\mathbf{A}\right). \tag{10}$$

Set $\mathbf{B} = \mathbf{K}^{-1}\mathbf{A}$; by submultiplicativity, and by inequality (8),

$$|\mathbf{B}| = \left|\mathbf{K}^{-1}\mathbf{A}\right| \leq \left|\mathbf{K}^{-1}\right| \, |\mathbf{A}| < 1.$$

Using $(9')$, we invert (10):

$$(\mathbf{K} - \mathbf{A})^{-1} = \left(\mathbf{I} - \mathbf{K}^{-1}\mathbf{A}\right)^{-1}\mathbf{K}^{-1} = \sum_{0}^{\infty} \left(\mathbf{K}^{-1}\mathbf{A}\right)^n \mathbf{K}^{-1}. \tag{10'}$$

This proves that $(\mathbf{K} - \mathbf{A})$ is invertible. $\qquad\qquad\square$

Definition. The *resolvent set* of $\mathbf{M}$ in $\mathcal{L}$ consists of those complex numbers λ for which

$$\lambda\mathbf{I} - \mathbf{M}$$

is invertible; the *spectrum* of $\mathbf{M}$ consists of those λ for which it is not. The resolvent set of $\mathbf{M}$ is denoted by $\rho(\mathbf{M})$, its spectrum by $\sigma(\mathbf{M})$.

In chapter 11, section 11.4, we defined the notion of an analytic function of a complex variable whose values lie in a Banach space over $\mathbb{C}$. Since a Banach algebra is, in particular, a Banach space over $\mathbb{C}$, we may speak of analytic functions whose values lie in a Banach algebra. As the reader may immediately verify, the product of two such analytic functions is analytic. All the standard paraphernalia of the theory of analytic functions—the Cauchy integral theorem, the Cauchy integral formula, power series, Laurent series, and so on—are meaningful and valid for functions that take their values in a Banach algebra.

Theorem 3.

(i) *The resolvent set $\rho(M)$ is an open subset of $\mathbb{C}$.*

(ii) *The resolvent of M, defined on $\rho(M)$ as $(\zeta I - M)^{-1}$, abbreviated as $(\zeta - M)^{-1}$, is an analytic function of ζ on $\rho(M)$.*

Proof. Suppose that λ is in $\rho(M)$; then by theorem 2 applied to $K = \lambda I - M$ and $A = hI$,

$$(\lambda - h)I - M = (\lambda I - M - hI)$$

is invertible for h small enough. This proves part (i).

By formula $(10')$,

$$((\lambda - h) - M)^{-1} = \sum_0^\infty (\lambda - M)^{n-1} h^n; \tag{11}$$

this shows that the resolvent can be expanded in a *power series* around each point λ of $\rho(M)$, convergent for $|h| < |(\lambda - M)^{-1}|^{-1}$, this proves analyticity, as asserted in part (ii). $\square$

The series (11) converges when $|h|$ is less than $|(\lambda - M)^{-1}|^{-1}$. From this we deduce

Corollary 3'. *For any λ in $\rho(M)$, denote by $d(\lambda)$ the distance of λ to the spectrum of M. Then*

$$|(\lambda - M)^{-1}| \geq d^{-1}. \tag{11'}$$

Theorem 4 (Gelfand).

(i) *The spectrum $\sigma(M)$ is a closed, bounded, nonempty set in $\mathbb{C}$.*

(ii) *The spectral radius of M, denoted as $|\sigma(M)|$, is defined as*

$$|\sigma(M)| = \max_{\lambda \text{ in } \sigma(M)} |\lambda|. \tag{12}$$

We claim that

$$|\sigma(M)| = \lim_{k \to \infty} |M^k|^{1/k}. \tag{12'}$$

Proof. Since $\rho(\mathbf{M})$ is open, its complement, $\sigma(\mathbf{M})$, is closed. Applying (9), (9′) to $\mathbf{A} = \zeta^{-1}\mathbf{M}$, we see that

$$(\zeta\mathbf{I} - \mathbf{M})^{-1} = \zeta^{-1}\left(\mathbf{I} - \mathbf{M}\zeta^{-1}\right)^{-1} = \sum_0^\infty \mathbf{M}^n \zeta^{-n-1} \tag{13}$$

converges for $|\zeta^{-1}\mathbf{M}| < 1$, that is, for $|\zeta| > |\mathbf{M}|$. This proves that every such ζ belongs to $\rho(\mathbf{M})$; it follows that every λ in $\sigma(\mathbf{M})$ satisfies $|\lambda| \le |\mathbf{M}|$. This proves that the spectrum is bounded.

Representation (13) is a Laurent series for the resolvent around ∞; the first term is $\zeta^{-1}\mathbf{I}$. Integrating (13) with respect to ζ around the contour $C : |\zeta| = c, c > |\mathbf{M}|$ gives

$$\oint (\zeta - \mathbf{M})^{-1}\, d\zeta = \mathbf{I}, \qquad \text{where } d\zeta = \frac{1}{2\pi i}\, d\zeta. \tag{14}$$

If the spectrum of $\mathbf{M}$ were empty, then $(\zeta - \mathbf{M})^{-1}$ would be, by part (ii) of theorem 3, an everywhere regular analytic function. Then by the Cauchy integral theorem, applicable to analytic functions in a Banach space, the integral on the left in (14) would be zero; since the right side isn't, this proves that $\sigma(\mathbf{M})$ is not empty, a result due to A. E. Taylor.

We investigate now more precisely the radius of convergence of the series (13). Let k be any integer; then we can decompose $n = kq + r, 0 \le r < k$. So $\mathbf{M}^n = \mathbf{M}^{kq+r} = (\mathbf{M}^k)^q \mathbf{M}^r$ from which we deduce that

$$|\mathbf{M}^n| \le |\mathbf{M}^r|\,|\mathbf{M}^k|^q.$$

This gives the estimate

$$\left|\sum \frac{\mathbf{M}^n}{\zeta^{n+1}}\right| \le \sum \frac{|\mathbf{M}^n|}{|\zeta|^{n+1}} \le \left(\sum_0^{k-1} \frac{|\mathbf{M}^r|}{|\zeta|^{r+1}}\right) \sum_q \left(\frac{|\mathbf{M}^k|}{|\zeta|^k}\right)^q.$$

Thus the series (13) converges absolutely if

$$1 > \frac{|\mathbf{M}|^k}{|\zeta|^k}, \qquad \text{that is, } |\zeta| > |\mathbf{M}^k|^{1/k}. \tag{15}$$

Thus every ζ satisfying (15) belongs to the resolvent set; it follows that every λ in $\sigma(M)$ satisfies $|\lambda| \le |\mathbf{M}^k|^{1/k}$. In view of definition (12) it follows that $|\sigma(\mathbf{M})| \le |\mathbf{M}^k|^{1/k}$; since this holds for all integers k,

$$|\sigma(\mathbf{M})| \le \liminf_{k\to\infty} |\mathbf{M}^k|^{1/k}. \tag{16}$$

We turn now again to representation (13) of the resolvent and express the coefficient of ζ^{-n-1} by the Cauchy integral formula:

$$\oint (\zeta - \mathbf{M})^{-1}\zeta^n\, d\zeta = \mathbf{M}^n. \tag{17}$$

As path of integration we may choose any contour C in the resolvent set of $\mathbf{M}$ that winds once around $\sigma(\mathbf{M})$. It follows from the definition (12) that $|\zeta| = |\sigma(\mathbf{M})| + \delta$ is such a contour. We may then estimate from (17):

$$|\mathbf{M}^n| \leq c\,(|\sigma(\mathbf{M})| + \delta)^{n+1}\,, \qquad c = \max_{|\zeta| = \sigma(\mathbf{M}) + \delta} |(\zeta\mathbf{I} - \mathbf{M})^{-1}|.$$

Take the nth root,

$$|\mathbf{M}^n|^{1/n} \leq c^{1/n}\,(|\sigma(\mathbf{M})| + \delta)^{1+1/n}$$

and then form the lim sup,

$$\limsup_{n\to\infty} |\mathbf{M}^n|^{1/n} \leq |\sigma(\mathbf{M}| + \delta.$$

Since this is true for any $\delta > 0$, it is true for $\delta = 0$;

$$\limsup |\mathbf{M}^n|^{1/n} \leq |\sigma(\mathbf{M})|. \tag{18}$$

Comparing (16) and (18), we conclude that the lim inf and lim sup are equal, and we obtain Gelfand's formula (12′) for the spectral radius. □

17.2 FUNCTIONAL CALCULUS

Since $\mathcal{L}$ is an algebra, we can form any *polynomial* p of an element $\mathbf{M}$ of $\mathcal{L}$ by setting

$$p(\mathbf{M}) = \sum^{N} a_j \mathbf{M}^j. \tag{19}$$

(19) defines a mapping from the algebra of polynomials into the algebra $\mathcal{L}$ that is, clearly, a homomorphism. This homomorphism can be extended to a larger class of functions than polynomials; for instance, we can define

$$e^{\mathbf{M}} = \sum \frac{\mathbf{M}^n}{n!}.$$

More generally, we can define

$$f(\mathbf{M}) = \sum a_n \mathbf{M}^n \tag{20}$$

for any *entire* function

$$f(\zeta) = \sum a_n \zeta^n. \tag{20′}$$

Still more generally it follows from (12′) that we can define (20) for any function $f(\zeta)$ whose power series converges in a circle whose radius exceeds $|\sigma(\mathbf{M})|$. We propose now a still further extension:

Definition. Let $\mathbf{M}$ be an element of $\mathcal{L}$, $f(\zeta)$ a function analytic in a domain G containing $\sigma(\mathbf{M})$. Let C be a contour in $G \cap \rho(\mathbf{M})$ that winds once around every point in $\sigma(\mathbf{M})$ but winds zero time around any point of the complement of G. We define

$$f(\mathbf{M}) = \oint (\zeta - \mathbf{M})^{-1} f(\zeta) d\zeta. \qquad (21)$$

By the Cauchy integral theorem, (21) is independent of the choice of the contour.

Theorem 5.

(i) *For f a polynomial, definitions (21) and (19) are the same.*

(ii) *The mapping (21) from the algebra of functions analytic on an open set containing $\sigma(\mathbf{M})$ into $\mathcal{L}$ is a homomorphism.*

(iii)

$$\sigma(f(\mathbf{M})) = f(\sigma(\mathbf{M})). \qquad (22)$$

(iv) *Let f be analytic on an open set containing $\sigma(\mathbf{M})$, and g analytic on an open set containing $f(\sigma(\mathbf{M}))$. Denote their composite by h,*

$$h(\zeta) = g(f(\zeta)); \qquad (23)$$

then

$$h(\mathbf{M}) = g(f(\mathbf{M})). \qquad (23')$$

Proof. (i) Replacing $f(\zeta)$ by a polynomial in (21) and using formula (17) shows that (21) and (19) are the same. The same argument shows that (21) is the same as (20) for f analytic in a disk of radius $> \sigma|(\mathbf{M})|$.

(ii) For any pair of complex numbers ζ and ω,

$$(\zeta\mathbf{I} - \mathbf{M}) - (\omega\mathbf{I} - \mathbf{M}) = (\zeta - \omega)\mathbf{I}.$$

Suppose that both ζ and ω belong to $\rho(\mathbf{M})$. Multiply the above identity by $(\zeta - \mathbf{M})^{-1}(\omega - \mathbf{M})^{-1}(\zeta - \omega)^{-1}$:

$$(\zeta - \omega)^{-1}[(\omega - \mathbf{M})^{-1} - (\zeta - \mathbf{M})^{-1}] = (\zeta - \mathbf{M})^{-1}(\omega - \mathbf{M})^{-1}. \qquad (24)$$

Relation (24) is called the *resolvent identity.*

The mapping $f \to f(\mathbf{M})$ given by (21) is obviously linear. We show now that it is multiplicative. Let f, g both be functions analytic in an open set $G \supset \sigma(\mathbf{M})$. We choose two contours C and D, both in $G \cap \rho(\mathbf{M})$ so that they have no point in common, and so that D lies inside C. That is, C winds once around every point ω of D, while D winds zero times around every point ζ of C. Using definition (21) for

f and g with contours C and D, we write $f(\mathbf{M})g(\mathbf{M})$ as a product of two integrals, which we express as a double integral. Then we use the resolvent identity (24):

$$
\begin{aligned}
f(\mathbf{M})g(\mathbf{M}) &= \oint\oint (\zeta - \mathbf{M})^{-1}(\omega - \mathbf{M})^{-1} f(\zeta)g(\omega)\,d\zeta\,d\omega \\
&= \oint\oint (\zeta - \omega)^{-1}[(\omega - \mathbf{M})^{-1} - (\zeta - \mathbf{M})^{-1}]f(\zeta)g(\omega)\,d\zeta\,d\omega \\
&= \oint\left[\oint (\zeta - \omega)^{-1}f(\zeta)\,d\zeta\right](\omega - \mathbf{M})^{-1}g(\omega)\,d\omega \\
&\quad - \oint\left[\oint(\zeta - \omega)^{-1}g(\omega)\,d\omega\right](\zeta - \mathbf{M})^{-1}f(\zeta)\,d\zeta.
\end{aligned}
\tag{25}
$$

Since C winds once around every point ω of D, the integral with respect to ζ above in the first term is, by the Cauchy integral formula, $f(\omega)$. Since D does not wind around any point ζ on C, the ω integration in the second term above is zero; so we conclude from (25) that

$$
f(\mathbf{M})g(\mathbf{M}) = \oint (\omega - \mathbf{M})^{-1} f(\omega)g(\omega)\,d\omega,
$$

which by (21) is $h(\mathbf{M})$, where $h(\omega) = f(\omega)g(\omega)$. This proves that the mapping (21) is multiplicative.

(iii) We have to show that μ belongs to the spectrum of $f(\mathbf{M})$ if and only if μ is of the form

$$
\mu = f(\lambda), \qquad \lambda \text{ in } \sigma(\mathbf{M}).
\tag{26}
$$

If μ is not of form (26), then $f(\zeta) - \mu$ does not vanish on $\sigma(\mathbf{M})$. Therefore $(f(\zeta) - \mu)^{-1} = g(\zeta)$ is analytic in an open set containing $\sigma(\mathbf{M})$, thus we may define $g(\mathbf{M})$ by formula (21). According to part (ii), $[f(\mathbf{M}) - \mu - \mathbf{I}]\,g(\mathbf{M}) = h(\mathbf{M})$, where $h(\zeta) = (f(\zeta) - \mu)\,g(\zeta) \equiv 1$. Thus $h(\mathbf{M}) = \mathbf{I}$, and $g(\mathbf{M})$ is the inverse of $f(\mathbf{M}) - \mu\mathbf{I}$. This proves that μ does not lie in $\sigma(f(\mathbf{M}))$.

On the other hand, suppose that μ is of form (26). Define the function $k(\zeta)$ by

$$
k(\zeta) = \frac{f(\zeta) - f(\lambda)}{\zeta - \lambda}.
$$

Clearly, k is analytic in an open set containing $\sigma(\mathbf{M})$, so $k(\mathbf{M})$ can be defined by (21). Since $(\zeta - \lambda)\,k(\zeta) = f(\zeta) - f(\lambda)$, it follows from part (ii) that

$$
(\mathbf{M} - \lambda\mathbf{I})\,k(\mathbf{M}) = f(\mathbf{M}) - f(\lambda)\mathbf{I}.
\tag{27}
$$

Since λ belongs to $\sigma(\mathbf{M})$, the first factor is not invertible. We appeal now to part (ii) of theorem 1, according to which the product $f(\mathbf{M}) - f(\lambda)\mathbf{I}$ is not invertible either.

(iv) By assumption, $g(\omega)$ is analytic on $f(\sigma(\mathbf{M}))$. Since by part (iii) the spectrum of $f(\mathbf{M})$ is $f(\sigma(\mathbf{M}))$, it follows that formula (21) can be applied to g in place of f, and $f(\mathbf{M})$ in place of $\mathbf{M}$ and D in place of C:

$$g(f(\mathbf{M})) = \oint (\omega - f(\mathbf{M}))^{-1} g(\omega) \, d\omega, \tag{28}$$

For ω on D, $(\omega - f(\zeta))^{-1}$ is an analytic function on $\sigma(\mathbf{M})$; therefore applying formula (21) once more, we get

$$(\omega \mathbf{I} - f(\mathbf{M}))^{-1} = \oint (\zeta - \mathbf{M})^{-1} (\omega - f(\zeta))^{-1} d\zeta, \tag{29}$$

provided that the contour C does not wind around any of the points ω on D. Now set (29) into (28):

$$g(f(\mathbf{M})) = \oint \oint (\zeta - \mathbf{M})^{-1} (\omega - f(\zeta))^{-1} g(\omega) d\zeta \, d\omega. \tag{30}$$

We reverse the order of integration; since C does not wind around points of D, it follows that D winds around every point ζ of C. By the Cauchy integral formula,

$$\oint (\omega - f(\zeta))^{-1} g(\omega) d\omega = g(f(\zeta)) = h(\zeta),$$

where we have used (23). Setting this in (30) on the right we get, by (21), $h(\mathbf{M})$, as asserted in (23′). □

Definition (21) and properties listed in theorem 5 are called the *functional calculus* for operators. Relation (22) is called the *spectral mapping theorem.*

Suppose that the spectrum of $\mathbf{M}$ can be decomposed as the union of n *pairwise disjoint closed components:*

$$\sigma(\mathbf{M}) = \sigma_1 \cup \cdots \cup \sigma_N, \qquad \sigma_j \cap \sigma_k = \phi. \tag{31}$$

For each j, denote by C_j a contour in the resolvent of $\mathbf{M}$ that winds once around each point of σ_j but not σ_k, $k \neq j$. We define

$$\mathbf{P}_j = \oint_j (\zeta - \mathbf{M})^{-1} d\zeta. \tag{32}$$

Theorem 6.

(i) The $\mathbf{P}_j$ are disjoint projections, that is,

$$\mathbf{P}_j^2 = \mathbf{P}_j \quad and \quad \mathbf{P}_j \mathbf{P}_k = 0 \quad for \ j \neq k. \tag{33}$$

(ii)

$$\sum_j \mathbf{P}_j = \mathbf{I}. \tag{34}$$

(iii) $\mathbf{P}_m \neq 0$ if σ_n is not empty.

Proof. Relations (33) are corollaries of part (ii) of theorem 5. Since $C = \sum C_j$ winds once around every point of $\sigma(M)$, (34) follows by summing (32) over all j, and using (14). We leave the proof of part (iii) to the diligent reader. $\square$

Exercise 1. Show that if $\mathbf{P}$ is a nonzero projection, that is, satisfies $\mathbf{P}^2 = \mathbf{P} \neq 0$, then

$$|\mathbf{P}| \geq 1. \tag{35}$$

Exercise 2. Show that the spectral radius $|\sigma(\mathbf{M})|$ depends upper semicontinuously on $\mathbf{M}$ in the norm topology, namely, that if $\lim \mathbf{M}_n = \mathbf{M}$, then

$$\lim \sup |\sigma(\mathbf{M}_n)| \leq |\sigma(\mathbf{M})|.$$

Exercise 3. Show that $|\exp \mathbf{M}| \leq \exp |\mathbf{M}|$.

Exercise 4. Show that if 0 does not belong to the $\sigma(\mathbf{M})$, and if 0 can be connected to ∞ by curve that lies in $\rho(\mathbf{M})$, then $\log(\mathbf{M})$ can be defined so that

$$\exp \log(\mathbf{M}) = \mathbf{M}.$$

Exercise 5. Define $\mathcal{L}_{\mathbf{M}}$ to be the closure of the algebra generated by $\mathbf{M}$ and $(\zeta - \mathbf{M})^{-1}$, ζ in $\rho(\mathbf{M})$. Show that $\mathcal{L}_{\mathbf{M}}$ is a commmutative subalgebra of $\mathcal{L}$.

NOTE. For the history of the spectral theory of operators in a Banach space, see pp. 607–609 of Dunford and Schwartz.

The term "spectrum" is due to Hilbert, a remarkable anticipation of its meaning in quantum mechanics.

BIBLIOGRAPHY

Dunford, N. Spectral theory I, Convergence to projections. *Trans. AMS*, **54** (1943): 185–217.

Gelfand, I. M. Normierte Ringe. *Mat. Sbornik, N.S.*, **51** (1941): 3–24.

Lorch, E. R. The spectrum of linear transformations. *Trans. AMS*, **54** (1942): 238–248.

Nagumo, M. Einige analytische Untersuchungen in linearen metrischen Ringen. *Jap. J. Math.*, **13** (1936): 61–80.

Riesz, F. Sur certaines systemes singuliers d'équation integrales. *An. École Norm. Sup. (3)*, **28** (1911): 33–62.

Taylor, A. E. The resolvent of a closed transformation. *Bull. AMS*, **44** (1938): 70–74.

Wiener, N. Note on a paper of M. Banach. *Fund. Math.*, **4** (1923): 136–143.

18

GELFAND'S THEORY OF COMMUTATIVE BANACH ALGEBRAS

A Banach algebra, defined in chapter 17, is an associative, complete normed algebra over $\mathbb{C}$. The algebras $\mathcal{L}$ we deal with in this chapter are assumed to have a *unit*, denoted as $\mathbf{I}$, with $|\mathbf{I}| = 1$, and to be *commmutative:*

$$\mathbf{M}\mathbf{N} = \mathbf{N}\mathbf{M} \qquad \text{for all } \mathbf{N}, \mathbf{M} \text{ in } \mathcal{L}. \tag{1}$$

The main topic, as in chapter 17, is invertibility, approached here through two concepts that turn out to be equivalent: *multiplicative functionals* and *maximal ideals*.

Definition. A *multiplicative functional* p in a Banach algebra $\mathcal{L}$ is a *homomorphism* of $\mathcal{L}$ into $\mathbb{C}$.

Although defined purely algebraically, homomorphisms of a commutative Banach algebra with a unit have the following analytic property:

Theorem 1. *Every homomorphism p of a commutative Banach algebra with unit into $\mathbb{C}$ is a contraction, that is, satisfies*

$$|p(\mathbf{M})| \leq |\mathbf{M}|. \tag{2}$$

Proof. Since $\mathbf{M} = \mathbf{I}\mathbf{M}$ for every $\mathbf{M}$, and since p is a homomorphism,

$$p(\mathbf{M}) = p(\mathbf{I}\mathbf{M}) = p(\mathbf{I})\, p(\mathbf{M}).$$

It follows from this that, unless $p \equiv 0$, a trivial case,

$$p(\mathbf{I}) = 1. \tag{3}$$

Let $\mathbf{K}$ be an invertible element of $\mathcal{L}$, that is, $\mathbf{K}\,\mathbf{N} = \mathbf{I}$. Then, by (3),

$$p(\mathbf{K})\,p(\mathbf{N}) = p(\mathbf{K}\,\mathbf{N}) = p(\mathbf{I}) = 1;$$

this proves

Lemma 2. *If $\mathbf{K}$ is invertible, $p(\mathbf{K}) \neq 0$.*

Suppose now that contrary to (2), $|p(\mathbf{M})| > |\mathbf{M}|$ for some $\mathbf{M}$; then

$$\mathbf{B} = \frac{\mathbf{M}}{p(\mathbf{M})} \qquad (4)$$

satisfies

$$|\mathbf{B}| < 1. \qquad (4')$$

It follows then from (9), (9') of theorem 2 in chapter 17 that

$$\mathbf{K} = \mathbf{I} - \mathbf{B}$$

is invertible. On the other hand, by (4) and (2),

$$p(\mathbf{K}) = p(\mathbf{I}) - p\left(\frac{\mathbf{M}}{p(\mathbf{M})}\right) = 1 - 1 = 0;$$

this contradicts the observation in lemma 2: if $\mathbf{K}$ is invertible, $p(\mathbf{K}) \neq 0$. Hence (2) holds for all $\mathbf{M}$. $\qquad\qquad\square$

The main result of this chapter is the converse of lemma 2.

Theorem 3. *An element $\mathbf{K}$ of a commutative Banach algebra $\mathcal{L}$ with a unit is invertible if and only if*

$$p(\mathbf{K}) \neq 0 \qquad (5)$$

for all homomorphisms p of $\mathcal{L}$ into $\mathbb{C}$.

Proof. As observed already in lemma 2, if $\mathbf{K}$ is invertible, then $p(\mathbf{K}) \neq 0$ for all homomorphisms p. What remains to be shown is the converse: if $\mathbf{K}$ is not invertible, there is a homomorphism $p : \mathcal{L} \to \mathbb{C}$ such that

$$p(\mathbf{K}) = 0. \qquad (6)$$

To construct such a p, we need some algebraic and analytic notions.

Definition. Let $\mathcal{L}$ be a commutative algebra with unit. A subset $\mathcal{I}$ of $\mathcal{L}$ is called an *ideal* if it has these three properties:

(i) $\mathcal{I}$ is a linear subspace of $\mathcal{L}$.

(ii) For any $\mathbf{M}$ in $\mathcal{L}$, $\mathbf{M}\mathcal{I} \subset \mathcal{I}$.

(iii) $\mathcal{I}$ is nontrivial, meaning that it is neither $\{0\}$ nor all of $\mathcal{L}$.

Note that an ideal cannot contain an invertible element $\mathbf{N}$. For then it would follow from (ii) that $\mathcal{I}$ contains every element of $\mathcal{L}$, contrary to (iii). In particular, $\mathcal{I}$ does not contain $\mathbf{I}$.

Lemma 4. *Let $\mathcal{L}$ and $\mathcal{A}$ be commutative algebras with unit over the same field, q a homorphic map of $\mathcal{L}$ onto $\mathcal{A}$. Suppose that q is nontrivial in this sense:*

(i) q is not an isomorphism.

(ii) $q(\mathcal{L})$ does not consist of $\{0\}$ only.

The kernel of the homomorphism q, consisting of all $\mathbf{K}$ in $\mathcal{L}$ mapped into 0 by q, is an ideal in $\mathcal{L}$. Conversely, every ideal $\mathcal{I}$ in $\mathcal{L}$ is the kernel of some nontrivial homomorphism.

Proof. It is easy to see that the kernel of q is an ideal. To show the converse, define $\mathcal{A}$ to be

$$\mathcal{A} = \mathcal{L}(\bmod \mathcal{I}) = \frac{\mathcal{L}}{\mathcal{I}},$$

meaning that $\mathcal{A}$ consists of *equivalence classes* of elements of $\mathcal{L}$, two elements $\mathbf{M}$ and $\mathbf{M}'$ being equivalent mod $\mathcal{I}$ if their difference belongs to $\mathcal{I}$:

$$\mathbf{M} \equiv \mathbf{M}' \bmod \mathcal{I} \quad \text{if } \mathbf{M} - \mathbf{M}' \in \mathcal{I}.$$

Addition and multiplication of equivalence classes is performed by picking arbitrary representatives of each class, adding or multiplying them, and then forming the class to which they belong.

The mapping q is taken as the natural assignment to each element $\mathbf{M}$ of $\mathcal{L}$ the class of all $\mathbf{M}'$ congruent to $\mathbf{M}$ mod $\mathcal{I}$. Clearly, the kernel of q is $\mathcal{I}$. $\qquad\square$

Lemma 4'. *$\mathcal{L}$, $\mathcal{A}$ and q as in lemma 4, $\mathcal{J}$ an ideal in $\mathcal{A}$. The inverse image of $\mathcal{J}$ is an ideal in $\mathcal{L}$.*

Proof is obvious.

We have noted earlier that an ideal contains no invertible elements. Conversely:

Lemma 5. *Every nonzero element $\mathbf{K}$ of $\mathcal{L}$ that is not invertible belongs to some ideal.*

Proof. That ideal is the *principal ideal* $\mathbf{K}\mathcal{L}$ generated by $\mathbf{K}$. It is easy to verify properties (i) and (ii); since $\mathbf{K}\mathcal{L}$ does not contain the identity $\mathbf{I}$, property (iii) holds too. $\qquad\square$

Definition. A *maximal ideal* is an ideal that is not contained in any other ideal.

Lemma 6. *Every ideal is contained in some maximal ideal.*

Proof. The ideals in $\mathcal{L}$ are *partially ordered* by inclusion. Let $\{\mathcal{I}_\alpha\}$ be a collection of *totally ordered* ideals. We claim that their *union* is an ideal: it is easy to see that properties (i) and (ii) hold. Since the identity $\mathbf{I}$ is not contained in any of the $\mathcal{I}_\alpha$, it is not contained in their union, either; this proves property (iii). We appeal now to Zorn's lemma to conclude that among all ideals containing a given one there is one that is maximal. $\square$

Combining lemma 5 and lemma 6 we deduce lemma 7.

Lemma 7. *A noninvertible element* $\mathbf{K}$ *of* $\mathcal{L}$ *belongs to some maximal ideal* $\mathcal{M}$.

Lemma 8. *Let* $\mathcal{M}$ *denote a maximal ideal of* $\mathcal{L}$. *Then* $\mathcal{A} = \mathcal{L}/\mathcal{M}$ *is a division algebra; that is, every nonzero element of* $\mathcal{A}$ *is invertible.*

Proof. If $\mathcal{A}$ contained a nonzero noninvertible element $\mathbf{C}$, $\mathcal{J} = \mathbf{C}\mathcal{A}$ would be an ideal contained in $\mathcal{A}$. Now consider the natural inclusion map $q : \mathcal{L} \to \mathcal{L}/\mathcal{M} = \mathcal{A}$. The inverse image $\mathcal{T}$ of $\mathcal{J}$ would be, according to lemma 4′, an ideal in $\mathcal{L}$, and it would properly contain $\mathcal{M}$, the inverse image of 0 in $\mathcal{A}$. Since $\mathcal{M}$ was assumed maximal, this is not possible. $\square$

We turn now to some analytical results:

Lemma 9. *The closure* $\overline{\mathcal{I}}$ *of an ideal* $\mathcal{I}$ *in a commutative Banach algebra* $\mathcal{L}$ *with unit is an ideal.*

Proof. It is easy to verify that $\overline{\mathcal{I}}$ has properties (i) and (ii), and that it does not consist of 0 alone. We claim that $\overline{\mathcal{I}}$ does not contain $\mathbf{I}$. Since $\mathcal{I}$ contains no invertible elements, and since by theorem 2 of chapter 17, all $\mathbf{N}$ contained in the open unit ball centered at $\mathbf{I}$ are invertible, $\mathbf{I}$ does not belong to $\overline{\mathcal{I}}$. $\square$

Lemma 10. *Every maximal ideal* $\mathcal{M}$ *in a commutative Banach algebra* $\mathcal{L}$ *with unit is closed.*

Proof. If $\mathcal{M}$ were not closed, its closure $\mathcal{M}$ would, by lemma 9, be an ideal, properly containing $\mathcal{M}$. This contradicts maximality. $\square$

Lemma 11. $\mathcal{L}$ *as above,* $\mathcal{I}$ *a closed ideal in* $\mathcal{L}$. *Then* $\mathcal{A} = \mathcal{L}/\mathcal{I}$ *is a Banach algebra in the natural norm of the quotient algebra.*

Exercise 1. Prove lemma 11.

The following result, due to Mazur, is the keystone to the sequence of lemmas above.

Theorem 12. *Let $\mathcal{A}$ be a Banach algebra with a unit that is a division algebra; then $\mathcal{A}$ is isomorphic to the field of complex numbers.*

Proof. As defined in chapter 17, the *spectrum* of an element K in $\mathcal{A}$ is the set of complex numbers ζ such that $\zeta \mathbf{I} - \mathbf{K}$ is not invertible, $\mathbf{I}$ being the unit of $\mathcal{A}$. According to theorem 4 of chapter 17, the spectrum of any $\mathbf{K}$ is *not empty*. This means that there is a complex number, call it κ, such that $\kappa \mathbf{I} - \mathbf{K}$ is *not* invertible. Since $\mathcal{A}$ is assumed to be a division algebra, this can only be if $\kappa \mathbf{I} - \mathbf{K}$ is the zero element in $\mathcal{A}$, in which case $\kappa \mathbf{I} = \mathbf{K}$. Thus every element $\mathbf{K}$ is a multiple of the unit; the mapping

$$\mathbf{K} \longrightarrow \kappa \tag{7}$$

is the isomorphism of $\mathcal{A}$ with $\mathbb{C}$. □

We are now ready to prove theorem 3; that is, given any noninvertible element $\mathbf{K}$ of $\mathcal{L}$, we construct a homomorphism $p : \mathcal{L} \to \mathbb{C}$ such that $p(\mathbf{K}) = 0$.

According to lemma 7, $\mathbf{K}$ belong to some maximal ideal $\mathcal{M}$; according to lemma 10, $\mathcal{M}$ is closed. According to lemma 11, $\mathcal{L}/\mathcal{M}$ is a Banach algebra, and according to lemma 8, $\mathcal{L}/\mathcal{M}$ is a division algebra. Then, by theorem 12, $\mathcal{L}/\mathcal{M}$ is isomorphic to $\mathbb{C}$. The composition

$$p_{\mathcal{M}} : \mathcal{L} \to \frac{\mathcal{L}}{\mathcal{M}} \to \mathbb{C} \tag{8}$$

is a homormophism of $\mathcal{L}$ onto $\mathbb{C}$, whose nullspace is $\mathcal{L}$. Since $\mathbf{K}$ belongs to $\mathcal{L}$,

$$p_{\mathcal{M}}(\mathbf{K}) = 0,$$

as asserted in theorem 3. □

We restate relation (8) as follows:

Theorem 13. *Let $\mathcal{L}$ denote a commutative Banach algebra with unit. To each maximal ideal $\mathcal{M}$ in $\mathcal{L}$ there corresponds a homomorphism $p_{\mathcal{M}}$ of $\mathcal{L} \to \mathbb{C}$ whose nullspace is $\mathcal{M}$:*

$$p_{\mathcal{M}}(\mathbf{K}) = 0 \quad \textit{iff } \mathbf{K} \textit{ in } \mathcal{M}.$$

Conversely, the null space of every homomorphism of $\mathcal{L}$ onto $\mathbb{C}$ is a maximal ideal.

We remark that the converse statement is a purely algebraic fact. We draw now some consequences of theorem 3:

Theorem 14. $\mathcal{L}$ *as above,* **N** *any element of* $\mathcal{L}$. *The spectrum of* **N** *is*

$$\sigma(\mathbf{N}) = \{p(\mathbf{N})\} \tag{9}$$

as p *ranges over all homomorphisms of* $\mathcal{L}$ *into* $\mathbb{C}$.

Proof. By definition of spectrum, ζ belongs to $\sigma(\mathbf{N})$ iff $\zeta\mathbf{I} - \mathbf{N}$ is not invertible. According to theorem 3 this is the case iff $p(\zeta\mathbf{I} - \mathbf{N}) = 0$ for some p. Since, by (3), $p(\mathbf{I}) = 1$, it follows that ζ lies in $\sigma(\mathbf{N})$ iff $\zeta = p(\mathbf{N})$ for some p. $\qquad\square$

We show next how to use the characterization of the spectrum contained in theorem 14 to give a new proof of the spectral mapping theorem. We recall the functional calculus developed in chapter 17; formula (21) there defines $f(\mathbf{M})$ as

$$f(\mathbf{M}) = \oint (\zeta - \mathbf{M})^{-1} f(\zeta) d\zeta \tag{10}$$

for every f analytic in an open set containing $\sigma(\mathbf{M})$. According to part (iii) of theorem 4 in chapter 17, equation (22),

$$\sigma(f(\mathbf{M})) = f(\sigma(\mathbf{M})). \tag{11}$$

The integral (10) is the limit in the sense of the norm of the usual partial sums employed in defining a Riemann integral. It follows that for any bounded linear map $\ell : \mathcal{L} \to \mathbb{C}$ we may apply ℓ inside the integral on the right in (10):

$$\ell(f(\mathbf{M})) = \oint \ell\left((\zeta - \mathbf{M})^{-1}\right) f(\zeta) d\zeta. \tag{10'}$$

In particular, (10') holds for every homomorphism p of $\mathcal{L} \to \mathbb{C}$:

$$p(f(\mathbf{M})) = \oint p\left((\zeta - \mathbf{M})^{-1}\right) f(\zeta) d\zeta. \tag{12}$$

Since p is a homomorphism,

$$p\left((\zeta - \mathbf{M})^{-1}\right) = (\zeta - p(\mathbf{M}))^{-1}.$$

Setting this into (12) gives

$$p(f(\mathbf{M})) = \oint (\zeta - p(\mathbf{M}))^{-1} f(\zeta) d\zeta. \tag{13}$$

It follows from theorem 14 that $p(\mathbf{M})$ belongs to $\sigma(\mathbf{M})$. By construction, the contour C winds once around every point of $\sigma(\mathbf{M})$; so by the Cauchy integral formula the right side of (13) equals $f(p(\mathbf{M}))$. This implies that

$$p(f(\mathbf{M})) = f(p(\mathbf{M})). \tag{14}$$

According to theorem 14, as p runs through all homomorphisms the left side of (14) fills up the spectrum of $f(M)$ while the right side fills up $f(\sigma(\mathbf{M}))$. So we conclude that (11) holds. □

The homomorphisms p constructed in this chapter can be regarded as functions of the associated maximal ideal $\mathcal{M}$ and of the element $\mathbf{N}$ of the algebra $\mathcal{L}$:

$$p = p(\mathcal{M}, \mathbf{N}). \tag{15}$$

For fixed $\mathbf{N}$, p is a function on the *space J* of *maximal ideals*.

Definition. The functions p defined above constitute the *Gelfand representation* of the commutative Banach algebra $\mathcal{L}$ on the space J of its maximal ideals.

Theorem 15.

  (i) *The Gelfand representation is a homomorphism of $\mathcal{L}$ into an algebra of complex-valued functions on the set J.*

  (ii) *The representation is a contraction:* $|p(\mathcal{M}, \mathbf{N})| \leq |\mathbf{N}|$.

 (iii) *The spectrum of $\mathbf{N}$ is the range of the function representing $\mathbf{N}$.*

 (iv) *The unit $\mathbf{I}$ is represented by $p(\mathcal{M}, I) \equiv 1$.*

  (v) *The functions p separate points of J; that is, given two distinct maximal ideals $\mathcal{M}$ and $\mathcal{M}'$, there is an $\mathbf{N}$ such that*

$$p(\mathcal{M}, \mathbf{N}) \neq p(\mathcal{M}', \mathbf{N}).$$

Proof. Part (i) expresses the fact that each p is a homomorphism; part (ii) restates (2), part (iii) restates theorem 14, part (iv) restates (3), and part (v) follows from theorem 13.

Definition. The *natural topology* on the space J of maximal ideals is the coarsest one in which all functions $p(\mathcal{M}, \mathbf{N})$, $\mathbf{N}$ fixed, are continuous. This is called the *Gelfand topology*.

Theorem 16. *J is compact in the Gelfand topology.*

Proof. Consider the product space

$$P = \prod_{\mathcal{L}} D_{|\mathbf{N}|}, \tag{16}$$

where D_r denotes the disk $|\zeta| \leq r$ in $\mathbb{C}$. Each disk is compact; therefore by Tychonov's theorem so is their product, P, in the product topology. By (15), $p(\mathcal{M}, \mathbf{N})$ lies in the disk $D_{|\mathbf{N}|}$. We map J into P by assigning to each $\mathcal{M}$ of J the point

$$\prod_{\mathcal{L}} p(\mathcal{M}, \mathbf{N}). \tag{17}$$

By (v) of theorem 15, (17) is an embedding of J in P. Clearly, the Gelfand topology is the same as that induced by the embedding. Since P is compact, the compactness of J would follow from knowing that the image of J under (17) is closed. Let $t = \prod t_N$ be a point of the closure of (17). We claim that $p(N) = t_N$ is a homomorphism of $\mathcal{L} \to \mathbb{C}$, namely that $t_{N+M} = t_N + t_M$, $t_{NM} = t_N t_M$, and $t_{cN} = ct_N$. These relations are, according to part (i) of theorem 15, satisfied by the points (17). Since these relations involve only two factors at a time, they remain true on the closure of (17). $\qquad\qquad\qquad\qquad\qquad\qquad\qquad\qquad\qquad\qquad\qquad\qquad\qquad$ $\square$

BIBLIOGRAPHY

Gelfand, I. M. Normierte Ringe. *Mat. Sbornik, N.S.*, **9** (1941): 3–24.

19

APPLICATIONS OF GELFAND'S THEORY OF COMMUTATIVE BANACH ALGEBRAS

19.1 THE ALGEBRA $C(S)$

S is a compact Hausdorff space, $\mathcal{L} = C(S)$, the algebra of continuous, complex-valued functions on S, normed by the maximum norm:

$$|f| = \max_s |f(s)|. \tag{1}$$

Given any point r on S, we can associate with r the homomorphism $p_r \colon \mathcal{L} \to \mathbb{C}$:

$$p_r(f) = f(r). \tag{2}$$

As noted in theorem 13 of chapter 18, the kernel of p_r is a maximal ideal

$$\mathcal{M}_r = \{f : f(r) = 0\}. \tag{3}$$

Theorem 1. *Every maximal ideal $\mathcal{M}$ in $C(S)$ is of form (3).*

Exercise 1. Prove theorem 1.

Theorem 1 shows that the maximal ideal space of $C(S)$ can be identified with S itself; the abstract theory gives nothing new!

19.2 GELFAND COMPACTIFICATION

S is a locally compact Hausdorff space, $C_b(S)$ the algebra of all complex valued, *bounded*, continuous functions on S, normed by

$$|f| = \sup_s |f(s)|. \tag{4}$$

Given any point r in S, $\mathcal{M}_r$ defined by (3) is a maximal ideal. We claim that if S is not compact there are others. We demonstrate this when $S = \mathbb{R}$:

Let $\{s_n\}$ be a sequence of points tending to ∞. Define $\mathcal{I}$ to consist of all f such that

$$\lim_{n \to \infty} f(s_n) = 0. \tag{5}$$

Clearly, $\mathcal{I}$ is an ideal, and equally clearly, the functions in $\mathcal{I}$ have no zeros in common. As observed in chapter 18, $\mathcal{I}$ is contained in some (in fact many) maximal ideals $\mathcal{M}$; yet clearly, $\mathcal{M}$ is not of form (3).

Although theorem 1 fails in the noncompact case, the following is true:

Theorem 2. *The set of maximal ideals $\mathcal{M}_r$ of form (3) is a dense subset, in the Gelfand topology, of the space of all maximal ideals of $C_b(S)$ when S is a locally compact Hausdorff space.*

Proof. Let $\mathcal{M}_\infty$ denote a maximal ideal. Open sets in the Gelfand topology that contain $\mathcal{M}_\infty$ contain $\mathcal{M}$ that satisfy for some $\epsilon > 0$

$$\left| p(\mathcal{M}, h_j) - p(\mathcal{M}_\infty, h_j) \right| < \epsilon, \qquad 1 \le j \le k, \tag{6}$$

where h_j are elements of the algebra, and $p(\mathcal{M})$ is defined by equation (15) of chapter 18. Set $h_j = f_j + c_j, c_j = p(\mathcal{M}_\infty, h_j)$; we can write (6) in the form

$$\left| p(\mathcal{M}, f_j) \right| < \epsilon, \quad p(\mathcal{M}_\infty, f_j) = 0, \qquad 1 \le j \le k. \tag{7}$$

We claim that every open set of form (7) contains some $\mathcal{M}_r$. For suppose not, then for every r in S, $\mathcal{M} = \mathcal{M}_r$ does not satisfy (7). Since by definition (14), chapter 18, and definition (3) of $\mathcal{M}_r$, $p(\mathcal{M}_r, f_j) = f_j(r)$, violating (7) means that for all r,

$$\max_j |f_j(r)| \ge \epsilon. \tag{8}$$

Since by (7), $p(\mathcal{M}_\infty, f_j) = 0$, f_j belongs to $\mathcal{M}_\infty$, $j = 1, \ldots, k$. Since $\mathcal{M}_\infty$ is an ideal, it follows that for arbitrary bounded, continuous functions g_j, $\sum g_j f_j$ also belongs to $\mathcal{M}_\infty$. Choosing $g_j = \overline{f}_j$, we conclude that

$$f = \sum |f_j|^2 \tag{9}$$

belongs to $\mathcal{M}_\infty$. From (8) we conclude that

$$f(r) \ge \epsilon^2 \tag{10}$$

for every r in S. This shows that f is an invertible element, and thus cannot belong to an ideal. This contradiction shows that every neighborhood of $\mathcal{M}_\infty$ contains some point $\mathcal{M}_r$. □

The maximal ideal space of $C_b(S)$ in the Gelfand topology, is called the *Gelfand compactification* of S. This space contains a dense subspace homeomorphic to S; the restriction of the space of continuous functions on the Gelfand compactification to this subspace is the space of all bounded continuous functions on S.

The next example is more fun.

19.3 ABSOLUTELY CONVERGENT FOURIER SERIES

$\mathcal{L}$ is the algebra of all complex-valued functions $f(\theta)$ on the unit circle S^1 that have an absolutely convergent Fourier series, namely of the form

$$f(\theta) = \sum c_n e^{in\theta}, \tag{11}$$

$$|f| = \sum |c_n| < \infty. \tag{12}$$

It is easy to verify that the norm (12) is submultiplicative: $|fg| \le |f|\,|g|$. So $\mathcal{L}$ is a Banach algebra. The function $f \equiv 1$ is its unit; its norm equals 1. For each point ω of S^1, the mapping

$$p_\omega(f) = f(\omega) \tag{13}$$

is a homomorphism of $\mathcal{L} \to \mathbb{C}$. Conversely:

Theorem 3. *Every homomorphism into $\mathbb{C}$ of the algebra $\mathcal{L}$ of functions on S^1 with norm (12) is of form (13).*

Proof. According to theorem 1 of chapter 18, every homomorphism p of a Banach algebra $\to \mathbb{C}$ has norm 1. Since $e^{i\theta}$ and $e^{-i\theta}$ both have norm 1, it follows that

$$\left|p\left(e^{i\theta}\right)\right| \le 1, \quad \left|p\left(e^{-i\theta}\right)\right| \le 1. \tag{14}$$

Since p is a homomorphism,

$$p\left(e^{i\theta}\right) p\left(e^{-i\theta}\right) = p(1) = 1. \tag{15}$$

Combining (14) and (15), we conclude that $|p(e^{i\theta})| = 1$; therefore we can write

$$p\left(e^{i\theta}\right) = e^{i\omega}, \quad \omega \text{ real.} \tag{16}$$

Since p is a homomorphism, $p(e^{in\theta}) = e^{in\omega}$ for all integers n, and for all finite sums

$$p\left(\sum c_n e^{in\theta}\right) = \sum c_n e^{in\omega}. \tag{17}$$

Since p is continuous and $\sum |c_n| < \infty$, (17) holds for infinite sums as well which converge in the sense of the norm (12). This proves that p is of form (13).   □

According to the main result of chapter 18, theorem 3, an element f of a Banach algebra $\mathcal{L}$ is invertible if $p(f) \neq 0$ for all homomorphisms p of $L \to \mathbb{C}$. In view of theorem 3 above we conclude

Theorem 4. *If a function f defined on the unit circle has absolutely convergent Fourier series, and doesn't vanish at any point of S^1, then its reciprocal f^{-1} also has absolutely convergent Fourier series.*

This celebrated theorem, due to Norbert Wiener, is astonishing, for there is no obvious relation between the Fourier series of a function f and of its reciprocal.
The situation is similar in more variables:

$$f(\theta) = \sum c_n e^{in\cdot\theta}, \qquad \theta = (\theta_1, \ldots, \theta_k), \ n = n_1, \ldots, n_k,$$
$$\sum |c_n| < \infty.$$

The analogues of theorems 3 and 4 hold, with similar proofs.

19.4 ANALYTIC FUNCTIONS IN THE CLOSED UNIT DISK

$\mathcal{A}$ is the algebra of all functions $f(z)$ analytic in the open unit disk $|z| < 1$, and continuous up to the boundary $|z| = 1$. Clearly, $\mathcal{A}$ is a Banach algebra under the norm

$$|f| = \max_{|z|\leq 1} |f(z)|. \tag{18}$$

Given any point w in $|z| \leq 1$, the mapping

$$p_w(f) = f(w) \tag{19}$$

is a homomorphism of $\mathcal{A} \to \mathbb{C}$. Conversely:

Theorem 5. *Every homomorphism of the algebra $\mathcal{A} \to \mathbb{C}$ is of form (19).*

Proof. By theorem 1, chapter 18, such a homomorphism p has norm ≤ 1. Since according to (18), $f(z) = z$ has norm $= 1$, it follows that

$$|p(z)| \leq 1. \tag{20}$$

Denote the value of $p(z)$ by w; since p is a homomorphism, $p(z^n) = w^n$ and for all finite sums

$$p\left(\sum_0^n a_j z^j\right) = \sum a_j w^j. \tag{21}$$

We can express (21) so: when $f(z)$ is a polynomial, $p(f) = f(w)$. Since every f in $\mathcal{A}$ can be approximated uniformly on $|z| \leq 1$ by polynomials, and since p is continuous, this relation holds for all f in $\mathcal{A}$, as asserted in theorem 5. □

Exercise 2. Show that every function in $\mathcal{A}$ can be approximated uniformly on the unit disk by polynomials.

Theorem 6. *Let $f_1, \ldots, f_m$ be a collection of m functions in $\mathcal{A}$ that have no common zero in the disk. Then there are functions $g_1, \ldots, g_m$ in $\mathcal{A}$ such that*

$$\sum g_j f_j \equiv 1. \tag{22}$$

Proof. Consider the set $\mathcal{I}$ of all functions of form

$$f = \sum h_j f_j, \quad h_j \text{ in } \mathcal{A}. \tag{23}$$

Unless $\mathcal{I}$ is all of $\mathcal{A}$, it is an ideal in $\mathcal{A}$, and therefore it is contained in some maximal ideal $\mathcal{M}$. According to theorem 13 of chapter 18, $\mathcal{M}$ is the null set of a homomorphism. By theorem 5, all homomorphisms are of form (19); therefore a maximal ideal is a collection of all functions in $\mathcal{A}$ that vanish at some point w. Since the functions f_j, $j = 1, \ldots, m$ were assumed to have no common zero, they cannot belong to the same ideal. This means that (23) cannot be an ideal, therefore the functions f of form (23) are all of $\mathcal{A}$. In particular, $f \equiv 1$ is of that form; this proves (22). □

The situation is similar for analytic functions of k complex variables, defined in the polydisk

$$\prod(|z_j| < 1)$$

and continuous up to the boundary. The analogues of theorems 5 and 6 hold, with analogous proofs.

19.5 ANALYTIC FUNCTIONS IN THE OPEN UNIT DISK

The algebra $\mathcal{B}$ of bounded analytic functions in the open unit disk $|z| < 1$. These form a Banach algebra under the norm

$$|f| = \sup_{|z|<1} |f(z)|. \tag{24}$$

Every mapping of form (19), $|w| < 1$, is a homomorphism: $\mathcal{B} \to \mathbb{C}$; therefore the set of f satisfying

$$f(w) = 0, \qquad |w| < 1 \tag{25}$$

is a maximal ideal $\mathcal{M}_w$. Not all maximal ideals in $\mathcal{B}$ are of this form. However, the following is true:

Theorem 7. *The set of maximal ideals $\mathcal{M}_w$ of form (25) is a dense subset, in the Gelfand topology, of the space of all maximal ideals.*

The analysis presented in the proof of theorem 2 shows that the theorem is equivalent to the following proposition:

Theorem 7′. *Let $f_1, \ldots, f_m$ be a collection of functions in $\mathcal{B}$ that have the property that*

$$\sum |f_j(z)| > 1 \tag{26}$$

for every z in $|z| < 1$. Then there exist m functions $g_j \in \mathcal{B}$ such that

$$\sum g_j\, f_j \equiv 1. \tag{27}$$

Theorem 7 is called the *corona theorem*. The implication (26) $\Rightarrow$ (27) is subtle and deep. It was shown to be true by Lennart Carleson, using function theoretic arguments. Tom Wolff succeeded 1979 in giving an entirely different proof, using methods of partial differential equations (e.g., see Koosis).

19.6 WIENER'S TAUBERIAN THEOREM

The next application deals with a Banach algebra without a unit. This is easily remedied by adjoining a unit $\mathbf{I}$ in a purely formal fashion. That is, if $\mathcal{A}$ is a Banach algebra without a unit, the enlarged algebra $\mathcal{L}$ consists of elements of the form $\lambda \mathbf{I} + \mathbf{M}$, λ a complex number, $\mathbf{M}$ an element of $\mathcal{A}$. Addition is defined componentwise, and multiplication according to the distributive law:

$$(\mu \mathbf{I} + \mathbf{N})\,(\lambda \mathbf{I} + \mathbf{M}) = \mu\lambda\mathbf{I} + \mu\mathbf{M} + \lambda\mathbf{N} + \mathbf{N}\mathbf{M}.$$

Note that $\mathcal{A}$ is a maximal ideal in $\mathcal{L}$.

Norm in the enlarged algebra is defined as follows:

$$|\lambda \mathbf{I} + \mathbf{M}| = |\lambda| + |\mathbf{M}|.$$

Clearly, this new norm is subadditive and submultiplicative, and the unit $\mathbf{I}$ has norm 1.

We take $\mathcal{A}$ to be the space L^1 of complex-valued integrable functions on $\mathbb{R}$, which we denote by lowercase letters. We define multiplication to be *convolution*:

$$(f * g)(s) = \int_{\mathbb{R}} f(s - u)\, g(u)\, du. \tag{28}$$

The change of variable of integration $s - u = v$ shows that convolution is *commutative*.

Lemma 8. *Convolution is submultiplicative in the L^1-norm, that is,*

$$|f * g|_{L^1} \le |f|_{L^1}\, |g|_{L^1}. \tag{29}$$

Proof. Suppose that both f and g are continuous and of compact support. Then so is $f * g$, and it is the L^1-limit as $\Delta \to 0$ of finite sums of the form

$$\sum_j f(s - j\Delta)\, g(j\Delta)\, \Delta. \tag{29'}$$

Note that the L^1-norm of a translate of f is the same as the L^1-norm of f; therefore, using the subadditivity of norm, we conclude that the L^1-norm of $(29')$ is bounded by

$$|f|_{L^1} \sum |g(j\Delta)|\, \Delta. \tag{30}$$

We let Δ tend to zero; the sum in $(29')$ tends to $f * g$, and (30) tends to $|f|_{L^1}\, |g|_{L^1}$, so we obtain in the limit inequality (29). $\square$

Actually the proof presented above applies to *any* norm for the function f that is translation invariant, leading to the following important inequality:

$$|f * g| \le |f||g|_{L^1}, \tag{31}$$

where $|\ \ |$ stands for *any translation invariant norm* for functions on $\mathbb{R}$. From now on all norms in this section denote the L^1-norm.

We denote by $\mathcal{L}$ the convolution algebra of L^1 functions augmented formally by a unit, which we denote by e. We proceed now to determine all maximal ideals of $\mathcal{L}$, or rather, what is equivalent, the multiplicative linear functionals p whose nullspaces they are. One of the maximal ideals in the convolution algebra is L^1 itself. For any other p, there is an f in L^1 such that $p(f) \ne 0$; we normalize f so that

$$p(f) = 1. \tag{32}$$

Let t be any real number, and denote by f_t the translate of f:

$$f_t(s) = f(s - t). \tag{33}$$

Lemma 9. f_t *depends continuously on* t *in the sense of the norm, that is,*

$$\lim_{h \to 0} |f_{t+h} - f_t| = 0. \tag{34}$$

Proof. This is clearly true when f is continuous and has compact support; since every f in L^1 can be approximated in norm by such functions, (34) holds for all f. □

We define now the function $\chi(t)$ as

$$\chi(t) = p(f_t). \tag{35}$$

Since p is a continuous linear functional, it follows from lemma 9 that χ is a continuous function. Since the norm of f_t is independent of t, it follows that $\chi(t)$ is uniformly bounded by $|f|$ for all real t.

Let t and r be two real numbers; we claim that

$$f_{t+r} * f = f_t * f_r. \tag{36}$$

To see this, we apply the definition (28) of convolution to the left side:

$$(f_{t+r} * f)(s) = \int f(v - t - r) f(s - v) dv.$$

Introducing $v - r = u$ as new variable of integration transforms this to

$$\int f(u - t) f(s - r - u) du;$$

this is $f_t * f_r$, as asserted in (36).

Now let p act on (36). Since p is multiplicative, we get that

$$p(f_{t+r}) p(f) = p(f_t) p(f_r).$$

Using the definition (35) of χ and the normalization (32), we deduce that χ satisfies the functional equation

$$\chi(t + r) = \chi(t) \chi(r). \tag{37}$$

We have already seen that χ is continuous and bounded; according to a well-known result of analysis, all such solutions of (37) are purely imaginary exponentials:

$$\chi(t) = e^{i\xi t}, \qquad \xi \text{ real}. \tag{38}$$

Having determined the action of p on f and its translates, we want to determine its action on all g in the algebra. We start by remarking that since p is continuous, and $p(f) \neq 0$, p does not vanish in a ball of small enough radius around f. Since f can be approximated arbitrarily closely in norm by continuous functions with compact support, there is such a function for which p does not vanish. This shows that we may take f to be continuous and of compact support.

Let g be any continuous function of compact support. Then the sum (29') tends in the sense of the L^1-norm to $f * g$. Using the notation (33), we can rewrite (29') as

$$\sum f_{j\Delta} g(j\Delta)\Delta. \tag{39}$$

Let p act on the preceding sum; since p is linear, we get $\sum p(f_{j\Delta}) g(j\Delta)\Delta$. Combining (35) and (38), we get the following approximation to $p(f * g)$:

$$\sum e^{i\xi j\Delta} g(j\Delta)\Delta. \tag{40}$$

As $\Delta \to 0$, (40) tends to

$$\tilde{g}(\xi) = \int_{\mathbb{R}} e^{i\xi v} g(v)\, dv, \tag{41}$$

that is, to the Fourier transform of g except for the factor $1/\sqrt{\pi}$. Since (39) tends to $f * g$, and since p is continuous, it follows that $p(f * g) = \tilde{g}(\xi)$. Since p is multiplicative, and $p(f) = 1$, it follows that

$$p(f * g) = p(g) = \tilde{g}(\xi) \tag{42}$$

for all continuous g of compact support. Since both p and the Fourier transform at ξ are continuous functions of g, it follows that (42) holds for all g in L^1. We summarize:

Theorem 10. *Every multiplicative linear functional of the convolution algebra L^1 is of the form (42), ξ some real number. Conversely, for every ξ, (42) is a multiplicative linear functional.*

Proof. The first part has been demonstrated above; the second part paraphrases the well-known fact that the Fourier transform of $f * g$ is the ordinary product of the Fourier transforms of f and g:

$$\widetilde{f * g} = \tilde{f}\, \tilde{g}. \qquad \square$$

Theorem 11 (Wiener). *Let f be an L^1 function on $\mathbb{R}$ whose Fourier transform $\tilde{f}(\xi)$ is nonzero for any ξ. Then the translates of f span all of L^1, that is, any L^1 function can be approximated in norm by linear combinations of translates of f.*

REMARK 1. The condition of the nonvanishing of $\tilde{f}$ is necessary, for if $\tilde{f}$ vanishes at η, so does the Fourier transform of any translate of f, of their linear combinations and L^1 limits of them.

Proof. When g is continuous and of compact support, $f * g$ is the L^1 limit of the linear combination (29') of translates of f. It follows from lemma 8 that for any g

in L^1, $f * g$ is the L^1-limit of $f * g_n$, where g_n are continuous, of compact support, and tend to g in the L^1-norm. Therefore, to prove the theorem, it is sufficient to show that the space of functions $f * g$, g in L^1, is dense in L^1. $\qquad\square$

Lemma 12. *Let f be a function as in theorem 11, and m an L^1 function whose Fourier transform $\tilde{m}$ has compact support. Then m can be written as*

$$m = f * g, \quad g \text{ in } L^1. \tag{43}$$

Note that the L^1 functions whose Fourier transforms have compact support are dense in L^1; therefore theorem 11 follows from lemma 12.

Exercise 3. Show that the set of L^1 functions m whose Fourier transform has compact support are dense in L^1.

Proof of Lemma 12. Choose a compact interval I so large that it contains the support of $\tilde{m}$. Construct an auxiliary function h in L^1 whose Fourier transform is real and has the following properties:

$$\tilde{h}(\xi) = \begin{cases} 1 & \text{on } I \\ \leq 1 & \text{everywhere.} \end{cases} \tag{44}$$

Define f^c to be the conjugate of f, in the following sense:

$$f^c(s) = \overline{f(-s)}. \tag{45}$$

As is well known, the Fourier transform of the conjugate of f is the complex conjugate of that of f:

$$\widetilde{f^c}(\xi) = \overline{\tilde{f}}(\xi). \tag{45'}$$

We make use now of the Banach algebra $\mathcal{L}$ obtained from the convolution algebra L^1 by formally adjoining a unit e. We may regard e as the Dirac δ of distribution theory, see Appendix B. Elements of this algebra are of the form

$$\lambda e + k, \quad k \text{ in } L^1, \lambda \text{ in } \mathbb{C}.$$

The functional $p_0(\lambda e + k) = \lambda$ is, obviously, multiplicative. It follows from theorem 10 that all others are of the form

$$p(\lambda e + k) = \lambda + \tilde{k}(\xi), \quad \xi \text{ real.}$$

Take, in particular, the element $e - h + f * f^c$. According to the above, $p_0(e - h + f * f^c) = 1$, clearly nonzero. For any other p,

$$p(e - h + f * f^c) = 1 - \tilde{h} + |\tilde{f}|^2. \tag{46}$$

It follows from (44) and the nonvanishing of $\tilde{f}$ that (46) is positive for all ξ. Thus $e - h + f * f^c$ does not belong to the nullspace of any multiplicative linear functional.

We claim that $e - h + f * f^c$ is invertible in $\mathcal{L}$; for according to theorem 3 of chapter 18, an element of a Banach algebra is invertible iff it does not belong to the nullspace of any multiplicative linear functional.

Denote by d the inverse of $e - h + f * f^c$:

$$\left(e - h + f * f^c\right) * d = e.$$

Multiply this relation by m:

$$\left(e - h + f * f^c\right) * d * m = m. \tag{47}$$

We claim that $(e - h) * m$ is zero; To see this regard e as the Dirac δ-distribution. The Fourier transform of $(e - h) * m$ is $(1 - \tilde{h})\tilde{m}$, see section B.5. According to the construction (44), $1 - \tilde{h}$ is zero in I while $\tilde{m}$ is zero on the complement of I. This shows that the Fourier transform in the sense of distributions of $(e - h) * m$ is zero; therefore so is $(e - h) * m$ itself. Setting this in (47) gives

$$f * f^c * d * m = m;$$

this is the desired relation (43), with $g = f^c * d * m$. $\qquad\qquad\qquad\qquad\qquad$ □

We give now an indication how theorem 11 is used in applications.

Let n be a *bounded* function on $\mathbb{R}$, which has a limit as s tends to ∞:

$$\lim_{s \to \infty} n(s) = a. \tag{48}$$

Let f be an L^1 function, normalized so that

$$\int f(u)\, du = 1. \tag{49}$$

It follows then easily from (48) and (49) that

$$\lim_{s \to \infty} (f * n)(s) = a. \tag{50}$$

The question is: can one deduce (48) from (50)? A result of this kind is called a *Tauberian theorem*.

Theorem 13. *Let n be a bounded function on $\mathbb{R}$, and f an L^1 function normalized as in (49). Suppose that (50) holds, that is, that the convolution of f and n tend to a number a as s tends to ∞. Suppose that the Fourier transform of f is nowhere zero. Then $n(s)$ tends to a in the mean in the following sense: for every value of d*

$$\lim_{s \to \infty} \frac{1}{d} \int_s^{s+d} n(u)\, du = a. \tag{51}$$

Proof. (50) implies that for any t,

$$\lim_{s \to \infty} (f_t * n)(s) = a. \tag{50'}$$

Taking linear combinations of (50'), we deduce that for

$$h = \sum c_j f_{t_j}, \tag{52}$$

$$\lim_{s \to \infty} (h * n)(s) = a \sum c_j = a \int h \, du. \tag{53}$$

Clearly, (53) holds for any function h that is the L^1-limit of a sequence of functions of form (52).

If the Fourier transform of f is nowhere zero, then it follows from theorem 11 that (53) holds for all h in L^1. Take, in particular,

$$h(u) = \begin{cases} 1/d & \text{for } 0 < s < d \\ 0 & \text{elsewhere;} \end{cases} \tag{54}$$

for this h, (53) becomes (51). □

REMARK 2. (51) is not quite (48) but close to it. If, for example, we know that n is uniformly continuous on $\mathbb{R}_+$, then (51) implies (48).

Exercise 4. Suppose that n is *slowly increasing* in the sense that

$$\sup_{u-1 < v < u} n(u) - n(v) \tag{55}$$

tends to 0 as u tends to ∞. Show that then (51) and (55) imply (48).

NOTE. Wiener showed how to use his Tauberian theorem to prove the prime number theorem.

Exercise 5. Let f be a function of class L^2 on $\mathbb{R}$. Show that the translates of f span L^2 if and only if $\tilde{f}$ does not vanish on a set of positive Lebesgue measure.

REMARK 3. A. Beurling has shown that if f belongs to all L^p, $1 \le p$, and if $\tilde{f}$ vanishes on a set of positive Hausdorff measure α, $0 \le \alpha < 1$, then the translates of f do not span L^p for $p < 2/(2 - \alpha)$.

19.7 COMMUTATIVE B^*-ALGEBRAS

Gelfand theory is particularly useful for studying commutative algebras of operators over a complex Hilbert space H. We recall from chapter 15 the notion of the transpose of an operator $\mathbf{A}$ mapping a Banach space X into X. When $\mathbf{A}$ maps a complex

Hilbert space H into itself, its conjugate transpose, called in this context its *adjoint*, is another operator $\mathbf{A}^* : H \to H$ so that the pair satisfies

$$(\mathbf{A}x, y) = (x, \mathbf{A}^*y) \tag{56}$$

for all x, y in H. Every bounded operator $\mathbf{A}$ has an adjoint $\mathbf{A}^*$; the following algebraic properties follow directly from the definition:

$$(\mathbf{A} + \mathbf{B})^* = \mathbf{A}^* + \mathbf{B}^*, \quad (k\mathbf{A})^* = \bar{k}\mathbf{A}^*,$$
$$\mathbf{A}^{**} = \mathbf{A}, \quad (\mathbf{AB})^* = \mathbf{B}^*\mathbf{A}^*. \tag{57}$$

We denote the operator norm of $\mathbf{A}$ as $\|\mathbf{A}\|$.

Theorem 14. *For a bounded, linear mapping* $\mathbf{A} : H \to H$, H *a Hilbert space,*

$$\|\mathbf{A}\| = \|\mathbf{A}^*\|, \tag{58}$$
$$\|\mathbf{A}^*\mathbf{A}\| = \|\mathbf{A}\|^2. \tag{59}$$

Proof. Take the supremum of the absolute value of each side of (56) for all x, y of unit length: $\|x\| = \|y\| = 1$. On the left take the supremum first with respect to y, then with respect to x; the result is $\|\mathbf{A}\|$. Reverse the order on the right, obtaining $\|\mathbf{A}^*\|$; this proves (58).

To prove (59), take $y = \mathbf{A}x$ in (56); estimating the right hand by the Schwarz inequality gives

$$\|\mathbf{A}x\|^2 \le \|x\| \, \|\mathbf{A}^*\mathbf{A}x\| \le \|x\|^2 \|\mathbf{A}^*\mathbf{A}\|.$$

Taking the supremum over all unit vectors x gives $\|\mathbf{A}\|^2 \le \|\mathbf{A}^*\mathbf{A}\|$. Since by submultiplicity and (58), $\|\mathbf{A}^*\mathbf{A}\| \le \|\mathbf{A}^*\| \, \|\mathbf{A}\| = \|\mathbf{A}\|^2$, (59) follows. $\qquad\square$

Definition. A complete normed algebra with a $*$ operation that has properties (57), (58), and (59) is called a $\mathcal{B}^*$-*algebra*.

An element $\mathbf{A}$ of a $\mathcal{B}^*$-algebra is called *self-adjoint* if $\mathbf{A}^* = \mathbf{A}$; $\mathbf{B}$ is called *anti-self-adjoint* if $\mathbf{B}^* = -\mathbf{B}$.

Theorem 15.

  (i) *The spectrum of a self-adjoint* $\mathbf{A}$ *of a commutative* $\mathcal{B}^*$-*algebra is real.*
  (ii) *The spectrum of an anti-self-adjoint element* $\mathbf{B}$ *is imaginary.*

Proof. According to Gelfand's theory the spectrum of $\mathbf{A}$ is the set of numbers $p(\mathbf{A})$, p being any multiplicative linear functional. We claim that for $\mathbf{A}$ self-adjoint, $p(\mathbf{A})$ is real; to see this, write

$$p(\mathbf{A}) = a + ib. \tag{60}$$

Let t be any real number; set $\mathbf{T} = \mathbf{A} + it\mathbf{I}$. Then $\mathbf{T}^* = \mathbf{A} - it\mathbf{I}$, and

$$\mathbf{T}^*\mathbf{T} = \mathbf{A}^2 + t^2\mathbf{I}. \tag{61}$$

From (60), $p(\mathbf{T}) = a + i(b + t)$, so

$$|p(\mathbf{T})|^2 = a^2 + (b + t)^2. \tag{62}$$

According to theorem 1 of chapter 18, every multiplicative linear functional is a contraction:

$$|p(\mathbf{T})|^2 \leq \|\mathbf{T}\|^2. \tag{63}$$

Using (62) on the left, and (59) and (61) on the right, we deduce from (63) that

$$a^2 + (b + t)^2 \leq \|\mathbf{T}^*\mathbf{T}\| \leq \|\mathbf{A}\|^2 + t^2.$$

If $b \neq 0$, this is clearly false for t large and of the same sign as b. So $b = 0$; this proves part (i) of theorem 15.

If $\mathbf{B}$ is anti-self-adjoint, it follows from (57) that $i\mathbf{B}$ is self-adjoint. Thus part (ii) follows from part (i). $\qquad\square$

Theorem 16. *Every multiplicative linear functional p on a commutative $\mathcal{B}^*$-algebra satisfies*

$$p(\mathbf{T}^*) = \overline{p(\mathbf{T})}. \tag{64}$$

Proof. Define $\mathbf{A}$ and $\mathbf{B}$ as

$$\mathbf{A} = \frac{\mathbf{T} + \mathbf{T}^*}{2}, \quad \mathbf{B} = \frac{\mathbf{T} - \mathbf{T}^*}{2}; \tag{65}$$

it follows from (57) that $\mathbf{A}^* = \mathbf{A}$, $\mathbf{B}^* = -\mathbf{B}$. Decompose $\mathbf{T}$ and $\mathbf{T}^*$ as

$$\mathbf{T} = \mathbf{A} + \mathbf{B}, \quad \mathbf{T}^* = \mathbf{A} - \mathbf{B}.$$

Since p is linear,

$$p(\mathbf{T}) = p(\mathbf{A}) + p(\mathbf{B}), \quad p(\mathbf{T}^*) = p(\mathbf{A}) - p(\mathbf{B}). \tag{66}$$

According to theorem 15, $p(\mathbf{A})$ is real and $p(\mathbf{B})$ imaginary; therefore (64) follows from (66).

Theorem 17. *For every $\mathbf{T}$ in a commutative $\mathcal{B}^*$ algebra,*

$$\|\mathbf{T}\| = |\sigma(\mathbf{T})|. \tag{67}$$

Proof. We first prove (67) for the self-adjoint elements $\mathbf{A}$ of the algebra. When $\mathbf{A}^* = \mathbf{A}$, (59) becomes

$$\|\mathbf{A}^2\| = \|\mathbf{A}\|^2. \tag{68}$$

It follows from (57) that if $\mathbf{A}$ is self-adjoint, so is any power of $\mathbf{A}$; therefore it follows from (68) that

$$\|\mathbf{A}^{2n}\| = \|\mathbf{A}^n\|^2.$$

Applying this to $n = 1, 2, 4, 8, \ldots, 2^k = m$ and combining these identities, we deduce that

$$\|\mathbf{A}^m\| = \|\mathbf{A}\|^m.$$

Taking the mth root, we conclude that

$$\lim_{k \to \infty} \|\mathbf{A}^m\|^{1/m} = \|\mathbf{A}\|, \qquad m = 2^k. \tag{69}$$

According to theorem 4 of chapter 17, the limit on the left of (69) is equal to the spectral radius of $\mathbf{A}$; this proves (67) for $\mathbf{T} = \mathbf{A}$ self-adjoint.

We turn now to arbitrary $\mathbf{T}$; it follows from (64) that for any p,

$$p\left(\mathbf{T}^*\mathbf{T}\right) = p(\mathbf{T}^*)p(\mathbf{T}) = |p(\mathbf{T})|^2. \tag{70}$$

Since every point in $\sigma(\mathbf{T})$ is of the form $p(\mathbf{T})$, it follows from (70) that the spectral radius of $\mathbf{T}^*\mathbf{T}$ is the square of the spectral radius of $\mathbf{T}$:

$$|\sigma(\mathbf{T}^*\mathbf{T})| = |\sigma(\mathbf{T})|^2. \tag{71}$$

By (57), $\mathbf{T}^*\mathbf{T}$ is self-adjoint; since we have already proved (67) in the symmetric case,

$$\|\mathbf{T}^*\mathbf{T}\| = |\sigma(\mathbf{T}^*\mathbf{T})|. \tag{72}$$

Combining this with (71) gives

$$\|\mathbf{T}^*\mathbf{T}\| = |\sigma(\mathbf{T})|^2$$

By (59), $\|\mathbf{T}^*\mathbf{T}\| = \|\mathbf{T}\|^2$; this yields (67) for all $\mathbf{T}$. □

NOTE. Gelfand invented his theory of commutative Banach algebras to derive the results presented in section 7. The application to Wiener's theorem was an afterthought.

HISTORICAL NOTE. Tauberian theorems were named so by Hardy and Littlewood after the obscure Austrian mathematician Alfred Tauber, 1866–1942, who wrote the first paper on the subject in 1897. His major contributions were to actuarial science. In 1942 he was deported to the Theresienstadt concentration camp.

BIBLIOGRAPHY

Beurling, A. On a closure problem, *Arkiv Mat.*, **1**, (19xx): 301–303.

Carleson, L. Interpolation by bounded analytic functions and the corona problem. *An. Math.*, **76** (1962): 547–559.

Gelfand, I. M. Normierte Ringe. *Mat. Sbornik, N.S.*, **9** (51) (1941): 3–24.

Koosis, P. *Introduction to H_p Spaces*. Cambridge Tracts in Mathematics, **115**. Cambridge University Press, Cambridge, 1998.

Wiener, N. Tauberian theorems. *An. Math.*, (2), **33** (1932): 1–100.

20

EXAMPLES OF OPERATORS AND THEIR SPECTRA

In this chapter we discuss and illustrate phenomena in the spectral theory of operators that go beyond the mere fact that operators belong to the Banach algebra of bounded linear maps $\mathcal{L}(X, X)$; we exploit the fact that these are operators that act on elements of a Banach space.

20.1 INVERTIBLE MAPS

First we address the operation of *invertibility* of a map $\mathbf{M}$ in $\mathcal{L}(X, X)$. By definition, $\mathbf{M}$ is invertible iff it maps X onto X in a one-to-one fashion. The inverse is necessarily bounded, thanks to the closed graph theorem; see theorem 11 of chapter 15. Thus there are only two ways for $\mathbf{M}$ to *fail* to be invertible:

(a) $\mathbf{M}$ is not one-to-one.
(b) $\mathbf{M}$ is not onto.

It follows that if the product $\mathbf{MK}$ of two maps is invertible, then $\mathbf{K}$ is one-to-one, and $\mathbf{M}$ is onto. If $\mathbf{M}$ and $\mathbf{K}$ commute, meaning that $\mathbf{MK} = \mathbf{KM}$, we deduce from invertibility of the product that both $\mathbf{M}$ and $\mathbf{K}$ are both one-to-one and onto, and therefore both $\mathbf{M}$ and $\mathbf{K}$ are invertible.

We have seen in theorem 2, chapter 17, that if an element $\mathbf{K}$ of a Banach algebra is invertible, then so are all nearby elements $\mathbf{K} - \mathbf{A}$, $|\mathbf{A}| <$ const. For maps we have the following additional result:

Theorem 1. *X a Banach space*, $\mathbf{K} : X \rightarrow X$ *a bounded linear map that maps X onto itself. Then all nearby maps, namely those of the form*

$$\mathbf{K} - \mathbf{A}, \qquad |\mathbf{A}| < \epsilon, \text{ with } \epsilon \text{ small enough}, \tag{1}$$

also maps X onto itself.

Proof. It follows from theorem 9 of chapter 15 that there is a constant k such that for any z in X there is an x in X such that

$$\mathbf{K}x = z, \quad |x| \leq k|z|. \tag{2}$$

We claim now that for any linear map $\mathbf{A} : X \to X$ whose norm satisfies $|\mathbf{A}| < 1/k$, $\mathbf{K} - \mathbf{A}$ maps X onto itself. Furthermore we claim that for any u there is an x such that

$$(\mathbf{K} - \mathbf{A})x = u, \quad |x| \leq \frac{k}{1 - k|\mathbf{A}|}|u|. \tag{3}$$

This x will be constructed as the limit of a sequence of approximations $\{x_n\}$, defined recursively as follows:

$$\mathbf{K}x_{n+1} = \mathbf{A}x_n + u, \quad x_0 = 0. \tag{4}$$

By (2), for $n = 1$ equation (4) has a solution x_1 satisfying $|x_1| \leq k|u|$. For $n > 1$ we construct x_{n+1} by substracting two consecutive equations (4):

$$\mathbf{K}(x_{n+1} - x_n) = \mathbf{A}(x_n - x_{n-1}). \tag{5}$$

By (2), (5) has a solution $(x_{n+1} - x_n)$ satisfying

$$|x_{n+1} - x_n| \leq k|\mathbf{A}| \, |x_n - x_{n-1}|. \tag{5'}$$

Since $k|\mathbf{A}| < 1$, it follows from (5') that the sequence $\{x_n\}$ is convergent. Letting $n \to \infty$ in (4) gives (3), with $x = \lim x_n$. Since $x_0 = 0$, $x = \sum_0^\infty (x_{n+1} - x_n)$;

$$|x| \leq \sum |x_{n+1} - x_n| \leq \sum (k|\mathbf{A}|)^n |x_1|,$$

which is $\leq k/(1 - k|\mathbf{A}|)|u|$, as asserted in (3). $\qquad\qquad\square$

Recall from chapter 17 that the spectrum $\sigma(\mathbf{M})$ of a bounded linear map $\mathbf{M}$ is the set of all complex numbers λ for which $(\lambda \mathbf{I} - \mathbf{M})$ is not invertible.

Theorem 2. *Let* $\mathbf{M} : X \to X$ *be a bounded linear operator,* λ *a boundary point of the spectrum of* $\mathbf{M}$. *Then the range of* $\lambda - \mathbf{M}$ *is not all of* X.

Proof. Since λ is assumed to be a boundary point of $\sigma(\mathbf{M})$, it is the limit of a sequence λ_n in the resolvent set of $\mathbf{M}$. According to corollary 3' of theorem 3 in chapter 17, the norm of the resolvent is $\geq$ the reciprocal of the distance to some point of the spectrum:

$$|(\zeta - \mathbf{M})^{-1}| \geq |\zeta - \lambda|^{-1}. \tag{6}$$

This implies that there is a u, depending on ζ, such that $(\zeta - \mathbf{M})^{-1}u = x$ is large compared to u:

$$(\zeta - \mathbf{M})x = u, \quad |x| > \frac{1}{2|\zeta - \lambda|}|u|. \tag{7}$$

Now in theorem 1 take $\mathbf{K} = \lambda - \mathbf{M}$, $\mathbf{A} = (\lambda - \zeta)\mathbf{I}$. If, contrary to theorem 2, the range of $\mathbf{K} = \lambda - \mathbf{M}$ is all of X, then we can take $\zeta = \lambda_n$ so close to λ that $|\mathbf{A}| = |\lambda - \zeta|$ is small enough for $(\mathbf{K} - \mathbf{A})x = u$ to have a solution x whose norm satisfies (3). This is in direct contradiction to relation (7), which asserts that the unique solution of

$$(\mathbf{K} - \mathbf{A})x = (\zeta - \mathbf{M})x = u$$

has a norm $|x|$ that is very large compared to $|u|$.   □

Theorem 3. *Let $\mathbf{M}$ be a bounded linear map: $X \to X$, $\mathbf{M}' : X' \to X'$ its transpose. Then*

$$\sigma(\mathbf{M}') = \sigma(\mathbf{M}). \tag{8}$$

Proof. The proof is based on this simple observation:
$\mathbf{K} : X \to X$ is invertible iff its transpose $\mathbf{K}' : X' \to X'$ is invertible. Now, if $\mathbf{K}$ is invertible, it has an inverse $\mathbf{L}$:

$$\mathbf{KL} = \mathbf{LK} = \mathbf{I}. \tag{9}$$

The transpose of these relations is

$$\mathbf{L}'\mathbf{K}' = \mathbf{K}'\mathbf{L}' = \mathbf{I}', \tag{9'}$$

which shows that $\mathbf{L}'$ is the inverse of $\mathbf{K}'$. When X is reflexive, the relation of $\mathbf{K}$ and $\mathbf{K}'$ is symmetric, so the proof is complete. In the nonreflexive case an additional argument is needed. Suppose that $\mathbf{K}'$ is invertible; then (9') holds, and so by transposition

$$\mathbf{K}''\mathbf{L}'' = \mathbf{L}''\mathbf{K}'' = \mathbf{I}''. \tag{9''}$$

Since $\mathbf{K}''$ and $\mathbf{I}''$, when restricted to X, are $\mathbf{K}$ and $\mathbf{I}$, respectively, it follows from (9'') that the nullspace of $\mathbf{K}$ is trivial. It follows that $\mathbf{K}$ is one-to-one, and by (9'') that $\mathbf{L}''$ restricted to the range of $\mathbf{K}$ is inverse to $\mathbf{K}$. Since $\mathbf{L}''$ is a continuous map, it follows that the range of $\mathbf{K}$ is closed. We claim that the range of $\mathbf{K}$ is all of X, for if not, there would be, by Hahn-Banach, a linear functional $\ell \neq 0$ annihilating $R_\mathbf{K}$, but such an ℓ, according to theorem 5 of chapter 15, belongs to the nullspace of $\mathbf{K}'$. Since $\mathbf{K}'$ is assumed invertible, this cannot be.

Theorem 3 follows from the observation by setting $\mathbf{K} = \lambda - \mathbf{M}$.   □

Corollary 1. *Let* **M** *be a bounded linear map of a Hilbert space* H *into* H, **M*** *its adjoint. Then*

$$\sigma(\mathbf{M}^*) = \sigma\overline{(\mathbf{M})}.$$

Exercise 1. Prove corollary 3.

   The discussion at the beginning of this chapter shows that λ can enter the spectrum of a map **M** in two ways:

   (a) $(\lambda\mathbf{I} - \mathbf{M})$ has a nontrivial nullspace in X.
   (b) The range of $(\lambda\mathbf{I} - \mathbf{M})$ is a proper subspace of X.

   A nonzero vector in the nullspace of $\lambda\mathbf{I}-\mathbf{M}$ is called an *eigenvector* of **M**. Since in an infinite-dimensional space alternative (b) could hold without (a) being the case, a point λ in the spectrum of **M** need not be an eigenvalue of **M**. This is the reason why eigenvectors play a less prominent role in the general theory of linear operators than in the finite-dimensional case. They do play a role, as we will see in the examples below, and in the chapters on compact operators.
   We turn now to some examples:

20.2 SHIFTS

Take X to be ℓ^2, consisting of vectors x with complex components:

$$x = (a_0, a_1, \ldots) \qquad \sum |a_j|^2 < \infty. \tag{10}$$

The right shift **R** and left shift **L** are defined by

$$\mathbf{R}x = (0, a_0, a_1, \ldots,), \qquad \mathbf{L}x = (a_1, a_2, \ldots). \tag{11}$$

Clearly, $\mathbf{LR} = \mathbf{I}$, $\mathbf{RL} \neq \mathbf{I}$, so neither **L** nor **R** is invertible. A moment's calculation shows that **R** and **L** are transposes of each other:

$$\mathbf{R}' = \mathbf{L}, \qquad \mathbf{R} = \mathbf{L}', \tag{12}$$

Theorem 4. *The spectrum of* **R** *and of* **L** *consists of the unit disk* $|\lambda| \leq 1$.

   Proof. Obviously **L** is a *contraction*, that is, $\|\mathbf{L}x\| \leq \|x\|$. Since equality holds for some x, it follows that $\|\mathbf{L}\| = 1$, and similarly that for any positive integer n, $\|\mathbf{L}^n\| = 1$. It follows from this that

$$\lim \|\mathbf{L}^n\|^{1/n} = 1, \tag{13}$$

so according to theorem 4 of chapter 17, the spectral radius of **L** equals 1. This shows that *no complex number* ζ, $|\zeta| > 1$, *belongs to the spectrum of* **L**.

Next we determine the eigenvalues and eigenvectors of $\mathbf{L}$. Suppose that $\mathbf{L}x = \lambda x$; by definition (11) this means that $(a_1, a_2, a_3, \ldots) = \lambda(a_0, a_1, a_2, \ldots)$, which implies that

$$a_n = \lambda^n a_0. \tag{14}$$

Since x belongs to ℓ^2, $\sum |a_n|^2 < \infty$; this is satisfied by (14) iff $|\lambda| < 1$. So we see that the eigenvalues of $\mathbf{L}$ are all complex numbers λ, $|\lambda| < 1$. All eigenvalues λ belong to $\sigma(\mathbf{L})$; since the spectrum is closed, *all λ in the unit disk belong to $\sigma(\mathbf{L})$.*

Since $\mathbf{R}$ and $\mathbf{L}$ are adjoint to each other, it follows from theorem 3 that $\sigma(\mathbf{L}) = \sigma(\mathbf{R})$; from this and the previous two statements theorem 4 follows.   □

Exercise 2. Show that $\mathbf{R}$ has no eigenvalues.

Exercise 3. Show that the spectrum of $\mathbf{R}$ and $\mathbf{L}$ acting on the spaces, ℓ^p, $1 \le p \le \infty$, consists of all points of the unit disk.

20.3 VOLTERRA INTEGRAL OPERATORS

Take X to be $C[0, 1]$, and $\mathbf{V}$ to be integration:

$$(\mathbf{V}x)(s) = \int_0^s x(r)\, dr. \tag{15}$$

Theorem 5. *The spectrum of the operator (15) acting on $C[0, 1]$ consists of the single point $\lambda = 0$.*

Proof. The n-fold iterate of $\mathbf{V}$ is given by the formula

$$(\mathbf{V}^n x)(s) = \frac{1}{(n-1)!} \int_0^s (s - r)^{n-1} x(r)\, dr, \tag{15'}$$

as may be verified by induction, using integration by parts. For any s in $[0, 1]$

$$|\mathbf{V}^n x(s)| \le \frac{1}{(n-1)!} \int_0^s (s-r)^n |x|\, dr \le \frac{|x|}{n!}.$$

So for any x in $C[0, 1]$, $|\mathbf{V}^n x| \le |x|/n!$, which implies, by definition of norm, that $|\mathbf{V}|^n \le 1/n!$. It follows from this that

$$\lim_{n \to \infty} |\mathbf{V}^n|^{1/n} = 0,$$

and so by theorem 4 of chapter 17, the spectral radius of $\mathbf{V}$ is zero. Since the spectrum is nonempty, theorem 5 follows.   □

Example 1. $X = \ell^2$; $\{\lambda_n\}$ a given bounded sequence of complex numbers. For x given by (10), define

$$\mathbf{M}x = (\lambda_0 a_0, \lambda_1 a_1, \ldots, \lambda_n a_n) \cdots . \tag{16}$$

Exercise 4. Prove that the spectrum of $\mathbf{M}$ as defined by (16) is the closure of the set $\{\lambda_n\}$.

Exercise 5. Take X to be ℓ^p, $1 \leq p \leq \infty$, and define $\mathbf{M}$ by (16). Prove that $\sigma(\mathbf{M})$ is the closure of the set $\{\lambda_n\}$.

Exercise 6. Suppose that the kernel of the integral operator

$$\mathbf{K}f(x)a = \int_0^s \mathbf{K}(s, t) f(t) \, dt \tag{17}$$

is a continuous function of s, t in $t \leq s$.

(a) Show that $\mathbf{K}$ maps $C[0, 1]$ into $C[0, 1]$.
(b) Show that the spectrum of $\mathbf{K}$ consists of the single point 0.

Operators of form (17) are called *Volterra operators*, after Vito Volterra, who first investigated their theory.

20.4 THE FOURIER TRANSFORM

Denote the Fourier transform by $\mathbf{F}$:

$$(\mathbf{F}f)(u) = \frac{1}{\sqrt{2\pi}} \int f(x)e^{ixu} dx = \tilde{f}(u).$$

We take it as known from the theory of the Fourier transform, see Appendix B, that $\mathbf{F}$ is an invertible, norm-preserving map of $L^2(\mathbb{R})$ onto $L^2(\mathbb{R})$. The inverse is given by

$$f(x) = \frac{1}{\sqrt{2\pi}} \int \tilde{f}(u)e^{-ixu} \, du.$$

Replacing x by $-x$ gives

$$f(-x) = \frac{1}{\sqrt{2\pi}} \int \tilde{f}(u)e^{ixu} \, du.$$

Denote the mapping $f(x) \to f(-x)$ by $\mathbf{R}$, it follows from the formula above that $\mathbf{F}^2 = \mathbf{R}$. Since $\mathbf{R}^2 = \mathbf{I}$, it follows that

$$\mathbf{F}^4 = \mathbf{I}.$$

It follows then from the spectral mapping theorem that the spectrum of $\mathbf{F}$ lies on the set consisting of the fourth roots of $1 : \pm 1, \pm i$.

Exercise 7.

    (a) Show that $\mathbf{F}$ maps the space of functions of the form $p(x)e^{-x^2/2}$, p a polynomial of degree $\leq n$, into itself.

    (b) Show that $\mathbf{F}$ has eigenfunctions of form $p(x)e^{-x^2/2}$.

    (c) Show that the eigenfunctions in (b) span all of $L^2(\mathbb{R})$. (Hint: See section 9.1).

BIBLIOGRAPHY

Volterra, V. Sulla inversione degli integrali definiti. *An. Mat.* (2), **25** (1897): 139–178.

21

COMPACT MAPS

The notion of compact maps, and their properties, are the bread and butter of functional analysis.

We recall that a subset S of a complete metric space is called *precompact* if its closure is compact. The following are useful criteria for precompactness:

(a) S is precompact if and only if every sequence of points of S contains a Cauchy subsequence.

(b) S is precompact iff for every $\epsilon > 0$ it can be covered by a finite number of balls of radius ϵ.

We turn to precompact subsets of Banach spaces. The following are easily deduced from (a) or (b).

(c) If C_1 and C_2 are precompact subsets of a Banach space X, then $C_1 + C_2$ is precompact.

(d) If C is a precompact set in a Banach space, so is its convex hull.

(e) If C is a precompact subset of a Banach space X, $\mathbf{M}$ a linear, bounded map of X into another Banach space U, then $\mathbf{M} C$ is a precompact subset of U.

Exercise 1. Prove statements (c), (d) and (e).

21.1 BASIC PROPERTIES OF COMPACT MAPS

Definition. X and U denote Banach spaces. A linear map $\mathbf{C} : X \to U$ is called *compact* if the image $\mathbf{C} B$ of the unit ball B in X is precompact in U.

Theorem 1.

(i) *The sum of two compact maps:* $X \to U$ *is compact.*

(ii) *The scalar multiple of a compact map is compact.*

(iii) *Let V be a Banach space, $\mathbf{M} : U \to V$ a bounded linear map, $\mathbf{C} : X \to U$ compact. Then the product $\mathbf{M} \mathbf{C} : X \to V$ is compact.*

(iv) *Let Z denote a Banach space, $\mathbf{N} : Z \to X$ a bounded linear map, $\mathbf{C} : X \to U$ compact. Then $\mathbf{C}\,\mathbf{N} : Z \to U$ is compact.*

(v) *Let $\mathbf{C}_n : X \to U$ be a sequence of compact maps that converge uniformly to $\mathbf{C}$,*

$$\lim |\mathbf{C}_n - \mathbf{C}| = 0. \tag{1}$$

Then $\mathbf{C}$ is compact.

Proof. Denote by $\mathbf{C}_1$, $\mathbf{C}_2$ two compact linear maps of $X \to U$; that means that the images $\mathbf{C}_1 B = C_1$ and $\mathbf{C}_2 B = C_2$ of the unit ball B in X are precompact. According to (c) above, $C_1 + C_2$ is then precompact. Since $(\mathbf{C}_1 + \mathbf{C}_2) B$ is contained in $C_1 B + C_2 B$, and since a subset of a precompact set is precompact, part (i) follows.

Part (ii) is a special case of (iii). Part (iii) in turn follows from property (e) above of precompact sets.

(iv) Since the bounded map $\mathbf{N}$ carries the unit ball of Z into some ball in X, $\mathbf{C}\mathbf{N}\,B$ is precompact.

(v) Given any $\epsilon > 0$, choose n so large that $|\mathbf{C}_n - \mathbf{C}| < \epsilon$. Since $\mathbf{C}_n$ is a compact map, $\mathbf{C}_n B$ can be covered by a finite number of balls of radius ϵ; but then $\mathbf{C}\,B$ is covered by balls of radius 2ϵ around the same centers. $\square$

Suppose that $U = X$; in the language of algebra, theorem 1 says that the compact maps form a *closed two-sided ideal* in $\mathcal{L}(X)$. According to a theorem of Calkin, this is the *only* closed two-sided ideal when X is a Hilbert space.

Theorem 2. *X and U are Banach spaces, $\mathbf{C} : X \to U$ a compact linear map. Let Y be a closed subspace of X, and V the closure in U of $\mathbf{C}\,Y$.*

(i) *The restriction of $\mathbf{C}$ to $Y \to V$ is a compact map.*

(ii) *Suppose that $U = X$, and the closed subspace Y is invariant under $\mathbf{C}$, namely is mapped into itself by $\mathbf{C}$. Then $\mathbf{C} : X/Y \to X/Y$ is compact.*

Proof. Part (i) is utterly obvious; so is part (ii), once the definitions of the norms in the quotient spaces are put in place.

Exercise 2. Prove that a *degenerate* bounded linear map $\mathbf{D}(\dim R_{\mathbf{D}} < \infty)$ is compact.

In the rest of this chapter we present a theory of compact operators due to F. Riesz. We start by restating lemma 7 from chapter 5, on the geometry of normed linear spaces. This lemma will be used over and over again:

Lemma 3. *X is a normed linear space, Y a closed linear subspace of X properly contained in X. Then there is an x in X such that*

$$|x| = 1 \quad and \quad d(x, Y) = \inf_{y \text{ in } Y} |x - y| \geq \tfrac{1}{2}. \tag{2}$$

In chapter 5 this lemma was used to prove theorem 6, which asserts that *the unit ball* in a normed linear space *is compact iff the space is finite dimensional.* Here we use it to study of compact maps **C** of a Banach space X into itself. The next three theorems are the basic results concerning these maps.

Theorem 4. *Let* **C** *be a compact map of a Banach space* $X \to X$; *denote by* **I** *the identity map* $X \to X$; *set*

$$\mathbf{T} = \mathbf{I} - \mathbf{C}. \tag{3}$$

(i) *The nullspace* $N_\mathbf{T}$ *of* **T** *is finite-dimensional.*

(ii) *Denote by* N_j *the nullspace of* $\mathbf{T}^j$,

$$N_j = N_{\mathbf{T}^j}. \tag{4}$$

There is an integer i *such that*

$$N_k = N_i \quad for\ k > i. \tag{5}$$

(iii) *The range* $R_\mathbf{T}$ *of* **T** *is closed.*

Proof. (i) By definition (3) of **T**, y lies in $N_\mathbf{T}$ if $y = \mathbf{C}\,y$. Since **C** is assumed compact, it follows that the unit ball in $N_\mathbf{T}$ is precompact. But then according to the result quoted above, $N_\mathbf{T}$ is finite-dimensional.

(ii) Assume, on the contrary, that (5) fails for all i, that is, that N_{i-1} is a proper subset of N_i for all i. By lemma 3, there would be for every i a vector y_i such that

$$y_i \text{ in } N_i, \quad |y_i| = 1, \quad d(y_i, N_{i-1}) > \tfrac{1}{2} \tag{6}$$

Take $m < n$; by definition (3) of **T**,

$$\mathbf{C}\,y_n - \mathbf{C}\,y_m = y_n - \mathbf{T}\,y_n - y_m + \mathbf{T}\,y_m. \tag{7}$$

The last three terms on the right belong to N_{n-1}, so, by (6), their sum differs from y_n by $\tfrac{1}{2}$ at least. This proves that $|\mathbf{C}\,y_n - \mathbf{C}\,y_m| > \tfrac{1}{2}$. Clearly, the sequence $\{\mathbf{C}\,y_n\}$ contains no Cauchy subsequence. Since each $|y_n| = 1$, this contradicts compactness of **C**.

(iii) We have to show that if $\{y_k\}$ is a convergent sequence of points in $R_\mathbf{T}$,

$$\lim y_k = y, \quad y_k = \mathbf{T}\,x_k, \tag{8}$$

then their limit y also belongs to $R_\mathbf{T}$. Denote by d_k the distance of x_k from $N_\mathbf{T}$:

$$d_k = \inf_{z \text{ in } N_\mathbf{T}} |x_k - z|. \tag{9}$$

We claim that the sequence d_k is bounded. Indeed, we can choose z_k in $N_\mathbf{T}$ so that $w_k = x_k - z_k$ satisfies

$$|w_k| = |x_k - z_k| < 2d_k. \tag{9'}$$

Clearly, since $\mathbf{T}z_k = 0$,

$$\mathbf{T}\,w_k = \mathbf{T}\,x_k - \mathbf{T}\,z_k = y_k. \tag{10}$$

Suppose that the d_k were unbounded. Since it follows from (8) that the $|y_k|$ are bounded, we divide (10) by d_k and conclude that

$$\mathbf{T}\frac{w_k}{d_k} = \frac{y_k}{d_k} \to 0. \tag{11}$$

Set $u_k = w_k/d_k$. It follows from (9') that $|u_k| < 2$. Using (11) and the definition (3) of $\mathbf{T}$, we see that $u_k - \mathbf{C}\,u_k \to 0$. Since $\mathbf{C}$ is compact, the second term has a convergent subsequence; but then so does the first term:

$$u_k \to u. \tag{12}$$

Since $\mathbf{T}$ is continuous, it follows from (11) that the limit u satisfies $\lim \mathbf{T}\,u_k = \mathbf{T}\,u = 0$, that is, u belongs to $N_{\mathbf{T}}$. On the other hand, it follows from (9) that $|w_k - z| \geq d_k$ for all z in $N_{\mathbf{T}}$. Dividing by d_k and using $u_k = w_k/d_k$, we deduce that $|u_k - z| \geq 1$ for all z in $N_{\mathbf{T}}$. Since we may take $z = u$, this contradicts (12) and shows that the sequence d_k is bounded.

Using the definition (3) of $\mathbf{T}$, we deduce from (10) and (8) that

$$w_k - \mathbf{C}\,w_k = y_k \to y. \tag{13}$$

It follows from (9') and the boundedness of d_k that the sequence w_k is bounded. Then by the compactness of $\mathbf{C}$, the second term on the left in (13) has a convergent subsequence. It follows from (13) that then the same subsequence of w_k converges to a limit w, and since $\mathbf{T}$ is continuous, that w satisfies

$$w - \mathbf{C}\,w = \mathbf{T}\,w = y.$$

This proves that the range of $\mathbf{T}$ is closed, as asserted.   □

The next result states that for $\mathbf{C}$ compact the range of $\mathbf{T} = \mathbf{I} - \mathbf{C}$ has finite codimension, equal to the dimension of the nullspace of $\mathbf{T}$. Recalling from chapter 2 the notion of the *index* of a map as the difference of the two, we state the result as follows:

Theorem 5. *Let $\mathbf{C}$ be a compact map of a Banach space $X \to X$. Then $\mathbf{T} = \mathbf{I} - \mathbf{C}$ satisfies*

$$\operatorname{ind} \mathbf{T} = \dim N_{\mathbf{T}} - \operatorname{codim} R_{\mathbf{T}} = 0. \tag{14}$$

Proof. We start with the special case that $N_{\mathbf{T}}$ is trivial:

$$\dim N_{\mathbf{T}} = 0. \tag{15}$$

We show that then codim $R_\mathbf{T} = 0$, which means that $R_\mathbf{T}$ is the whole space X. Now suppose, on the contrary, that $R_\mathbf{T} = X_1$ is a *proper* subspace of X. Then, since by assumption (15), $\mathbf{T}$ is one-to-one, it follows that $\mathbf{T} X_1 = X_2$ is a *proper* subspace of X_1. Define X_k as $\mathbf{T}^k X$. We deduce similarly that $X \supset X_1 \supset X_2 \supset \cdots$, and that all inclusions are proper.

According to part (iii) of theorem 4, X_1, the range of $\mathbf{T}$, is closed. We claim that every subspace X_k is closed. Indeed X_k is the range of $\mathbf{T}^k$, and $\mathbf{T}^k = (\mathbf{I} - \mathbf{C})^k = \mathbf{I} + \sum_1^k (-1)^j \binom{k}{j} \mathbf{C}^j$ is, according to theorem 1, of the form $\mathbf{I}$ plus a compact operator. Therefore we can conclude from theorem 4 that X_k is closed. We appeal now to lemma 3; we can choose x_k in X_k so that

$$|x_k| = 1, \quad \text{dist}(x_k, X_{k+1}) > \tfrac{1}{2}. \tag{16}$$

Let m and n be two distinct indices, $m < n$. Then, using the definition (3) of $\mathbf{T}$,

$$\mathbf{C} x_m - \mathbf{C} x_n = x_m - \mathbf{T} x_m - x_n + \mathbf{T} x_n.$$

The last three terms on the right all belong to X_{m+1}; therefore, by (16), $|\mathbf{C} x_m - \mathbf{C} x_n| > \tfrac{1}{2}$. This contradicts the assumption that $\mathbf{C}$ maps the unit ball into a precompact set, and completes the proof of (14) under the assumption (15).

We turn now to the case when $\mathbf{T}$ has a nontrivial nullspace; according to theorem 4 there is an index i such that

$$N_{i+1} = N_i, \tag{17}$$

where N_j is defined by formula (4). It follows from that definition that $N = N_i$ is an invariant subspace of $\mathbf{T}$, and therefore also of $\mathbf{C}$. So we may apply part (ii) of theorem 2 and conclude that $\mathbf{C} : X/N \to X/N$ is a compact mapping. We claim that $\mathbf{T} : X/N \to X/N$ has trivial nullspace; for a nontrivial nullvector would imply that for some x *not* in N is mapped into N. Since N is the nullspace of $\mathbf{T}^i$, this would put x into N_{i+1}, the nullspace of $\mathbf{T}^{i+1}$. This of course contradicts (17). Thus $\mathbf{T}$ on X/N satisfies assumption (15); it follows then that $\mathbf{T}$ maps X/N one-to-one onto itself. This means that for any y in X there is an x in X, z in N such that

$$\mathbf{T}x = y + z. \tag{18}$$

We can express this as

$$X = R_\mathbf{T} + N, \tag{19}$$

meaning that every y in X can be expressed as the sum of a vector in $R_\mathbf{T}$ and of one in N. These spaces have, for $i > 1$, a nonempty intersection, consisting of those vectors n in N that lie in the range of $\mathbf{T}$; that is, the vectors n are of the form

$$n = \mathbf{T}z. \tag{20}$$

It follows from (17) that z in (20) belongs to N. According to a basic theorem of linear algebra, the dimension of the nullspace of $\mathbf{T}$ in N equals the codimension of the range of T in N. It follows that the dimension of the space of vectors of form (20) is

$$\dim N - \dim N_T. \tag{21}$$

Combining this with (19) shows that the codimension of R_T in X is equal to dim N_T, as asserted in (14). $\square$

21.2 THE SPECTRAL THEORY OF COMPACT MAPS

Theorem 6 (F. Riesz). *X denotes a Banach space, C a compact linear map of $X \to X$.*

   (i) The spectrum of C consists of an at most denumerable set of complex numbers $\{\lambda_n\}$ that accumulate only at 0. If $\dim X = \infty$, 0 belongs to $\sigma(C)$.

   (ii) Each nonzero λ_j is a point eigenvalue of C, of finite multiplicity; that is, for each $\lambda = \lambda_j$

      the nullspace of $C - \lambda$ is finite-dimensional,
      there is an integer i such that the nullspace of $(C - \lambda)^k$ is the same as the nullspace of $(C - \lambda)^i$ for all $k > i$.

  (iii) The resolvent $(\zeta - C)^{-1}$ has a pole at each nonzero λ_j.

Proof. Define T for $\zeta \neq 0$, as $T = I - \zeta^{-1}C$. It follows from theorem 5 that if the nullspace of T is trivial, meaning that dim $N_T = 0$, then the range of T is all of X. This shows that every nonzero point of the spectrum of T is an eigenvalue. It follows from part (ii) of theorem 4 that the multiplicty of an eigenvalue is finite; this proves part (ii) of theorem 6.

We turn to (i). To show that the eigenvalues λ_n of C can accumulate only at 0, consider an infinite sequence $\{\lambda_n\}$, $\lambda_n \neq \lambda_m$ for $n \neq m$, of eigenvalues, with corresponding eigenvectors x_n:

$$Cx_n = \lambda_n x_n. \tag{22}$$

Define Y_n to be the linear space spanned by $x_1, \ldots, x_n$. Since eigenvectors pertaining to distinct eigenvalues are linearly independent, Y_{n-1} is a proper subset of Y_n. We apply now lemma 3, with $X = Y_n$ and $Y = Y_{n-1}$; there is a y_n in Y_n such that

$$|y_n| = 1, \quad |y_n - y| > \tfrac{1}{2} \quad \text{for all } y \text{ in } Y_{n-1}. \tag{23}$$

By definition of Y_n, y_n is of the form

$$y_n = \sum_1^n a_j x_j.$$

Thus

$$\mathbf{C}y_n - \lambda_n y_n = \sum_1^n (\lambda_j - \lambda_n) a_j x_j \in Y_{n-1}. \tag{24}$$

This shows that for $n > m$,

$$\mathbf{C}y_n - \mathbf{C}y_m = \lambda_n y_n - y, \qquad y \text{ in } Y_{n-1}.$$

So by (23),

$$|\mathbf{C}y_n - \mathbf{C}y_m| \ge \frac{|\lambda_n|}{2}. \tag{25}$$

Since each y_n is a unit vector, and since $\mathbf{C}$ is assumed to map the unit ball into a precompact set, there can be only a finite number of λ_n with $|\lambda_n| > \delta$.

We can restate (25) in a quantitative form. Recall the definition of the *capacity function* $C(\epsilon, K)$ of a precompact set K in a metric space: it is the *maximum number of points* $z_1, \ldots, z_C$ in K such that the distance of any two distinct z_j is at least ϵ:

$$d(z_n, z_m) \ge \epsilon, \qquad n \ne m.$$

In terms of the function C the inequality (25) leads to the following estimate of the number $N(\epsilon)$ of eigenvalues $\ge \epsilon$:

$$N(\epsilon) \le C(\epsilon, 2\mathbf{C}(B)), \tag{26}$$

where B is the unit ball in X.

Exercise 3. Show that the factor 2 on the right in (26) can be omitted.

(iii) To show that the resolvent of $\mathbf{C}$ has poles at λ_j, take ζ to be near but $\ne \lambda_j$. By definition of the resolvent, $(\zeta - \mathbf{C})^{-1}x = u$ means that

$$x = \zeta u - \mathbf{C}u. \tag{27}$$

We will solve this equation for u in two stages. Choose i so large that $N_{i+1} = N_i$, where N_i is $N_{(\lambda - \mathbf{C})^i}$, and denote N_i as N. Since N is an invariant subspace of $\mathbf{C}$, $\mathbf{C}$ can be interpreted as a map

$$\mathbf{C} : X/N \longrightarrow X/N. \tag{28}$$

According to part (ii) of theorem 2, $\mathbf{C}$ in (28) is compact. We claim that λ belongs to the resolvent set of $\mathbf{C}$ over X/N. If it did not, λ would, by part (ii), belong to the point spectrum of $\mathbf{C}$ over X/N; that means that some point y in X, $\ne 0 \mod N$, would be mapped by $\mathbf{C}$ into N. But such a y would belong to N_{i+1} and not to N_i, contrary to our choice of i. It follows then, since $\mathbf{C}$ is compact over X/N, that $\lambda - \mathbf{C}$

is invertible on X/N. Since the collection of invertible maps is open, it follows that $\zeta - \mathbf{C}$ is invertible for $|\zeta - \lambda|$ sufficiently small, and that

$$|(\zeta - \mathbf{C})^{-1}| \leq \text{const.} \tag{29}$$

on X/N for all such ζ.

The first step in solving (27) is to solve the congruence

$$\zeta v - \mathbf{C}v \equiv x \bmod N; \tag{30}$$

according to (29), (30) has a unique solution v, and $|v| \leq \text{const.}\,|x|$. The congruence (30) means that

$$\zeta v - \mathbf{C}v = x - n, \qquad n \in N. \tag{31}$$

The second step is to find a solution z in N of the equation

$$\zeta z - \mathbf{C}z = n. \tag{32}$$

Adding (32) to (31), we obtain $u = v + z$ as the solution of (27).

Solving (32) for z in N is a problem in linear algebra. By definition of λ as an eigenvalue of $\mathbf{C}$, we know that λ belongs to the spectrum of $\mathbf{C}$ over N. Since N is finite dimensional, it follows that no ζ sufficiently close to λ and $\neq \lambda$ belongs to the spectrum of $\mathbf{C}$ over N. Thus (32) has a unique solution for such ζ, and since the resolvent over a finite-dimensional space is a rational function, that solution z satisfies $|z| \leq \text{const.}\,|\zeta - \lambda|^{-i}|n|$. It follows from (31) that $|n| \leq \text{const.}\,|x|$, so $|z| \leq \text{const.}\,|\zeta - \lambda|^{-i}|x|$. Combining this with $|v| \leq \text{const.}\,|x|$, we get

$$|u| = |(\zeta - \mathbf{C})^{-1}x| = |v + z| \leq |v| + |z| \leq \text{const.}\,|\zeta - \lambda|^{-i}|x|.$$

This inequality can be expressed so: for ζ near enough to λ but $\neq \lambda$,

$$|(\zeta - \mathbf{C})^{-1}| \leq \text{const.}\,|\zeta - \lambda|^{-i}.$$

In words, near an eigenvalue λ of a compact map $\mathbf{C}$, the resolvent blows up at most like the minus ith power of the distance to λ. From this it is easy to deduce, as in classical function theory, that the resolvent of $\mathbf{C}$ has a pole of order i. This completes the proof of theorem 6. $\square$

Note that the proof of part (iii) gives another proof of part (ii).

Exercise 4. Show that the resolvent of $\mathbf{C}$ has a Laurent expansion around λ of the form

$$(\zeta - \mathbf{C})^{-1} = \sum_{-i}^{\infty} \mathbf{A}_j(\zeta - \lambda)^j. \tag{33}$$

Show that $\mathbf{A}_{-1}$ is a projection, namely that $\mathbf{A}^2_{-1} = \mathbf{A}_{-1}$; see theorem 6 of chapter 17. Show that the range of this projection is N_i; show that

$$\mathbf{A}_{-j} = (\mathbf{C} - \lambda)^{j-1}\mathbf{A}_{-1}, \quad j = 2, \ldots, i. \tag{34}$$

Exercise 5. Show that a compact operator on a Banach space X, $\dim X = \infty$, is not invertible.

In what follows, X denotes a Banach space, and $\mathbf{B}$ a bounded linear operator: $X \to X$ such that for some integer n

$$\mathbf{B}^n = \mathbf{C} \tag{35}$$

is a compact operator. The basic properties of compact operators, theorems 4, 5, and 6, are true for operators whose power is compact.

Theorem 4′. *Let $\mathbf{B} : X \to X$ be a map with compact power; set*

$$\mathbf{S} = \mathbf{I} - \mathbf{B}. \tag{36}$$

(i) *The nullspace $N_{\mathbf{S}}$ of $\mathbf{S}$ is finite-dimensional.*
(ii) *There is an integer i such that*

$$N_{\mathbf{S}^k} = N_{\mathbf{S}^i} \quad \text{for all } k > i. \tag{37}$$

(iii) *The range $R_{\mathbf{S}}$ of $\mathbf{S}$ is closed.*

Theorem 5′. *For $\mathbf{S}$ as defined in (36).*

$$\operatorname{ind}\mathbf{S} = \dim N_{\mathbf{S}} - \operatorname{codim} R_{\mathbf{S}} = 0. \tag{38}$$

Theorem 6′. $\mathbf{B} : X \to X$ *a map with compact power.*

(i) *The spectrum of $\mathbf{B}$ consists of an at most denumerable set of complex numbers $\{\beta_n\}$ that accumulate only at 0. If $\dim X = \infty$, 0 belongs to $\sigma(\mathbf{B})$.*
(ii) *Each β_j is of finite multiplicity and finite index.*
(iii) *The resolvent of $\mathbf{B}$ has poles at β_j.*

Proof. We start with the identity

$$\mathbf{I} - \mathbf{B}^n = (\mathbf{I} - \mathbf{B})(\mathbf{I} + \mathbf{B} + \cdots + \mathbf{B}^{n-1}). \tag{39}$$

Denote $\mathbf{I} - \mathbf{B}^n = \mathbf{T}, \mathbf{I} + \mathbf{B} + \cdots + \mathbf{B}^{n-1} = \mathbf{Q}$. Then (39) can be rewritten as

$$\mathbf{T} = \mathbf{SQ} = \mathbf{QS}. \tag{40}$$

From this we deduce that

$$N_\mathbf{T} \supset N_\mathbf{S}, \qquad R_\mathbf{T} \subset R_\mathbf{S}, \tag{41}$$

By assumption $\mathbf{B}^n = \mathbf{C}$ is compact; it follows from theorems 4 and 5 that

$$\dim N_\mathbf{T} < \infty, \qquad \operatorname{codim} R_\mathbf{T} < \infty. \tag{42}$$

Combining (42) and (41), it follows that $\dim N_\mathbf{S} < \infty$, $\operatorname{codim} R_\mathbf{S} < \infty$. By theorem 4, $R_\mathbf{T}$ is closed; from this and (41) it follows that so is $R_\mathbf{S}$.

Raising (40) to the kth power gives

$$\mathbf{T}^k = \mathbf{S}^k \mathbf{Q}^k = \mathbf{Q}^k \mathbf{S}^k,$$

from which we deduce that

$$N_{\mathbf{T}^k} \supset N_{\mathbf{S}^k}, \qquad R_{\mathbf{T}^k} \subset R_{\mathbf{S}^k}. \tag{43}$$

By assumption, $\mathbf{B}^n = \mathbf{C}$ is compact, by part (ii) of theorem 4, $N_{\mathbf{T}^k}$ is independent of k for $k > i$. It follows then from (42) that the nullspaces $N_{\mathbf{S}^k}$, $k = 1, 2, \ldots$ are all contained in the finite-dimensional space $N_{\mathbf{T}^i}$. From this it follows that the spaces $N_{\mathbf{S}^k}$, ordered by inclusion, are independent of k for k large enough. This completes the proof of theorem $4'$. $\square$

We turn now to theorem $6'$. According to the spectral mapping theorem (see theorem 5 of chapter 17), and since $\mathbf{C} = \mathbf{B}^n$,

$$\sigma(\mathbf{C}) = \sigma(\mathbf{B}^n) = \sigma(\mathbf{B})^n. \tag{44}$$

By assumption, $\mathbf{C}$ is compact; so according to theorem 6, $\sigma(\mathbf{C})$ consists of a denumerable set of points accumulating only at 0. It follows from (44) that the same is true of the spectrum of $\mathbf{B}$; this proves (i).

Exercise 6. Estimate the number of eigenvalues β of $\mathbf{B}$ that satisfy $|\beta| > \epsilon$ in terms of the capacity $C(\epsilon, \mathbf{B}^n B)$.

We have already proved part (ii) of theorem $6'$; we leave part (iii) as the next exercise.

Exercise 7. Prove part (iii) of theorem $6'$. $\square$

We turn now to theorem $5'$. If the operator $\mathbf{Q}$ defined in (39) is invertible, then it follows from the factorizations (40) that the nullspaces of $\mathbf{T}$ and $\mathbf{S}$ have the same dimension, and their ranges have the same codimension. Thus ind $\mathbf{T} =$ ind $\mathbf{S}$; since according to theorem 5, ind $\mathbf{T}$ is zero, ind $\mathbf{S} = 0$ also, as asserted by theorem $5'$.

To ascertain that $\mathbf{Q}$ is invertible, we note that $\mathbf{Q}$ is a polynomial in $\mathbf{B}$, and we appeal to the spectral mapping theorem, theorem 4 of chapter 17. According to it, the spectrum of $\mathbf{Q}$ consists of complex numbers of the form $1 + \beta + \cdots + \beta^n$, β in the spectrum of B. It follows that if $\sigma(\mathbf{B})$ contains no nth root of unity other than 1, $\mathbf{Q}$ is invertible. If $\sigma(B)$ does contain nth roots of unity, we perturb slightly the product (40). $\qquad\square$

Exercise 8. Show by an example that if $\mathbf{M}$ is the *strong* limit of a sequence of compact operators, $\mathbf{M}$ need not be compact.

Exercise 9. Show that if $\mathbf{C}$ is compact and $\{\mathbf{M}_n\}$ tends *strongly* to $\mathbf{M}$, then $\mathbf{CM}_n$ and $\mathbf{M}_n\mathbf{C}$ tend *uniformly* to $\mathbf{CM}$ and $\mathbf{MC}$, respectively.

Theorem 7 (Schauder). *The transpose $\mathbf{C}'$ of a compact map $\mathbf{C} : X \to U$ is compact, and conversely.*

Proof. We have to show that the image of the unit ball in U' under $\mathbf{C}'$ is precompact. According to criterion a for precompactness, given any sequence $\{\ell_n\}$ in U, $|\ell_n| \leq 1$, we have to show that $\{\mathbf{C}'\ell_n\}$ has a Cauchy subsequence. Denote by K the closure of $\mathbf{C}B$, B the unit ball in X. Since $\mathbf{C}$ is assumed compact, K is a compact subset of U. The functions (ℓ_n, u) of u are uniformly bounded and are equicontinuous on K:

$$|(\ell_n, u) - (\ell_n, v)| = |(\ell_n, u - v)| \leq |u - v|.$$

According to the theorem of Arzela-Ascoli, a *uniformly bounded, equicontinuous* sequence of functions on a *compact* set K has a uniformly convergent subsequence:

$$|(\ell_n, u) - (\ell_m, u)| < \epsilon \tag{45}$$

for all $n, m > N$, for all u in K. Since every u of the form $u = \mathbf{C}x$, $|x| \leq 1$, belongs to K, we conclude from (45) that

$$|(\ell_n - \ell_m, \mathbf{C}x)| = |(\mathbf{C}'\ell_n - \mathbf{C}'\ell_m, x)| \leq \epsilon \tag{46}$$

for all x, $|x| \leq 1$. By definition of the norm in U', this proves that

$$|\mathbf{C}'\ell_n - \mathbf{C}'\ell_m| < \epsilon \tag{47}$$

for $n, m > N$, namely that $\{\mathbf{C}'\ell_n\}$ is a Cauchy sequence.

Conversely, if $\mathbf{C}'$ is a compact mapping, then by what we have proved above, $\mathbf{C}''$ is a compact map. Since $\mathbf{C}$ is the restriction of $\mathbf{C}''$ to X, it follows from part (i) of theorem 2 that $\mathbf{C}$ is a compact map. This completes the proof of theorem 7. $\qquad\square$

Theorem 8. *Let $\mathbf{C} : X \to X$ be a compact map, and $\mathbf{T} = \mathbf{I} - \mathbf{C}$.*

  (i) A vector u belongs to the range of $\mathbf{T}$ iff $(u, \ell) = 0$ for every ℓ in the nullspace of $\mathbf{T}'$.
  (ii) $\dim N_{\mathbf{T}'} = \dim N_{\mathbf{T}}$.

Proof. The defining relation between $\mathbf{T}$ and its transpose is

$$(\mathbf{T}x, \ell) = (x, \mathbf{T}'\ell).$$

It follows that the nullspace of $\mathbf{T}'$ is the annihilator $R_{\mathbf{T}}^{\perp}$ of the range of $\mathbf{T}$.

(i) Since according to theorem 4, $R_{\mathbf{T}}$ is closed, it follows from theorem 8 of chapter 8 that every vector u that satisfies $(u, \ell) = 0$ for all ℓ in $R_{\mathbf{T}}^{\perp}$ belongs to $R_{\mathbf{T}}$.

(ii) We saw in chapter 8, exercise 2, that for a closed subspace R of X, the annihilator of R is isomorphic with the dual of X/R. Therefore $\dim R^{\perp} = \dim(X/R)' = \dim X/R = \operatorname{codim} R$. Apply this to $R = R_{\mathbf{T}}$. Since the annihilator of $R_{\mathbf{T}}$ is the nullspace of $\mathbf{T}'$, $\dim N_{\mathbf{T}'} = \operatorname{codim} R_{\mathbf{T}}$. Since, by theorem 5, $\operatorname{ind}_{\mathbf{T}} = 0$, we deduce that $\operatorname{codim} R_{\mathbf{T}} = \dim N_{\mathbf{T}}$. It follows that $\dim N_{\mathbf{T}'} = \dim N_{\mathbf{T}}$. Theorem 8 is called the Fredholm alternative. □

Theorem 9. *A compact map* $\mathbf{C} : X \to U$ *maps every weakly convergent sequence into one that converges strongly.*

Exercise 10.

(i) Prove theorem 9.

(ii) What about its converse?

NOTE. F. Riesz's paper was, and is, fundamental for the theory of compact operators in a Banach space. There he took Hilbert's definition of a compact operator, called in those days completely continuous, and showed how to extend this notion to Banach spaces. His accomplishment is all the more remarkable, since 1918 predates Banach's fundamental paper by a good five years! And unlike Banach, he treats normed linear spaces over the complex field, which is essential for spectral theory.

HISTORICAL NOTE. Julius Schauder (1899–1943) was the most brilliant of the Polish mathematicians of his time, Schauder bases, the Schauder fixed point theorem, the Leray-Schauder degree of a mapping, as well as many fundamental results in the theory of elliptic and hyperbolic partial differential equations, are his creation. Being Jewish, he was killed during the Nazi occupation of Poland. Such things were so routine, nobody knows when or where.

BIBLIOGRAPHY

Calkin, J. W. Two-sided ideals and congruences in the ring of bounded operator in Hilbert space. *An. Math.* (2), **42** (1941): 839–873.

Riesz, F. Über lineare Funktionalgleichungen. *Acta Math.*, **41** (1918): 71–98.

Schauder, J. Über lineare, vollstätige Funktionaloperationen. *Studia Math.* **2** (1930): 183–196.

22

EXAMPLES OF COMPACT OPERATORS

22.1 COMPACTNESS CRITERIA

We start by giving some useful criteria in various topologies for sets of functions to be compact. The first theorem is the famous Arzela-Ascoli criterion.

Definition. S is a Hausdorff space; a collection $\{g\}$ of complex-valued functions g on S is called *equicontinuous if* for every point s of S and every $\epsilon > 0$ there is an open set N containing s such that for every r in N and every g of the collection

$$|g(r) - g(s)| < \epsilon. \tag{1}$$

Theorem 1. S *is a compact Hausdorff space,* $\{g\}$ *a collection of complex-valued functions on* S *satisfying these conditions:*

  (i) The collection $\{g\}$ *is equicontinuous.*
  (ii) $\{g\}$ *is uniformly bounded:*

$$|g(s)| \leq M \tag{1'}$$

  for all s *in* S *and every* g *in the collection.*

Assertion. *The collection* $\{g\}$ *is precompact in the maximum norm on* S.

For a proof, we refer to any text on real variables, such as that by Royden.

Exercise 1. Show that conditions (i) and (ii) of theorem 1 are necessary for a collection to be precompact in the maximum norm.

Exercise 2. Show that if Q is an open bounded set in $\mathbb{R}^n$ such that any two points of Q can be connected by a path of length $\leq \ell$, then a family of functions $\{g\}$ on Q is precompact in the maximum norm if the functions g and their first partial derivatives are uniformly bounded in Q:

$$|g| \leq M, \quad |\partial_i g| \leq M \quad \text{in } Q.$$

Exercise 3. Formulate a version of theorem 1 for functions whose values lie in a metric space.

Another equally—or even more—important compactness criterion is due to Rellich:

Theorem 2. *Q is a domain in $\mathbb{R}^m$, open and bounded, whose boundary is smooth. Suppose that $\{u\}$ is a collection of functions in Q such that the functions u and their first derivatives are uniformly bounded in the $L^2(Q)$ norm:*

$$\|u\| \leq M, \quad \|\partial_i u\| \leq M, \quad i = 1, \dots, m. \tag{2}$$

Then the collection $\{u\}$ is precompact in the $L^2(Q)$ norm.

Proof of Rellich's criterion is based on Poincaré's inequality: for any smooth function u defined over a smoothly bounded domain Q,

$$\int_Q |u(x)|^2 \, dx \leq \left| \int u \, dx \right|^2 + d^2 \sum \int_Q |\partial_i u|^2 \, dx, \tag{3}$$

where d is the diameter of Q. For details we refer to Courant-Hilbert.

We remark that Poincaré's inequality, and likewise Rellich's criterion, do not hold for arbitrary bounded domains.

22.2 INTEGRAL OPERATORS

We turn now to integral operators, see equation (1), chapter 16, where we have defined $g = \mathbf{K}f$ by

$$g(s) = \int_T K(s, t) f(t) \, dn(t), \tag{4}$$

$s \in S, t \in T$, both compact metric spaces. We will investigate conditions on the kernel $K(s, t)$ that make $\mathbf{K}$ compact in various topologies.

Theorem 3. *The integral operator (4) is compact as a map $L^1 \to C$ if the kernel $K(s, t)$ is a continuous function of s and t.*

Proof. To show that $\mathbf{K}$ a compact operator with respect to these norms, we have to verify that the functions

$$g = \mathbf{K}f, \quad |f|_{L^1} \leq 1, \tag{5}$$

satisfy the compactness criteria (1), (1′). Since S and T are compact spaces, the kernel is uniformly bounded, so by theorem 1 of chapter 16, $\mathbf{K}$ is bounded; thus condition (1′) is satisfied. To verify (1), we study $g(r) - g(s)$:

$$|g(r) - g(s)| = \left| \int [K(r,t) - K(s,t)] f(t) \, dn \right| \leq \sup_t |K(r,t) - K(s,t)| \, |f|_{L^1}. \tag{6}$$

Since $T \times S$ is compact, the kernel is uniformly continuous, so the right side of (6) tends to zero as r tends to s. This shows that the functions g in (5) are equicontinuous.
□

Using an analogous argument we can prove

Theorem 3′. *The integral operator* (4) *is compact as a map* $C \to C$ *if the kernel* $K(s,t)$ *is a continuous function of* s *in the* L^1 *norm with respect to* t.

Next we recall theorem 2 of chapter 16, which says that an integral operator $\mathbf{K}$ whose kernel K is square integrable is a bounded mapping: $L^2 \to L^2$, and that

$$\|\mathbf{K}\|^2 \leq \iint |K(s,t)|^2 \, dm \, dn. \tag{7}$$

Condition (7) implies more:

Theorem 4. *An integral operator with a square integrable kernel is a compact map of* $L^2 \to L^2$.

Proof. Let $\{u_j\}$ be an orthonormal basis of $L^2(S, dm)$. We expand $K(s,t)$ for fixed t into a series

$$K(s,t) = \sum_j K_j(t) \, u_j(s). \tag{8}$$

Using the Parseval relation for $\{u_j\}$ gives for almost all values of t,

$$\int |K(s,t)|^2 \, dm(s) = \sum_j |K_j(t)|^2. \tag{9}$$

Integrating (9) gives

$$\iint |K(s,t)|^2 \, dm \, dn = \sum \int |K_j(t)|^2 \, dn(t). \tag{10}$$

Now define

$$K_N(s,t) = \sum_{j \leq N} K_j(t) \, u_j(s), \tag{11}$$

and denote by $\mathbf{K}_N$ the integral operator with kernel K_N. Clearly, $\mathbf{K}_N$ is a degenerate operator, meaning its range is finite dimensional; then, as observed in exercise 2 of chapter 21, $\mathbf{K}_N$ is compact. We claim that as $N \to \infty$, $\mathbf{K}_N$ tends in norm to $\mathbf{K}$. For applying inequality (7) to $\mathbf{K} - \mathbf{K}_N$, we get

$$\|\mathbf{K} - \mathbf{K}_N\|^2 \leq \iint |K - K_N|^2 \, dm \, dn = \sum_{j > N} \int K_j^2(t) \, dn(t). \tag{12}$$

It follows from (10) that the right side of (12) tends to 0 as $N \to \infty$. As noted in theorem 1, part (v), chapter 21, the uniform limit of compact maps is compact. $\quad\square$

Exercise 4. Construct an example of an integral operator whose kernel satisfies Holmgren's condition (see theorem 3 of chapter 16) that is not compact.

In chapter 20, theorem 4, we have shown that the operation of integration, defined as

$$(\mathbf{V}x)(s) = \int_0^s x(t) \, dt, \tag{13}$$

regarded as mapping of $C[0, 1]$ to $C[0, 1]$, has the spectrum consisting of 0. Here we give another proof of this fact. The kernel of the integral operator $\mathbf{V}$ in (15) is

$$K(s, t) = \begin{cases} 1 & \text{for } t < s \\ 0 & \text{for } t > s. \end{cases} \tag{14}$$

Clearly, T is a continuous function of s in the L^1 norm in t. So according to theorem 3′, $\mathbf{V}$ is a compact map of $C[0, 1]$ into itself.

According to theorem 5, chapter 21, the spectrum of a compact operator consists of 0 and eigenvalues $\lambda \neq 0$. We show now that $\mathbf{V}$ has no eigenvalues, for suppose that x were an eigenfunction, λ an eigenvalue $\neq 0$,

$$\mathbf{V}x = \int_0^s x(t) \, dt = \lambda x(s). \tag{15}$$

The left side is a differentiable function of s, therefore so is the right. Differentiating (15), we get

$$x(s) = \lambda x'(s).$$

All nonzero solutions of this are of the form $x(s) = c e^{s/\lambda}$, $c \neq 0$. In particular, $x(0) = c \neq 0$; setting $s = 0$ in (15), we get $0 = \lambda x(0)$, in contradiction to what we have shown above. This proves that $\mathbf{V}$ has no eigenvalue $\lambda \neq 0$, and so

$$\sigma(\mathbf{V}) = \{0\}. \tag{16}$$

$\square$

22.3 THE INVERSE OF ELLIPTIC PARTIAL DIFFERENTIAL OPERATORS

We turn now to a class of operators that are defined by solving differential equations. Let Q be a bounded domain in $\mathbb{R}^n$ with a smooth boundary; denote by Δ the Laplace operator. It is well known, see section 7.2, that the boundary value problem

$$\Delta u = f \text{ in } Q, \quad u = 0 \text{ on } \partial Q, \tag{17}$$

has a unique solution u for every f in $C^\infty(Q)$.

Definition. Denote the solution u of (17) as

$$u = Sf. \tag{18}$$

Theorem 5. S *is a compact map of* $L^2(Q)$ *into* $L^2(Q)$.

Proof. Multiply (17) by u, and integrate over Q, integrating by parts:

$$-\int_Q \sum |\partial_j u|^2 ds = \int_Q fu\, ds. \tag{19}$$

As noted in lemma 1 of chapter 7, for all functions u that vanish on ∂Q,

$$\|u\|_0 \le d\|u\|_1, \tag{20}$$

where $\|u\|_0$ denotes the L^2-norm of u over Q, $\|u\|_1^2$ the sum of the squares of the L^2-norms of the first derivatives of u and d the diameter of Q. Using the Schwarz inequality on the right of (19), and then (20), we get

$$\|u\|_1^2 \le \|f\|_0\|u\|_0 \le d\|f\|_0\|u\|_1,$$

so

$$\|u\|_1 \le d\|f\|_0, \quad \|u\|_0 \le d^2\|f\|_0. \tag{21}$$

The image of the unit ball in $L^2(Q)$ under S consists of solutions u of (17) corresponding to f with $\|f\|_0 \le 1$. It follows from (21) that these satisfy

$$\|u\|_1 \le d, \quad \|u\|_0 \le d^2. \tag{21'}$$

According to theorem 2, a set of functions satisfying (21') is precompact in the $L^2(Q)$ norm. This proves that S is a compact operator. □

Exercise 5. Show that theorem 5 is true even when Q is not smoothly bounded. (Hint: Use lemma 2 in chapter 7.)

Theorem 5 can be extended with practically no change to all second order elliptic operators with Dirichlet boundary condition. Under the Neumann boundary condition $u_n = 0$ it is necessary to require Q to be smoothly bounded.

22.4 OPERATORS DEFINED BY PARABOLIC EQUATIONS

Our next example concerns solutions of parabolic equations of the form

$$u_t = \Delta u \tag{22}$$

for functions $u(s, t)$, $s \in Q$ as in the previous example, and $t \geq 0$. It is well known, and will be proved in the chapter on semigroups, that the initial boundary value problem for the parabolic equation has a unique solution; that is, there is a unique solution of (23) satisfying

$$u(s, 0) \text{ given in } Q, \qquad u = 0 \text{ on } \partial Q \text{ for all } t > 0. \tag{23'}$$

We denote by $S(T)$ the operator relating $u(s, 0)$ to $u(s, T)$, $t \geq 0$:

$$S(T) : u(s, 0) \longrightarrow u(s, T). \tag{23''}$$

Theorem 6. $S(T) : L^2(Q) \to L^2(Q)$ *is a compact operator for* $T > 0$.

Proof. Multiply equation (23) by u, integrate with respect to s over Q, with respect to t from 0 to T. We get, after integration by parts, that

$$\frac{1}{2} \int u^2 ds \Big|_0^T = \int_0^T \int u \Delta u \, ds \, dt = -\iint \sum |\partial_j u|^2 \, ds \, dt \leq 0. \tag{24}$$

This proves that

$$\int u^2(s, T) \, ds \leq \int u^2(s, 0) \, ds,$$

namely that $\|u(t)\|_0$ is a *decreasing* function of t. In terms of the solution operator S defined in (23'') this can be expressed as follows:

$$\|S(T)\| \leq 1. \tag{25}$$

Next multiply (23) by $t \Delta u$ and integrate with respect to s over Q and with respect to t from 0 to T. We get

$$\int_0^T \int t \, u_t \, \Delta u \, ds \, dt = \int_0^T \int t (\Delta u)^2 ds \, dt. \tag{26}$$

Integrate by parts on the left with respect to x and t; using the abbreviation u_j for the partial derivatives of u, we get the following expression for the left side of (26):

$$-\frac{T}{2} \int \sum u_j^2(T)\, ds + \frac{1}{2} \iint \sum u_j^2\, dt\, ds.$$

Since the right side of (26) is positive, we deduce that

$$\frac{T}{2} \int \sum u_j^2(T)\, ds \leq \frac{1}{2} \iint \sum u_j^2\, ds\, dt.$$

We can use (24) to estimate the right side above; we obtain

$$T \int \sum u_j^2(T)\, ds \leq \frac{1}{2} \int u^2(0)\, ds.$$

Using Rellich's compactness criterion, theorem 2, and exercise 6, we conclude that for $T > 0$, the image of $\int u^2(0)\, ds \leq 1$ under $\mathbf{S}(T)$ is precompact.   □

22.5 ALMOST ORTHOGONAL BASES

The following result due to Paley and Wiener shows that a not too large perturbation of an orthonormal basis is a basis:

Theorem 7. *Let H denote a Hilbert space, and $\{x_n\}$ an orthonormal basis of H. Let $\{y_n\}$ a collection of elements that doesn't differ too much from $\{x_n\}$ in the sense that*

$$\sum \|x_n - y_n\|^2 < \infty. \tag{27}$$

We further assume that the $\{y_n\}$ are linearly independent in the sense that no y_n lies in the closed linear span of the other y_k.

Assertion. *The $\{y_n\}$ form a basis in the sense that every u in H can be written uniquely as a linear combination of $\{y_n\}$.*

Proof. (Birkhoff-Rota, and Sz. Nagy) Since the $\{x_n\}$ are an orthonormal basis, every u in H can be expanded as

$$u = \sum a_n x_n, \quad a_n = (u, x_n), \quad \text{and} \quad \|u\|^2 = \sum |a_n|^2. \tag{28}$$

Define the linear map $\mathbf{B} : H \to H$ as follows: for u given by (28),

$$\mathbf{B}u = \sum a_n y_n. \tag{29}$$

The series on the right converges; for write $\mathbf{B}u - u = \sum a_n(y_n - x_n)$. Then, by the triangle inequality and Schwarz's inequality,

$$\|\mathbf{B}u-u\| \leq \sum |a_n| \, \|y_n-x_n\| \leq \left(\sum |a_n|^2\right)^{1/2} \left(\sum \|y_n - x_n\|^2\right)^{1/2} \leq \|u\| \text{ const.} \tag{30}$$

where in the last step we have used assumption (27), and (28). This proves that $\mathbf{B} - \mathbf{I}$ is a bounded linear map. We claim that $\mathbf{B} - \mathbf{I}$ is a compact map; for write

$$\mathbf{B}u - u = \sum_{0}^{N} a_n(y_n - x_n) + \sum_{N+1}^{\infty} = \mathbf{G}_N u + \mathbf{R}_N u. \tag{31}$$

We estimate $\mathbf{R}_N u$ as in

$$\|\mathbf{R}_N u\| \leq \|u\| \left(\sum_{N+1}^{\infty} \|y_n - x_n\|^2\right)^{1/2}. \tag{32}$$

It follows from (32) and (27) that

$$\lim_{N\to\infty} \|\mathbf{R}_N\| = 0.$$

This and the decomposition (31) show that $\mathbf{B} - \mathbf{I}$ is the uniform limit of $\mathbf{G}_N$. The mapping $\mathbf{G}_N$ are degenerate, therefore compact; so $\mathbf{B} - \mathbf{I}$ is the uniform limit of compact maps. According to theorem 2 of chapter 21, $\mathbf{B} - \mathbf{I}$ itself is compact.

The nullspace of $\mathbf{B}$ is trivial; for if $\mathbf{B}u$ were $= 0$ for some $u \neq 0$, by (29) $0 = \sum a_n y_n$ would give a nontrivial linear relation among the y_n, which is excluded by assumption (ii). We apply now theorem 5 of chapter 21 to $\mathbf{B} = \mathbf{I} + (\mathbf{B} - \mathbf{I})$ to conclude that the range of $\mathbf{B}$ is all of H, as asserted in theorem 7. $\qquad\qquad\square$

BIBLIOGRAPHY

Birkhoff, G. and Rota, G.-C. On the completeness of Sturm-Lionville expansions. *Am. Math. Monthly*, **67** (1960): 835–841.

Courant, R. and Hilbert, D. *Methoden der Mathematischen Physik*. Springer, Berlin, 1993; see p. 488.

Paley, R. E. A. C. and Wiener, N. *Fourier Transforms in the Complex Domain*. AMS Coll Publ., New York, 1934.

Poincaré, H. *Rend. Circ. Mat.* Palermo (1894).

Rellich, F. Ein Satz über mittlere Kowvergenz. *Ges. Wiss. Gött.*, Nachrichten (1930).

Sz.-Nagy, B. Expansion theorems of Paley-Wiener type. *Duke Math. J.*, **14** (1947): 975–978.

23

POSITIVE COMPACT OPERATORS

23.1 THE SPECTRUM OF COMPACT POSITIVE OPERATORS

A classic and important result of linear algebra, due to Perron (e.g., see Lax, p. 196) asserts that a matrix with all positive entries has a positive eigenvalue that is the largest in absolute value among all eigenvalues. In this chapter we present an infinite-dimensional generalization.

Theorem 1. *Let Q be a compact Hausdorff space, $X = C(Q)$, $\mathbf{K}$ a linear map $C(Q) \to C(Q)$, mapping real-valued functions into real-valued functions. We assume that*

   (i) $\mathbf{K}$ *is strictly positive in the sense that if p is any nonnegative function on Q, $p \not\equiv 0$, then $\mathbf{K}p$ is positive on Q.*

   (ii) $\mathbf{K}$ *is compact.*

We claim that then

   $\mathbf{K}$ *has a positive eigenvalue σ of multiplicity one and index one, with positive eigenfunction.*

   All other eigenvalues μ of K are smaller in absolute value than σ:

$$|\mu| < \sigma. \tag{1}$$

Proof. The strict positivity of the operator $\mathbf{K}$ implies that $\mathbf{K}$ is *strictly monotone*, that is,

$$\text{if} \quad x \leq y \quad \text{and} \quad x \not\equiv y, \quad \text{then} \quad \mathbf{K}x < \mathbf{K}y. \tag{2}$$

We consider now the action of $\mathbf{K}$ on nonnegative functions. We consider the set of all positive numbers κ such that there is a nonnegative function x such that at all

POSITIVE COMPACT OPERATORS

points of Q,

$$\kappa x \leq \mathbf{K}x, \qquad 0 \leq x. \tag{3}$$

We denote by $|x|$ the maximum norm of x and by $|\mathbf{K}|$ the corresponding norm of $\mathbf{K}$. It is easy to see that the set of κ is bounded, for (3) implies that $\kappa|x| \leq |\mathbf{K}x| \leq |\mathbf{K}|\,|x|$, which shows that no κ can exceed $|\mathbf{K}|$. We show now that the set of κ satisfying (3) is not empty. Indeed, take $x(t) \equiv 1$; it follows from the strict positivity of the $\mathbf{K}$ that $\mathbf{K}x$ is positive. Inequality (3) holds with $\kappa = \min \mathbf{K}x$.

Next we use the monotonicity of $\mathbf{K}$; combining (2) and (3), we get $\kappa \mathbf{K}x \leq \mathbf{K}^2 x$, which together with (3) gives $\kappa^2 x \leq \mathbf{K}^2 x$. Using this argument recursively we get that for any natural number n

$$\kappa^n x \leq \mathbf{K}^n x, \tag{4}$$

Since the function x is ≥ 0, we deduce the norm inequality

$$\kappa^n |x| \leq |\mathbf{K}^n x| \leq |\mathbf{K}^n|\,|x|,$$

so

$$\kappa^n \leq |\mathbf{K}^n|. \tag{5}$$

We take the nth root and use the formula for the spectral radius (see theorem 4 of chapter 17):

$$\kappa \leq \lim |\mathbf{K}^n|^{1/n} = |\sigma(\mathbf{K})|. \tag{6}$$

Since the set of κ is nonempty, it follows that $\mathbf{K}$ has a positive spectral radius. Since $\mathbf{K}$ is compact, it follows that the *set of eigenvalues of* $\mathbf{K}$ *is not empty.*

Next we show the converse of inequality (6). Since, according to theorem 6 of chapter 21, the nonzero spectrum of $\mathbf{K}$ is an isolated point spectrum, there is an eigenvalue λ and eigenfunction z such that

$$\mathbf{K}z = \lambda z, \quad |\lambda| = |\sigma(\mathbf{K})|. \tag{7}$$

We claim the following inequality for $|z(s)| = y(s)$, $\sigma = |\sigma(\mathbf{K})|$:

$$\sigma y \leq \mathbf{K}y. \tag{8}$$

Take any point q in the space Q; multiply the function z by a complex number of absolute value 1 so that $\lambda z(q)$ is real and positve. Decompose z as $z = u + iv$. Separating the real part of (7), we get

$$\lambda z(q) = (\mathbf{K}u)(q). \tag{9}$$

Since the operator $\mathbf{K}$ is monotone, and since $u \leq y$, we deduce from (9) that

$$|\lambda| y(q) = (\mathbf{K}u)(q) \leq (\mathbf{K}y)(q). \tag{10}$$

This proves (8), and a little more; since $\mathbf{K}$ is strictly monotone, the sign of equality holds in (10) only if z is real and positive, and λ is real and positive.

We show now that in (8) the sign of equality holds. Indeed, suppose that for some q,

$$\sigma y(q) < \mathbf{K}y(q). \tag{11}$$

By continuity, if inequality (11) holds then and there is a positive δ such that in some neighborhood N around q

$$\sigma y(s) + \delta \leq \mathbf{K}y(s), \qquad s \text{ in } N. \tag{12}$$

We construct now a function p that is positive inside N, zero outside N. Since $\mathbf{K}$ is strictly positive, $\mathbf{K}p > 0$ everywhere. We define the function x by

$$x = y + \epsilon p, \qquad \epsilon > 0. \tag{13}$$

We claim that for ϵ small enough there is a constant c such that

$$(\sigma + c\epsilon)x \leq \mathbf{K}x. \tag{14}$$

We verify this fact first for s in N. Clearly, the right side of (14), $\mathbf{K}y + \epsilon\mathbf{K}p$ is $> \mathbf{K}y$, and the left side differs from λy by $O(\epsilon)$. It follows therefore from (12) that (14) holds in N for ϵ small enough and $c \leq 1$. For s outside N, the function $p(s) = 0$, so inequality (14) asserts that

$$(\sigma + c\epsilon)y \leq \mathbf{K}y + \epsilon\mathbf{K}p.$$

In view of (8) this would follow from

$$cy \leq \mathbf{K}p. \tag{15}$$

Since $\mathbf{K}$ is a strictly positive and p is nonnegative, it follows that $\mathbf{K}p > 0$ on Q. Clearly, c can be chosen so small that (15) holds. This completes the proof of (14).

Inequality (14) shows that if (12) were to hold, $\kappa = \sigma + c\epsilon$ and $x = y + \epsilon p$ would satisfy in inequality (3). But according to (6), no such κ can exceed σ; this is a contradiction, caused by assuming that in (8) the sign $<$ holds at point. It follows that in (8) equality holds everywhere. We showed earlier that then the eigenfunction z of $\mathbf{K}$ with eigenvalue λ whose absolute value is $|\sigma(\mathbf{K})|$ is positive when multiplied by a constant, and that λ is real and positive. This proves part (ii) of theorem 1.

We claim that σ has multiplicity 1. For if not, $\mathbf{K}$ would have two linearly independent eigenfunctions. According to what was shown above, both of these can be chosen to be positive, but some linear combination of them would change sign, a contradiction. This proves the first part of (i) in theorem 1.

There remains to show that $\mathbf{K}$ has no generalized eigenfunctions with eigenvalues σ. Denote by $\mathbf{K}'$ the transpose of $\mathbf{K}$; $\mathbf{K}'$ acts on Borel measures of finite total mass; it is related to $\mathbf{K}$ by

$$(\mathbf{K}x, m) = (x, \mathbf{K}'m).\tag{16}$$

It follows from (16) that $\mathbf{K}'$ too is strictly positive; that is, that if m is a nonnegative measure $\neq 0$, then $\mathbf{K}'m$ is a positive measure. $\mathbf{K}'$ and $\mathbf{K}$ have the same spectrum. Therefore $\mathbf{K}'$ has the same dominant eigenvalue σ as $\mathbf{K}$. Since $\mathbf{K}$ is compact, so is $\mathbf{K}'$. The eigen measure m associated with the eigenvalue σ can be shown to be positive by the same argument as used above.

We are ready to show that $\mathbf{K}$ has no generalized eigenfunction with eigenvalue σ, for suppose that w were such generalized eigenfunction:

$$(\mathbf{K} - \sigma)w = v, \quad (\mathbf{K} - \sigma)v = 0.\tag{17}$$

From (16) we deduce that

$$((\mathbf{K} - \sigma)w, m) = (w, (\mathbf{K}' - \sigma)m)\tag{17'}$$

Since $(\mathbf{K} - \sigma)w = v$ and $(\mathbf{K}' - \sigma)m = 0$, we deduce from (17) that

$$(v, m) = 0.$$

But since v is a positive function and m a positive measure, this is impossible. This completes the proof of theorem 1.   □

23.2 STOCHASTIC INTEGRAL OPERATORS

We consider now an application of theorem 1.

Theorem 2. *Let $K(s, t)$ be continuous positive function on $0 \le s, t \le 1$, satisfying*

$$\int_0^1 K(s, t)\, ds = 1 \quad \text{for all } t.\tag{18}$$

Denote by $\mathbf{K}$ the integral operator with kernel K; then

(i) 1 is an eigenvalue of $\mathbf{K}$, with corresponding eigenfunction y that is positive.
(ii) x is any function satisfying

$$\int x(t)\, dt = 1.\tag{19}$$

Then

$$\lim_{n \to \infty} \mathbf{K}^n x = y\tag{20}$$

provided that we normalize the eigenfunction y so that

$$\int y(t)\,dt = 1. \tag{21}$$

Proof. We have shown in chapter 22 that an integral operator with a continuous kernel over a compact set such as $[0, 1]$ is a compact operator. Since the kernel $K(x, y)$ is positive, the operator $\mathbf{K}$ is strictly positive. It follows from the first part (ii) of theorem 1, that $\mathbf{K}$ has a positive eigenvalue σ with positive eigenfunctions:

$$\int K(s, t)\,y(t)\,dt = \sigma\,y(s). \tag{22}$$

Integrating this with respect to s and using (18), we get

$$\int y(t)\,dt = \sigma \int y(s)\,ds.$$

Since $y(t) > 0$, $\int y\,dt \neq 0$, so it follows that $\sigma = 1$. This proves the first part of (ii).

To prove the second part, consider the space Z of all continuous functions z with mean value zero:

$$\int z(t)\,dt = 0. \tag{23}$$

Z is a closed subspace of $X = C[0, 1]$. We claim that Z is *invariant* under $\mathbf{K}$, that is, if z belongs to Z, so does $u = \mathbf{K}z$. To see this integrate both sides of

$$u(s) = \int K(s, t)\,z(t)\,dt$$

with respect to s. Using (18) gives, after changing the order of integration,

$$\int u(s)\,ds = \int \int K(s, t)\,z(t)\,dt\,ds = \int z(t)\,dt = 0.$$

Clearly, the spectrum of $\mathbf{K}$ over Z is a point spectrum consisting of the spectrum of $\mathbf{K}$ over X, with the eigenvalue $\sigma = 1$ removed, since $\sigma = 1$ has multiplicity 1, and the corresponding eigenfunction y does not belong to Z. According to the first part of (ii), theorem 1, all the remaining eigenvalues μ are < 1 in absolute value, and so the spectral radius η of $\mathbf{K}$ over Z, abbreviated as $\mathbf{K}_y$, satisfies

$$|\sigma(\mathbf{K}_y)| = \eta < 1.$$

According to formula (12′) in theorem 4 of chapter 17,

$$\lim_{n \to \infty} |\mathbf{K}_y^n|^{1/n} = \eta.$$

In particular, for any z in Z,

$$\lim \mathbf{K}^n z = 0. \tag{24}$$

We apply this now to $z = x - y$. Since x satisfies (19), and y is normalized by (21), z satisfies (23) and so belongs to Z. So (24) applies:

$$\lim_{n \to \infty} \mathbf{K}^n (x - y) = 0.$$

Since y is an eigenfunction, $y = \mathbf{K}y = \cdots = \mathbf{K}^n y$; this proves (17). $\quad\square$

Note that the *rate* at which $K^n x$ tends to y depends on the size of the *second largest eigenvalue* η *of* K. Interesting estimates of the second largest eigenvalue are derived in Lawler and Sokal.

A nonnegative function $x(t)$ that satisfies (19) can be interpreted as a *probability density*. The kernel $K(s, t)$ is the probability density of transition from state t to state s. Relation (18) means that with probability 1 an occupant of state t makes a transition to some state s. The operator $\mathbf{K}$ then models a random process that changes a random variable whose distribution has density x into one whose distribution has density $\mathbf{K}x$. The eigenfunction y normalized by (21) is a probability density that is *invariant* under the random process described by $\mathbf{K}$.

Given a probability distribution x, $\mathbf{K}^n x$ represents the probability distribution of the system after it has been subjected n times to the random process $\mathbf{K}$. The probabilistic meaning of the limiting relation (17) is that the n-fold application of the random process $\mathbf{K}$ turns, as $n \to \infty$, any distribution x into the invariant distribution y.

23.3 INVERSE OF A SECOND ORDER ELLIPTIC OPERATOR

Theorem 3. *Let* $\mathbf{L}$ *be a differential operator of form*

$$\mathbf{L} = -\Delta + \sum b_i \partial_i + c \tag{25}$$

acting on periodic functions of $s_1, \ldots, s_m$. *Here* $\partial_i = \partial / \partial s_i$, $\Delta = \sum \partial_i^2$, b_i, *and* c *smooth periodic functions of* s, c *positive. Then for any smooth periodic function* f, *the equation*

$$\mathbf{L}u = f \tag{26}$$

has a unique smooth periodic solution u. *We denote this solution as* $\mathbf{K}f = u$. *The operator* $\mathbf{K}$ *regarded as a mapping of the space* C *of continuous, periodic functions into* C *has the following properties:*

  (*i*) $\mathbf{K}$ *is a bounded map.*

 (*ii*) $\mathbf{K}$ *is strictly positive.*

(*iii*) $\mathbf{K}$ *is compact.*

Proof. We sketch those parts of the proof that do not require technicalities. Let $s_{\max}$ and $s_{\min}$ denote the points where the function u takes on its maximum and minimum values, respectively. At such points the first derivatives of u are zero, and the second derivatives are nonpositive, respectively nonnegative. Therefore we conclude from equations (25) and (26) that

$$
\begin{aligned}
c(s_{\max})u_{\max} &\leq f(s_{\max}), \\
c(s_{\min})u_{\min} &\geq f(s_{\min}).
\end{aligned}
\tag{27}
$$

Since we have assumed that the function c is positive, we conclude that $|u|_{\max} \leq$ const.$|f|_{\max}$, which proves that $\mathbf{K}$ is bounded in the maximum norm, as asserted in (i).

It follows from (27) that if f is positive for all s, then so is u. We omit the additional argument to prove what is claimed in part (ii): if f is nonnegative but $\not\equiv 0$, u is positive.

We forgo a proof of part (iii), except to remind the reader that an analogous but weaker result—the compactness of $\mathbf{K}$ in the L^2-norm—was proved in theorem 5 of chapter 22. $\qquad\square$

By theorem 1, an operator $\mathbf{K}$ having the properties above has a positive eigenvalue λ that dominates all others, with the corresponding eigenfunction strictly positive. Since $\mathbf{K}$ is the inverse of $\mathbf{L}$, its eigenvalues are the reciprocals of those of $\mathbf{L}$. Therefore we conclude from theorem 3:

Theorem 4. *The second-order elliptic partial differential operator* $\mathbf{L}$ *defined in* (25) *has a positive eigenvalue that is smaller than the absolute value of all other eigenvalues. The eigenfunction corresponding to this eigenvalue is positive.*

Theorems 3 and 4 hold for operators $\mathbf{L}$ of form (25) where the Laplacian Δ is replaced by any *second* order elliptic operator $\sum a_{ij}\partial_i\partial_j$, (a_{ij}) a positive definite matrix.

Theorem 1 for integral operators has been derived by Jentsch. Krein and Rutman have further extended theorem 1 to spaces where the set of positive functions are replaced by a convex cone in a Banach space.

Estimates for the second largest eigenvalue have been derived by E. Hopf.

BIBLIOGRAPHY

Jentsch, R. Über Integralgleichungen mit positivem Kern. *J. Reine Angew. Math.* **141** (1912): 235–244.

Hopf, E. An inequality for positive operators. *J. of Math. and Mech.*, **12** (1963): 683–692, 889–892.

Krein, M. G. and Rutman, M. A. Linear operators leaving invariant a cone in Banach space. *Usp. Mat. Nauk 3* (1), **23** (1948): 3–95, *AMS Transl.*, **26** (1950).

Lawler, G. F. and Sokal, A. D. Bounds on the L^2 spectrum for Markov chains and Markov processes; a generalization of Cheeger's inequality. *Trans. AMS,* **309**, (1988): 557–580.

Lax, P. D. *Linear Algebra.* Wiley, New York, 1997.

24

FREDHOLM'S THEORY OF INTEGRAL EQUATIONS

The historically first general theory dealing with the solution of linear equations in infinite-dimensional spaces is due to Ivar Fredholm in the year 1900. Its importance was immediately recognized, and it spurred a great deal of further work by Hilbert, Schmidt, F. Riesz, Banach, and many others. These newer theories, set in what was formalized as Hilbert and Banach space, have completely replaced Fredholm's theory. Unlike the new, abstract theories, Fredholm dealt with integral operators, and his central notion was the determinant associated with such operators. Since this determinant appears in some modern theories (inverse scattering, completely integrable systems), it is time to resurrect it.

24.1 THE FREDHOLM DETERMINANT AND THE FREDHOLM RESOLVENT

$$u(x) + \int_0^1 K(x, y)u(y)\, dy = f(x), \tag{1}$$

where f is a given continuous function on the interval $[0,1]$ and u an unknown function to be determined. The kernel $K(x, y)$ is assumed here to be continuous; Fredholm's treatment allowed some singularities.

The question of when equation (1) can be solved can be settled by theorem 5 of chapter 21. To put equation (1) in that context, we will regard both functions f and u as elements of the space of continuous functions C[0,1]. As shown in chapter 22, the integral operator on the left of (1) is a compact operator. According to theorem 5 of chapter 21, the dimension of the nullspace of the operator acting on u on the left side of (1) equals the codimension of its range; by theorem 8, the range is characterized as those f that are orthogonal to the nullspace of the *transpose* of the operator, namely with kernel $K'(x, y) = K(y, x)$. This result was proved by Fredholm for operators of form (1) and is known to this day as the *Fredholm alternative*.

Fredholm's approach was to replace the integral in (1) by a Riemann sum over n intervals of length h. This yields a system of n linear equations for the values u_j of u at the n nodes j/n of the subdivision. Fredholm expressed the solution of these equations as ratios of determinants and took the continuum limits of these determinants as $n \to \infty$.

The discretized form of (1) is

$$u_i + h \sum K_{ij} u_j = f_i, \qquad i = 1, \ldots, n, \tag{1'}$$

where $f_i = f(ih), h = 1/n$ and $K_{ij} = K(ih, jh)$. Denote by $D(h)$ the determinant of the matrix acting on the vector u in (1'):

$$D(h) = \det(I + hK_{ij}). \tag{2}$$

Clearly, $D(h)$ is a polynomial in h:

$$D(h) = \sum_0^n a_m h^m. \tag{2'}$$

The coefficients a_m can be determined conveniently as Taylor coefficients:

$$a_m = \frac{1}{m!} \left(\frac{d}{dh} \right)^m D(h)|_{h=0}. \tag{2''}$$

To differentiate a determinant, we use the rule

$$\frac{d}{dh} \det(C_1, \ldots, C_n) = \sum_\ell \det\left(C_1, \ldots, \frac{d}{dh} C_\ell, \ldots, C_n \right). \tag{3}$$

We use the same rule to find the mth derivative. The resulting formula is simple because each column C_j is a *linear* function of h. A further simplification is possible because at $h = 0, C_j(0) = E_j$, the jth unit vector. So using (2) in (2''), we get an expression involving determinants of principal minors of K_{ij}:

$$D(h) = 1 + h \sum_i K_{ii} + \frac{h^2}{2} \sum_{i,j} \det \begin{pmatrix} K_{ii} & K_{ij} \\ K_{ji} & K_{jj} \end{pmatrix} + \cdots. \tag{4}$$

We now set $h = 1/n$ and let n tend to ∞. The kth term in (4), a sum with respect to k parameters, tends to a k-fold integral. To write these in a compact form, Fredholm introduced the following convenient abbreviation:

$$K \begin{pmatrix} x_1, \ldots, x_k \\ y_1, \ldots, y_k \end{pmatrix} = \det K(x_i, y_j), \qquad 1 \leq i, j \leq k. \tag{5}$$

The formal limit as $n \to \infty$ of the finite sum (4), $h = 1/n$, is the infinite series

$$D = \sum_0^\infty \frac{1}{k!} \int \cdots \int K \begin{pmatrix} x_1, \ldots, x_k \\ x_1, \ldots, x_k \end{pmatrix} dx_1 \ldots dx_k. \tag{6}$$

Definition. D is called the Fredholm determinant of the operator acting on the left in (1).

Lemma 1. *The series (6) converges.*

Proof. To prove convergence, Fredholm relied on an inequality due to Hadamard for determinants to estimate the terms of the series (6):

$$|\det (C_1, \ldots, C_k)| \leq \prod \|C_j\|, \tag{7}$$

where $\|C\|$ denotes the Euclidean length of the vector C. The geometric interpretation of (7) is that the volume of the parallelopiped whose vertices are sums of vectors C_j is $\leq$ the volume of the *rectangular* parallelopiped whose sides C'_j have the same lengths as the C_j;

$$\|C'_j\| = \|C_j\|,$$

where the C'_j are orthogonal to each other.

Exercise 1. Furnish an analytic proof of inequality (7).

The kernel K, being continuous, is bounded:

$$|K(x, y)| \leq M \qquad \text{for all } x, y,$$

so the length of each column vector of the $k \times k$ matrix (5) is $\leq M\sqrt{k}$. Therefore, according to (7)

$$\left| K \begin{pmatrix} x_1, \ldots, x_k \\ y_1, \ldots, y_k \end{pmatrix} \right| \leq M^k k^{k/2}. \tag{8}$$

The kth term in series (6) is then less $M^k k^{k/2}/k!$, which, by Stirling's formula, is $\leq (Me)^k k^{-k/2}$. This estimate shows that the series (6) converges. $\qquad\square$

Fredholm showed how to get a better estimate than (8) in the case where the kernel K satisfies a Hölder condition in the variable y;

$$|K(x, y) - K(x, z)| \leq M|y - z|^\alpha, \tag{9}$$

In the matrix on the right in (5) we subtract the $(i + 1)$st column from the ith, $i = 1, \ldots, k - 1$. This new matrix has the same determinant as the original one, and its ith column has length $\leq M|y_{i+1} - y_i|^\alpha \sqrt{k}$. It follows from (7) that

$$\left| K \begin{pmatrix} x_1, \ldots, x_k \\ y_1, \ldots, y_k \end{pmatrix} \right| \leq k^{k/2} M^k \left(\prod |y_{i+1} - y_i| \right)^\alpha.$$

By the arithmetic-geometric mean inequality

$$\prod |y_{i+1} - y_i| \le \left(\frac{1}{k}\right)^k,$$

so we deduce that

$$\left| K \left(\begin{matrix} x_1, \dots, x_k \\ y_1, \dots, y_k \end{matrix} \right) \right| \le (k^k)^{(1/2)-\alpha} M^k. \tag{8'}$$

The same inequality holds if $K(x, y)$ satisfies a Hölder condition in the variable x.

We return now to equations (1'). To solve such a system of equations, we multiply (1') by the inverse of the matrix acting on the unknowns. The elements of the inverse matrix can be represented as determinants of the minors of co-order 1 (i.e., one less than the order of the matrix to be inverted) divided by the determinant of the matrix. We apply this procedure to the system (1') and obtain a formula analogous to (2), which can be brought to a form analogous to (4). Fredholm, by passing to the limit $h \to 0$, has determined the continuous analogue R of these determinants. Using the notation (5), we can write R in the form

$$R(x, y) = K(x, y) + \int K \left(\begin{matrix} x, x_1 \\ y, x_1 \end{matrix} \right) dx_1 + \cdots$$

$$= \sum_{0}^{\infty} \frac{1}{k!} \int \cdots \int K \left(\begin{matrix} x, x_1, \dots, x_k \\ y, x_1, \dots, x_k \end{matrix} \right) dx_1 \dots dx_k. \tag{10}$$

Inequality (8) shows that the series on the right in (10) converges uniformly for all x and y. This shows that $R(x, y)$ is a continuous function of x, y.

We show now how to use the kernel $R(x, y)$ and the determinant D to solve equation (1). Expand the determinants in (10), defined by (5), according to their first rows:

$$K \left(\begin{matrix} x, x_1, \dots, x_k \\ y, x_1, \dots, x_k \end{matrix} \right) = K(x, y) \, K \left(\begin{matrix} x_1, \dots, x_k \\ x_1, \dots, x_k \end{matrix} \right)$$

$$- K(x, x_1) \, K \left(\begin{matrix} x_1, x_2, \dots, x_k \\ y, x_2, \dots, x_k \end{matrix} \right)$$

$$+ K(x, x_2) \, K \left(\begin{matrix} x_1, x_2, \dots, x_k \\ y, x_1, x_3, \dots, x_k \end{matrix} \right) - \cdots \tag{11}$$

Integrate (11) over the unit cube in $x_1, \dots, x_k$ space. We claim that the integrals of the last k terms on the right are all equal; this can be seen by interchanging in the jth integral

$$(-1)^j \int \cdots \int K(x, x_j) \, K \left(\begin{matrix} x_1, x_2, & \cdots, & x_k \\ y, x_1, \dots, x_{j-1}, x_{j+1}, \dots, x_k \end{matrix} \right) dx_1 \dots dx_k$$

the names of the variables x_1 and x_j, and then performing one row permutation and $j - 2$ column permutations. The result is

$$\int \ldots \int K \begin{pmatrix} x, x_1, \ldots, x_k \\ y, x_1, \ldots, x_k \end{pmatrix} dx_1 \ldots dx_k = K(x, y) \int \ldots \int K \begin{pmatrix} x_1, \ldots, x_k \\ x_1, \ldots, x_k \end{pmatrix}$$

$$dx_1 \ldots dx_k - k \int \ldots \int K(x, x_1) \begin{pmatrix} x_1, \ldots, x_k \\ y, x_2, \ldots, x_k \end{pmatrix} dx_1, \ldots dx_k.$$

Divide this by $k!$ and sum. Recalling the definition (10) of $R(x, y)$ and (6) of D, we can write the resulting relation as

$$R(x, y) = K(x, y)D - \int K(x, x_1)R(x_1, y)\,dx_1,$$

which we rewrite as

$$R(x, y) + \int K(x, z)R(z, y)\,dz - DK(x, y) = 0. \tag{12}$$

If, instead of using the first row, we expand the determinants in (10) according to the first column, we get the analogous identity

$$R(x, y) + \int K(z, y)R(x, z)\,dz - DK(x, y) = 0. \tag{12'}$$

We return now to the integral equation (1). Fredholm regarded the left side as an *operator* acting on the unknown u; let's denote the integral operator with kernel $K(x, y)$ by the symbol $\mathbf{K}$:

$$(\mathbf{K}u)(x) = \int K(x, y)u(y)\,dy. \tag{13}$$

Similarly $\mathbf{R}$ is the integral operator whose kernel is R. Equation (1) can be abbreviated as

$$(\mathbf{I} + \mathbf{K})u = f. \tag{13'}$$

Fredholm observed that operators of form (13') form a semigroup, namely that the product of two such operators is of the same form:

$$(\mathbf{I} + \mathbf{H})(\mathbf{I} + \mathbf{K}) = \mathbf{I} + \mathbf{L}, \tag{14}$$

where the kernel of $\mathbf{L}$ is

$$L(x, y) = K(x, y) + H(x, y) + \int H(x, z)K(z, y)\,dz. \tag{14'}$$

Theorem 2. *Let K be a continuous kernel, and suppose that $D \neq 0$. Then the operator $\mathbf{I} + \mathbf{K}$ is invertible, and its inverse is $\mathbf{I} - D^{-1}\mathbf{R}$, where D and R are defined by (6) and (10).*

Proof. Relations (12) and (12′) mean in operator language that

$$\mathbf{R} + \mathbf{KR} - D\mathbf{K} = 0,$$
$$\mathbf{R} + \mathbf{RK} - D\mathbf{K} = 0. \tag{15}$$

Since D is assumed $\neq 0$, we can rewrite these as

$$(\mathbf{I} + \mathbf{K})(\mathbf{I} - D^{-1}\mathbf{R}) = \mathbf{I},$$
$$(\mathbf{I} - D^{-1}\mathbf{R})(\mathbf{I} + \mathbf{K}) = \mathbf{I}. \tag{15′}$$

$\square$

The converse is also true.

Theorem 3. *Let K be a continuous kernel such that $D = 0$; then the operator $\mathbf{I} + \mathbf{K}$ has a nontrivial null-space and so is not invertible.*

Proof. In $R(x, y)$ fix the value of y, and denote the resulting function of x by r:

$$R(\cdot, y) = r(\cdot).$$

Then equation (12) can be written as follows:

$$r + \mathbf{K}r = 0;$$

this shows that r belongs to the nullspace of $\mathbf{I} + \mathbf{K}$. The flaw in this argument is that $R(x, y)$ could be zero for all x and y, so that r is the zero function. So we must argue differently.

Let λ denote a complex parameter. Replacing K by λK in formula (6) gives a power series in λ, which we denote as

$$D(\lambda) = \sum \frac{\lambda^k}{k!} \iint \cdots \int K \begin{pmatrix} x_1, \ldots, x_k \\ x_1, \ldots, x_k \end{pmatrix} dx_1 \ldots dx_k. \tag{16}$$

Similarly we define the function $R(x, y; \lambda)$ by replacing K with λK in (10):

$$R(x, y; \lambda) = \sum \frac{\lambda^{k+1}}{k!} \int \cdots \int K \begin{pmatrix} x, x_1, \ldots, x_k \\ y, x_1, \ldots, x_k \end{pmatrix} dx_1 \ldots dx_k. \tag{17}$$

Lemma 4. *$D(\lambda)$ and $R(x, y; \lambda)$ are entire analytic functions of λ, that is, analytic for all complex values of λ.*

Exercise 2. Prove lemma 4.

If we set $\lambda = 1$ in (16) and (17), we get back D and R defined in (6) and (10). So, if $D = 0$, $D(\lambda)$ has a zero at $\lambda = 1$. Since $D(\lambda)$ is analytic and not $\equiv 0$, this zero is of finite order.

Lemma 5. *Suppose that $D(\lambda)$ has a zero of order m at $\lambda = 1$. Then there is a value of x such that $R(x, x; \lambda)$ has a zero of order $< m$ at $\lambda = 1$.*

Proof. Set $y = x$ in (17) and integrate with respect to x:

$$\int R(x, x; \lambda)\, dx = \sum \frac{\lambda^{k+1}}{k!} \int \cdots \int K \begin{pmatrix} x, x_1, \ldots, x_k \\ x, x_1, \ldots, x_k \end{pmatrix} dx\, dx_1 \ldots dx_k.$$

The right side equals the derivative of (16) with respect to λ and then multiplied by λ:

$$\int R(x, x; \lambda)\, dx \equiv \lambda \frac{d}{d\lambda} D(\lambda).$$

By definition of m, the right side has zero of order $m - 1$ at $\lambda = 1$; therefore so does the left side. But then $R(x, x; \lambda)$ cannot have a zero of order $\geq m$ for every x. $\quad\square$

Denote by ℓ the largest number such that $R(x, y; \lambda)$ has a zero of order ℓ at $\lambda = 1$ for every x and y. According to lemma 5, $\ell < m$. We write

$$R(x, y; \lambda) = g(x, y)(\lambda - 1)^\ell + O(\lambda - 1)^{\ell+1}. \tag{18}$$

By definition of ℓ, $g(x, y) \not\equiv 0$. We take now relation (12) for $\lambda \neq 1$:

$$R(x, y; \lambda) + \int \lambda K(x, z) R(z, y; \lambda)\, dz = \lambda K(x, y) D(\lambda).$$

Divide both sides by $(\lambda - 1)^\ell$ and let λ tend to 1. Since $D(\lambda)$ has a zero of order $m > \ell$, the right side tends to zero. According to (18), $R(x, y; \lambda)(\lambda - 1)^\ell$ tends to g, so we obtain

$$g(x, y) + \int K(x, z) g(z, y)\, dz = 0. \tag{19}$$

Since $g \not\equiv 0$, there is a value y_0 such that $g(x, y_0) \neq 0$ for some x. Clearly, $u(x) = g(x, y_0)$ belongs to the nullspace of $\mathbf{I} + \mathbf{K}$, as asserted in theorem 3. $\quad\square$

Exercise 3. Suppose that the kernel $K(x, y)$ is degenerate; that is, $K(x, y) = \sum_1^n k_i(x) h_i(y)$. Show that then $D(\lambda)$ is a polynomial of degree $\leq n$.

We connect now the zeros of the analytic function $D(\lambda)$ to the eigenvalues of $\mathbf{K}$.

Theorem 6. *The complex number κ is an eigenvalue of the integral operator $\mathbf{K}$ iff $\lambda = -1/\kappa$ is a zero of $D(\lambda)$.*

Proof. Theorems 2 and 3 say that $\kappa = -1$ belongs to the spectrum of the integral operator $\mathbf{K}$ iff $D(1) = 0$. Replacing K by λK yields theorem 6. $\quad\square$

Theorem 6 shows that $D(\lambda)$ plays the role of the characteristic polynomial for the operator $\mathbf{K}$. The *algebraic multiplicity* of an eigenvalue κ of an operator $\mathbf{K}$ is the dimension of the space of its generalized eigenvectors, that is, the union of the nullspaces of $(\kappa\mathbf{I}-\mathbf{K})^i$, i any positive integer. According to theorem 6 of chapter 21, the geometric multiplicity of a nonzero eigenvalue of a compact operator is finite.

Theorem 7. *Let κ be a nonzero eigenvalue of the operator $\mathbf{K}$ defined in (13). According to theorem 6, $\lambda = -1/\kappa$ is a zero of the function $D(\lambda)$ defined in (16). Denote by m the multiplicity of this zero. Claim: m is the algebraic multiplicity of the eigenvalue κ.*

Proof. The proof is the same as that in the finite-dimensional case, employing minors of cofinite order of the determinant D. We omit the details.   □

The next result is a generalization to operators of two important matrix relations.

Theorem 8. *Denote by $\kappa_1, \kappa_2, \ldots$ the eigenvalues of an integral operator $\mathbf{K}$ whose kernel $K(x, y)$ is Hölder continuous in x or y with Hölder exponent $> \frac{1}{2}$. Then*

$$\int K(x, x)\,dx = \sum \kappa_j, \tag{20}$$

$$D = \prod(1 + \kappa_j). \tag{20'}$$

The series and the product converge absolutely.

The quantity on the left in (20) is called the *trace* of the integral operator $\mathbf{K}$; (20) is called the *trace formula*. In chapter 30 we will establish such a trace formula for a large, invariantly defined class of operators in Hilbert space.

The proof of theorem 8 relies on *Hadamard's factorization theorem*. We state it for the case that is needed:

Theorem 9. *Let $f(\lambda)$ be an entire analytic function of λ, defined for all complex λ of order < 1, that is, it satisfies the growth condition*

$$|f(\lambda)| \leq \text{const.} \,(\exp|\lambda_1|^\rho), \qquad \rho < 1. \tag{21}$$

Denote the roots of f, counted with multiplicity, by $\{\lambda_j\}$, and assume that $\lambda = 0$ is not a root, that is, that $f(0) \neq 0$. Then $\sum |\lambda_j|^{-1}$ converges, and f can be factored as

$$f(\lambda) = f(0) \prod \left(1 - \frac{\lambda}{\lambda_j}\right). \tag{22}$$

This result is not deep; for a proof, we refer the reader to Ahlfors's text on complex variables. We will apply (22) to $f(\lambda) = D(\lambda)$:

Lemma 10. *If the kernel K satisfies a Hölder condition with exponent α, with respect to either the x or the y variable, then $D(\lambda)$ satisfies the growth condition*

$$|D(\lambda)| \leq \text{const.} \left(\exp |\lambda|^{2/(1+2\alpha)} \right).$$

Exercise 4. Prove lemma 10, using the definition (16) of $D(\lambda)$ and inequality (8′).

Clearly, for $\alpha > \frac{1}{2}$ the function $D(\lambda)$ is of order less than 1, and so can be factored as in (22). Using the connection established in theorem 6 between the zeros of D and the eigenvalues κ_j of $\mathbf{K}$, we can write this factorization as

$$D(\lambda) = \prod (1 + \kappa_j \lambda). \tag{22′}$$

Here we have made use of the fact, evident from (16), that $D(0) = 1$. We differentiate (22′) and set $\lambda = 0$. On the left side in (22′) we obtain $dD(\lambda)/d\lambda|_{\lambda=0}$, which according to the power series expansion (16) is equal to $\int K(x, x)\, dx$. The infinite product on the right is for $|\lambda| \leq 1$ the uniform limit of finite products; therefore its derivative is the limit of the derivative of the finite products. This shows that the derivative of the right side of (22′) at $\lambda = 0$ equals $\sum \kappa_j$. This proves (20). Setting $\lambda = 1$ in (22′) is identity (20′). This completes the proof of theorem 8. $\square$

24.2 THE MULTIPLICATIVE PROPERTY OF THE FREDHOLM DETERMINANT

Next we present Fredholm's extension of the *multiplicative property* of determinants to operators. We will denote the determinant of $\mathbf{I} + \mathbf{K}$ by D_K, of $\mathbf{I} + \mathbf{H}$ by D_H, and so on. Similarly we will denote the inverse of $\mathbf{I} + \mathbf{K}$ by $\mathbf{I} - D_K^{-1}\mathbf{R}_K$, and the kernel of $\mathbf{R}_K$ as $R_K(x, y)$.

Theorem 11. *Let $\mathbf{H}$ and $\mathbf{K}$ be integral operators with continuous kernels, and set $(\mathbf{I} + \mathbf{H})(\mathbf{I} + \mathbf{K}) = \mathbf{I} + \mathbf{L}$. Then*

$$D_L = D_H D_K. \tag{23}$$

Proof. We present Fredholm's beautiful argument based on computing the variation of D_K. We define

$$\delta D_K = \frac{d}{d\epsilon} D_{K + \epsilon \delta K}|_{\epsilon=0} \tag{24}$$

To compute δD_K as function of δK, we first calculate the variation of the determinant (5). Using formula (3) for differentiating a determinant, we get a sum of k determinants:

$$\delta K \begin{pmatrix} x_1, \ldots, x_k \\ x_1, \ldots, x_k \end{pmatrix} = \sum_\ell \det K_\ell \tag{25}$$

where the lth column of K_ℓ is $\delta K(x_i, y_\ell)$. Expand $\det K_\ell$ with respect to the lth column:

$$\delta K \begin{pmatrix} x_1, \ldots, x_k \\ x_1, \ldots, x_k \end{pmatrix} = \sum (-1)^{\ell+m} K \begin{pmatrix} x_1, \ldots, (x_m), \ldots, x_k \\ x_1, \ldots, (x_\ell), \ldots, x_k \end{pmatrix} \delta K(x_m, x_\ell), \quad (25')$$

where the parentheses indicate that the lth column and the mth row are to be omitted. Integrate $(25')$ over the k-dimensional unit cube. All k terms with $\ell = m$ are equal; denoting $x_\ell = x_m = x$ and relabeling the remaining variables $x_1, \ldots, x_{k-1}$, we get the following expression for the integral of the sum of all terms in $(25')$ with $\ell = m$:

$$k \int \ldots \int K \begin{pmatrix} x_1, \ldots, x_{k-1} \\ x_1, \ldots, x_{k-1} \end{pmatrix} dx_1 \ldots dx_{k-1} \int \delta K(x, x) \, dx. \quad (26)$$

The remaining $k(k-1)$ integrals with $x_\ell \neq x_m$ are also equal; relabel $x_m = x$, $x_\ell = y$, and the remaining variables as $x_1, \ldots, x_{k-2}$. Suppose that $\ell < m$; then $\ell - 1$ row transpositions and $m - 2$ column transpositions bring all terms in $(25')$ with $\ell \neq m$ into the same form so that the sum of all these terms is

$$-k(k-1) \int \ldots \int K \begin{pmatrix} y, x_1, \ldots, x_{k-2} \\ x, x_1, \ldots, x_{k-2} \end{pmatrix} \delta K(x, y) \, dx_1 \ldots dx_{k-2} \, dx \, dy. \quad (26')$$

To obtain δD_K, we form termwise the variation of the series (6) defining D_K. We get the sum of (26) and $(26')$ divided by $k!$ Using the definition (6) of D_K, we write the sum coming from (26) as

$$D_K \int \delta K(x, x) \, dx,$$

and the sum coming from $(26')$ in terms of $R_K(x, y)$ defined by (10), as

$$- \iint R_K(y, x) \delta K(x, y) \, dx \, dy.$$

Altogether we get

$$\delta D_K = D_K \int \delta K(x, x) \, dx - \iint R_K(y, x) \delta K(x, y) \, dx \, dy.$$

Assume that $D_K \neq 0$ and divide by D_K; we get

$$\delta \log D_K = \int \delta K(x, x) \, dx - D_K^{-1} \iint R_K(y, x) \delta K(x, y) \, dx \, dy. \quad (27)$$

According to formula $(15')$ the operator $\mathbf{I} - D_K^{-1} \mathbf{R}_K$ is the inverse of the operator $\mathbf{I} + \mathbf{K}$. Therefore we can regard the x integration on the right in (27) as an application of $(\mathbf{I} + \mathbf{K})^{-1}$ to $\delta K(\cdot, y)$, and rewrite (27) in operator notation as

$$\delta \log D_K = \int (\mathbf{I} + \mathbf{K})^{-1} \delta K(\cdot, y) \, dy. \quad (28)$$

Alternatively we can regard the y integration on the right in (27) as the application of the *transpose* of $\mathbf{I} - D_K^{-1}\mathbf{R}_K$ to the function $\delta K(x, \cdot)$. Since the transpose of the inverse is the inverse of the transpose, we can also rewrite (27) as follows:

$$\delta \log D_K = \int (\mathbf{I} + \mathbf{K}')^{-1} \delta K(x, \cdot) \, dx, \tag{28'}$$

where $\mathbf{K}'$ denotes the transpose of the operator $\mathbf{K}$.

Let K and H be any two continuous kernels, $D_K \neq 0, D_H \neq 0$. Denote, as before, the product of the operators $\mathbf{I} + \mathbf{H}$ and $\mathbf{I} + \mathbf{K}$ as $\mathbf{I} + \mathbf{L}$. By theorem 2, $\mathbf{I} + \mathbf{K}$ and $\mathbf{I} + \mathbf{H}$ are invertible; therefore so is their product $\mathbf{I} + \mathbf{L}$. Then by theorem 3, $D_L \neq 0$.

To compute the variation of D_L in terms of δH and δK, we express the kernel of L in terms of H and K. Since $\mathbf{L} = \mathbf{K} + \mathbf{H} + \mathbf{HK}$,

$$L(x, y) = K(x, y) + H(x, y) + \int H(x, z) K(z, y) \, dz,$$

and so

$$\delta L(x, y) = \delta K(x, y) + \delta H(x, y) + \int \delta H(x, z) K(z, y) \, dz + \int H(x, z) \delta K(z, y) \, dz. \tag{29}$$

We can express the integrations on the right in (29) as the actions of the operators $\mathbf{K}'$ and $\mathbf{H}$:

$$\delta L(x, y) = (\mathbf{I} + \mathbf{K}') \delta H(x, \cdot) + (\mathbf{I} + \mathbf{H}) \delta K(\cdot, y). \tag{29'}$$

We set this expression for δL into formula (27) applied to the kernel L. We express the terms involving δH as in (28') and those involving δK as in (28); we obtain

$$\delta \log D_L = \int (\mathbf{I}+\mathbf{L}')^{-1}(\mathbf{I}+\mathbf{K}') \delta H(x, \cdot) \, dx + \int (\mathbf{I}+\mathbf{L})^{-1}(\mathbf{I}+\mathbf{H}) \delta K(\cdot, y) \, dy. \tag{30}$$

We recall that $\mathbf{I} + \mathbf{L}$ is defined as $(\mathbf{I} + \mathbf{H})(\mathbf{I} + \mathbf{K})$; taking the inverse and then the transpose, we get

$$(\mathbf{I} + \mathbf{L})^{-1} = (\mathbf{I} + \mathbf{K})^{-1}(\mathbf{I} + \mathbf{H})^{-1}, \quad (\mathbf{I} + \mathbf{L}')^{-1} = (\mathbf{I} + \mathbf{H}')^{-1}(\mathbf{I} + \mathbf{K}')^{-1}.$$

Setting these into (30) gives

$$\delta \log D_L = \int (\mathbf{I} + \mathbf{H}')^{-1} \delta H(x, \cdot) \, dx + \int (\mathbf{I} + \mathbf{K})^{-1} \delta K(\cdot, y) \, dy. \tag{30'}$$

Comparing this with (28) and (28'), we get

$$\delta \log D_L = \delta \log D_H + \delta \log D_K. \tag{31}$$

Now deform K and H into zero so that $D_K \neq 0$, $D_H \neq 0$ during this deformation. This is easily done by setting $K(t) = \lambda(t)K$, $H(t) = \lambda(t)H$, where the complex-valued function $\lambda(t)$ avoids zeros of both $D_K(\lambda)$ and $D_H(\lambda)$. Formula (31) shows that

$$\frac{d}{dt}[\log D_{L(t)} - \log D_{K(t)} D_{H(t)}] = 0.$$

Since $L(0) = K(0) = H(0) = 0$ and $D_0 = I$, we deduce that $\log D_L - \log D_K D_H = 0$; this proves the multiplicative property of the determinant when $D_K \neq 0$, $D_H \neq 0$. When $D_H = 0$, $I + H$ is not onto, and when $D_K = 0$, $I + K$ is not one-to-one. In either case $(I + H)(I + K)$ is not invertible, so $D_L = 0$. Thus $D_L = D_H D_K$ in all cases.   □

Exercise 5. Justify the calculation of δD_K termwise in the infinite series defining D_K.

24.3 THE GELFAND-LEVITAN-MARCHENKO EQUATION AND DYSON'S FORMULA

So far we have taken the integration in the integral equation (1) over the unit interval. It could of course be any interval $[a, b]$, even an infinite interval $[a, \infty]$, provided that the kernel K tends to zero fast enough as its arguments tend to ∞.

Exercise 6. Show that if

$$|K(x, y)| \leq M(x)M(y),$$

where $M(x)$ is a decreasing function and integrable,

$$\int_a^\infty M(x)\, dx < \infty,$$

then the series (6) and (10) defining the Fredholm determinants converge.

Let $K(x, y)$ be a continuous kernel defined for all real x and y; suppose that as x, y tends to ∞, the kernel tends to zero with sufficient rapidity. For any real a we define the Fredholm operator $I + K_a$, where K_a is the integral operator over the interval (a, ∞) with kernel K:

$$(K_a u)(x) = \int_a^\infty K(x, y)u(y)\, dy, \qquad a \leq x. \tag{32}$$

Denote by $D(a)$ the Fredholm determinant of $I + K_a$. We will study how $D(a)$ depends on a; we assume that $D(a) \neq 0$.

Let h be any real number; the simple transformation $y \to y + h$ carries the operator K_{a+h} into the integral operator with kernel $K(x + h, y + h)$ over the *fixed* interval (a, ∞). The derivative of this kernel with respect to h at $h = 0$ is $\delta K = K_x + K_y$. Therefore the derivative of $D(a)$ with respect to a can be computed from formulas (27), (28) for the variation of D. To carry this out, we need a few identities connecting the kernels K and R. We derive these from formulas (12) and (12'), which express the fact that $\mathbf{I} + \mathbf{K}$ and $\mathbf{I} - D^{-1}\mathbf{R}$ are inverses of each other:

$$R(x, y) + \int K(x, z)R(z, y)\,dz - DK(x, y) = 0, \tag{12}$$

$$R(x, y) + \int K(z, y)R(x, z)\,dz - DK(x, y) = 0. \tag{12'}$$

Differentiate (12) with respect to x:

$$R_x(x, y) + \int K_x(x, z)R(z, y)\,dz - DK_x(x, y) = 0.$$

Setting $y = x$ in this relation gives

$$R_x(x, x) + \int K_x(x, z)R(z, x)\,dz - DK_x(x, x) = 0. \tag{33}$$

Similarly

$$R_y(y, y) + \int K_y(z, y)R(y, z)\,dz - DK_y(y, y) = 0. \tag{33'}$$

We take now formula (27) for the variation of $\log D$ and set $\delta K = K_x + K_y$:

$$\delta \log D(a) = \int [K_x(x, x) + K_y(x, x)]\,dx - D^{-1} \iint R(y, x)[K_x(x, y)$$
$$+ K_y(x, y)]\,dx\,dy. \tag{34}$$

We use relations (33) and (33') to replace the two double integrals on the right in (34) by single integrals:

$$\delta \log D(a) = \int [K_x(x, x) + K_y(x, x)]\,dx + \int [D^{-1}R_x(x, x) - K_x(x, x)]\,dx$$
$$+ \int [D^{-1}R_y(y, y) - K_y(y, y)]\,dy$$
$$= D^{-1} \int_a^\infty [R_x(x, x) + R_y(x, x)]\,dx$$
$$= -D^{-1}R(a, a).$$

Since the variation of $\log D(a)$ is its derivative with respect to a,

$$\frac{d}{da} \log D(a) = -D^{-1}(a)R(a, a), \quad \text{which gives} \quad \frac{d}{da}D(a) = -R(a, a). \tag{35}$$

Exercise 7. Derive formula (35) by differentiating formula (6) with respect to the lower limit of integration.

We return now to formula (12) and note that it has the following interpretation. Let y be any fixed number, $a \leq y$; then the solution w' of the integral equation

$$w' + \mathbf{K}_a w' = K(\cdot, y)$$

is $w'(x) = D^{-1}(a) R(x, y)$. Setting, in particular, $y = a$, $x = a$, we get $w'(a) = D^{-1}(a) R(a, a)$. Using the relation (35), we have therefore

Theorem 12. *Denote by $w = w(x; a)$ the solution of the integral equation*

$$w(x) + \int_a^\infty K(x, y) w(y)\, dy = -K(x, a). \tag{36}$$

Then

$$w(a; a) = \frac{d}{da} \log D(a). \tag{37}$$

We show now how to apply this result to the so-called *inverse problem* of scattering theory concerning the differential operator

$$-\partial_x^2 + q, \tag{38}$$

where $q = q(x)$ is a potential that rapidly tends to zero as $x \to \pm\infty$. The inverse problem is to determine the potential q from the so-called *reflection coefficient*, to be defined in chapter 37. This problem arises in physics, in situations where the potential q is inaccessible to direct measurement whereas the reflection coefficient is.

Gelfand, Levitan, and Marchenko have devised an integral equation from whose solution q can be determined. Denote by $Z(x)$ the Fourier transform of the reflection coefficient; in the case where the operator (37) has point eigenvalues, these have to be included in Z. The G-L-M equation for an auxiliary function w is

$$w(x; a) + \int_a^\infty Z(x + y) w(y; a)\, dy = -Z(x + a). \tag{39}$$

According to G-L-M (e.g., see Lax, pp. 84–91) the potential q is related to the solution w of the G-L-M equation by

$$q(a) = -2\frac{d}{da} w(a, a). \tag{40}$$

Equation (39) is a special case of equation (36), with $K(x, y) = Z(x + y)$. We can therefore appeal to theorem 12 to deduce from (40) the following formula due to Freeman Dyson:

$$q(a) = -2\frac{d^2}{da^2} \log D(a), \tag{41}$$

where $D(a)$ is the Fredholm determinant of the G-L-M operator acting on w on the left in (39).

BIBLIOGRAPHY

Ahlfors, L. V. *Complex Analysis*. McGraw-Hill, New York, 1979.

Dyson, F. J. Fredholm determinants and inverse scattering problems. *Comm. Math. Phys.*, **47** (1976): 171–183.

Fredholm, I. Sur une classe d'équations fonctionnelles. *Acta Math.*, **27** (1903): 365–390.

Gelfand, I. M. and Levitan, B. M. On the determination of a differential equation by its spectral function. *Izv. Akad. Navk SSSR Ser. Mat.*, **15** (1951): 309–360; AMS *Transl.*, **1** (1955): 254–304.

Lax, P. D. *Outline of a Theory of the KdV Equation. Recent Mathematical Methods in Nonlinear Wave Propagation*, Lecture Notes in Mathematics, **1640**. Springer Verlag, 1996, pp. 70–102.

Marchenko, V. A. Concerning the theory of differential operators of second order. *Dokl. Akad. Nauk SSSR*, **72** (1950): 457–460.

25

INVARIANT SUBSPACES

Let X denote a Banach space, $\mathbf{M}$ a bounded, linear map: $X \to X$. We recall from linear algebra the notion of an *invariant subspace*: it is a subspace Y of X that is mapped into itself under $\mathbf{M}$. Y is called nontrivial if it is not the whole space and consists of more than the zero vector. In the context of normed linear spaces we are interested in *closed invariant subspaces Y* under $\mathbf{M}$.

Given a closed invariant subspace Y of $\mathbf{M}$, there is the possibility of inverting $\mathbf{M}$ in two stages. The task is, given u in X, to find an x_0 in X such that $\mathbf{M}x_0 = u$. First we solve this equation mod Y, that is, find an x_1 that satisfies $\mathbf{M}x_1 \equiv u(\mathrm{mod}\,Y)$; then $z = u - \mathbf{M}x_1$ belongs to Y. Put $x_0 = x_1 + y$, and choose y in Y so that it satisfies $\mathbf{M}y = z$. The first step amounts to inverting $\mathbf{M}$ on the quotient space X/Y; this task can be handled adequately by analytical tools only if X/Y has a normed structure, and that is possible only if Y is a closed subspace of X.

It follows from the algorithm presented above that if $\mathbf{M}$ is invertible on X/Y and on Y, then it is invertible on X. The converse is false, as shown in chapter 2.

Exercise 1. Prove that if $\mathbf{M} : X \to X$ is invertible, and Y a finite-dimensional subspace of X invariant under $\mathbf{M}$, then $\mathbf{M}$ is invertible on Y and X/Y.

Exercise 2. Show that the subspaces of X invariant under an operator $\mathbf{M}$ form a *lattice* in the sense that the *intersection* of two invariant subspaces is an invariant supspace, and the *closure of the sum* of two invariant subspace is an invariant subspace.

25.1 INVARIANT SUBSPACES OF COMPACT MAPS

Of course every subspace spanned by eigenvectors of $\mathbf{M}$ is invariant under $\mathbf{M}$, but there are plenty of operators without any eigenvalues. As we will show below, many of these possess nontrivial invariant subspaces. The following result is due to von Neumann for Hilbert space, and to Aronszajn and Smith for Banach spaces. The proof given below is due to Lomonosov, with a simplification due to Hilden.

Theorem 1. *A compact map* **C** *of a complex Banach space* X *of dimension greater than 1 into itself has a nontrivial closed invariant subspace.*

Proof. Suppose that $\mathbf{C} \neq 0$; normalize $\mathbf{C}$ so that

$$|\mathbf{C}| = 1. \tag{1}$$

Since $\mathbf{C} \neq 0$, we can choose x_0 in X so that

$$|\mathbf{C}x_0| > 1, \quad |x_0| > 1. \tag{2}$$

Denote the closed unit ball around x_0 by B:

$$B = \{x \,\|\, |x - x_0| \leq 1\}. \tag{3}$$

It follows from (2) that the origin does not belong to B. Denote by K the closure of the image of B under $\mathbf{C}$:

$$K = \overline{\mathbf{C}B}. \tag{4}$$

Since $\mathbf{C}$ is a compact map, K is a compact set. It follows from (2) and (1) that K does not contain the origin.

We will prove the existence of an invariant subspace by arguing indirectly: assume that $\mathbf{C}$ has *no* nontrivial closed invariant subspace. Now for any $y \neq 0$ the set $\{p(\mathbf{C})y\}$, where p is any polynomial, is a linear subspace of X invariant under $\mathbf{C}$. The closure of this subspace is a closed invariant subspace of $\mathbf{C}$, and since $\mathbf{C}$ is not supposed to have any nontrivial ones, this one must be trivial, meaning all of X. So it follows from our supposition that for any $y \neq 0$, the set $\{p(\mathbf{C})y\}$, p any polynomial, is dense in X.

Since we have seen before that K does not contain 0, we can apply the above to any y of K. In particular, for any y in K there is a polynomial p such that

$$|p(\mathbf{C})y - x_0| < 1. \tag{5}$$

Since (5) is a strict inequality, the set of y that satisfies (5) with given p is an open set O_p. Such open sets cover K; since K is compact, a finite collection suffices. Thus there is a finite collection of polynomials $p_1, \ldots, p_N$ such that for every y in K inequality (5) holds with some $p = p_i$. We introduce the abbreviation $p_i(\mathbf{C}) = \mathbf{C}_i$, $i = 1, \ldots, N$; the state of affairs can be expressed as follows: for every y in K,

$$|\mathbf{C}_i y - x_0| < 1 \tag{6}$$

for at least one i.

The point x_0 belongs to B; according to definition (4) of K, $\mathbf{C}x_0 \in K$. Therefore we may in (6) set $y = \mathbf{C}x_0$ and conclude that for some $i = i_1$, $|\mathbf{C}_{i_1}\mathbf{C}x_0 - x_0| < 1$. By definition (3) of B, this means that the point $\mathbf{C}_{i_1}\mathbf{C}x_0$ belongs to B, and so, by definition (4) of K, $\mathbf{C}\mathbf{C}_{i_1}\mathbf{C}x_0 \in K$. Therefore we may in (6) set $y = \mathbf{C}\mathbf{C}_{i_1}\mathbf{C}x_0$ and conclude that for some $i = i_2$,

$$|C_{i_2}CC_{i_1}Cx_0 - x_0| < 1.$$

Repeating this argument, we conclude recursively that

$$\left|\prod_1^n (C_{i_k}C)x_0 - x_0\right| < 1.$$

We obtain from this by the triangle inequality that

$$\left|\prod_1^n (C_{i_k}C)x_0\right| \geq |x_0| - 1; \qquad (7)$$

note that according to (2), the right side is positive.

Since the C_i are polynomials in C, they commute with each other and C, so (7) can be rewritten as

$$\left|\left(\prod_1^n C_{i_k}\right) C^n x_0\right| \geq |x_0| - 1. \qquad (7')$$

Denote the largest of the norms of C_i by c: $|C_i| \leq c, i = 1, \ldots, N$. Then it follows from $(7')$ that

$$c^n |C^n| |x_0| \geq |x_0| - 1.$$

Taking the nth root and letting $n \to \infty$, we conclude that

$$\lim_{n \to \infty} |C^n|^{1/n} \geq \frac{1}{c}.$$

According to spectral theory (see theorem 4, chapter 17) the quantity on the left is the *spectral radius* of C. Since $|\sigma(C)| > 0$, the spectrum of C contains points other than 0. But according to the spectral theory of compact operators (theorem 6 of chapter 21) such points are eigenvalues of C of finite multiplicity. The corresponding eigenspace is invariant under C, in contradiction to our supposition. This proves theorem 1. ☐

25.2 NESTED INVARIANT SUBSPACES

Once we have proved the existence of a single invariant subspace Y, we can, by invoking that theorem repeatedly, prove the existence of infinitely many invariant subspaces. Thus Y itself has an invariant subspace, and so does X/Y, which gives rise to an invariant subspace $Z, Y \subset Z \subset X$.

In an interesting paper, Ringrose has explored such families of invariant subspaces.

Definition. Let X be a complex Banach space. A collection of closed subspaces $\{M\}$ of X is called a *nest* if the subspaces M are nested, meaning *totally ordered* by inclusion; a nest will be denoted by the symbol $\mathcal{N}$.

Definition. Let X be a complex Banach space, $\mathcal{N}$ a nest of closed subspaces of X, $\mathbf{C}$ a compact map of X into itself. If each subspace in the nest $\mathcal{N}$ is invariant under $\mathbf{C}$, then $\mathcal{N}$ is called an *invariant nest* for $\mathbf{C}$.

Theorem 2. *Let $\mathbf{C}$ be a compact map of a complex Banach space X into itself.*

(i) *There exists a maximal invariant nest $\mathcal{N}$ for $\mathbf{C}$.*

(ii) *$\mathcal{N}$ contains the trivial subspaces $\{0\}$ and X.*

(iii) *Let $\mathcal{N}_0$ be a subset of $\mathcal{N}$. The subspace $N = \cap\{L: L \text{ in } \mathcal{N}_0\}$ belongs to $\mathcal{N}$.*

Proof. The existence of a maximal invariant nest follows from Zorn's lemma. The space N described in (iii) is obviously closed and invariant under $\mathbf{C}$. Take any K in $\mathcal{N}$; if some L in $\mathcal{N}_0$ is contained in K, so is N. Otherwise, every L in $\mathcal{N}_0$ contains K, but then so does N. Therefore, since $\mathcal{N}$ is maximal, N must belong to $\mathcal{N}$.   □

Theorem 2′ (Ringrose). *Let $\mathbf{C}$ be a compact map of a complex Banach space X into itself, $\mathcal{N}$ a maximal invariant nest of $\mathbf{C}$. Denote by M any of the closed subspaces contained in the nest $\mathcal{N}$, and denote by M_- the subspace*

$$M_- = \overline{\cup\{L \text{ in } \mathcal{N}: L \text{ properly contained in } M\}}. \tag{8}$$

By maximality, M_- belongs to the nest $\mathcal{N}$.

(i) *The quotient space M/M_- has dimension 0 or 1.*

(ii) *Suppose that $\dim M/M_- = 1$. Then $\mathbf{C}$ maps M/M_- into itself as multiplication by a scalar; denote this scalar by μ. If $\mu \neq 0$, μ is an eigenvalue of $\mathbf{C}$.*

(iii) *Conversely, every nonzero eigenvalue γ of $\mathbf{C}$ occurs as μ for some M in $\mathcal{N}$. The number of times γ occurs as μ in $\mathcal{N}$ equals the algebraic multiplicity of γ as eigenvalue of $\mathbf{C}$, that is,*

$$\max_i \ \dim \text{ nullspace of } (\gamma\mathbf{I} - \mathbf{C})^i.$$

Proof. (i) Since both M and M_- are invariant under $\mathbf{C}$, $\mathbf{C}$ maps M/M_- into itself and is a compact map. If $\dim M/M_-$ were > 1, by the Aronszajn-Smith theorem $\mathbf{C}$ would have a nontrivial invariant subspace in M/M_-. This would correspond to an invariant subspace L of $\mathbf{C}$ in X, where L properly contains M_- and is properly contained in M. Since $\mathcal{N}$ is maximal, L belongs to $\mathcal{N}$; but this contradicts the definition (8) of M_-.

(ii) $\mu\mathbf{I} - \mathbf{C}$ is the zero mapping of M/M_- into itself. This means that $\mu\mathbf{I} - \mathbf{C}$ maps M into M_-. Since $\mathbf{C}$ restricted to M is compact, and $\mu \neq 0$, it follows from theorem 5 of chapter 21, that the nullspace of $(\mu\mathbf{I} - \mathbf{C})$ in M has the same dimension as the codimension of $(\mu\mathbf{I} - \mathbf{C})M$ in M. We have seen that $(\mu\mathbf{I} - \mathbf{C})M$ is contained in M_-, and therefore has codimension at least 1. This proves that the nullspace of $(\mu\mathbf{I} - \mathbf{C})$ contains nonzero vectors, namely that μ is an eigenvalue of $\mathbf{C}$.

(iii) Suppose that γ is an eigenvalue of $\mathbf{C}$, and x a corresponding eigenvector. Define $\mathcal{A}_x$ to be the collection of all subspaces in $\mathcal{N}$ which contain x. Define $M = \cap\{K: K \text{ in } \mathcal{A}_x\}$; by theorem 2, M belongs to $\mathcal{A}_x$. Define M_- by (8); suppose that M_- is properly contained in M. Then M_- does not contain x, since by definition M is the smallest space in $\mathcal{N}$ that contains x. So M/M_- is represented by x; the action of $\mathbf{C}$ on x is to multiply it by γ. It follows that $\mu = \gamma$, as asserted in part (iii).

To complete this proof, we have to show that M_- is properly contained in M. According to the Fredholm alternative, theorem 8 of chapter 21, a vector y in X belongs to the range of $\mathbf{C} - \gamma\mathbf{I}$ iff $(y, m) = 0$ for every m in the nullspace of the transpose $\mathbf{C}' - \gamma\mathbf{I}$. Take for simplicity the case that γ is an eigenvalue of multiplicity 1. We claim that the corresponding eigenvector x satisfies $(x, m) \neq 0$. If it did not, x would be in the range of $\mathbf{C} - \gamma\mathbf{I}$, namely $(\mathbf{C} - \gamma\mathbf{I})u = x$ for some u in x. Applying $(\mathbf{C} - \gamma\mathbf{I})$ to this relation gives

$$(\mathbf{C} - \gamma\mathbf{I})^2 u = 0,$$

which would make u a generalized eigenvector, and γ an eigenvector of multiplicity at least 2, contrary to our assumption.

Let M be as above, the smallest subspace in the nest $\mathcal{N}$ that contains x. Let L be subspace in the nest $\mathcal{N}$ that is properly contained in M; such an L does not contain x. Since γ has multiplicity 1, the nullspace of $\mathbf{C} - \gamma\mathbf{I}$ over X is spanned by the single eigenvector x. It follows that the nullspace of $\mathbf{C} - \gamma\mathbf{I}$ over L contains only the zero vector. Therefore $\mathbf{C} - \gamma\mathbf{I}$ is a one-to-one map of L into L; according to the Fredholm alternative, $\mathbf{C} - \gamma\mathbf{I}$ maps L into itself. It follows that the every vector y in L belongs to the range of $\mathbf{C} - \gamma\mathbf{I}$ and therefore satisfies the compatibility condition $(y, m) = 0$.

We are ready to show that x does not belong to M_- as defined by (8). If it did, it would be the limit of a sequence of vectors y_n in some L_n properly contained in M. As we have shown above, every such y_n satisfies the compatibility condition $(y_n, m) = 0$, but then, passing to the limit, we would deduce that $(x, m) = 0$. This contradicts our previous finding that $(x, m) \neq 0$. $\square$

Exercise 3. Modify the argument above to include eigenvalues of arbitrary algebraic multiplicity.

Constructing a nest of invariant subspaces for a compact operator is the Banach space analogy of bringing a finite matrix into upper triangular form. An account of the theory of such triangular forms can be found in K. R. Davidson's monograph on nest algebras.

We give now an example: $X = C[0, 1]$, $\mathbf{V}$ is integration,

$$(\mathbf{V}f)(t) = \int_0^t f(s)\, ds. \tag{9}$$

We saw in chapter 22, equation (15), that this map is compact but has no eigenvectors; therefore its spectrum consists of the point 0. On the other hand, the closed subspaces C_a of C consisting of those f that vanish on $[0, a]$ are obviously invariant for **V**.

Exercise 4. Prove that the subspaces C_a, $0 \le a \le 1$ form a *maximal* nest.

Brodsky and Donoghue have, independently, proved in the L^2 topology the converse of the propostion above:

Theorem 3. *Define H to be the Hilbert space $L^2[0, 1]$, and define H_a to consist of those functions in H that are zero on the subinterval $[0, a]$, $0 \le a \le 1$. The integration operator **V** defined by (9) maps H into itself, and every H_a into itself. Conversely, **V** has no other invariant subspaces.*

Proof. (Donoghue) We start with a brief discussion of the convolution of L^1 functions on the real line. The convolution $f * g$ of two L^1 functions on the real line is defined by

$$(f * g)(t) = \int f(s) g(t - s) \, ds. \tag{10}$$

$\square$

We have shown in section 6 of chapter 19 that the convolution of two L^1 functions is L^1, and that convolution is associative and commutative.

Suppose that both $f(s)$ and $g(s)$ are zero for s large enough negative, and denote by ℓ_f, respectively ℓ_g the lower end of the support of f and g:

$$\ell_f = \sup\{\ell: f(s) = 0 \text{ for } s < \ell\}. \tag{11}$$

It follows from this definition that the integrand on the right in the definition (10) of convolution is zero when $t < \ell_f + \ell_g$, for either one or the other factor is zero. This shows that $(f * g)(t) = 0$ when $t < \ell_f + \ell_g$, from which it follows that $\ell_{f*g} \ge \ell_f + \ell_g$. According to the Titchmarsh convolution theorem the sign of equality holds:

$$\ell_{f*g} = \ell_f + \ell_g. \tag{12}$$

A proof of this remarkable theorem will be given in chapter 38.

The mapping **V** defined by (9) it can be expressed as a convolution:

$$Vf = h * f, \tag{13}$$

where f is set to zero outside the interval $[0,1]$, and the function h is the Heaviside function

$$h(s) = \begin{cases} 0 & \text{for } s < 0 \\ 1 & \text{for } 0 < s. \end{cases} \tag{14}$$

Note that the right side of (13) is defined on all of $\mathbb{R}$; its restriction to $[0,1]$ equals $\mathbf{V}f$. It follows from the associative property for convolution that for any natural number n,

$$\mathbf{V}^n f = h^{(n)} * f, \tag{13'}$$

where $h^{(n)}$ is the n-fold convolution of h with itself. An easy calculation, already performed in chapter 20, shows that

$$h^{(n)}(s) = \begin{cases} 0 & \text{for } s < 0 \\ s^n/n! & \text{for } 0 < s. \end{cases} \tag{14'}$$

The nontrivial part of theorem 3 is that the only invariant subspaces of $\mathbf{V}$ consist of all L^2 functions that vanish on an interval $[0, a]$. This follows from

Lemma 4. *Denote by f any function in $L^2[0, 1]$. The set of functions f, $\mathbf{V}f$, $\mathbf{V}^2 f$, ... span $L^2(\ell, 1)$, where $\ell = \ell_f$.*

Proof. Suppose that g in $L^2[0, 1]$ is orthogonal to $\mathbf{V}^n f$, $n = 0, 1, \ldots$. We can, using (13'), write this condition as

$$(h^{(n)} * f, g) = 0, \qquad n = 0, 1, \ldots \tag{15}$$

where $(,)$ denotes the scalar product in $L^2[0, 1]$. Now define the function g_- in $L^2(-1, 0)$ by

$$g_-(s) = \bar{g}(-s). \tag{16}$$

For any function k in $L^2[0, 1]$, we can write the L^2 scalar product as

$$(k, g) = (k * g_-)(0). \tag{17}$$

Using associativity of convolution and (17), we can write condition (15) as

$$(h^{(n)} * f * g_-)(0) = 0;$$

using (17) once more, we find that this is equivalent to

$$(h^{(n)}_-, f * g_-) = 0. \tag{15'}$$

Note that $f * g_-$ is supported in $[-1, 1]$, and $h^{(n)}_-(s) = s^n/n!$ on the interval $[-1, 0]$ and zero for $s > 0$. Since by the Weierstrass approximation theorem polynomials are dense in $L^2[-1, 0]$, it follows from $(15')$ that $f * g_- = 0$ on $[-1, 0]$. We can express this by the inequality $\ell_{f*g_-} \geq 0$. According to the Titchmarsh convolution theorem, it follows that

$$\ell_f + \ell_{g_-} \geq 0,$$

which implies that $g_-(s) = 0$ for $s < -\ell_f$. By definition (16) of g_- this is the same as

$$g(s) = 0 \qquad \text{for} \quad s > \ell_f.$$

Since g is any function orthogonal to all $\mathbf{V}^n f$, this shows that the orthogonal complement of the span of $\{\mathbf{V}^n f\}$ is contained in $L^2[0, \ell]$. It follows that the span of $\{\mathbf{V}^n f\}$ contains $L^2[\ell, 1]$. On the other hand, since each function $\mathbf{V}^n f$ is zero on $[0, \ell]$, it follows that the set $\{\mathbf{V}^n f\}$ spans $L^2[\ell, 1]$.   □

Theorem 3 follows easily from lemma 4. To see this, let Y be a closed invariant subspace of $\mathbf{V}$ in $L^2[0, 1]$. Given any f in Y, $\mathbf{V}f$, $\mathbf{V}^2 f$, and so on, also belongs to Y, and so by lemma 4, Y contains $L^2[\ell_f, 1]$. It follows that Y contains $L^2[a, 1]$, where

$$a = \inf_{f \text{ in } Y} \ell_f.$$

On the other hand, by definition of a, $f(t) = 0$ for $t < a$ for any f in Y. This shows that $Y = L^2[a, 1]$.   □

The Aronszajn-Smith theorem has been extended in several directions. Robinson and Bernstein have shown, using Robinson's nonstandard analysis, that an operator $\mathbf{T}$ has an invariant subspace, provided that there is a polynomial p such that $p(\mathbf{T})$ is compact. Lomonosov has shown that every operator that commutes with a compact operator has an invariant subspace. Whether every noncompact operator has an invariant subspace has been an open problem, until Enflo exhibited a Banach space X, especially constructed for this purpose, and a linear mapping of X into X that is irreducible, meaning that it has *no* nontrivial invariant subspace. Subsequently irreducible linear maps were exhibited for more familiar spaces but none in reflexive spaces. It is in particular an open question if there are irreducible operators in Hilbert space, and it is an open question whether this question is interesting.

In chapter 38 we will present Beurling's description of all invariant subspace of the right shift operator $\mathbf{R}$ acting on ℓ^2.

BIBLIOGRAPHY

Aronszajn, N. and Smith, K. T. Invariant subspaces of completely continuous operators. *An. Math.*, **60** (1954): 345–350.

Bernstein, A. R. and Robinson, A. Solution of an invariant subspace problem of K. T. Smith and P. R. Halmos. *Pacific J. Math.*, **16** (1966): 421–431.

Brodskii, M. S. On a problem of I. M. Gelfand. *Uspekhi Mat. Nauk* (N.S.), **12** (1957): 129–132.

Davidson, K. R. *Nest Algebras*. Pitman Research Notes in Math, **191**. Longman Scientific and Technical, Essex, England, 1988.

Donoghue, W. F. The lattice of invariant subspaces of a completely continuous quasinilpotent transformations. *Pacific J. Math.*, **7** (1957): 1031–1935.

Enflo, P. On the invariant subspace problem for Banach spaces. *Acta Math.*, **158** (1987): 213–313.

Lomonosov, V. I. Invariant subspaces for the family of operators which commute with a completely continuous operator. *Funct. Anal. Appl.*, **7** (1973); 213–214.

Radjavi, H. and Rosenthal, P. *Invariant Subspaces*. Springer, New York, 1973.

Ringrose, J. R. *Compact Non-self-adjoint Operators*. Van Nostrand Reinhold, New York, 1971.

Ringrose, J. R. Superdiagonal forms for compact linear operators. *Proc. London Math. Soc.* (3), **12** (1962): 367–384.

26

HARMONIC ANALYSIS
ON A HALFLINE

In this chapter we refine the technique described in chapter 21 for the study of compact operators, and use it to deduce the exponential decay of a class of functions.

26.1 THE PHRAGMÉN-LINDELÖF PRINCIPLE
FOR HARMONIC FUNCTIONS

The maximum principle of the theory of analytic functions has a classical generalization due to Phragmén and Lindelöf for analytic functions defined on domains that stretch to infinity. The assumptions are that the function is bounded on the boundary of the domain, and that it is of limited growth as z approaches infinity. The conclusion is that the function is bounded in the whole domain. This principle has an analogue for harmonic functions which we now state and prove:

Let $h(x, y)$ be a harmonic function defined in the half-strip $-1 \leq x \leq 1$, $0 \leq y$, and continuous up to the boundary. We assume that

(i) $h(\pm 1, y) = 0$ for $y \geq 0$.

(ii) $|h(x, y)| \leq$ const. e^{ly} in the half-strip, where l is some positive number less than $\pi/2$.

We claim that then $|h(x, y)| \leq$ const. $e^{-my} k$ in the half-strip, where m is any number less than $\pi/2$.

Proof. We want to compare h to the auxiliary harmonic function

$$k_\epsilon(x, y) = A \cos mx e^{-my} + \epsilon \cos mx e^{my}, \qquad (1)$$

where m is some positive number,

$$\ell < m < \frac{\pi}{2}. \qquad (1')$$

Because of the upper bound placed on m, the function $\cos mx$ is positive on $-1 \leq x \leq 1$; choose A in (1) so large that for all x on $[-1, 1]$,

$$|h(x, 0)| \leq A \cos mx. \tag{2}$$

We claim that $|h(x, y)| \leq k_\epsilon(x, y)$ for all x, y in the half-strip, for any positive choice for ϵ. To see this, we note that

$$-k_\epsilon(x, y) < h(x, y) < k_\epsilon(x, y) \tag{3}$$

for $y = 0, x$ in $[-1, 1]$, since A has been so chosen. Furthermore (3) holds for $y \geq 0, x = \pm 1$, since by hypothesis (i) h is zero there, while k_ϵ is positive. Finally we note that (3) holds for $y = Y$ large enough, $|x| \leq 1$, since by hypothesis (ii) h grown at most as e^{ly}, while k_ϵ grows as e^{my}, and $m > \ell$. Applying the classical maximum principle to the rectangle $0 \leq y \leq y, -1 \leq x \leq 1$, we conclude that (3) holds inside the rectangle as well. Since Y is arbitrary, (3) holds in the entire half-strip for any positive value of ϵ. Letting ϵ tend to zero, we obtain that

$$|h(x, y)| \leq Ae^{-my}$$

in the entire half-strip. This proves that h decays exponentially as y tends to ∞. $\square$

26.2 AN ABSTRACT PRAGMÉN-LINDELÖF PRINCIPLE

This section is devoted to the statement and proof of a sweeping generalization due to the author of the Phragmén-Lindelöf theorem stated above. It is based on a functional analytical abstraction of the space of harmonic functions.

Definition. Let X be a Banach space over $\mathbb{C}$, S a linear space of locally integrable functions $u(y)$ on the positive reals $y > 0$ whose values belong to the Banach space X. We impose the following two conditions on the space S:

(i) S is *translation invariant*, that is, if $u(y)$ belongs to S, so does $u(y + t)$, t any positive number.
(ii) Define the norms $|u|_a^b$ as

$$|u|_a^b = \int_{-a}^{b} |u(y)| dy, \tag{4}$$

where $0 < a < b$. We require that the unit ball in S in the $|u|_a^b$ norm be *precompact* in the $|u|_{a_0}^{b_0}$-norm whenever $[a_0, b_0]$ is a compact subinterval of $[a, b]$, that is, $a < a_0 < b_0 < b$.

A similar definition can be given in terms of the L^p-norm

$$|u|_a^b = \left(\int_a^b |u(y)|^p dy \right)^{1/p} \tag{4'}$$

in place of the L^1-norm.

Property (ii) is called *interior compactness*.

If we adjoin to S all limits of sequences of functions in S that converge in the L^1 sense on compact subsets of $\mathbb{R}_+$, the extended space retains both properties (i) and (ii). Therefore we may assume without loss of generality that (iii) S *is closed in the sequential topology described above.*

Example 1. Take X to be the space of continuous functions on $[-1, 1]$ that vanish at the endpoints. Take S to consist of all functions $u(y) = h(x, y)$, where h is any harmonic function in the half-strip $x \in [-1, 1]$, $y \geq 0$, continuous up to the boundary and equal zero for $x = \pm 1$. It is not hard to show that the space S defined in example 1 has properties (i) and (ii).

Example 2. Let **L** be a linear elliptic operator of order $2n$, whose coefficients are independent of one of the variables, call it y. Let G be a domain in the space of the remaining variables, call them x, whose closure is compact and whose boundary is smooth. Take the Banach space X to be the space H_0^n of functions in G with square integrable derivatives of order n, whose derivatives of order $i = 0, 1, \ldots, n - 1$ are zero on the boundary of G. Take S to consists of all functions of the form $u(y) = h(x, y)$, where $h(x, y)$ is any solution of $\mathbf{L}h = 0$ in the half-cylinder $G \times \mathbb{R}_+$ whose values for fixed y belong to X.

Using the theory of elliptic equations we can show that a space S defined above is interior compact.

Example 3. $X = \mathbb{C}$, S the space spanned on $\mathbb{R}_+$ by a set of exponential functions $\{e^{-\mu_n y}\}$, where the μ_n are real, positive, and $\Sigma \mu_n^{-1} < \infty$. According to Müntz's theorem (see chapter 9) S is a proper subspace of the space of continuous functions on $\mathbb{R}_+$ that vanish at ∞. These spaces were studied by Laurent Schwartz; the interior compactness of such a space follows from properties established by Schwartz.

We state now our *abstract Phragmen-Lindelöf principle.*

Theorem 1. *Let S be a translation invariant, interior compact space of vector-valued functions on $\mathbb{R}_+$. Then there exists a positive number c, depending only on the space S, such that every function u in S that satisfies*

$$\int_0^\infty |u(y)| dy < \infty$$

also satisfies

$$\int_0^\infty |u(y)|e^{cy}dy < \infty.$$

Proof. Denote by K the subspace of S consisting of integrable function $u(y)$; K is a Banach space under the L^1-norm

$$|u| = \int_0^\infty |u(y)|dy. \tag{5}$$

Denote by $\mathbf{T}(t)$ the translation operator acting on K :

$$(\mathbf{T}(t)u)(y) = u(y + t). \tag{6}$$

Clearly, the operators $\mathbf{T}(t)$ are contractions, that is of norm ≤ 1. Therefore the spectrum of each $\mathbf{T}(t)$ lies in the unit disk. The key to theorem 1 is to show that the spectrum of $\mathbf{T}(t), t > 0$, lies *inside* the unit disk.

The proof is somewhat elaborate and is based on 12 lemmas and 2 propositions. In what follows $\mathbf{T}$ denotes any of the operators $\mathbf{T}(t), t > 0$.

Proposition A. *Every nonzero boundary point λ of the spectrum of $\mathbf{T}$ that lies inside the unit disk: $|\lambda| < 1$ and $\lambda \neq 0$, is an eigenvalue of finite multiplicity. That is,*

(i) $(\mathbf{T} - \lambda)$ has a nontrivial nullspace of finite multiplicity.

(ii) The space of all generalized eigenfunctions is finite dimensional.

(iii) Denote by N the space of all generalized eigenfunctions. The mapping $\mathbf{T} - \lambda$ of K/N into K/N is invertible.

(iv) λ is an isolated point of the spectrum of $\mathbf{T}$ over K.

Proof. We need a series of lemmas:

Lemma 1. *Let λ be a complex number inside the unit disc, $|\lambda| < 1$, and $\lambda \neq 0$. Let $\{u_n\}$ be a bounded sequence of functions in K such that for some positive integer k,*

$$\lambda - \lim_{n \to \infty} (\mathbf{T} - \lambda)^k u_n = v. \tag{7}$$

Then a subsequence of u_n converges strongly in K, and the limit u satisfies

$$(\mathbf{T} - \lambda)^k u = v.$$

Proof. It suffices to prove the result for $k = 1$. We have assumed that the sequence $\{u_n\}$ is bounded: $|u_n| \leq M$. Therefore

$$|u_n|_0^{3t} \leq |u_n| \leq M.$$

Since the functions u_n lie in an interior compact space, there is a subsequence, again denoted as $\{u_n\}$, that converges in the $|u|_t^{2t}$-norm:

$$|u_n - u_m|_t^{2t} \to 0. \tag{8}$$

We claim that the sequence $\{u_n\}$ converges in the $|u|_0^a$-norm for any a. For according to (7) with $k = 1$, $\mathbf{T}u_n - \lambda u_n$ converges in the L^1-norm. Since $(\mathbf{T}u_n)(y) = u_n(y+t)$ and since $\lambda \neq 0$, it follows from this and (8) that $\{u_n\}$ converges in the $|\ |_0^t$-norm; it further follows that $\{u_n\}$ converges in the $|\ |_{2t}^{3t}$-norm. Repeating this argument N times we conclude that $\{u_n\}$ converges in the $|\ |_0^{Nt}$-norm, as asserted.

To complete the proof of lemma 1, we show that the functions u_n have uniformly small L^1 norms near ∞. We take the algebraic identity

$$\mathbf{T}^m = \lambda^m + \sum_{j=1}^{m} \lambda^{m-j} \mathbf{T}^{j-1} (\mathbf{T} - \lambda)$$

and apply it to u_n. Denoting $(\mathbf{T} - \lambda)u_n = v_n$, we get

$$\mathbf{T}^m u_n = \lambda^m u_n + \sum_{1}^{m} \lambda^{m-j} \mathbf{T}^{j-1} v_n. \tag{9}$$

By hypothesis (7), $v_n = v + e_n$, where $|e_n| \to 0$. We rewrite the identity above as

$$\mathbf{T}^m u_n = \lambda^m u_n + \sum_{1}^{m} \lambda^{m-j} \mathbf{T}^{j-1} v + \sum_{1}^{m} \lambda^{m-j} \mathbf{T}^{j-1} e_n. \tag{9'}$$

We claim that as m and n tend to ∞, the right side of (9') tends to zero in the norm of K. The last group of terms can be estimated as

$$\left| \sum \lambda^{m-j} \mathbf{T}^{j-1} e_n \right| \le \sum_{1}^{m} |\lambda|^{m-j} |e_n| \le \frac{|e_n|}{1 - |\lambda|}.$$

To estimate the next group of terms, we note that $|\mathbf{T}^{j-1} v| = d_j$ tends to zero as j tends to ∞; therefore

$$\left| \sum \lambda^{m-j} \mathbf{T}^{j-1} v \right| \le \sum_{1}^{m} |\lambda|^{m-j} d_j$$

tends to zero as m tends to ∞. Finally $|\lambda^m u_n| \le |\lambda|^m M$ tends to zero as m tends to ∞. So we conclude from (9') that given any positive ϵ,

$$|\mathbf{T}^m u_n| < \epsilon$$

for all n and m large enough. Since

$$|\mathbf{T}^m u| = \int_{mt}^{\infty} |u(y)| \, dy,$$

this proves that the functions u_n are uniformly small at ∞. Hence the subsequence $\{u_n\}$ converges in the L^1-norm not only over finite subintervals but on the whole positive axis. $\square$

Lemma 2. *Let* **T** *be any bounded map of a Banach space* K *over* $\mathbb{C}$ *into* K, *and let* λ *be a boundary point of the spectrum of* **T**.

(i) *There exists a sequence of unit vector* u_n *such that*

$$|(\mathbf{T} - \lambda)u_n| \to 0. \tag{10}$$

(ii) *The image of* K *under* $\mathbf{T} - \lambda$ *is a proper subspace of* K.

Lemma 2 is a restatement of theorem 2 of the chapter 20. $\square$

We turn now to proving part (i) of proposition A. Let λ be a boundary point of the spectrum of **T**, $|\lambda| < 1$ and $\lambda \neq 0$. According to part (i) of lemma 2, there is a sequence of functions $\{u_n\}$ in K, $|u_n| = 1$, that satisfies (10). According to lemma 1, a subsequence converges to a limit u that satisfies $(\mathbf{T} - \lambda)u = 0$. Since a limit of unit vector is a unit vector, u is an eigenvector; this shows that λ is an eigenvalue. Denote by N_1 the space of eigenvectors with eigenvalue λ; we claim that it is finite dimensional. To see this, we note that $(\mathbf{T} - \lambda)u = 0$ means that

$$u(y + t) = \lambda u(y).$$

It follows that for an eigenfunction u,

$$|u|_0^s = \int_0^{3t} |u(y)|dy = (|\lambda|^{-1} + 1 + |\lambda|) \int_t^{2t} |u(y)|dy,$$
$$= (|\lambda|^{-1} + 1 + |\lambda|)|u|_t^{2t}.$$

This shows that for eigenvectors the $| \ |_0^{3t}$ norm, and the $| \ |_t^{2t}$ norms are equivalent. But the unit sphere in the former norm is by interior compactness precompact in the latter norm; that makes the unit ball of N_1 in the $| \ |_0^{3t}$-norm precompact. But then according to theorem 6 of chapter 5, N_1 is finite dimensional. This completes the proof of part (i) of proposition A.

We turn now to part (ii) concerning generalized eigenfunctions, defined as the nullspace N_k of the operator $(\mathbf{T} - \lambda)^k$, $k = 1, 2, \ldots$. As already noted in chapter 2, theorem 2, if N_1 is finite dimensional, then all the nullspaces N_k are finite dimensional, and furthermore their dimensions satisfy the inequality

$$\dim N_k - \dim N_{k-1} \geq \dim N_{k+1} - \dim N_k. \tag{11}$$

This inequality derives from the fact that $(\mathbf{T} - \lambda)$ maps N_{k+1}/N_k one-to-one into N_k/N_{k-1}.

Lemma 3. *There is an index i such that the sign of equality holds in (11) for $k \geq i$:*

$$\dim N_{k+1}/N_k = \dim N_k/N_{k-1}, \qquad k \geq i. \tag{11'}$$

Proof. A nonincreasing sequence of nonnegative integers such as $\dim N_{k+1}/N_k$ eventually consists of the same numbers. □

REMARK 1. It follows from (11') that for $k \geq i$, $\mathbf{T} - \lambda$ maps N_{k+1}/N_k one-to-one onto N_k/N_{k-1}.

Lemma 4. *With N_k defined as above denote their union by N : $N = \cup N_k$.*

(i) $\mathbf{T} - \lambda$ *maps N/N_i one-to-one onto N/N_{i-1}.*

(ii) $(\mathbf{T} - \lambda)^{-1}$ *is a bounded map of N/N_{i-1} onto N/N_i.*

Proof. Part (i) is purely linear algebra:

Exercise 1. Prove part (i) of lemma 4.

(ii) If $(\mathbf{T} - \lambda)^{-1}$ were unbounded, there would be a sequence U_n in N/N_i, $|U_n| = 1$ such that $(\mathbf{T} - \lambda)U_n = V_n$ is a nullsequence in N/N_{i-1}. Since $|U_n| = 1$, there is an element u_n in the coset U_n satisfying $|u_n| \leq 2$. On the other hand, $|V_n| \to 0$ means that we can write

$$(\mathbf{T} - \lambda)u_n = v_n + z_n, \tag{12}$$

where $z_n \in N_{i-1}$, $|v_n| \to 0$. Applying $(\mathbf{T} - \lambda)^{i-1}$ to (12), we get

$$(\mathbf{T} - \lambda)^i u_n = (\mathbf{T} - \lambda)^{i-1} v_n. \tag{12'}$$

Since $|v_n| \to 0$ and $(\mathbf{T} - \lambda)^{i-1}$ is a bounded map, the norm of the right side of (12') tends to zero; therefore so does the left side:

$$\lim_{n \to \infty} |(\mathbf{T} - \lambda)^i u_n| = 0. \tag{13}$$

We appeal now to lemma 1: since $|u_n| \leq 2$ and (13) is relation (7) with $v = 0$, it follows that subsequence of u_n converges a limit u that satisfies

$$(\mathbf{T} - \lambda)^i u = 0.$$

Such a u belongs to N_i. Recalling the definition of norm in a coset mod N_i, we conclude that

$$|U_n| \leq |u_n - u| \to 0$$

for a subsequence. But this contradicts our assumption that $|U_n| = 1$ for all n; therefore assuming that $(\mathbf{T} - \lambda)^{-1}$ is unbounded leads to a contradiction. $\quad\square$

Denote the closure of N in K by $\bar{N}$. Since N_{i-1} and N_i are finite dimensional and thereby closed, $\bar{N}/N_{i-1}$ and $\bar{N}/N_i$ are the completion of N/N_{i-1} and N/N_i, respectively. Since $(\mathbf{T} - \lambda)^{-1}$ is bounded, it can be extended as a bounded mapping of $\bar{N}/N_{i-1}$ to $\bar{N}/N_i$. This shows that $(\mathbf{T} - \lambda)$ maps $\bar{N}/N_i$ *one-to-one onto* $\bar{N}/N_{i-1}$.

Lemma 5. λ *is not in the spectrum of* $\mathbf{T}$ *as a map of* $K/\bar{N}$ *into* $K/\bar{N}$.

Proof. We argue indirectly and assume the contrary. We have assumed that λ is a boundary point of the spectrum of $\mathbf{T}$ over K. That means that there is a sequence of complex numbers $\mu_n \to \lambda$ that belong to the resolvent set of $\mathbf{T}$ over K. Suppose that infinitely many of these belong to the resolvent set of $\mathbf{T}$ over $K/\bar{N}$; then if λ belonged to the spectrum of $\mathbf{T}$ over $K/\bar{N}$, there would be, according to part (i) of lemma 3, a sequence $\{U_n\}$ of cosets mod $\bar{N}$ such that $|U_n| = 1, |(\mathbf{T} - \lambda)U_n| \to 0$. Let u_n be an element of the coset U_n, $|u_n| < 2$; then

$$(\mathbf{T} - \lambda)u_n = v_n + z_n, \tag{14}$$

where z_n belongs to $\bar{N}$ and $|v_n| \to 0$. It follows from (14) that $|z_n| < 3$. As observed above, $(\mathbf{T} - \lambda)^{-1}$ maps $\bar{N}/N_{i-1}$ boundedly onto $\bar{N}/N_i$; this means that there is an element w_n in $\bar{N}$ and a constant c such that $|w_n| \leq c|z_n| \leq 3c$, and

$$(\mathbf{T} - \lambda)w_n = z_n(\bmod N_{i-1}).$$

Subtracting this form (14) yields

$$(\mathbf{T} - \lambda)(u_n - w_n) = v_n(\bmod N_{i-1}).$$

Applying $(\mathbf{T} - \lambda)^{i-1}$ to both sides, we have

$$(\mathbf{T} - \lambda)^i(u_n - w_n) = (\mathbf{T} - \lambda)^{i-1}v_n. \tag{14$'$}$$

Since $(\mathbf{T} - \lambda)^{i-1}$ is a bounded operator, the norm of the right side of (14$'$) tends to zero; therefore so does the left side:

$$\lim_{n\to\infty} |(\mathbf{T} - \lambda)^i(u_n - w_n)| = 0. \tag{15}$$

Since $|u_n - w_n| \leq |u_n| + |w_n|$ is uniformly bounded, we can appeal to lemma 1 and conclude that a subsequence of $u_n - w_n$ converges to a limit u that satisfies

$$(\mathbf{T} - \lambda)^i u = 0.$$

Such a u belongs to N_i; according to the definition of the norm of cosets

$$|U_n| \leq |u_n - w_n - u| \to 0,$$

in contradiction to $|U_n| = 1$ for all n. Therefore it must be the case that all but a finite number of the μ_n belong to the spectrum of $\mathbf{T}$ over $K/\bar{N}$. Since μ_n belongs to the resolvent set of $\mathbf{T}$ over K, $\mathbf{T} - \mu_n$ maps K onto K, and therefore $K/\bar{N}$ onto $K/\bar{N}$. So the only way μ_n can belong to the spectrum of $\mathbf{T}$ over $K/\bar{N}$ is for μ_n to be an eigenvalue. That is, there exists a coset U_n, $|U_n| = 1$, such that $(\mathbf{T} - \mu_n)U_n = 0$. Since $\mu_n \to \lambda$, it follows that $|(\mathbf{T} - \lambda)U_n| \to 0$; so we are back where we were before, heading into a contradiction. We got into this contradiction by supposing— erroneously—that that λ belongs to the spectrum of $\mathbf{T}$ over $K/\bar{N}$. $\square$

Lemma 6. λ *belongs to the resolvent set of* $\mathbf{T}$ *over* K/N_{i-1}.

Proof. Lemma 5 shows $\mathbf{T} - \lambda$ maps K onto $K/\bar{N}$; previously we have deduced from lemma 4 that $\mathbf{T} - \lambda$ maps $\bar{N}$ onto $\bar{N}/N_{i-1}$. Combining these statements, we conclude that $\mathbf{T} - \lambda$ maps K onto K/N_{i-1}, and therefore that $\mathbf{T} - \lambda$ maps K/N_{i-1} *onto* itself.

We have assumed that λ is a boundary point of the spectrum of $\mathbf{T}$ over K and therefore that λ is a limit of a sequence of points $\mu_n \neq \lambda$ in the resolvent set of $\mathbf{T}$ over K. Since N_{i-1} is finite dimensional, it follows from exercise 1 of chapter 25 that μ_n belongs to the resolvent set of $\mathbf{T}$ over K/N_{i-1} as well; therefore if λ belonged to the spectrum of $\mathbf{T}$ over K/N_{i-1}, it would be a boundary point of the spectrum. According to part (ii) of lemma 2 the image of K/N_{i-1} under $\mathbf{T}-\lambda$ would be a proper subspace of K/N_{i-1}, but this contradicts the fact established above that $\mathbf{T} - \lambda$ maps K/N_{i-1} *onto* itself. $\square$

Since λ is in the resolvent set of $\mathbf{T}$ over K/N_{i-1}, it follows that λ is not an eigenvalue, that is, that $\mathbf{T}-\lambda$ maps no element of K not in N_{i-1} into N_{i-1}. This shows that $N_i = N_{i-1}$; thus the space of generalized eigenvectors N_i is of finite dimensions. This proves parts (ii) and (iii) of proposition A.

Part (iv), that λ is an isolated point of the spectrum of $\mathbf{T}$ over K, now follows as in the classical Riesz theory. Since the resolvent set of $\mathbf{T}$ over K/N_{i-1} is open, all complex numbers μ sufficiently close to λ are in the resolvent set of $\mathbf{T}$ over K/N_{i-1}. On the other hand, the spectrum of $\mathbf{T}$ over N_{i-1} consists of the single point λ. Since the resolvent set of $\mathbf{T}$ over K is the intersection of its resolvent set over N_i and K/N_i, λ is an isolated point of the spectrum. This completes the proof of proposition A. $\square$

Proposition B. *The spectrum of* $\mathbf{T}$ *is a discrete set of points accumulating only at the origin.*

The proof rests on four lemmas.

Lemma 7. *Let* λ *be a complex number in the spectrum of* $\mathbf{T}$ *over* K *with absolute value 1:* $|\lambda| = 1$. *Then there is an eigenfunction* v *in* S *satisfying*

$$v(y + t) = \lambda v(y). \tag{16}$$

The eigenfunction v *lies in* S, *not in* K.

Proof. Since the spectrum of **T** over K lies in the closed unit disk, every point λ of absolute value 1 in the spectrum of **T** is a boundary point of the spectrum. Therefore according to part (i) of lemma 2, given any nullsequence $\{\epsilon_n\}$ of positive numbers, say $\epsilon_n = 1/n^2$, there is a sequence $\{u_n\}$ of functions in K for which the quotient

$$\frac{|(\mathbf{T} - \lambda)u_n|}{|u_n|} \leq \epsilon_n. \tag{17}$$

For any positive number a we can rewrite (17) as

$$\frac{\sum_k |(\mathbf{T} - \lambda)u_n|_{ka}^{(k+1)a}}{\sum_k |u_n|_{ka}^{(k+1)a}} \leq \epsilon_n.$$

It follows that for each n there is an integer k_n such that

$$\frac{|(\mathbf{T} - \lambda)u_n|_{k_n a}^{(k_n+1)a}}{|u_n|_{k_n a}^{(k_n+1)a}} \leq \epsilon_n. \tag{17'}$$

We define v_n as

$$v_n(y) = c_n u_n(y + k_n a),$$

c_n a constant so chosen that

$$|v_n|_0^t = 1, \tag{18}$$

where t is the amount of translation by $\mathbf{T}(t)$. (17′) can be rewritten as

$$|(\mathbf{T} - \lambda)v_n|_0^a \leq \epsilon_n |v_n|_0^a. \tag{17''}$$

We choose now $a = nt$, and denote

$$A_n = |v_n|_0^{nt}.$$

We claim that for all integers $k \leq n$,

$$|v_n|_0^{kt} \leq k + (k - 1)\epsilon_n A_n. \tag{18'}$$

For $k = 1$ this is the normalization (18); for $k > 1$ it follows inductively form (17″). So we have for $k = n$ that

$$A_n = |v_n|_0^{nt} \leq n + (n - 1)\epsilon_n A_n.$$

Since we have chosen $\epsilon_n = 1/n^2$, it follows that $A_n \leq 2n$. Setting this into (17″), we get

$$|(\mathbf{T} - \lambda)v_n|_0^{nt} \leq \frac{2}{n}. \tag{18''}$$

Now we proceed as in the proof of the first part of lemma 1 and use (17'') and (18')
to select a subsequence of $\{v_n\}$ that converges to a limit v in the norm $\mid \mid_0^b$, b any
positive number. Since S is closed in this topology, v belongs to S and satisfies (16).
It follows from the normalization (18) that $v \neq 0$.   □

Lemma 8. *For t small enough, $\lambda = -1$ does not belong to the spectrum of $\mathbf{T}(t)$.*

Proof. Suppose, on the contrary, that -1 lies in the spectrum of $\mathbf{T}(t)$. Then according to lemma 7 there is v_t in S satisfying

$$v_t(y + t) = -v_t(y).$$

As t tends to zero, these functions v_t oscillate more and more rapidly, and clearly
violate interior compactness. Hence $\lambda = -1$ does not belong to the spectrum of
$\mathbf{T}(t)$ for t small enough.   □

Lemma 9. *The spectrum of $\mathbf{T}(t)$ in $0 < |\lambda| < 1$ is a discrete set of points accumu-
lating at most at the origin or on the unit circle.*

Proof. We have shown that for t small enough $\lambda = -1$ belongs to the resolvent
set of $\mathbf{T}(t)$. Therefore some open set around $\lambda = -1$ belongs to the resolvent set.
  According to proposition A, the boundary points of the spectrum of $\mathbf{T}(t)$ inside
the unit circle form a discrete set of points that acumulate at most at the origin or
on the unit circle. We claim that all points of the spectrum inside the unit circle
are boundary points. Suppose on the contrary that there were a point that is not a
boundary point. Such a point could be connected to $\lambda = -1$ by a polygonal path that
avoids the discrete set of boundary points, but that is a contradiction, for such a path
must contain a boundary point. This proves lemma 9 for t small enough; for any t it
follows from $\mathbf{T}(t) = \mathbf{T}^m(t/m)$ and the spectral mapping theorem.
  Let p be any positive number; denote by $S^{(p)}$ the set of functions $u^{(p)}$ of the form

$$u^{(p)}(y) = e^{-py}u(y), \qquad u \text{ in } S. \tag{19}$$

Clearly, $S^{(p)}$ is translation invariant, and it is interior compact, since for a and b
finite,

$$e^{-pb}|u|_a^b \leq |u^{(p)}|_a^b \leq e^{-pa}|u|_a^b. \tag{19'}$$

  We define the space $K^{(p)}$ as the subspace of $S^{(p)}$ consisting of integrable functions
$u^{(p)}$; $K^{(p)}$ is a Banach space under the L^1-norm.
  From here on $\mathbf{T}$ denotes the *unit translation* $\mathbf{T}(1)$.

Lemma 10. *If λ belongs to the spectrum of $\mathbf{T}$ acting on K, and p is a positive
number, then λe^{-p} belongs to the spectrum of $\mathbf{T}$ acting on $K^{(p)}$.*

Proof. Recall that every λ in the spectrum of $\mathbf{T}$ over K is an eigenvalue in the sense that there is an eigenfunction u satisfying

$$u(y + 1) = \lambda u(y),$$

For $|\lambda| < 1$ the eigenfunction u belongs to K, and for $|\lambda| = 1$, u belongs to S. Clearly, $u^{(p)} = e^{-py}u$ belongs to $K^{(p)}$, and satisfies

$$u^{(p)}(y + 1) = e^{-p(y+1)}u(y + 1) = \lambda e^{-p}u^{(p)}(y).$$

This shows that $u^{(p)}$ is an eigenfunction of $\mathbf{T}$ over $K^{(p)}$ with eigenvalue λe^{-p}. $\quad\square$

According to lemma 9 applied to the space $K^{(p)}$, the spectrum of $\mathbf{T}$ acting on $K^{(p)}$ has no points of accumulation inside the unit circle other than zero. Combining this with lemma 10, we conclude that the spectrum of $\mathbf{T}$ acting on K has no point of accumulation on the unit circle, as asserted in Proposition B. $\quad\square$

We are now ready to prove theorem 1. We only need a couple of lemmas, the first an extension of lemma 10:

Lemma 11'. *Denote by* $\Sigma^{(p)}$ *the spectrum of* $\mathbf{T}$ *acting on the space* $K^{(p)}$. *Let* p *and* q *be two positive numbers,* $p < q$. *Then*

$$e^{(p-q)}\Sigma^{(p)} \subset \Sigma^{(q)}. \tag{20}$$

The proof is the same as that of lemma 10 after we observe that $S^{(q)} = e^{(p-q)y}S^{(p)}$.

According to proposition B, $\Sigma^{(q)}$ is a discrete pointset in the unit disk accumulating only at the origin; it follows from (20) that $\Sigma^{(p)}$ contains a point *on* the unit circle for only a *discrete* set of values of p, $p < q$. Choose p to be different from these discrete set of values, and arrange the eigenvalues λ_j of $\mathbf{T}$ acting on $K^{(p)}$ in decreasing order in absolute value:

$$1 > |\lambda_1| \geq |\lambda_2| \geq \ldots \to 0.$$

With each eigenvalue λ_j we can define a *projection* $\mathbf{P}_j$,

$$\mathbf{P}_j = \frac{1}{2\pi i}\int_{C_j}(\zeta - \mathbf{T})^{-1}d\zeta,$$

where C_j is a circle around λ_j that contains no other point of the spectrum of $\mathbf{T}$.

Lemma 12.

(i) *The operators* $\mathbf{P}_j$ *are pairwise disjoint projections:*

$$\mathbf{P}_j^2 = \mathbf{P}_j, \mathbf{P}_j\mathbf{P}_k = 0 \quad for \ j \neq k.$$

(ii) Each $\mathbf{P}_j$ commutes with $\mathbf{T}$.

(iii) The range of $\mathbf{P}_j$ is the space of generalized eigenfunctions of $\mathbf{T}$ with eigenvalue λ_j.

Exercise 2. Prove lemma 11.

Note that according to proposition A the range of each $\mathbf{P}_j$ is finite dimensional.

Denote by $K_m^{(p)}$ the intersection of the nullspaces of $\mathbf{P}_1, \ldots, \mathbf{P}_{m-1}$; $K_m^{(p)}$ is invariant under $\mathbf{T}$. Denote by $\mathbf{T}_{p,m}$ the operator $\mathbf{T}$ acting on $K_m^{(p)}$. The spectrum of $\mathbf{T}_{p,m}$ consists of the eigenvalues $\lambda_m, \lambda_{m+1} \ldots$. According to the formula for spectral radius (see theorem 4 of chapter 17),

$$\lim_{k \to \infty} |\mathbf{T}_{p,m}^k|^{1/k} = |\lambda_m|.$$

It follows that given any positive ϵ, for k large enough

$$|\mathbf{T}_{p,m}^k| < (|\lambda_m| + \epsilon)^k. \tag{21}$$

Every $u^{(p)}$ in $K^{(p)}$ can be decomposed as

$$u^{(p)} = \sum_1^{m-1} f_j + v, \tag{22}$$

where $f_j = \mathbf{P}_j u^p$, and v belongs to $K_m^{(p)}$. For any u in K, $e^{-py} u = u^{(p)}$ belongs to $K^{(p)}$; using the decomposition (22), we can write

$$u = \Sigma e^{py} f_j + e^{py} v. \tag{22'}$$

Now choose the integer m so large that $|\lambda_m| < \frac{1}{4} e^{-p}$, and choose $\epsilon < \frac{1}{4} e^{-p}$. It follows from inequality (21) that for k large

$$\int_k^{k+1} e^{py} |v(y)| dy \leq e^{p(k+1)} \int_k^{\infty} |v(y)| dy \leq e^{p(k+1)} |\mathbf{T}_{p,m}^k v|$$

$$\leq e^{p(k+1)} \left(\frac{1}{2} e^{-p} \right)^k |v| = e^p \left(\frac{1}{2} \right)^k |v|. \tag{23}$$

This shows that the last term on the right in (22') is integrable. We claim that the rest of the terms $e^{py} f_j$ decay exponentially, for they are eigenfunctions or generalized eigenfunctions of $\mathbf{T}$ with eigenvalues $\lambda_j e^p$. All of these eigenvalues must be less than 1, for otherwise they and their sum would not be integrable. This would contradict the fact that u belongs to the space K.

Clearly, we can choose c positive and so small that the conditions

$$e^c |\lambda_j| e^p < 1, \qquad \frac{1}{2} e^c < 1,$$

are fulfilled for all λ_j that satisfy $|\lambda_j| e^p < 1$. It follows then from (22') and (23) that $e^{cy} u(y)$ is integrable. This completes the proof of theorem 1. $\qquad\square$

26.3 ASYMPTOTIC EXPANSION

We show now how to use the results derived for the proof of theorem 1 to give an asymptotic description of functions contained in the space K.

Lemma 13.

  (i) *There is a basis for the eigenfunctions of* $\mathbf{T}(1)$ *over* K *consisting of exponential functions, that is, functions of the form*

$$e^{\mu y} w, \qquad \mathrm{Re}\, \mu < 0. \tag{24}$$

  (ii) *There is a basis of the generalized eigenfunctions of* $\mathbf{T}(1)$ *over* K *consisting of exponential polynomials, that is, a sum of functions of the form*

$$y^k e^{\mu y} w_k. \tag{25}$$

Proof. The translation operators $\mathbf{T}(t)$ commute. It follows that an eigenspace of $\mathbf{T}(1)$ is an invariant subspace of all the translations $\mathbf{T}(t)$. It follows from this that there is a basis for the eigenfunctions of $\mathbf{T}(1)$ consisting of eigenfunctions v for all the translations:

$$\mathbf{T}(t)v = \lambda(t)v. \tag{26}$$

Since v belongs to K, it is an integrable function of y. Therefore $\mathbf{T}(t)v$ is a continuous function of t in the L^1-norm. This shows that $\lambda(t)$ is continuous. Since translations satisfy

$$\mathbf{T}(s+t) = \mathbf{T}(s)\mathbf{T}(t),$$

it follows from (26) that

$$\lambda(s+t) = \lambda(s)\lambda(t).$$

The only continuous solutions of this equation are $\lambda(t) = e^{\mu t}$. Since the operator $\mathbf{T}(t)$ is norm decreasing, it follows that $\mathrm{Re}\, \mu < 0$. This proves part (i) of lemma 12; part (ii) can be argued similarly. $\qquad\square$

Denote by $\{\mu_j\}$ the set of all μ that appear in an eigenfunction (24) of $\mathbf{T}(1)$. It follows from proposition B that the real parts of μ_j tend to $-\infty$.

Theorem 2. *Let S be as in theorem 1, and K the space of those functions in S that are integrable. Every function $u(y)$ in K has an asymptotic expansion of the form*

$$u(y) \approx \sum e^{\mu_j y} e_j. \tag{27}$$

where each e_j is a finite sum of the form

$$e_j = \sum y^k w_{j,k}.\tag{27'}$$

Proof. Decompose u as in equation (22), with $p = 0$. Use the formula for the spectral radius proves the asymptotic character of the expansion (27). □

We return to example 2 of elliptic operators of order $2n$ of form

$$\mathbf{L} = \sum_0^{2n} \mathbf{A}_j \partial_y^{2n-j},\tag{28}$$

where $\mathbf{A}_j = \mathbf{A}_j(x, \partial x)$ are linear partial differential operators of order j in the variables $x_1, \ldots, x_m$, with coefficients that may depend on x but not on y. Ellipticity means that

$$L(\xi, \eta) = \sum A_j(x, \xi)\eta^{2n-j} > c(|\xi|^{2n} + \eta^{2n})$$

for all x, c some positive constant. A typical example is

$$\mathbf{L}_0 = \partial_y^{2n} + \Delta_x^n.$$

Let G be a smoothly bounded domain in x-space, whose closure is compact. We take S to be the space of all solutions $u = u(x, y)$ of $\mathbf{L}u = 0$ in the half-cylinder $G \times \mathbb{R}_+$, which satisfy so-called coercive boundary conditions on ∂G. We take these boundary conditions the same for all y; the simplest example is Dirichlet boundary conditions. We regard these solutions as functions $u(y)$ whose values belong to the Banach space of functions of x in G, satisfying the boundary conditions, normed by the Sobolev norm

$$\int_G \sum_{a < 2n} |\partial_x^a u|^2 dx.$$

It follows from the theory of elliptic equations that the space S of these functions $u(y)$ form an interior compact space.

Consider all solutions of $\mathbf{L}u = 0$ in the half cylinder $G \times \mathbb{R}_+$ whose dependence of y is exponential:

$$u(x, y) = e^{\mu y} w(x).\tag{29}$$

Setting this in (28), we conclude that w satisfies the equation

$$\left(\sum \mu^{2n-j} A_j\right)w = 0,\tag{30}$$

and w satisfies the prescribed boundary conditions. Applying theorem 2 to this concrete situation yields the following result.

Theorem 3.

(i) *The set of values μ_k for which (30) has a nontrivial solution w satisfying the boundary conditions is a discrete set in the sense that each strip*

$$a < \operatorname{Re} \mu_k < b$$

contains only a finite number of μ_k.

(ii) *Each solution of $\mathbf{L}u = 0$ that is integrable in the half cylinder has an asymptotic expansion of the form*

$$u \approx \sum e^{\mu_k y} w_k(y), \qquad \operatorname{Re} \mu_k < 0,$$

where $w_k(y)$ is a polynomial in y.

Exercise 3. Take

$$\mathbf{L} = (\partial_y^2 + \partial_x^2)^2,$$

G the interval $[0, \pi]$, and u subject to Dirichlet boundary conditions

$$u = u_x = 0 \qquad \text{when } x = 0 \text{ or } \pi,$$

Show that μ in (30) satisfies the equation $\tan^2 \mu\pi = (\mu\pi)^2/[1 + (\mu\pi)]^2$.

BIBLIOGRAPHY

Lax, P. D. A Phragmén-Lindelöf theorem in harmonic analysis and its applications to some questions in the theory of elliptic equations. *CPAM*, **10** (1957): 361–389.

Schwartz, L. Etude des sommes d'exponentielles, 2nd ed. Actualités scientifiques et industrielles 959, Hermann, Paris, 1959.

27
INDEX THEORY

We start with stating the main results of index theory in chapter 2. Let U, V be linear spaces, in general, infinite dimensional. A linear map $\mathbf{T} : U \to V$ is said to have *finite index* if it has these properties:

  (i) The nullspace $N_{\mathbf{T}}$ of $\mathbf{T}$, is a finite-dimensional subspace of U.
  (ii) The quotient space $V / R_{\mathbf{T}}$, $R_{\mathbf{T}}$ the range of $\mathbf{T}$, is finite dimensional.

For such an operator we define the *index* as

$$\operatorname{ind} \mathbf{T} = \dim N_{\mathbf{T}} - \operatorname{codim} R_{\mathbf{T}}. \tag{1}$$

A map $\mathbf{G}$ from one linear space into another is called *degenerate* if its range is finite dimensional. We recall from chapter 2 the following results:

Theorem A. *A linear map* $\mathbf{T} : U \to V$ *has finite index iff* $\mathbf{T}$ *has a pseudoinverse, that is, a linear map* $\mathbf{S} : V \to U$ *such that*

$$\mathbf{ST} = \mathbf{I} + \mathbf{G}, \quad \mathbf{TS} = \mathbf{I} + \mathbf{H}, \tag{2}$$

where $\mathbf{I}$ *denotes the identity in* U *and* V *respectively, and* $\mathbf{G}$, $\mathbf{H}$ *are degenerate maps.*

Theorem B. *Let* $\mathbf{T} : U \to V$ *and* $\mathbf{R} : V \to W$ *be linear maps with finite index. Then their product* $\mathbf{RT}$ *has finite index, and*

$$\operatorname{ind} \mathbf{RT} = \operatorname{ind} \mathbf{R} + \operatorname{ind} \mathbf{T}. \tag{3}$$

Theorem C. *Let* $\mathbf{T} : U \to V$ *be a linear map with finite index, and* $\mathbf{G} : U \to V$ *a degenerate linear map. Then* $\mathbf{T} + \mathbf{G}$ *has finite index, and*

$$\operatorname{ind} (\mathbf{T} + \mathbf{G}) = \operatorname{ind} \mathbf{T}. \tag{4}$$

27.1 THE NOETHER INDEX

In this chapter we present a corresponding analytic theory dealing with *bounded* linear maps of one *Banach space* U into another V. As before, $\mathbf{T}$ has finite index if it has properties (i) and (ii). Condition (ii) implies, according to the closed graph theorem, theorem 14 of chapter 15, that the range of $\mathbf{T}$ is closed.

The natural concept of pseudoinverse in this context is the following:

Definition. Two bounded linear maps $\mathbf{T} : U \to V$ and $\mathbf{S} : V \to U$ are called *pseudoinverses* of each other if

$$\mathbf{ST} = \mathbf{I} + \mathbf{K}, \quad \mathbf{TS} = \mathbf{I} + \mathbf{H}, \tag{5}$$

where $\mathbf{K}$ and $\mathbf{H}$ are *compact* maps of U, respectively V, into themselves.

The analogues of theorems A, B, and C hold in the Banach space context, with very much the same proofs. There are three additional results with no counterparts in the linear algebra context.

We will use repeatedly the basic result theorem 5 of chapter 21:

Theorem 0. *Let* $\mathbf{K} : U \to U$ *be a compact map. Then* $\mathbf{I} + \mathbf{K}$ *has finite index, and*

$$\operatorname{ind}(\mathbf{I} + \mathbf{K}) = 0 \tag{6}$$

The Banach space analogues of theorems A, B, and C are theorems 1, 2, and 3:

Theorem 1. *A bounded map* $\mathbf{T} : U \to V$ *has finite index iff* $\mathbf{T}$ *has a pseudoinverse in the sense of definition (5).*

Proof. It is easy to verify that the pseudoinverse $\mathbf{S}$ constructed in chapter 2 for theorem A can be chosen to be bounded in case $\mathbf{T}$ is a bounded map. Since degenerate maps are compact par excellence, (2) implies (5). Conversely, if (5) is satisfied, it follows that the nullspace of $\mathbf{T}$ is contained in the nullspace of $\mathbf{I} + \mathbf{K}$, and that the range of $\mathbf{T}$ contains the range of $\mathbf{I} + \mathbf{H}$. Since according to theorem 0, $\mathbf{I} + \mathbf{K}$ and $\mathbf{I} + \mathbf{H}$ have finite indices,

$$\dim N_{\mathbf{T}} \le \dim N_{\mathbf{I}+\mathbf{K}} < \infty,$$

and

$$\operatorname{codim} R_{\mathbf{T}} \le \operatorname{codim} R_{\mathbf{I}+\mathbf{H}} < \infty. \qquad \square$$

A special case of theorem B is

Theorem 2. *Let* $\mathbf{T}: U \to V$ *and* $\mathbf{R}: V \to W$ *be bounded maps with finite index. Then their product* $\mathbf{RT}$ *has finite index, equal to the sum of the indices of* $\mathbf{R}$ *and of* $\mathbf{T}$.

From theorem 2 we can deduce

Theorem 2'. *If* **T** *and* **S** *are pseudoinverses to each other,*

$$\text{ind } \mathbf{T} = -\text{ind } \mathbf{S}. \tag{7}$$

Proof. Applying the multiplicative law (3) to (5) we deduce that

$$\text{ind } \mathbf{T} + \text{ind } \mathbf{S} = \text{ind } (\mathbf{I} + \mathbf{K}).$$

According to theorem 0, ind $(\mathbf{I} + \mathbf{K}) = 0$, so (7) follows. □

The next result is due to Yood:

Theorem 3. *Suppose that* **T** $: U \rightarrow V$ *has finite index, and* **L** *is a compact linear map* $U \rightarrow V$. *Then* **T** + **L** *has finite index, and*

$$\text{ind } (\mathbf{T} + \mathbf{L}) = \text{ind } \mathbf{T}.$$

Proof. Since **T** has finite index, it has pseudoinverse **S**. Obviously **S** is also pseudoinverse to **T** + **L**, for

$$\mathbf{S}(\mathbf{T} + \mathbf{L}) = \mathbf{ST} + \mathbf{SL} = \mathbf{I} + \mathbf{K} + \mathbf{SL},$$

and since **L** is compact, so is **SL**. So, by (7),

$$\text{ind } (\mathbf{T} + \mathbf{L}) = -\text{ind } \mathbf{S} = \text{ind } \mathbf{T}. \qquad \square$$

Every bounded linear map **T** $: U \rightarrow V$ has a *transpose* **T**' $: V' \rightarrow U'$, defined by the relation

$$(\mathbf{T}u, \ell) = (u, \mathbf{T}'\ell). \tag{8}$$

Theorem 4. *Let* **T** $: U \rightarrow V$ *have finite index; then so does its transpose* **T**', *and*

$$\text{ind } \mathbf{T}' = -\text{ind } \mathbf{T}. \tag{9}$$

Proof. Let **S** be a pseudoinverse of **T**; taking the transpose of (5) gives

$$\mathbf{T}'\mathbf{S}' = \mathbf{I}' + \mathbf{K}', \quad \mathbf{S}'\mathbf{T}' = \mathbf{I}' + \mathbf{H}'.$$

Since according to theorem 7 of chapter 21 the transpose of a compact operator is compact, it follows that **T**' and **S**' are pseudoinverse to each other. This proves that **T**' has finite index. Next we show that

$$\dim N_{\mathbf{T}'} = \text{codim } R_{\mathbf{T}} \quad \text{and} \quad \text{codim } R_{\mathbf{T}'} = \dim N_{\mathbf{T}}. \tag{10}$$

It follows from (8) that the nullspace of $\mathbf{T}'$ is the annihilator of the range of $\mathbf{T}$; this proves the first part of (10). Similarly (8) shows that the nullspace of $\mathbf{T}$ is the annihilator in U of the range of $\mathbf{T}'$. This would prove the second part of (10) if U were reflexive; otherwise, it only shows that

$$\operatorname{codim} R_{\mathbf{T}'} \geq \dim N_{\mathbf{T}}. \tag{10'}$$

Subtracting the first part of (10) from (10') gives

$$-\operatorname{ind} \mathbf{T}' \geq \operatorname{ind} \mathbf{T}. \tag{11}$$

By (11), applied to $\mathbf{S}$,

$$-\operatorname{ind} \mathbf{S}' \geq \operatorname{ind} \mathbf{S}. \tag{11'}$$

Add this relation to (11):

$$-\operatorname{ind} \mathbf{T}' - \operatorname{ind} \mathbf{S}' \geq \operatorname{ind} \mathbf{T} + \operatorname{ind} \mathbf{S}.$$

According to (7), both sides of the above inequality are zero. However, this can only be if equality holds in both (11) and (11'), and therefore also in (10'). This completes the proof of (10); relation (9) follows. □

The next result, due to Dieudonné, is called the *stability* of the index:

Theorem 5. *If* $\mathbf{T} : U \to V$ *has finite index, there is* $\epsilon > 0$ *such that for all bounded linear maps* $\mathbf{M} : U \to V$ *satisfying* $|\mathbf{M}| < \epsilon$,

$$\operatorname{ind}(\mathbf{T} + \mathbf{M}) = \operatorname{ind} \mathbf{T}. \tag{12}$$

Proof. Let $\mathbf{S}$ be a pseudoinverse to $\mathbf{T}$; then by (5)

$$\mathbf{S}(\mathbf{T} + \mathbf{M}) = \mathbf{ST} + \mathbf{SM} = \mathbf{I} + \mathbf{K} + \mathbf{SM}. \tag{13}$$

Choose $\epsilon = |\mathbf{S}|^{-1}$; then $|\mathbf{SM}| < 1$, so by theorem 2 of chapter 17, $\mathbf{I} + \mathbf{SM}$ is invertible. Multiply (13) by $(\mathbf{I} + \mathbf{SM})^{-1}$ on the left:

$$(\mathbf{I} + \mathbf{SM})^{-1} \mathbf{S}(\mathbf{T} + \mathbf{M}) = \mathbf{I} + (\mathbf{I} + \mathbf{SM})^{-1} \mathbf{K}.$$

This shows that $(\mathbf{I} + \mathbf{SM})^{-1} \mathbf{S}$ is pseudoinverse to $\mathbf{T} + \mathbf{M}$; so, by (7),

$$\operatorname{ind}(\mathbf{T} + \mathbf{M}) = -\operatorname{ind}(\mathbf{I} + \mathbf{SM})^{-1} \mathbf{S}. \tag{13'}$$

Since $(\mathbf{I} + \mathbf{SM})^{-1}$ is invertible, multiplication by it doesn't change the index:

$$\operatorname{ind}(\mathbf{I} + \mathbf{SM})^{-1} \mathbf{S} = \operatorname{ind} \mathbf{S}.$$

Setting this into (13)' and using (7) gives (12). □

Theorem 5 can be reformulated as follows:

Theorem 5′. *If* **T** *has finite index and* $\{\mathbf{T}_n\}$ *a sequence that tends to* **T** *in the norm topology, then for n large enough,* $\mathbf{T}_n$ *has finite index, and*

$$\lim_{n\to\infty} \text{ind }\mathbf{T}_n = \text{ind }\mathbf{T}.$$

A direct consquence of theorem 5′ is

Theorem 5″. *Let* $\mathbf{T}(t)$ *be a one-parameter family of mappings* $U \to V$, $0 \le t \le 1$. *Suppose that for each* t, $\mathbf{T}(t)$ *has finite index, and that* $\mathbf{T}(t)$ *depends continuously on* t *in the norm topology. Then* ind $\mathbf{T}(t)$ *is independent of* t; *in particular,*

$$\text{ind }\mathbf{T}(0) = \text{ind }\mathbf{T}(1).$$

This result, called the *homotopy invariance of index*, is enormously useful in calculating the index of mappings. We will give examples later on.

It is easy to see that the index is *not* continuous in the *strong* topology for maps. As an example, take $U = V = \ell^2$. For $u = (a_1, a_2, \ldots)$ define

$$\mathbf{T}_n u = (a_1, a_2, \ldots, a_n, 0, a_{n+1}, \ldots).$$

Clearly, $N_{\mathbf{T}_n} = \{0\}$, $R_{\mathbf{T}_n} = \{v | v_{n+1} = 0\}$, so ind $\mathbf{T}_n = -1$. But $s - \lim \mathbf{T}_n = \mathbf{I}$, whose index is zero. However, Hörmander has shown the continuity of the index for strongly convergent sequences of maps under an additional assumption that is easily verified in many cases:

Theorem 6 (Hörmander). *Let* $\mathbf{T}_n : U \to V$ *and* $\mathbf{S}_n : V \to U$ *be sequences of bounded linear maps such that*

$$\mathbf{S}_n\mathbf{T}_n = \mathbf{I} + \mathbf{K}_n, \quad \mathbf{T}_n\mathbf{S}_n = \mathbf{I} + \mathbf{H}_n, \tag{14}$$

where $\{\mathbf{K}_n\}$ *and* $\{\mathbf{H}_n\}$ *are uniformly compact sequences of mappings, in this sense: the vectors*

$$\{\mathbf{K}_n u, |u| \le 1, n = 1, 2, \ldots\} \tag{15}$$

belong to a compact subset of U, *and similarly the vectors*

$$\{\mathbf{H}_n v, |v| \le 1, n = 1, 2, \ldots\} \tag{15′}$$

belong to a compact subset of V. *Suppose that the sequences* $\{\mathbf{T}_n\}$ *and* $\{\mathbf{S}_n\}$ *converge strongly:*

$$s - \lim \mathbf{T}_n = \mathbf{T}, \quad s - \lim \mathbf{S}_n = \mathbf{S}. \tag{16}$$

Then

$$\lim_{n \to \infty} \text{ind} \, \mathbf{T}_n = \text{ind} \, \mathbf{T}. \tag{17}$$

For a proof, see Hörmander, theorem 19.1.10.
The following is an immediate consequence of theorem 6:

Theorem 6'. *Let* $\mathbf{T}(t) : U \to V$ *and* $\mathbf{S}(t) : V \to U$ *be bounded linear maps depending on a parameter* t, $0 \le t \le 1$, *continuously in the strong topology. Assume that* $\mathbf{S}(t)$ *and* $\mathbf{T}(t)$ *are pseudoinverses of each other:*

$$\mathbf{S}(t)\mathbf{T}(t) = \mathbf{I} + \mathbf{K}(t), \quad \mathbf{T}(t)\mathbf{S}(t) = \mathbf{I} + \mathbf{H}(t),$$

where the maps $\mathbf{K}(t)$ *and* $\mathbf{H}(t)$ *are uniformly compact in the sense of (15), (15′). Then* ind $\mathbf{T}(t)$ *is independent of* t; *in particular,*

$$\text{ind} \, \mathbf{T}(0) = \text{ind} \, \mathbf{T}(1).$$

NOTE. Operators with finite index are usually called Fredholm operators. There is no historical justification for this. It would be more appropriate to call operators with index $\neq 0$ Noether operators, and their index the Noether index, for Fritz Noether had given the first example of such an operator, defined the notion of index, and proved its stability, in the context of singular integral operators; see Hörmander and Dieudonné. The important multiplicative property (3) is due to Atkinson and Gohberg.

HISTORICAL NOTE. Fritz Noether was the brother of Emmy Noether; he fled to the Soviet Union to escape Nazi terror, and there fell victim to Stalin's terror.

We turn now to some examples of operators whose index can be calculated explicitly. In chapter 30 we will show that in some cases the index can be expressed as the differences of the trace of two operators.

27.2 TOEPLITZ OPERATORS

We denote by $L^2(S^1)$ the space of square integrable complex-valued functions u on the unit circle. It forms a Hilbert space under the L^2-norm

$$\|u\|^2 = \int |u(\theta)|^2 \, d\theta. \tag{18}$$

The functions $e^{ik\theta}$ form an orthonormal basis; every u in L^2 can be expanded as

$$u(\theta) = \sum_{-\infty}^{\infty} u_k \, e^{ik\theta}, \tag{19}$$

where the Fourier coefficients are given by

$$u_k = \int u(\theta)\, e^{-ik\theta}\, d\theta. \tag{20}$$

The Parseval relation holds:

$$\|u\|^2 = \sum_{-\infty}^{\infty} |u_k|^2. \tag{21}$$

Definition. H_+ is the subspace of $L^2(S^1)$ consisting of functions u whose negative coefficients are zero:

$$u \text{ in } H_+ \quad \text{iff} \quad u_k = 0 \quad \text{for } k < 0. \tag{22}$$

Given u in $\mathbf{L}^2$, of form (19), we define its *projection* $\mathbf{P}_+$ onto H_+ by

$$\mathbf{P}_+ u = \sum_{0}^{\infty} u_k\, e^{ik\theta}, \tag{23}$$

where u_k are its Fourier coefficients. It follows from (21) that

$$\|\mathbf{P}_+\| = 1. \tag{24}$$

The space H_- and the projection onto it are defined similarly.

The space H_+ consists of boundary values of functions analytic in the unit disk. The elements of H_- are boundary values of functions antianalytic in the unit disk, that is, functions whose complex conjugate is analytic.

Definition. Let $s(\theta)$ be a continuous complex-valued function on the unit circle S^1. We associate with the function s the *Toeplitz operator* $\mathbf{T}_s$, mapping H_+ into H_+, given by the formula

$$\mathbf{T}_s u = \mathbf{P}_+(su), \qquad u \text{ in } H_+. \tag{25}$$

In words, $\mathbf{T}_s$ is multiplication by s, followed by projection into H_+. We call s the *symbol* of $\mathbf{T}_s$. Clearly, $\mathbf{T}_s$ depends linearly on its symbol: $\mathbf{T}_{s+r} = \mathbf{T}_s + \mathbf{T}_r$.

When we represent functions of class H_+ in terms of their Fourier coefficients, a Toeplitz operator becomes a truncated discrete convolution:

$$(\mathbf{T}_s u)_k = \sum_{j=0}^{\infty} s_{k-j} u_j, \qquad k = 0, 1, \ldots. \tag{25'}$$

Here s_n and u_n denote the nth Fourier coefficients of the functions s and u, respectively. The semiinfinite matrix in (25') has identical entries along each of its dexter diagonals $k - j =$ const. Such matrices are called *Toeplitz matrices*; they arise naturally in discretizations of partial differential operators (e.g., see S. Parter and S. Osher) and in statistical mechanics (see McCoy).

Theorem 7. *Let s be a continuous complex valued function on S^1, $\mathbf{T}_s$ the Toeplitz operator with symbol s. Then*

(i) $\mathbf{T}_s$ is a bounded map of $H_+ \to H_+$, and

$$\|\mathbf{T}_s\| \leq |s|_{\max}. \tag{26}$$

(ii) If s is nowhere zero on S^1, then $\mathbf{T}_s$ has finite index.
(iii)

$$\operatorname{ind} \mathbf{T}_s = -W(s), \tag{27}$$

where $W(s)$ is the winding number of $s(\theta)$ around the origin.

Recall that the winding number of a curve $s(\theta)$ around 0 is the increase in the argument of $s(\theta)$ as θ goes from 0 to 2π, divided by 2π. Analytically

$$W(s) = \operatorname{Im} \int_0^{2\pi} s(\theta)^{-1} \frac{ds}{d\theta} d\theta. \tag{28}$$

Proof. (i) Multiplication by s is a bounded operation, with bound $|s|_{\max}$, and $\mathbf{P}_+$ is bounded, with norm 1. Since $\mathbf{T}_s$ is the product of these two, (26) follows.
   (ii) We claim that $\mathbf{T}_{s^{-1}}$ is a pseudoinverse of $\mathbf{T}_s$. To verify this, we need

Lemma 8. *For s continuous,*

$$\mathbf{C} = \mathbf{P}_+ s - s \tag{29}$$

is a compact map of H_+ into L^2.

Proof. Since s is continuous, given any ϵ we can approximate s uniformly by a trigonometric polynomial s_ϵ so that

$$|s(\theta) - s_\epsilon(\theta)| < \epsilon \qquad \text{for all } \theta. \tag{30}$$

The mapping $\mathbf{C}_\epsilon = \mathbf{P}_+ s_\epsilon - s_\epsilon$ annihilates any function u in H_+ of the form $u = \sum_M^\infty u_k e^{ik\theta}$, M the degree of s_ϵ. Since these functions form a linear subspace of H_+ of codimension M, it follows that the range of $\mathbf{C}_\epsilon$ has dimension $\leq M$. In particular, each $\mathbf{C}_\epsilon$ is compact. It follows from (26) and (30) that $\mathbf{C}_\epsilon$ tends to $\mathbf{C}$ uniformly in the norm. Since the uniform limit of compact maps is compact, (29) is compact.

We can perform now the pseudoinversion; we form the product and get, using (29), that

$$\mathbf{T}_{s^{-1}}\mathbf{T}_s = \mathbf{P}_+s^{-1}\mathbf{P}_+s = \mathbf{P}_+s^{-1}(s + \mathbf{P}_+s - s) = \mathbf{I} + \mathbf{P}_+s^{-1}\mathbf{C}.$$

Since $\mathbf{C}$ is compact, it follows that $\mathbf{T}_{s^{-1}}\mathbf{T}_s$ differs by a compact operator from the identity. Since s and s^{-1} play symmetric roles, it follows that $\mathbf{T}_s$ and $\mathbf{T}_{s^{-1}}$ are pseudoinverses.

(iii) To actually compute the index of $\mathbf{T}_s$, we will deform it continuously by deforming s. We will make use of the following result of topology:

Lemma 9. *Two members of the class of continuous, complex-valued, nowhere zero functions s on S^1 can be continuously deformed into each other within that class if and only if they have the same winding number $W(s)$.*

Proof. That the winding numbers have to be the same follows from the invariance of the winding number under deformation. To prove the converse, take first the case that the winding number of s is zero. Such a function has a single-valued logarithm, $\log s(\theta)$. Deform this function to zero as $t \log s(\theta)$. Exponentiation yields

$$s(\theta, t) = e^{t \log s(\theta)}, \qquad 1 \geq t \geq 0,$$

a deformation of $s(\theta)$ into the constant function 1.

Given any s of winding number N, we write it as

$$s(\theta) = e^{iN\theta}(e^{-iN\theta} s(\theta)).$$

The second factor has winding number zero, and therefore can be deformed into the constant function 1. So $s(\theta)$ can be deformed into $e^{iN\theta}$, $N = W(s)$.

Analytically the simplest curve that winds N times around the origin is $e^{iN\theta}$. For N positive the Toeplitz operator $\mathbf{T}_N$ whose symbol is $e^{iN\theta}$ is just multiplication by $e^{iN\theta}$. Clearly, multiplication by $e^{iN\theta}$ has only the trivial nullspace, and its range in H_+ has codimension N, since it consists of functions of form $\sum_N^\infty u_k e^{ik\theta}$. Therefore

$$\operatorname{ind} \mathbf{T}_N = -N. \tag{31}$$

For $N < 0$ the mapping $\mathbf{T}_N = \mathbf{P}_+ e^{iN\theta}$ is onto H_+, and its nullspace consists of linear combinations of $1, e^{i\theta}, \ldots, e^{i(N-1)\theta}$; thus it has dimension N. Therefore (31) holds for $N < 0$ as well. $\qquad\square$

We have shown in lemma 9 that every $s(\theta)$ that is $\neq 0$ can be deformed into $e^{iN\theta}$; that is, there is a one-parameter family $s(\theta, t)$, continuous in θ, t, such that

$$s(\theta, t) \neq 0, \quad s(\theta, 0) = s(\theta), \quad s(\theta, 1) = e^{iN\theta}.$$

Since the winding number $W(s)$ is invariant under continuous deformation,

$$W(s) = W(s(0)) = W(s(1)) = N.$$

It follows from (26) that

$$\|\mathbf{T}_{s(t)} - \mathbf{T}_{s(t')}\| = \|\mathbf{T}_{s(t)-s(t')}\| \leq \|s(t) - s(t')\|_{\max};$$

since $s(\theta, t)$ depends continuously on t, $\mathbf{T}_{s(t)}$ depends continuously on t in the norm topology. Appealing to the homotopy invariance of the index, theorem 5″, we conclude that

$$\text{ind }\mathbf{T}_s = \text{ind }\mathbf{T}_N.$$

Combining this with (31) and the identification of N as the winding number of s we get (27). □

In the course of proving theorem 7, we have shown that for the special function $s_N(\theta) = e^{iN\theta}$, the dimension of the nullspace of $\mathbf{T}_N$ is either 0 or N, depending on the sign of N. This turns out to be true for all functions s.

Theorem 10. *Let s be a continuous, complex-valued, nowhere zero function on the unit circle S^1, $\mathbf{T}_s$ the Toeplitz operator with symbol s. Then*

(i) If $W(s) > 0$, then $\mathbf{T}_s$ is one-to-one, and its range has codimension $W(s)$.

(ii) If $W(s) < 0$, then $\mathbf{T}_s$ has a nullspace of dimension $W(s)$, and maps H_+ onto H_+.

(iii) If $W(s) = 0$, then $\mathbf{T}_s$ is invertible.

Proof. We first prove part (iii); when the winding number of s is zero, s has a single-valued logarithm:

$$s(\theta) = \exp \ell(\theta), \quad \ell(\theta) = \log s(\theta).$$

We split ℓ into its analytic and antianalytic parts:

$$\ell = \ell_+ + \ell_-, \quad \ell_+ \text{ in } H_+, \quad \ell_- \text{ in } H_-.$$

We assume at first that s is smooth, say C^∞; then so is ℓ and so are ℓ_+ and ℓ_-. Exponentiate:

$$s = e^\ell = e^{\ell_+ + \ell_-} = e^{\ell_+} e^{\ell_-} = s_+ s_-. \tag{32}$$

The function s_+ is the boundary value of an analytic funciton, s_- the boundary of value of an antianalytic function. Both are continuous up to the boundary, and nonzero in the closed unit disk. We show now how to invert $\mathbf{T}_s$ with the aid of s_+ and s_-. Write

$$\mathbf{T}_s u = \mathbf{P}_s u = f,$$

u, f in H_+. This equation means that

$$su = f + g_-, \quad g_- \text{ in } H_-.$$

Express s as $s_+ s_-$, and divide by s_-:

$$s_+ u = s_-^{-1} f + s_-^{-1} g_-.$$

Since $s_-^{-1} = \exp -\ell_-$, the product $s_-^{-1} g_-$ belongs to H_-; applying $\mathbf{P}_+$ gives $s_+ u = \mathbf{P}_+ s_-^{-1} f$. Divide by s_+:

$$u = s_+^{-1} \mathbf{P}_+ s_-^{-1} f. \tag{32'}$$

This shows that $s_+^{-1} \mathbf{P} s_-^{-1}$ is the inverse of $\mathbf{T}_s$.

We turn now to parts (i) and (ii). Denote the winding number of s by W. The function $s e^{-W\theta}$ has a winding number zero; therefore the mapping $u \to f$ given by

$$\mathbf{P}_+ s e^{-W\theta} u = f$$

is invertible. This is the same as saying that $\mathbf{T}_s$ maps $e^{-W\theta} H_+$ one-to-one onto H_+. From this (i) and (ii) follow.   □

We remark that formula (32') is not only a theoretical tool but a practical method for inverting $\mathbf{T}_s$.

We return now to the case when s is merely continuous, and $W(s) = 0$. For any $\epsilon > 0$ we can approximate s uniformly by a smooth function r so that

$$|r - s|_{\max} < \epsilon. \tag{33}$$

For ϵ small enough

$$|r^{-1} s - 1|_{\max} < \tfrac{1}{2}. \tag{33'}$$

We draw two conclusions from this inequality:

(i) It follows from (33') combined (26) that

$$\|\mathbf{T}_{r^{-1}s} - \mathbf{I}\| < \tfrac{1}{2}.$$

It follows from theorem 2 of chapter 17 that $\mathbf{T}_{r^{-1}s}$ is an invertible operator.

(ii) It follows from (33') that r and s have the same winding number. Since we have assumed that $W(s) = 0$, also $W(r) = 0$. Since r is smooth, it can be factored as in (32):

$$r = r_+ r_-,$$

where r_+ is the boundary value of an analytic function that is nowhere zero in the unit disk, and r_- the boundary value of a nowhere zero antianalytic function in the unit disk. Therefore $W(r_+) = 0 = W(r_-)$.

We claim that the operator $\mathbf{T}_{r^{-1}s}$ can be factored as follows:

$$\mathbf{T}_{r^{-1}s} = \mathbf{T}_{r_-^{-1}sr_+^{-1}} = \mathbf{P}_+ r_-^{-1} s r_+^{-1} = \mathbf{P}_+ r_-^{-1} \mathbf{P}_+ s \mathbf{P}_+ r_+^{-1} = \mathbf{T}_{r_-^{-1}} \mathbf{T}_s \mathbf{T}_{r_+^{-1}}.$$

This is so because the operator $\mathbf{P}_+$ to the left of r_+^{-1} acts as the identity; the operator $\mathbf{P}_+$ to the left of s removes an antianalytic function that would have been removed by the leftmost operator $\mathbf{P}_+$. As observed above, the operator $\mathbf{T}_{r^{-1}s}$ on the left is invertible, so are the two operators $\mathbf{T}_{r_-^{-1}}$ and $\mathbf{T}_{r_+^{-1}}$ on the right, because the winding numbers of r_+ and r_- are zero. It follows therefore that the third operator in the product on the right, $\mathbf{T}_s$, is invertible too. This proves part (iii) of theorem 10.   □

The proof above is due to Gohberg, who pointed out that it also applies to piecewise continuous functions s, provided that there is some continuous function r such that inequality (33′) is satisfied, with any constant on the right less than 1:

$$|r^{-1}s - 1|_{\max} < c, \qquad c < 1. \tag{33″}$$

The geometric meaning of the restriction (33″) becomes clear if we multiply both sides by $|r|$ and write it as $|r(\theta) - s(\theta)| < r(\theta)$, valid for all θ. If $r(\theta)$ is the position of a person taking a walk around the origin, and $s(\theta)$ the position of that person's dog, held on a leash of variable length but always less then $c|r(\theta)|$, no matter how busily the dog jumps around in the circle to which it is confined by the leash, it must circle the origin the same number of times as its master. For a full discussion we refer to Gohberg and Krupnik, as well as the last chapter in Douglas.

Krein and Gohberg have extended theorem 7 to $n \times n$ *matrix-valued* functions $\mathbf{S}(\theta)$, acting by multiplication on vector-valued functions $u(\theta)$. For fixed n, we denote by H_+ that subspace of the L^2 vector-valued functions on S^1 whose negative Fourier coefficients are zero. $\mathbf{P}_+$ is orthogonal projection onto H_+.

$\mathbf{S}(\theta)$ denotes a continuous matrix-valued function on S^1 whose entries are complex-valued functions. Since $\mathbf{S}(\theta)$ is a bounded function, the *matrix Toeplitz operator*

$$\mathbf{T_S} = \mathbf{P}_+ \mathbf{S} \tag{34}$$

is a bounded mapping of $H_+ \to H_+$.

Theorem 11. *Let* $\mathbf{S}(\theta)$ *denote a continuous, complex matrix valued function on* S^1, *invertible at all points of* S^1.

(i) $\mathbf{T_S} : H_+ \to H_+$ *defined by (34) has a pseudoinverse.*

(ii) *Since* $\mathbf{S}(\theta)$ *is invertible,* $s(\theta) = \det \mathbf{S}(\theta)$ *is nonzero on* S^1; *we claim that*

$$\operatorname{ind} \mathbf{T_S} = -W(\det \mathbf{S}).$$

Proof. $\mathbf{T_{S^{-1}}}$ is a pseudoinverse of $\mathbf{T_S}$; the proof is the same as in theorem 7. To calculate its index, we deform $\mathbf{S}$ into something simple, using the following topological result:

Lemma 12. *Two matrix-valued continuous functions on S^1 invertible at all points can be deformed continuously into each other within this class of functions iff the winding number of their determinants are equal.*

Starting with any $\mathbf{S}$, we deform it into $\mathbf{S}_N$ of diagonal form:

$$\mathbf{S}_N(\theta) = \begin{pmatrix} e^{iN\theta} & 0 \\ 0 & 1 \end{pmatrix}, \quad N = W(\det \mathbf{S}),$$

where all diagonal entries but the first are $= 1$. The matrix Toeplitz operator $\mathbf{T}_N$, whose symbol is $\mathbf{S}_N$ above, is the direct sum of scalar Toeplitz operators: its index can be computed componentwise by formula (31): ind $\mathbf{T}_N = N$. Theorem 11 now follows from the homotopy invariance of the index.   □

Exercise 1. Show that for

$$\mathbf{S}(\theta) = \begin{pmatrix} e^{i\theta} & 0 \\ 0 & e^{-i\theta} \end{pmatrix},$$

dim $N_{\mathbf{T}_\mathbf{S}} = 1$, codim $R_{\mathbf{T}_\mathbf{S}} = 1$.

The preceding example shows that theorem 10 is false for matrix-valued symbols. When $S(\theta)$ can be factored as $S = S_- S_+$, S_- antianalytic, S_+ analytic, both invertible at every point of the unit disk, then $\mathbf{T}_S = S_+^{-1} \mathbf{P}_+ S_-^{-1}$. Even when it exists, such a factorization can no longer be performed by taking logarithms. A method based on solving a system of PDE $-s$ is given in Lax.

In the proof of theorem 11 we have made essential use of topological notions and results. Conversely, notions and results from index theory are powerful tools in differential topology; a basic result of this kind is the Atiyah-Singer index theorem.

An important extension of the theory of Toeplitz operators, replacing S^1 by the real axis, has been given by Wiener and Hopf. An extension to functions of two variables has been given by Strang. Further generalizations of the notion of Toeplitz operators are due to L. Boutet de Monvel and V. Guillemin, and to C. Berger and Coburn. In the farthest reaching generalization the notion of dimension of nullspace and range is taken in the sense of the dimension function in von Neumann algebras.

27.3 HANKEL OPERATORS

A companion to Toeplitz operators can be constructed by replacing the projection into H_+ by projection into H_-:

Definition. Let $s(\theta)$ be a continuous function on the unit circle S^1. We define the *Hankel operator* $\mathbf{H}_s$ as the mapping of H_+ into H_- given by the formula

$$\mathbf{H}_s u = \mathbf{P}_-(su). \tag{35}$$

If we represent functions of class H_+ and H_- by their Fourier coefficients, a Hankel operator appears as

$$(\mathbf{H}_s u)_k = \sum_{j=0}^{\infty} s_{k+j} u_j, \qquad k = 0, 1, \dots \qquad (35')$$

Here s_n denotes the $(-n)$th Fourier coefficient of s, and u_j, $j = 0, 1, \dots$ the Fourier coefficients of u. Note that in order to maintain symmetry we have included the zeroth Fourier term in both H_+ and H_-. The semi-infinite matrix representing a Hankel operator has identical entries along each sinister diagonal $k + j = $ const. Such a matrix is called a Hankel matrix.

Exercise 2. Show that every Hankel operator is compact.

Exercise 3. Show that the norm of a Hankel operator $\mathbf{H}_s$ satisfies

$$\|\mathbf{H}_s\| \leq \inf |s - q|_{\max}, \qquad (36)$$

where q ranges over all analytic functions in the unit disk that are continuous on the unit circle S^1, and zero at $z = 0$. According to a theorem of Nehari, the sign of equality holds in (36).

For further reading turn to chapter 19 of Hörmander.

BIBLIOGRAPHY

Atkinson, F. V. The normal solubility of linear operators in normed space. *Mat. Sbornik, N.S.*, **28** (1951): 3–14.

Berger, C. A. and Coburn, L. A. Toeplitz operators and quantum mechanics. *Funct. Anal.*, **68** (1986): 273–299.

Boutet de Monvel, L. and Guillemin, V. The spectral theory of Toeplitz operators. *An. Math. Studies,* **99** (1981).

Böttcher, A. and Grudsky, S. M. *Toeplitz Matrices, Asymptotic Linear Algebra and Functional Analysis,* Birkhäuser, Boston, 2000.

Dieudonné, J. Sur les homomorphismes d'espace normeés. *Bull. Sci. Math.* (2), **67** (1943): 72–84.

Dieudonné, J. The index of operators in Banach spaces. *Integral Eq. Oper. Theory,* **8** (1985): 580–589.

Douglas, R. G. *Banach Algebra Techniques in Operator Theory,* 2nd ed. Graduate Texts in Mathematics, **179**. Springer, New York, 1988.

Gohberg, I. C. On linear equations in normed space. *Dokl. Akad. Nauk SSSR (N.S.),* **76** (1951): 477–480.

Gohberg, I. C. and Krein, M. G. Systems of integral equations on a half line with kernels depending on the difference of arguments. *AMS Trans.,* **14** (1960): 217–288.

Gohberg, I. C. and Krupnick, N. Ja. The algebra generated by Toeplitz matrices. *Funct. Anal. Appl.,* **3** (1969): 119–137.

Hörmander, L. *The Analysis of Linear Partial Differential Operators III,* Springer, New York, 1985.

Lax, P. D. On the factorization of matrix valued functions. *CPAM,* **29** (1976): 683–688.

McCoy, B. M. Introductory remarks to Szegö's paper. *On Certain Hermitean Forms Associated with the Fourier Series of a Positive Function*, Vol. 1. *Gábor Szegö's Collected Papers*. Birkhäuser, Boston, 1981.

Noether, F. Über eine Klasse singulären Integralgleichungen. *Math. Ann.*, **82** (1921): 42–63.

Osher, S. Systems of difference equations with general homogeneous boundary conditions. *Trans. AMS*, **137** (1969): 177–201.

Parter, S. V. On the eigenvalues of certain generalizations of Toeplitz matrices. *Arch. Rat. Mech. Anal.*, **11** (1962): 244–257.

Sarason, D. Toeplitz operators with piecewise quasicontinuous symbols. *Indiana U. Math. J.*, **26** (1977): 817–838.

Strang, G. Toeplitz operators in a quarter plane. *Bull. AMS*, **76** (1970): 1303–1307.

Yood, B. Properties of linear transformations preserved under addition of a completely continuous transformation. *Duke Math. J.*, **18** (1951): 599–612.

28

COMPACT SYMMETRIC OPERATORS IN HILBERT SPACE

One of the most beautiful—as well as the most useful—results of linear algebra is the spectral theory of hermitean symmetric matrices. We recall that a matrix $\mathbf{A}$ is hermitean symmetric if it is its own adjoint:

$$\mathbf{A}^* = \mathbf{A}.$$

The spectral theory of such matrices says that $\mathbf{A}$ has a complete set of orthogonal eigenvectors, and that the corresponding eigenvalues are real. This result has a perfect generalization, due to Hilbert, to hermitean symmetric operators in a Hilbert space that are compact. In this chapter we present this generalized theory, and give some concrete applications.

Definition. An operator $\mathbf{A}$ mapping a complex Hilbert space H into itself is called *hermitean symmetric*—symmetric for short—if it is its own adjoint, that is, if for all x and y in H

$$(\mathbf{A}x, y) = (x, \mathbf{A}y). \tag{1}$$

Exercise 1. Show that a symmetric operator $\mathbf{A}$ as above is closed. Show that $\mathbf{A}$ is bounded.

Theorem 1. *Let $\mathbf{A}$ be a symmetric operator:*

(i) *The (hermitean) quadratic form $(\mathbf{A}x, x)$ is real for all x in H.*

(ii) *The quadratic form (2) is not identically zero unless the operator $\mathbf{A} \equiv 0$.*

Proof. Set $y = x$ in (1):

$$(\mathbf{A}x, x) = (x, \mathbf{A}x). \tag{2}$$

315

Since the scalar product in a complex Hilbert space is skew symmetric, (2) is real. If $(\mathbf{A}x, x) = 0$ for all x, then by setting $x \pm y$ in place of x we deduce that the bilinear form $(\mathbf{A}x, y) = 0$ for all x and y in H. It follows that $\mathbf{A}x = 0$ for all x. $\qquad\square$

Definition. A symmetric operator $\mathbf{K}$ mapping a Hilbert space H into itself is called *positive* if the associated quadratic form $(\mathbf{K}x, x)$ is nonnegative for every x in H. This is denoted as $0 \leq \mathbf{K}$.

Definition. Let $\mathbf{A}$ and $\mathbf{B}$ denote two symmetric operators mapping a Hilbert space H into itself. The inequality $\mathbf{A} \leq \mathbf{B}$ ($\mathbf{A}$ less than $\mathbf{B}$, $\mathbf{B}$ greater than $\mathbf{A}$) means that $0 \leq \mathbf{B} - \mathbf{A}$.

Exercise 2. Show that the sum of two positive operator is positive.

Exercise 3. Show that if $\mathbf{A} \leq B_k$ and $\mathbf{C} \leq \mathbf{D}$, then $\mathbf{A} + \mathbf{C} \leq B_k + \mathbf{D}$.

Strict positivity, and strict inequality are defined analogously.

In this chapter we will study the spectral theory of *compact* symmetric operators. We recall from chapter 21 that an operator $\mathbf{A} : H \rightarrow H$ is called *compact* if it maps the unit ball in H into a compact set, that is, if the set $\{\mathbf{A}x, \|x\| \leq 1\}$ is *precompact*. We recall for convenience the notion of precompact; a set is *precompact* if its closure is compact. In a metric space this can be expressed in two equivalent ways:

Definition I. A subset R of a metric space is *precompact* if every sequence $\{z_n\}$ of vectors in R contains a convergent subsequence.

Definition II. A subset R is *precompact* if for any $\epsilon > 0$ the set R can be covered by a finite number of balls of radius ϵ.

Theorem 2. *A compact symmetric operator $\mathbf{A}$ maps a weakly convergent sequence $\{x_n\}$ into a sequence $\{\mathbf{A}x_n\}$ that is strongly convergent.*

Proof. It follows from (1) that if $\{x_n\}$ converges weakly, also $\{\mathbf{A}x_n\}$ converges weakly. A weakly convergent sequence is uniformly bounded: $\|x_n\| \leq$ const. It follows from the definition of a compact operator that $\{\mathbf{A}x_n\}$ lies in a precompact set; so by definition I a subsequence of $\{\mathbf{A}x_n\}$ converges strongly to some limit z. We claim that the whole sequence converges to z; for it not, there would be some ϵ-ball centered at z such that infinitely many $\mathbf{A}x_n$ lie outside it. By precompactness, this sequence has a subsequence that converges strongly to a limit z'; clearly, $\|z'-z\| \geq \epsilon$. But that contradicts the weak convergence of the whole sequence $\{\mathbf{A}x_n\}$. $\qquad\square$

We state now the main result of this chapter, the spectral theorem:

Theorem 3. $\mathbf{A}$ *denotes a compact symmetric operator mapping a complex Hilbert space H into itself. Then there is an orthonormal base $\{z_n\}$ for H consisting of* *eigenvector of* $\mathbf{A}$:

$$\mathbf{A}z_n = \alpha_n z_n. \tag{3}$$

The eigenvalues α_n are real and their only point of accumulation is 0.

Proof. If $\mathbf{A}$ is the zero operator, any orthonormal basis will do. Suppose that $\mathbf{A} \neq 0$. By theorem 2, the quadratic form $(\mathbf{A}x, x) \not\equiv 0$, so it takes on some nonzero value, say positive. Look at the values of the quadratic form on the unit sphere $\|x\| = 1$. Since $\mathbf{A}$ is a bounded operator, $(\mathbf{A}x, x)$ does not exceed $\|\mathbf{A}\|$ on the unit sphere; we claim that it achieves its maximum there. To see this, denote by m its supremum:

$$\sup_{\|x\|=1} (\mathbf{A}x, x) = m; \tag{4}$$

m is a positive quantity. Let $\{x_n\}$ be a maximizing sequence. Since the unit ball in a Hilbert space is weakly sequentially compact (see theorem 7 in chapter 10), a subsequence, also denoted as $\{x_n\}$, converges weakly to a limit we denote as z. We claim that z solves our maximum problem, for, by theorem 2, since $\mathbf{A}$ is compact, $\mathbf{A}x_n$ converges *strongly* to $\mathbf{A}z$. It follows from this that $(\mathbf{A}x_n, x_n)$ converges to $(\mathbf{A}z, z)$. Since $\{x_n\}$ is a maximizing sequence, the former converges to m. Therefore

$$(\mathbf{A}z, z) = m. \tag{4'}$$

To show that z maximizes (4) we have to verify that z is a unit vector. Since z is a weak limit of unit vector, its norm is ≤ 1 according to theorem 5 in chapter 10. Since m is positive, (4') shows that $z \neq 0$. Now define $y = z/|z|$; it is a unit vector, and

$$(\mathbf{A}y, y) = \frac{(\mathbf{A}z, z)}{\|z\|^2} = \frac{m}{\|z\|^2}.$$

If $\|z\|$ were less than 1, $(\mathbf{A}y, y)$ would be greater than m, contradicting the definition (4) of m as the supremum.

The homogeneous function

$$R_{\mathbf{A}}(x) = \frac{(\mathbf{A}x, x)}{\|x\|^2} \tag{5}$$

is called the *Rayleigh quotient*. Clearly, the vector z maximizes $R_{\mathbf{A}}(z)$ among all nonzero vectors, not just unit vectors. Let w be any vector in H, t any real number. The function $R(z + tw)$ as function of t achieves its maximum at $t = 0$; therefore by calculus its t-derivative is zero there. Differentiating (5) yields

$$\frac{(\mathbf{A}w, z) + (\mathbf{A}z, w)}{\|z\|^2} - (\mathbf{A}z, z) \frac{(w, z) + (z, w)}{\|z\|^4} = 0,$$

from which, using the symmetry (1) of $\mathbf{A}$ and (4'), we get

$$\mathrm{Re}(\mathbf{A}z - mz, w) = 0.$$

Since w is an arbitrary vector, $\mathbf{A}z - mz = 0$ i.e. z is an eigenvector of $\mathbf{A}$, with eigenvalue m.

Once we have proved the existence of one eigenvector, we can deduce the existence of a complete set. This is based on the observation that for a symmetric operator $\mathbf{A}$, the orthogonal complement of an eigenvector is invariant under $\mathbf{A}$. To see this, suppose that z is an eigenvector, and y is orthogonal to z; then $\mathbf{A}y$ is orthogonal to z, because

$$(\mathbf{A}y, z) = (y, \mathbf{A}z) = (y, \lambda z) = 0.$$

It follows that if y is orthogonal to a collection $\{z_m\}$ of eigenvector, so is $\mathbf{A}y$.

Now take $\{z_m\}$ to be the collection of *all* eigenvectors of $\mathbf{A}$; denote by Y their orthogonal complement. As shown above $\mathbf{A}$ maps Y into itself, and of course, $\mathbf{A}$ restricted to Y is symmetric. Therefore, as shown above, unless Y consists of the zero vector, it would contain an eigenvector of $\mathbf{A}$, a contradiction since Y is orthogonal to all eigenvector.   □

Exercise 4. Show, using theorem 2, that the sequence of eigenvalues tend to zero.

Exercise 5. Show that if $w - \lim x_n = x$, and if $\lim \|x_n\| = \|x\|$, then x_n converges strongly to x.

The argument used to prove the existence of the first eigenvector was constructive. The same constructive argument can be used to furnish all subsequent eigenvectors. Arrange the positive eigenvalues of $\mathbf{A}$ in decreasing order:

$$\mathbf{A}z_n = \alpha_n z_n, \qquad \alpha_1 \geq \alpha_2 \cdots > 0. \tag{6}$$

Then

$$\alpha_N = \max_{x \perp z_1, \ldots z_{N-1}} \frac{(\mathbf{A}x, x)}{\|x\|^2}. \tag{6'}$$

The vector that maximizes (6') is the Nth eigenvector. The negative eigenvalues of $\mathbf{A}$ can be characterized by similar minimum problems.

Suppose, as is often the case, that we are interested in the eigenvalues rather than the eigenvectors; then formula (6') is not so useful, for this maximum problem involves explicitly the unwanted eigenvectors. Fortunately there are formulas, two distinct ones, one due to E. Fischer, the other to R. Courant, that characterize the Nth eigenvalue of $\mathbf{A}$ without reference to eigenvectors of previous eigenvalues.

Theorem 4. *Let $\mathbf{A}$ be a compact symmetric operator; denote its positive eigenvalues, indexed in decreasing order, by α_k, $k = 1, 2, \ldots$; see (6). Denote by $R_\mathbf{A}(x)$ its Rayleigh quotient, defined by (5).*

(i) Fischer's principle:

$$\alpha_N = \max_{S_N} \min_{x \in S_N} R_A(x), \tag{7}$$

where S_N is any linear subspace of H of dimension N.
(ii) Courant's principle:

$$\alpha_N = \min_{S_{N-1}} \max_{x \perp S_{N-1}} R_A(x). \tag{8}$$

Proof. (i) Since S_N is N dimensional, it contains a nonzero vector y satisfying the $N - 1$ linear conditions $(y, z_k) = 0, k = 1, \ldots, N - 1$. For such a vector y it follows from (6′) that $R_A(y) \leq \alpha_N$. Since y belongs to S_N, it follows that

$$\min_{x \in S_N} R_A(x) \leq \alpha_N. \tag{9}$$

Inequality (9) holds for any subspaces of dimension N. On the other hand, if we take S_N to be the space spanned by the eigenvector $z_1, \ldots z_N$, the minimum of the Rayleigh quotient on S_N, reached for $x = z_n$, is α_N. This proves (7).

(ii) Given any subspace S_{N-1} of dimension $N - 1$, the N-dimensional space spanned by the first N eigenvectors contains a vector y that is perpendicular to S_{N-1}. Since for every vector y in the span of the first N eigenvectors $R_A(y) \geq \alpha_N$, it follows that for every subspace S_{N-1} of dimension $N - 1$,

$$\max_{x \perp S_{N-1}} R_A(x) \geq \alpha_N. \tag{10}$$

On the other hand, if we take S_{N-1} to be the space spanned by $z_1, \ldots z_{N-1}$, then according to (6′) the sign of equality holds in (10). This proves (8). □

Similar pairs of variational principles hold for the negative eigenvalues.

In a finite-dimensional space (7) and (8) hold for all eigenvalues, positive and negative. In this case the two principles are equivalent; (7) applied to $-A$ gives (8).

In an infinite-dimensional Hilbert space (7) and (8) are distinct; (7) can be used to give *lower bounds* for the Nth positive eigenvalue, whereas (8) can be used to give *upper bounds.*

Theorem 5. *Let* A *and* B *denote two compact symmetric operator;* $A \leq B$. *Denote their positive eigenvalues, indexed in decreasing order by* α_k *and* β_k, $k = 1, 2, \ldots$, *respectively. We claim that the kth eigenvalue of* A *is less than or equal the corresponding eigenvalue of* B:

$$\alpha_k \leq \beta_k. \tag{11}$$

Proof. It follows from the definition of inequality for symmetric operators that $A \leq B$ means that $(Ax, x) \leq (Bx, x)$ for all x. Then also $R_A(x) \leq R_B(x)$ for all x; the conclusion now follows from either Fisher's principle (7) or Courant's principle (8).   □

For negative eigenvalues the opposite inequality holds.

Exercise 6. Show that a compact positive symmetric operator has no negative eigenvalues.

In chapter 17 we have developed a functional calculus for elements of a Banach algebra, in particular, for bounded operators mapping a Banach space into itself. The class of functions f for which we were able to define $f(\mathbf{A})$ consisted of the set of functions analytic on an open set containing the spectrum of $\mathbf{A}$. We show now that thanks to the spectral theory developed in this chapter, we can for $\mathbf{A}$ symmetric and compact, define $f(\mathbf{A})$ for every f defined on the spectrum of $\mathbf{A}$.

Theorem 6. *Let $\mathbf{A}$ be a compact symmetric operator. We can assign to every bounded complex-valued function $f(\sigma)$ defined on the spectrum of $\mathbf{A}$ an operator that we denote as $f(\mathbf{A})$, so that*

> *(i) the operator assigned to the function $f(\sigma) \equiv 1$ is the identity $\mathbf{I}$.*
> *(ii) the operator assigned to the identity function $f(\sigma) = \sigma$ is $\mathbf{A}$.*
> *(iii) the assignment $f \rightarrow f(\mathbf{A})$ is a isomorphism of the ring of bounded function on $\sigma(\mathbf{A})$ into the algebra of bounded maps of H into H.*
> *(iv) this isomorphism is isometric:*
>
> $$\|f(\mathbf{A})\| = \sup_{\sigma \text{ in } \sigma(\mathbf{A})} |f(\sigma)|.$$
>
> *(v) when f is real valued, $f(\mathbf{A})$ is symmetric.*
> *(vi) when f is positive on the spectrum of $\mathbf{A}$, so is $f(\mathbf{A})$.*

Proof. The proof is shorter than the statement of the theorem. Denote by $\{z_n\}$ an orthonormal basis consisting of eigenvectors of $\mathbf{A}$, with eigenvalue α_n. Express x in H in terms of this basis:

$$x = \sum c_n z_n, \tag{12}$$

and define $f(\mathbf{A})$ to act as follows:

$$f(\mathbf{A})x = \sum f(\alpha_n) c_n z_n. \tag{13}$$

Properties (i) to (vi) are now obvious.   □

Corollary 1. *Suppose that the operator $\mathbf{A}$ is positive; then the spectrum of $\mathbf{A}$ lies on the nonnegative reals. Therefore $f(\lambda) = \sqrt{\lambda}$ is real and can be chosen positive on the spectrum of $\mathbf{A}$; $\sqrt{\mathbf{A}}$, called the positive square root of $\mathbf{A}$, is symmetric and positive.*

Exercise 7. Prove that the positive square root of $\mathbf{A}$ is unique, i.e. that there is no other positive operator whose square is $\mathbf{A}$.

Here is a no-cost extension of the spectral theorem for compact symmetric operators:

Theorem 7. *Let $\{\mathbf{A}_\gamma\}$ be a collection of symmetric operators mapping a Hilbert space H into itself that commute pairwise: $\mathbf{A}_\gamma \mathbf{A}_\delta = \mathbf{A}_\delta \mathbf{A}_\gamma$. Suppose that at least one of the $\mathbf{A}_\gamma$ is compact; then there an orthonormal basis $\{z_n\}$ consisting of common eigenvectors of all the $\mathbf{A}_\gamma$:*

$$\mathbf{A}_\gamma z_n = \alpha_n(\gamma) z_n. \tag{14}$$

Proof. Denote by $\mathbf{A}$ one of the compact operators in the collection, by α_n its eigenvalues. Denote by S_n the eigenspace corresponding to α_n, that is, the space of vectors z satisfying

$$\mathbf{A}z = \alpha_n z. \tag{14'}$$

It follows from theorem 3 that each S_n is finite dimensional, that they are orthogonal, and that they span H:

$$H = S_1 \oplus S_2 \oplus \cdots.$$

Each S_n is invariant under all the other $\mathbf{A}_\gamma$. To see this, let $\mathbf{A}_\gamma$ act on $(14')$; using commutativity, we get

$$\mathbf{A}_\gamma \mathbf{A}z = \mathbf{A}\mathbf{A}_\gamma z = \alpha_n \mathbf{A}_\gamma z.$$

Restricted to S_n, $\mathbf{A}_\gamma$ is a symmetric operator; therefore S_n is the orthogonal sum of eigenspaces of $\mathbf{A}_\gamma$. We take now another operator $\mathbf{A}_\mu$ of the collection and decompose each of these common eigenspaces of $\mathbf{A}$ and $\mathbf{A}_\gamma$ into an orthogonal sum of eigenspaces of $\mathbf{A}_\mu$, and so on. Since S_n is finite dimensional, this process must come to an end after a finite number of steps with a decomposition of S_n into a sum of eigenspaces for all operators in the collection. Then we turn to S_{n+1} and repeat the process, and so on. □

Exercise 8. Show, without recourse to theorem 3, that eigenvectors of a symmetric operator belonging to distinct eigenvalues are orthogonal.

Theorem 7 has the following important application to normal operators:

Definition. An operator $\mathbf{N}$ mapping a Hilbert space H into itself is called *normal* if $\mathbf{N}$ and its adjoint commute:

$$\mathbf{N}^*\mathbf{N} = \mathbf{N}\mathbf{N}^*.$$

Corollary 2. *Every compact normal operator has a complete set of orthonormal eigenvectors.*

Proof. Decompose **N** into the sum of its symmetric and antisymmetric parts:

$$\mathbf{N} = \mathbf{R} + \mathbf{J}, \qquad \text{where } \mathbf{R} = \frac{\mathbf{N} + \mathbf{N}^*}{2}, \ \mathbf{J} = \frac{\mathbf{N} - \mathbf{N}^*}{2}.$$

Clearly, **R** is symmetric, **J** antisymmetric, and $\mathbf{N}^* = \mathbf{R} - \mathbf{J}$. Since **N** and $\mathbf{N}^*$ commute, so do **R** and **J**. According to Schauder's theorem, theorem 7 in chapter 21, the adjoint $\mathbf{N}^*$ of the compact operator **N** is compact; therefore so are **J** and **R**.

We appeal now to theorem 7 and conclude that **R** and **J** have a complete orthonormal set of common eigenvectors; clearly, these are eigenvectors of **N** as well. □

Definition. An operator **U** is called *unitary* if it maps *H* onto itself isometrically, that is, $\|\mathbf{U}x\| = \|x\|$.

Exercise 9. Show that a unitary map **U** satisfies $\mathbf{U}^*\mathbf{U} = \mathbf{1}$.

Exercise 10. Let **U** be a unitary operator of form $\mathbf{I} + \mathbf{C}$, **C** compact. Show that **U** has a complete set of orthonormal eigenvectors, and that all eigenvalues have absolute value 1.

BIBLIOGRAPHY

Courant, R. Über die Eigenwerte bei den Differenzialgleichungen der Mathematischen Physik. *Math. Zeitschr.*, **7** (1920): 1–57.

Fischer, E. Über quadratische Formen mit reellen Koeffizienten. *Monatshefte Math. Phys.*, **16** (1905): 234–249.

Hilbert D. Grundzüge einer allgemeinen Theorie der linearen Integralgleichungen. *Nachr. Akad. Wiss. Göttingen. Math.–Phys. Kl* (1906): 157–227.

29
EXAMPLES OF COMPACT SYMMETRIC OPERATORS

The spectral theory developed in the previous chapter is one of the workhorses of analysis. In this chapter we present a few examples.

29.1 CONVOLUTION

We take the unit circle S^1 and form the Hilbert space $H = L^2(S^1)$, namely the square integrable functions. Let a denote any complex-valued function of class $L^1(S^1)$, and define the operator $\mathbf{A} : H \to H$ as convolution with a:

$$(\mathbf{A}u)(x) = \int_{S'} a(y)\, u(x-y)\, dy. \tag{1}$$

By a change of variables we can also write

$$(\mathbf{A}u)(x) = \int_{S^1} a(x-y)\, u(y)\, dy.$$

Theorem 1.

(i) *Every convolution operator commutes with translation:*

$$(\mathbf{T}_c u)(x) = u(x+c).$$

(ii) *Any two convolution operator commute.*

Exercise 1. Prove theorem 1.

Theorem 2.

(i) $\mathbf{A}$ *defined in (1) is a bounded operator, and*

$$\|\mathbf{A}\| \le |a|_{L^1}. \tag{2}$$

(ii) **A** *is a compact operator.*

(iii) **A** *is a normal operator.*

(iv) *If a satisfies*

$$a(-x) = \bar{a}(x), \tag{3}$$

then **A** *is symmetric.*

Proof. (i) Approximate the integral on the right of (1) by a sum

$$(\mathbf{A}u)(x) \simeq h \sum a(n\,h)\,u(x - n\,h).$$

Using the triangle inequality, we get that the right side is bounded in norm by

$$h \sum |a(n\,h)|\,\|u\|.$$

From this inequality (2) follows for smooth a and u by letting h tend to zero. For u in L^2 and a in L^1 we use approximation by smooth functions.

(ii) Suppose that a is in $L^2(S^1)$; then we can use the Schwarz inequality to estimate the modulus of continuity of $\mathbf{A}u$:

$$(\mathbf{A}u)(x) - (\mathbf{A}u)(z) = \int [a(x - y) - a(z - y)]\,u(y)\,dy$$

$$\leq \left[\int |a(x - y) - a(z - y)|^2\,dy\right]^{1/2} \|u\|$$

$$= \left[\int |a(y) - a(y + z - x)|^2\,dy\right]^{1/2} \|u\|.$$

As $z - x$ tends to zero, the integral on the right tends to zero; this proves that the image of the L^2 unit ball under **A** forms a set of equicontinuous, uniformly bounded set of functions. According to Arzela-Ascoli such a set of functions is precompact in the maximum norm. Therefore this set is perforce precompact in the coarser L^2-norm.

Any L^1 function a can be approximated in the L^1-norm by a sequence of L^2 functions. It follows from (2) that the corresponding operator $\mathbf{A}_n$ approximate **A** in the operator norm. Since the uniform limit of compact operators is compact, (ii) follows.

(iii) Let u and v be any pair of L^2 functions. Multiply (1) by $\bar{v}(x)$ and integrate over S^1; we get

$$(\mathbf{A}u, v) = \iint a(x - y)\,u(y)\,\bar{v}(x)\,dy\,dx. \tag{4}$$

Interchange the names of the integration variables given for the integral on the right of (4),

$$\iint a(y - x)\,\bar{v}(y)\,u(x)\,dy\,dx = (u, \mathbf{A}^*v),$$

where $\mathbf{A}^*$ is the operation of convolving with the function a^* defined as

$$a^*(x) = \bar{a}(-x). \tag{5}$$

Therefore the adjoint $\mathbf{A}^*$ of $\mathbf{A}$ too is a convolution operator. Since convolutions commute, $\mathbf{A}$ is a normal operator.

(iv) If a satisfies (3), then $a^* = a$ and so $\mathbf{A}^* = \mathbf{A}$. $\qquad\qquad\Box$

According to theorem 7 of chapter 28, a collection of normal operators that commute pairwise and one of which is compact have an orthonormal basis consisting of common eigenvectors. Take the collection consisting of all convolutions $\mathbf{A}$ and all translations $\mathbf{T}$; we showed above that the hypotheses of theorem 7, chapter 28 are fulfilled; therefore there exists an orthonormal basis $\{e_k\}$ of $L^2(S^1)$ such that each $e = e_k$ satisfies

$$a * e = \alpha e, \quad e(x + c) = \tau(c)\,e(x). \tag{6}$$

We have seen before that $e(x + c)$ depends continuously on c in the L^2-norm. It follows that the eigenvalue $\tau(c)$ defined in (6) is a continuous function of c. Clearly, $\tau(c) \neq 0$, for otherwise $e(x + c)$ would be zero for all x, impossible for an eigenfunction.

Interchange the role of x and c in (6):

$$e(x + c) = \tau(c)\,e(x) = \tau(x)\,e(c).$$

Dividing by $\tau(c)\,\tau(x)$, we get

$$\frac{e(x)}{\tau(x)} = \frac{e(c)}{\tau(c)}.$$

This shows that $e = $ const. τ; renorming e, we can make that constant $= 1$. So equation (6) can be rewritten as

$$e(x + c) = e(c)\,e(x). \tag{7}$$

The function e, being equal to the function τ, is continuous. It is well known that the only continuous solution of the functional equation (7) is the exponential function. Since e is continuous on S^1, it is periodic, so

$$e_k(x) = e^{ikx}, \quad k \text{ integer}. \tag{8}$$

This completes the proof that the exponentials (8) form a *complete* orthogonal system on S^1.

There are of course simpler ways of proving that the exponentials (8) are complete, but the proof presented above has the virtue that it can be generalized from S^1

to other compact commutative group. Thus in precisely this fashion Hermann Weyl succeeded in proving that the exponentials $e^{i\xi x}$, ξ real, are complete in the space of *almost periodic* functions. For details, see the F. Riesz and Sz.-Nagy text, *Functional Analysis*.

29.2 THE INVERSE OF A DIFFERENTIAL OPERATOR

Denote by **L** the second-order differential operator

$$\mathbf{L} = -\partial_x^2 + q, \quad \partial_x = \frac{d}{dx}, \tag{9}$$

acting on functions on the interval $[0, 2\pi]$ and vanishing at the endpoints. Here q is a continuous real valued function, say bounded from below by 1:

$$q(x) \geq 1. \tag{10}$$

It is not hard to show (e.g., see chapter 7) that the boundary value problem

$$\mathbf{L}u = f, \quad u(0) = 0, \quad u(2\pi) = 0, \tag{11}$$

has a unique solution u for any given continuous function f. We denote the dependence of this solution u on f by **A**:

$$\mathbf{A}f = u; \tag{12}$$

in words, **A** is the *inverse* of **L**. We will show below that **A** as defined above is bounded in the L^2-$(0, 2\pi)$ norm, and thus can be extended by continuity to the whole Hilbert space $H = L^2$.

Theorem 3. *The operator* **A** *as defined above is*

   (i) bounded.
   (ii) compact.
  (iii) symmetric.
  (iv) positive with respect to the L^2 scalar product.

Proof. For u twice continuously differentiable, multiply equation (11) by u and integrate over $[0, 2\pi]$. Using definition (9) of **L**, integration by parts gives

$$\int \left(u_x^2 + q u^2 \right) dx = \int uf \, dx. \tag{13}$$

We estimate the right side by Schwarz' inequality, and use the arithmetic geometric mean inequality and restriction (10) on the left; we get

$$\int u_x^2 + \frac{1}{2} \int u^2 \, dx \leq \frac{1}{2} \int f^2 \, dx.$$

The boundedness of $\mathbf{A}$ follows from this. It also follows that $\mathbf{A}$ maps the unit ball in the L^2 norm into a set of functions u for which $\|u_x\|^2 \leq 1$. According to Rellich's criterion (see chapter 22) this set is precompact; this proves that $\mathbf{A}$ is a compact operator.

(iii) The symmetry of $\mathbf{A}$ follows from that of $\mathbf{L}$: for u, v twice differentiable, we get by integration by parts that

$$(\mathbf{L}u, v) = (u, \mathbf{L}v).$$

Setting $\mathbf{L}u = f$, $\mathbf{L}v = g$, we get

$$(f, \mathbf{A}g) = (\mathbf{A}f, g).$$

(iv) Since the right side of (13) is $(\mathbf{A}f, f)$, positivity of $\mathbf{A}$ follows.

Since $\mathbf{A}$ as defined for bounded f is a bounded operator in the L^2-norm, it can be extended by continuity to all f in L^2. The extended operator retains all properties listed in theorem 3. □

Exercise 2. Show that for all f in L^2, $\mathbf{A}f$ is continuous and is zero at 0 and π. (Hint: Use the estimate $\|u_x\| \leq \frac{1}{2}\|f\|$.)

We can now apply the main result of chapter 28 to $H = L^2(0, 2\pi)$ and conclude that the operator $\mathbf{A}$ has a complete set of orthonormal eigenfunctions $\{e_n\}$ in L^2:

$$\mathbf{A}e_n = \alpha_n e_n. \tag{14}$$

Since $\mathbf{A}$ is positive, so are its eigenvalues.

Using the result of exercise 2, we conclude from (14) that each e_n is continuous. For such e, $\mathbf{L}$ is the inverse of $\mathbf{A}$, so we apply $\mathbf{L}$ to (14) to conclude that

$$\mathbf{L}e_n = \lambda_n e_n, \quad \lambda_n = \alpha_n^{-1}. \tag{14'}$$

The eigenvalues α_n of $\mathbf{A}$ tend to zero; it follows from (14)' that the eigenvalues λ_n of $\mathbf{L}$ tend to infinity.

29.3 THE INVERSE OF PARTIAL DIFFERENTIAL OPERATORS

The analysis in the last section can be applied, with a few changes, to partial differential operators. The simplest case is

$$\mathbf{L} = -\Delta, \tag{15}$$

the Laplace operator $\Delta = \sum \partial_j^2$, $\partial_j = \partial/\partial x_j$. Let G be a domain in $\mathbb{R}^n$, and consider the boundary value problem

$$\mathbf{L}\, u = f \text{ in } G, \qquad u = 0 \text{ on the boundary of } G. \tag{16}$$

It is a basic result of the theory of partial differential equations (see chapter 7) that for f sufficiently differentiable and G smoothly bounded, this boundary value problem has a unique solution u. As before, we denote this solution u as $\mathbf{A}\, f$. The analogue of theorem 3 holds in this case too and is proved by the same technique, namely by integration by parts. This allows the extension of $\mathbf{A}$ by continuity to all of $H = L^2(G)$, and the application of the spectral theorem of chapter 28 to conclude that $\mathbf{A}$ has a complete orthonormal set of eigenfunctions u_n, with positive eigenvalues α_n.

The argument that was used to prove that $\mathbf{A}$ is a compact operator shows that $\mathbf{A}$ is a smoothing operator. A not too technical argument shows that for a large enough k, $\mathbf{A}^k$ is highly smoothing, that is, it turns any $L^2(G)$ function into one that is as often differentiable as we wish. Applying the operator $\mathbf{A}$ many times to the eigenvalue equation $\mathbf{A}\, e = \alpha\, e$ gives $\mathbf{A}^k\, e = \alpha^k\, e$. It follows that the eigenfunction e is sufficiently differentiable; therefore $\mathbf{L}\,\mathbf{A}\, e = e$. Applying L to the eigenvalue equation gives therefore

$$\mathbf{L}\, e_n = \lambda_n\, e_n, \qquad \lambda_n = \alpha_n^{-1},$$

meaning that the e_n are eigenfunctions of the differential operator $\mathbf{L}$ as well.

This setup is applicable to more general elliptic operators. For instance, we can set

$$\mathbf{L} = -\sum \partial_i\, a_{ij}\, \partial_j + g,$$

where (a_{ij}) is a positive definite matrix whose entries are smooth functions of x and g is a smooth nonnegative function. The boundary condition $u = 0$ on the boundary may be replaced by others, for instance, the Neumann condition that the normal derivative of u vanish on the boundary. But note that whereas for the boundary condition $u = 0$ smoothness of the boundary may be relinquished, it cannot be for the Neumann condition.

The Fischer and Courant variational principals for the eigenvalues of $\mathbf{A}$ can be translated into variational principles for the eigenvalues of $\mathbf{L}$. They yield excellent asymptotic estimates for the size of the nth eigenvalue; see Hermann Weyl, and Courant quoted in chapter 28.

BIBLIOGRAPHY

Weyl, H. Über die asymptotische Verteilung der Eigenwerte. *Göttinger Nachr.* (1911): 110–117.

30

TRACE CLASS AND TRACE FORMULA

A remarkable result of linear algebra is the *trace formula*, which says that the sum of the eigenvalues of a square matrix equals the trace of the matrix defined as the sum of its diagonal elements. In 1959 Lidskii showed that this relation is valid also for a large class of compact operator in Hilbert space. This result is rather deep, and its proof correspondingly tricky. Lidskii's trace formula is a powerful tool in many branches of analysis.

30.1 POLAR DECOMPOSITION AND SINGULAR VALUES

Let H be a separable Hilbert space over C, and $\mathbf{T}$ some compact operator mapping H into itself. Denote the adjoint of $\mathbf{T}$ as $\mathbf{T}^*$; the product $\mathbf{T}^*\mathbf{T}$ is clearly a nonnegative, symmetric operator; according to the functional calculus described in chapter 28, $\mathbf{T}^*\mathbf{T}$ has a uniquely determined positive square root $\mathbf{A} = (\mathbf{T}^*\mathbf{T})^{1/2}$. For any u in H,

$$\|\mathbf{T}u\|^2 = (\mathbf{T}u, \mathbf{T}u) = (u, \mathbf{T}^*\mathbf{T}u) = (u, \mathbf{A}^2 u) = (\mathbf{A}u, \mathbf{A}u) = \|\mathbf{A}u\|^2. \qquad (1)$$

As we apply (1) to $u - v$, we deduce that if $\mathbf{A}u = \mathbf{A}v$, then $\mathbf{T}u = \mathbf{T}v$. This enables us to define the operator $\mathbf{U}$ on the range of $\mathbf{A}$ as follows:

$$\mathbf{U} : \mathbf{A}u \to \mathbf{T}u. \qquad (2)$$

It further follows from (1) that $\mathbf{U}$ is an isometry on the range of $\mathbf{A}$.

Denote the range of $\mathbf{A}$ by R, and define $\mathbf{U}$ to be zero on the orthogonal complement of R:

$$\mathbf{U}n = 0 \qquad \text{for } n \perp R.$$

Since $(\mathbf{U}n, v) = (n, \mathbf{U}^*v) = 0$ for $n \perp R$, and for all v, it follows that $\mathbf{U}^*$ maps H into the orthogonal complement of $R^\perp$; thus the range of $\mathbf{U}^*$ lies in the closure of $\overline{R}$ of R. We claim that

$$\mathbf{U}^*\mathbf{U}w = w \qquad \text{for } w \text{ in } \overline{R}. \qquad (3)$$

329

To see this take any z and w in $\overline{R}$. Since $\mathbf{U}$ is an isometry on R, it also preserves the scalar product of any two elements in $\overline{R}$:

$$(z, w) = (\mathbf{U}z, \mathbf{U}w) = (z, \mathbf{U}^*\mathbf{U}w).$$

It follows that $(z, \mathbf{U}^*\mathbf{U}w - w) = 0$. Since z is an arbitrary elements of $\overline{R}$,

$$\mathbf{U}^*\mathbf{U}w - w \perp R.$$

On the other hand, we have shown that $\mathbf{U}^*$ maps H into $\overline{R}$. So we see that for w in $\overline{R}$, $\mathbf{U}^*\mathbf{U}w - w$ both belongs to $\overline{R}$ and is perpendicular to it. Therefore (3) follows.

We summarize the information contained in (2) and (3):

Theorem 1. *Every compact operator* $\mathbf{T}$ *can be factored as*

$$\mathbf{T} = \mathbf{UA}, \tag{2'}$$

where $\mathbf{A}$ *is a positive symmetric operator, and* $\mathbf{U}^*\mathbf{U} = \mathbf{I}$ *on the range of* $\mathbf{A}$.

The operator $\mathbf{A}$ is called the *absolute value* of $\mathbf{T}$, and $(2')$ is called the *polar decomposition* of $\mathbf{T}$.

Theorem 1 is true not only for compact operators but all bounded operator. The only place in the proof where compactness was used is the construction of the square root of $\mathbf{T}^*\mathbf{T}$. As we will show in the next chapter, every bounded positive symmetric operator has a square root, not just the compact ones.

When $\mathbf{T}$ is compact, so is its absolute value $\mathbf{A}$. The nonzero eigenvalues of $\mathbf{A}$, denotes as $\{s_j\}$, are positive numbers that tend to zero; we index them in decreasing order. The numbers s_j are called the *singular values* of the operator $\mathbf{T}$, and denoted as $s_j(\mathbf{T})$.

Exercise 1. Show that for each j, $s_j(\mathbf{T})$ is a continuous function of $\mathbf{T}$ in the norm topology.

30.2 TRACE CLASS, TRACE NORM, AND TRACE

Definition. A compact map $\mathbf{T}$ of a Hilbert space H into H is in *trace class* when

$$\sum_1^\infty s_j(\mathbf{T}) < \infty. \tag{4}$$

This sum (4) is called the *trace norm* of $\mathbf{T}$:

$$\|\mathbf{T}\|_{\mathrm{tr}} = \sum s_j(\mathbf{T}). \tag{4'}$$

Exercise 2. Show that $\|T\| \le \|T\|_{\mathrm{tr}}$.

The next theorem enumerates the basic properties of the trace norm:

Theorem 2. *Let* **T** *be a trace class operator,* **B** *any bounded operator. Then*

    (i) $\|\mathbf{T}\|_{\mathrm{tr}} = \|\mathbf{T}^*\|_{\mathrm{tr}}$

    (ii) $\|\mathbf{BT}\|_{\mathrm{tr}} \le \|\mathbf{B}\|\,\|\mathbf{T}\|_{\mathrm{tr}}$

    (iii) $\|\mathbf{TB}\|_{\mathrm{tr}} \le \|\mathbf{B}\|\,\|\mathbf{T}\|_{\mathrm{tr}}.$

    (iv) *For any pair of trace class operators* **T** *and* **S**, **T** $+$ **S** *is trace class, and*

$$\|\mathbf{T} + \mathbf{S}\|_{\mathrm{tr}} \le \|\mathbf{T}\|_{\mathrm{tr}} + \|\mathbf{S}\|_{\mathrm{tr}}.$$

In words, the trace class is closed under adjointness, and is a two-sided ideal in the algebra of all bounded operators. The trace norm satisfies the triangle inequality.

Proof. (i) We will show that $s_j(\mathbf{T}^*) = s_j(\mathbf{T})$. The singular values of $\mathbf{T}^*$ are the positive eigenvalues of the square root of $\mathbf{T}^{**}\mathbf{T}^* = \mathbf{TT}^*$. We claim that $\mathbf{TT}^*$ and $\mathbf{T}^*\mathbf{T}$ have the same positive eigenvalues. To see this, let z be an eigenvector, λ an eigenvalue of $\mathbf{T}^*\mathbf{T}$:

$$\mathbf{T}^*\mathbf{T}z = \lambda z, \qquad \lambda \ne 0.$$

Let **T** act on both sides:

$$\mathbf{TT}^*\mathbf{T}z = \lambda \mathbf{T}z,$$

which shows that λ is an eigenvalue of $\mathbf{TT}^*$, with eigenvector $\mathbf{T}z$; $\mathbf{T}z$ is not zero because $\lambda \ne 0$. Since $\mathbf{A} = (\mathbf{T}^*\mathbf{T})^{1/2}$, the eigenvectors of $\mathbf{A}$ are those of $\mathbf{T}^*\mathbf{T}$, and the eigenvalues the square root of those of $\mathbf{T}^*\mathbf{T}$. This proves that $s_j(\mathbf{T}) = s_j(\mathbf{T}^*)$, and completes the proof of (i).

Exercise 3. Give an example of a bounded mapping such that $\mathbf{T}^*\mathbf{T}$ has zero for an eigenvalue but $\mathbf{TT}^*$ does not.

(ii) and (iii): We will show that $s_j(\mathbf{T}) \le \|\mathbf{B}\|\, s_j(\mathbf{T})$. To deduce this, we will verify that the absolute value square of $\mathbf{BT}$ is less than $\|\mathbf{B}\|^2$ times the absolute value square of $\mathbf{T}$. Clearly, the associated quadratic forms satisfy the inequality

$$(\mathbf{T}^*\mathbf{B}^*\,\mathbf{BT}\,u, u) = \|\mathbf{BT}u\|^2 \le \|\mathbf{B}\|^2\ \|\mathbf{T}u\|^2 = \|\mathbf{B}\|^2\,(\mathbf{T}^*\mathbf{T}u, u)\,;$$

this is the meaning of

$$(\mathbf{BT})^*\,\mathbf{BT} \le \|\mathbf{B}\|^2\,\mathbf{T}^*\mathbf{T}.$$

According to theorem 5 of chapter 28, the jth eigenvalue is a monotonic function, so

$$s_j^2(\mathbf{BT}) \le \|\mathbf{B}\|^2\,s_j^2(\mathbf{T}). \tag{5}$$

Taking the square root and summing over j, we obtain inequality (ii).

Since the singular values of adjoint operators are the same, we deduce from (5) that

$$s_j(\mathbf{TB}) = s_j(\mathbf{B}^*\mathbf{T}^*) \le \|\mathbf{B}^*\|^2 s_j(\mathbf{T}^*) = \|\mathbf{B}\| \, s_j(\mathbf{T}). \tag{5'}$$

Summing over j we deduce (iii).

To prove (iv), we establish the following characterization of the trace class and trace norm:

$$\|\mathbf{T}\|_{\text{tr}} = \sup \sum_n |(\mathbf{T}f_n, e_n)|, \tag{6}$$

where the supremum is taken over all pairs of orthonormal bases $\{f_n\}$ and $\{e_n\}$. We have to show that the right side of (6) never exceeds $\|\mathbf{T}\|_{\text{tr}}$, and equals it for appropriate choice of f_n and e_n.

Denote by z_j the normalized eigenvectors of the absolute value $\mathbf{A}$:

$$\mathbf{A}z_j = s_j z_j, \quad \|z_j\| = 1.$$

For any vector f, we can expand,

$$f = \sum_j (f, z_j) z_j, \quad \mathbf{A}f = \sum_j s_j(f, z_j) z_j.$$

Apply $\mathbf{U}$ to both sides; using the polar decomposition $\mathbf{T} = \mathbf{U}\mathbf{A}$, we get

$$\mathbf{T}f = \sum s_j(f, z_j) w_j, \tag{7}$$

where $w_j = \mathbf{U}z_j$. According to theorem 1, w_j form an orthogonal basis of the range of $\mathbf{A}$. We take the scalar product of (7) with e:

$$(\mathbf{T}f, e) = \sum s_j(f, z_j)(w_j, e). \tag{7'}$$

We set now $f = f_n$, $e = e_n$ and sum over n:

$$\sum (\mathbf{T}f_n, e_n) = \sum \sum s_j(f_n, z_j)(w_j, e_n). \tag{8}$$

We claim that the double series on the right converges absolutely and is $\le \|\mathbf{T}\|_{\text{tr}}$. To see this, we sum first with respect to n and apply the Schwarz inequality; we get the following estimate for the sum on the right in (8):

$$\sum_j s_j \left(\sum_n |(f_n, z_j)|^2 \sum_n |(w_j, e_n)|^2 \right)^{1/2}.$$

By the Parseval relation

$$\sum_n |(f_n, z_j)|^2 = \|z_j\|^2 = 1, \quad \sum_n |(w_j, e_n)|^2 = \|w_j\|^2 = 1.$$

This shows that the right side of (8) is bounded by $\sum s_j = \|\mathbf{T}\|_{\text{tr}}$.

To complete the proof, we choose $f_n = z_n, e_n = w_n$, supplemented by an arbitrary orthonormal basis on the orthogonal complement of the range of $\mathbf{A}$. Setting $f_n = z_n$, $e_n = w_n$, in $(7')$ we get, since $\mathbf{U}$ is an isometry on the range of $\mathbf{A}$, that

$$(\mathbf{T}f_n, e_n) = (\mathbf{UA}z_n, \mathbf{U}z_n) = s_n.$$

Summing over n, we get equality in (6).

The right side of formula (6) is a supremum of a sum of absolute value of linear function of $\mathbf{T}$. Therefore it is a subadditive function of $\mathbf{T}$; it follows that if it is finite for $\mathbf{S}$ and $\mathbf{T}$, it is finite for $\mathbf{S} + \mathbf{T}$ and satisfies the triangle inequality. $\qquad\square$

Exercise 4. Show that the trace class operators form of a complete linear space under the trace norm.

A bounded operator $\mathbf{T}$ in Hilbert space can be represented as an infinite matrix with respect to any orthonormal basis $\{f_n\}$. The mnth element of this matrix is $(\mathbf{T}f_n, f_m)$. Therefore the trace of this matrix is

$$\sum(\mathbf{T}f_n, f_n), \tag{9}$$

provided that this series converges.

Theorem 3. *For every trace class operator $\mathbf{T}$ the series (9) converges absolutely to a limit that is independent of the orthonormal basis chosen. It is called the trace of $\mathbf{T}$, and is denoted as* tr $\mathbf{T}$.

Proof. Set in (8) $e_n = f_n$, to obtain

$$\sum(\mathbf{T}f_n, f_n) = \sum\sum s_j(f_n, z_j)(w_j, f_n). \tag{10}$$

As we have already shown, the double series on the right converges, and its value is $\leq \|\mathbf{T}\|_{\mathrm{tr}}$.

To show that the trace is independent of choice of the orthonormal basis, we sum (10) first with respect to n. Using the Parseval relation

$$\sum_n (f_n, z)(w, f_n) = (w, z)$$

we can write (10) as

$$\mathrm{tr}\,\mathbf{T} = \sum s_j(\mathbf{T})(w_j, z_j), \tag{11}$$

which is clearly basis independent. $\qquad\square$

We state now some of the basic properties of trace.

Theorem 4. *Let $\mathbf{T}$ be a trace class operator.*

(i) $|\mathrm{tr}\,\mathbf{T}| \leq \|\mathbf{T}\|_{\mathrm{tr}}$

(ii) tr $\mathbf{T}$ *is a linear function of* $\mathbf{T}$.

(iii) tr $\mathbf{T}^* = \overline{\text{tr} \, \mathbf{T}}$.

(iv) *For any bounded operator* $\mathbf{B}$, tr $\mathbf{T}\,\mathbf{B} = $ tr $\mathbf{B}\,\mathbf{T}$.

Proof. Inequality (i) was derived in the course of proving the convergence of (8). Properties (ii) and (iii) follow from the definition (9) of trace. To prove (iv), we start with formula (7), and let $\mathbf{B}$ act on both side:

$$\mathbf{B}\mathbf{T} \, f = \sum s_j(f, z_j) \, \mathbf{B} \, w_j,$$

so

$$(\mathbf{B}\mathbf{T}f, f) = \sum s_j(f, z_j)(\mathbf{B} \, w_j, f).$$

Set $f = f_n$ and sum with respect to n. Reversing the order of summation and using the Parseval relation as in the derivation of (11), we get

$$\text{tr} \, \mathbf{B}\mathbf{T} = \sum s_j(\mathbf{T})(\mathbf{B} \, w_j, z_j). \tag{12}$$

On the other hand, replacing f by $\mathbf{B} \, f$ in (7) gives

$$\mathbf{T}\,\mathbf{B} \, f = \sum_j s_j(\mathbf{T})(\mathbf{B}f, z_j) \, w_j = \sum s_j(\mathbf{T})(f, \mathbf{B}^* z_j) \, w_j.$$

Proceeding as before, we get

$$\text{tr} \, \mathbf{T}\,\mathbf{B} = \sum s_j(\mathbf{T})(w_j, \mathbf{B}^* z_j) = \sum s_j(\mathbf{T})(\mathbf{B} \, w_j, z_j),$$

which is the same as formula (12) for tr $\mathbf{B}\mathbf{T}$.   □

30.3 THE TRACE FORMULA

The deepest, and most important, property of the trace was proved by Lidskii in 1959:

Theorem 5. *The trace of a trace class operator is the sum of its eigenvalues:*

$$\text{tr} \, \mathbf{T} = \sum \lambda_j(\mathbf{T}). \tag{13}$$

The identity (13) *is called the trace formula.*

Proof. When $\mathbf{T}$ is a normal operator of trace class, we can, according to corollary 7 of chapter 28, choose an orthonormal basis consisting of eigenvectors of $\mathbf{T}$. By (9),

$$\text{tr} \, \mathbf{T} = \sum (\mathbf{T} \, f_n, f_n) = \sum \lambda_n,$$

proving (13).

For $\mathbf{T}$ not normal, the eigenvector are in general not orthogonal, and there may be generalized eigenvectors:

$$\mathbf{T} w_n = \lambda_n w \quad \text{or} \quad \mathbf{T} w_n = \lambda_n w_n + w_{n-1}.$$

We can, by the Gram-Schmidt process, orthonormalize them; f_n is a linear combination of $w_1, \ldots, w_n$ so that

$$\mathbf{T} f_n = \lambda_n f_n + \text{linear combination of } f_1, \cdots, f_{n-1}.$$

Since the f_n have been chosen to be orthonormal,

$$(\mathbf{T} f_n, f_n) = \lambda_n.$$

Summing over all n would yield the trace formula, *provided* that the f_n form a basis for the whole Hilbert space. They do if the eigenvector and generalized eigenvector span the whole space, but if they don't, then the f_n have to be supplemented by an orthonormal basis h_m for the orthogonal complement of the eigenvector of $\mathbf{T}$. The expression of the trace of $\mathbf{T}$ now reads

$$\operatorname{tr} \mathbf{T} = \sum (\mathbf{T} f_n, f_n) + \sum (\mathbf{T} h_m, h_m) = \sum \lambda_n + \sum (\mathbf{T} h_m, h_m). \quad (14)$$

The task is to show that the second sum on the right is zero. For this we need some lemmas.

Lemma 6. *Let $\mathbf{T}$ be a compact operator on a Hilbert space H, K the orthogonal complement of its eigenvectors and generalized eigenvectors.*

(i) K is an invariant subspace of $\mathbf{T}^*$.
(ii) The spectrum of $\mathbf{T}^*$ over K consist of the single point $\lambda = 0$.

Proof. (i) Let e be an eigenvector, possibly generalized, of $\mathbf{T}$:

$$\mathbf{T} e = \lambda e + f,$$

f another generalized eigenvector, and suppose that u is orthogonal to e and f. We claim that so is $\mathbf{T}^* u$; for

$$(e, \mathbf{T}^* u) = (\mathbf{T} e, u) = (\lambda e + f, u) = \lambda(e, u) + (f, u) = 0.$$

(ii) According to Schauder's theorem, the adjoint $\mathbf{T}^*$ of a compact operator is compact. If λ were a nonzero eigenvalue of $\mathbf{T}^*$ on K, then $\bar{\lambda}$ would be an eigenvalue of $\mathbf{T}$ in H of finite multiplicity. According to theorem 6 of chapter 21, there is an integer i such that the nullspace of $(\mathbf{T}^* - \bar{\lambda})^i$ equals the nullspace of $(\mathbf{T}^* - \bar{\lambda})^{i+1}$, but is larger than the nullspace of $(\mathbf{T}^* - \bar{\lambda})^{i-1}$. Let u in K be a member of the nullspace of $(\mathbf{T}^* - \bar{\lambda})^i$ but not of $(\mathbf{T}^* - \bar{\lambda})^{i-1}$. Then the equation

$$(\mathbf{T}^* - \bar{\lambda}) v = u$$

has *no* solution; for a solution v would belong to the nullspace of $(\mathbf{T}^* - \overline{\lambda})^{i+1}$ but not to that of $(\mathbf{T}^* - \overline{\lambda})^i$. According to the Fredholm alternative, theorem 8 of chapter 21, there must be an eigenvector w of $\mathbf{T}$, $(\mathbf{T} - \lambda)\, w = 0$, that is *not* orthogonal to u. But this is a contradiction, for u belongs to K, and so u is orthogonal to all eigenvectors of $\mathbf{T}$. $\square$

If $\mathbf{T}$ is of trace class over H, so is $\mathbf{T}^*$. We claim that $\mathbf{T}^*$ restricted to its invariant subspace K is of trace class—this follows immediately from relation (6) characterizing trace class operators.

Back to formula (14); the second sum on the right can be rewritten as

$$\sum (h_m, \mathbf{T}^* h_m) = \overline{\sum (\mathbf{T}^* h_m, h_m)}.$$

Since h_m is an orthonormal basis of K, this sum is the complex conjugate of the trace of $\mathbf{T}^*$ over K. Its vanishing can be formulated so:

Lidskii's Lemma. *Let $\mathbf{T}$ be a trace class operator that has no eigenvalues, except zero; then* $\operatorname{tr} \mathbf{T} = 0$.

The rest of this section is devoted to proving this proposition. We start with an estimate of the eigenvalues of a compact operator in terms of its singular values.

Lemma 7. *Let $\mathbf{T}$ be a compact operator, with nonzero eigenvalues $\lambda_1, \lambda_2, \ldots$, arranged in decreasing order of their absolute value, including multiplicity. Denote as before the singular values of $\mathbf{T}$ as $s_j(\mathbf{T})$, arranged similarly. Then for any N,*

$$\prod_1^N |\lambda_j| \le \prod_1^N s_j(\mathbf{T}). \tag{15}$$

Proof. Denote by E_N the space spanned by the first N eigenvectors of $\mathbf{T}$, and denote by $\mathbf{P}_N$ orthogonal projection onto E_N. Denote by $\mathbf{T}_N$ the restriction of $\mathbf{T}$ to the invariant subspace E_N. Denote by $\mathbf{A}_N$ the absolute value of $\mathbf{T}_N$:

$$\mathbf{T}_N = \mathbf{U}_N \, \mathbf{A}_N. \tag{16}$$

Since the eigenvalues λ_j are nonzero, $\mathbf{T}_N$ is invertible. Then so is $\mathbf{U}_N$, and therefore $\mathbf{U}_N$ is unitary. Taking the determinant of (16) gives

$$|\det \mathbf{T}_N| = \det \mathbf{A}_N.$$

Since the determinant of a matrix is the product of its eigenvalues, we can rewrite this identity as

$$\prod_1^N |\lambda_j| = \prod_1^N \lambda_j(\mathbf{A}_N). \tag{17}$$

The operator $\mathbf{T}\mathbf{P}_N$ acts on E_N as the matrix $\mathbf{T}_N$; on the orthogonal complement of E_N, $\mathbf{T}\mathbf{P}_N = 0$. It follows that the absolute value of $\mathbf{T}\mathbf{P}_N$ is $\mathbf{A}_N$ on E_N, zero on $E_N^{\perp}$. It follows that $\lambda_j(\mathbf{A}_N) = s_j(\mathbf{T}\mathbf{P}_N)$, $j = 1, \ldots, N$. We appeal now to inequality (5)'; $s_j(\mathbf{T}\mathbf{B}) \leq \|\mathbf{B}\| \, s_j(\mathbf{T})$. Apply this to $\mathbf{B} = \mathbf{P}_N$; we obtain

$$s_j(\mathbf{T}\mathbf{P}_N) \leq s_j(\mathbf{T}). \tag{17'}$$

Since we have shown that $\lambda_j(\mathbf{A}_N) = s_j(\mathbf{T}\mathbf{P}_N)$, setting (17') into the right side of (17) yields (15). □

We can deduce further inequalities between $|\lambda_j|$ and s_j with the aid of the following simple principle:

Lemma 8. *Let* $a_1 \geq a_2 \geq \cdots$ *and* $b_1 \geq b_2 \geq \cdots$ *be two decreasing sequences of real numbers, satisfying for each* N

$$\sum_1^N a_j \leq \sum_1^N b_j. \tag{18}$$

Let F *be a convex function defined on* $\mathbb{R}$ *that tends to zero as its argument tends to* $-\infty$. *Then*

$$\sum_1^N F(a_j) \leq \sum_1^N F(b_j) \tag{18'}$$

for every N.

Proof. The set of functions F described above form a convex cone. It was shown in chapter 14, section 14.2, that the extreme rays of this convex cone are the piecewise linear one:

$$F(x) = \begin{cases} 0 & \text{for } x \leq z \\ x - z & \text{for } z \leq x, \end{cases}$$

z an arbitrary real number. For this choice of F inequality (18') can be reduced to

$$\sum_1^P (a_j - z) \leq \sum_1^Q (b_j - z), \tag{19}$$

where

$$a_j \geq z \quad \text{for } j \leq P, \qquad a_j < z \quad \text{for } j > P$$

and

$$b_j \geq z \quad \text{for } j \leq Q, \qquad b_j < z \quad \text{for } j > Q.$$

To verify (19), we observe that the right side can be characterized as

$$\max_{M} \sum_{1}^{M} (b_j - z).$$

For $M = P$, $\sum_{1}^{P}(b_j - z)$ is, according to (18), greater than the left side of (19), so even more so for the maximum.

It was shown in section 14.2 that every F in lemma 8 is a convex combination of the points on the extreme rays. Since both sides of inequality (18') are linear functions of F, and since (18') holds for every extreme F, it holds for all F. ☐

We apply lemma 8 to

$$a_j = \log |\lambda_j(\mathbf{T})|, \quad b_j = \log s_j(\mathbf{T});$$

taking the logarithm of (15) shows that inequalities (18) are satisfied for this choice. Choosing $F(x) = e^x$, we deduce from (18') that

$$\sum_{1}^{N} |\lambda_j(\mathbf{T})| \le \sum_{1}^{N} s_j(\mathbf{T}). \tag{20}$$

Choosing $F(x) = \log(1 + re^x), r > 0$, gives

$$\prod_{1}^{N}(1 + r|\lambda_j|) \le \prod_{1}^{N}(1 + r s_j). \tag{21}$$

To estimate the trace of $\mathbf{T}$, we approximate $\mathbf{T}$ by finite-dimensional projections. Let $\{h_n\}$ be an arbitrary orthonormal basis of the Hilbert space; denote by $\mathbf{P}_N$ orthogonal projection onto the span of $h_1, \cdots, h_N$. Denote by $\mathbf{T}_N$ the projection of $\mathbf{T}$ onto the range of $\mathbf{P}_N$:

$$\mathbf{T}_N = \mathbf{P}_N \mathbf{T} \mathbf{P}_N. \tag{22}$$

Lemma 9. *Suppose that $\mathbf{T}$ is a trace class operator that has no nonzero eigenvalue. Denote $\mathbf{T}_N$ as above. Then*

(i) $\mathbf{T}_N$ *approaches $\mathbf{T}$ uniformly:*

$$\lim_{N \to \infty} \|\mathbf{T}_N - \mathbf{T}\| = 0.$$

(ii) $\lim \operatorname{tr} \mathbf{T}_N = \operatorname{tr} \mathbf{T}.$

(iii) *denote the spectral radius of $\mathbf{T}_N$ by σ_N; σ_N tends to zero as $N \to \infty$.*

Proof. (i) is true for any compact operator $\mathbf{T}$, and (ii) is the definition of trace.

(iii) By assumption $\mathbf{T} - \lambda$ is invertible for every $\lambda \neq 0$. Given any $\delta > 0$, denote $m(\delta) = m$ the quantity

$$m = \max_{|\lambda| \geq \delta} \|(\mathbf{T} - \lambda)^{-1}\|.$$

By (i) we can choose $M(\delta)$ so large that for $N > M(\delta)$,

$$\|\mathbf{T}_N - \mathbf{T}\| < \frac{1}{m}.$$

For such N and $|\lambda| \geq \delta$, $(\mathbf{T}_N - \mathbf{T})(\mathbf{T} - \lambda)^{-1}$ has norm < 1, so

$$\mathbf{T}_N - \lambda = \mathbf{T}_N - \mathbf{T} + \mathbf{T} - \lambda = \left[(\mathbf{T}_N - \mathbf{T})(\mathbf{T} - \lambda)^{-1} + \mathbf{I}\right](\mathbf{T} - \lambda)$$

is invertible when $|\lambda| \geq \delta$. Therefore $\sigma_N < \delta$. $\qquad\qquad\square$

Denote the eigenvalues of $\mathbf{T}_N$ as $\lambda_j^{(N)}$, $j = 1, \ldots, N$. Denote by D_N the polynomial

$$D_N(\lambda) = \prod_1^N \left(1 - \lambda \lambda_j^{(N)}\right). \qquad (23)$$

Lemma 10.

$$\lim_{N \to \infty} D_N(\lambda) = e^{-\lambda \alpha}, \qquad \alpha = \operatorname{tr} \mathbf{T},$$

uniformly on every bounded set of complex numbers λ.

Proof. Take the logarithmic derivative of (23):

$$\frac{D_N'}{D_N} = -\sum_j \frac{\lambda_j^{(N)}}{1 - \lambda \lambda_j^{(N)}}, \qquad D_N' = dD_N/d\lambda.$$

Since each $|\lambda_j^{(N)}|$ is $\leq \sigma_N$, we can for $|\lambda| < 1/\sigma_N$ expand each term on the right as a geometric series:

$$\frac{D_N'}{D_N} = -\sum_j \sum_{k=1}^{\infty} \lambda^{k-1} \lambda_j^{(N)k} = -\sum_1^{\infty} S_k^{(N)} \lambda^{k-1}, \qquad (24)$$

where

$$S_k^{(N)} = \sum_{j=1}^N \lambda_j^{(N)k}.$$

For $k > 1$ we estimate $S_k^{(N)}$, crudely, as follows. Since each $|\lambda_j^{(N)}| \leq \sigma_N$, $|S_k^{(N)}| \leq \sigma_N^{k-1} \sum |\lambda_j^{(N)}|$. We apply now inequality (20) to $\mathbf{T}_N$ and deduce, using theorem 2, that

$$|S_k^{(N)}| \leq \sigma_N^{k-1} \|\mathbf{T}_N\|_{\mathrm{tr}} \leq \sigma_N^{k-1} \|\mathbf{T}\|_{\mathrm{tr}}, \qquad k > 1. \tag{25}$$

For $k = 1$ we have

$$S_1^{(N)} = \mathrm{tr}\,\mathbf{T}_N. \tag{25'}$$

We rewrite (24) as

$$\frac{D_N'}{D_N} + \mathrm{tr}\,\mathbf{T} = \mathrm{tr}\,\mathbf{T} - S_1^{(N)} - \sum_2^\infty S_k^{(N)} \lambda^{k-1}.$$

Taking absolute values and using (25) and (25'), we get, for $|\lambda| < 1/\sigma_N$ summing the geometric series,

$$\left| \frac{D_N'}{D_N} + \mathrm{tr}\,\mathbf{T} \right| \leq |\mathrm{tr}\,\mathbf{T} - \mathrm{tr}\,\mathbf{T}_N| + \frac{|\lambda|\sigma_N}{1 - |\lambda|\sigma_N} \|\mathbf{T}\|_{\mathrm{tr}}.$$

Now let $N \to \infty$. Using parts (ii) and (iii) of lemma 9, we conclude that

$$\lim_{N\to\infty} \left| \frac{D_N'}{D_N} + \mathrm{tr}\,\mathbf{T} \right| = 0,$$

uniformly for all λ in a compact set. Integrating this relation with respect to λ and using $D_N(0) = 1$, we deduce lemma 10. $\qquad\qquad\square$

We use now the definition (23) of D_N to estimate $|D_N(\lambda)|$ as follows:

$$|D_N(\lambda)| \leq \prod_1^N \left(1 + |\lambda|\,|\lambda_j^{(N)}| \right).$$

Using inequality (21) with $r = |\lambda|$ applied to the operator $\mathbf{T}_N$, we see that the right side above is less than

$$\prod \left(1 + |\lambda|\,s_j(\mathbf{T}_N) \right).$$

According to inequality (17'), $s_j(\mathbf{T}_N) \leq s_j(\mathbf{T})$, so we get the inequality

$$|D_N(\lambda)| \leq \prod_1^N \left(1 + |\lambda|\,s_j(\mathbf{T}) \right).$$

Letting $N \to \infty$ and using lemma 10, we obtain

$$\left| e^{-\lambda\alpha} \right| \leq \prod_1^\infty \left(1 + |\lambda|\,s_j(\mathbf{T}) \right).$$

Using the inequality $1 + r \le e^r$ on all but the first M factor on the right gives

$$\left| e^{-\lambda\alpha} \right| \le \prod_1^M \left(1 + |\lambda| \, s_j\right) \exp\left(|\lambda| \sum_{M+1}^\infty s_j\right) = P_M\left(|\lambda|\right) e^{|\lambda|\,\epsilon_M}, \qquad (26)$$

where P_M is a polynomial of degree M, and $\epsilon_M = \sum_{M+1}^\infty s_j$.

Now choose the argument of λ so that $-\lambda\,\alpha$ is positive, and let $|\lambda|$ tend to infinity. Since a polynomial grows more slowly than any exponential, we deduce from (26) that $|\alpha| \le \epsilon_M$. Since ϵ_M tends to zero as M tends to infinity, it follows that $\alpha = 0$. Using lemma 10, $\operatorname{tr} \mathbf{T} = -\alpha = 0$; this completes the proof of Lidskii's lemma, and thereby of the trace formula. $\square$

The proof presented above for the trace formula is due to Gohberg and Krein. Lemmas 7 and 8 have been derived by Hermann Weyl. Lidskii's proof relied on the Hadamard factorization theorem for entire functions of exponential type.

Lidskii defined trace class by forming the linear span with complex coefficients of self-adjoint trace class operators. In Dunford-Schwartz trace class operators are defined as the product of two Hilbert-Schmidt operators; see section 30.8. The trace formula appears in Dunford-Schwartz, but there is no reference to Lidskii. Under questioning, Jack Schwartz admitted that he discovered and proved the trace formula independently.

30.4 THE DETERMINANT

In this section we sketch the definition of the determinant of operators of the form $\mathbf{I} + \mathbf{T}$, $\mathbf{T}$ of trace class, and its fundamental properties. A full discussion is given in Gohberg, Goldberg, and Kaashoek.

For degenerate operators, $\mathbf{G}$, those with finite-dimensional range, the definition is taken from linear algebra. Let $\mathbf{G}$ act on a Hilbert space H, and K a finite-dimensional subspace of H that contains the range of $\mathbf{G}$. With respect to any orthonormal basis of K, $\mathbf{I} + \mathbf{G}$ can be expressed as a matrix; the determinant of this matrix is independent of the choice of the orthonormal basis, or of the subspace K. It is defined as the determinant of the operator $\mathbf{I} + \mathbf{G}$. The determinant has the usual properties:

$$\det (\mathbf{I} + \mathbf{G})(\mathbf{I} + \mathbf{F}) = \det (\mathbf{I} + \mathbf{G}) \det (\mathbf{I} + \mathbf{F}), \qquad (27a)$$

$$\det (\mathbf{I} + \mathbf{G}) = \prod_1^N (1 + \lambda_j), \qquad (27b)$$

where λ_j are the eigenvalues of $\mathbf{G}$ acting on K, including multiplicity. For different choices of K we may get a different number of eigenvalues that are zero; clearly that doesn't change the right side of (27b).

Every trace class operator $\mathbf{T}$ can be approximated in trace norm by degenerate operators; for instance, take the polar decomposition of $\mathbf{T} = \mathbf{UA}$, and approximate

A by $\mathbf{A}_N = \mathbf{A}\mathbf{P}_N$, where $\mathbf{P}_N$ is projection onto the space spanned by the first N eigenfunctions of $\mathbf{A}$. Clearly, by definition of trace norm, $\|\mathbf{T} - \mathbf{U}\mathbf{A}_N\|_{\text{tr}}$ tends to zero as N tends to ∞. The following results holds about such approximations:

Lemma. *Let* $\mathbf{T}$ *be a trace class operator, and* $\mathbf{T}_N$ *a sequence of degenerate operators tending to* $\mathbf{T}$ *in trace norm. Then* $\det(\mathbf{I} + \mathbf{T}_N)$ *tends to a limit that is independent of the choice of the sequence. This limit is defined as* $\det(\mathbf{I} + \mathbf{T})$.

The main result about determinants as defined above is that the two properties listed in (27) are retained. For proofs consult GGK.

In chapter 24 we presented Fredholm's theory, based on the notion of the determinant of operators of form $\mathbf{I} + \mathbf{K}$, $\mathbf{K}$ a one-dimensional integral operator with continuous kernel. As we will show in section 30.6, not all operator $\mathbf{K}$ of this form are of trace class; so the notion of determinant can be extended beyond the one sketched in this section.

30.5 EXAMPLES AND COUNTEREXAMPLES
OF TRACE CLASS OPERATORS

In this section we will study one-dimensional integral operators $\mathbf{K}$ of the form

$$(\mathbf{K}u)(s) = \int_0^1 K(s, t)u(t)\,dt, \tag{28}$$

acting on the Hilbert space $H = L^2[0, 1]$.

Many—one might say almost all—bounded operators that interest us are integral operators in one or several dimensions with kernels that may have singularities. In this section we will treat mostly kernels that are continuous functions. Recall that in chapter 24 we have shown that such integral operators are compact maps of $C[0, 1]$ into $C[0, 1]$.

Exercise 5. Show that an integral operator with continuous kernel is a compact map of $L^2[0, 1]$ into itself.

The adjoint of $\mathbf{K}^*$ of the operator (28) is another integral operator whose kernel K^* is the conjugate transpose of K:

$$K^*(s, t) = \overline{K}(t, s).$$

Clearly, $\mathbf{K}$ is a symmetric operator iff its kernel is skew symmetric, that is, if $K^* = K$.

For symmetric integral operators the spectral theory developed in chapter 28 is applicable: $\mathbf{K}$ has a complete set of orthonormal eigenfunctions e_j and real eigenvalues κ_j accumulating at zero:

$$\mathbf{K}e_j = \kappa_j e_j. \tag{29}$$

Since K maps L^2 functions into continuous functions, every eigenfunction e_j, with $\kappa_j \neq 0$, is a continuous function. When the kernel is real, the eigenfunctions can be chosen to be real valued.

The following remarkable result was proved early in the game, in 1909, by Mercer:

Theorem 11 (Mercer). *Let K be a real-valued symmetric, continuous function of s and t. Assume in addition that the operator $\mathbf{K}$ in (28) is positive in the usual sense:*

$$(\mathbf{K}u, u) \geq 0 \qquad \text{for all } u \text{ in } H.$$

Then the kernel K can be expanded in a uniformly convergent series

$$K(s,t) = \sum \kappa_j \, e_j(s) \, e_j(t), \tag{30}$$

where κ_j and e_j are the eigenvalues and normalized eigenfunctions of $\mathbf{K}$.

Proof. The key fact is the elementary observation that the kernel of a positive integral operator is nonnegative on the diagonal. To see this, suppose, on the contrary, that for some r, $K(r, r)$ were negative; then $K(s, t)$ would be negative for s, t close enough to r, and then

$$(\mathbf{K}u, u) = \iint K(s,t) \, u(t) \, u(s) \, ds \, dt$$

would surely be negative for all functions u that are nonnegative and whose support lies close enough to r.

Define the degenerate kernel K_N as the partial sum of the series on the right in (30):

$$K_N(s,t) = \sum_1^N \kappa_j \, e_j(s) \, e_j(t),$$

and denote by $\mathbf{K}_N$ the integral operator with kernel K_N. Clearly, the difference $\mathbf{K} - \mathbf{K}_N$ is a positive operator, for its eigenvectors are e_j, and its eigenvalues κ_j $j > N$, and zero. Therefore its kernel $K - K_N$ is nonnegative on the diagonal:

$$0 \leq K(s,s) - \sum_1^N \kappa_j \, e_j^2(s). \tag{31}$$

This proves that the partial sums of the infinite series

$$\sum \kappa_j \, e_j^2(s) \tag{30'}$$

are uniformly bounded by $K(s, s)$. Since each term is nonnegative, it follows that the series (30') converges for each s. Since the partial sums form an increasing sequence of functions, by Dini's theorem convergence is uniform for all s in $[0, 1]$. Using (31), we can, by the Schwarz inequality, estimate the remainder of the series on the right in (30) and prove its convergence uniformly for all s and t.

Call this limit K_∞; we claim that $K_\infty = K$. To see why, denote by $\mathbf{K}_\infty$ the integral operator whose kernel is K_∞. From the definition of K_∞ as the right side of (30), we see that e_j is an eigenfunction of $\mathbf{K}_\infty$, with eigenvalues κ_j. Thus $\mathbf{K}$ and $\mathbf{K}_\infty$ act the same way on all e_j, and therefore on all their linear combinations. Since both $\mathbf{K}$ and $\mathbf{K}_\infty$ map functions orthogonal to all e_j into zero, it follows that $\mathbf{K}u = \mathbf{K}_\infty u$ for all functions u. But then $\mathbf{K}$ and $\mathbf{K}_\infty$ have the same kernel.   □

Exercise 6. Show that an integral operator whose kernel is continuous and $\neq 0$ is $\neq 0$.

Set in equation (30) $s = t$ and integrate; we get

$$\int K(s, s)\, ds = \sum \kappa_j. \qquad (32)$$

Since the eigenvalues of a symmetric positive operator are its singular values, we conclude

Corollary 11A. *An integral operator that satisfies the hypotheses of Mercer's theorem is of trace class.*

Corollary 11B. *The trace of an integral operator that satisfies the hypotheses of Mercer's theorem equals the integral of its kernel along the diagonal.*

Formula (32) holds much more generally:

Theorem 12. *Let* $\mathbf{K}$ *be an integral operator of form* (28), *of trace class, with a continuous kernel. Then the trace of* $\mathbf{K}$ *equals the integral of its kernel along the diagonal.*

Proof. We take first the case that the kernel is not merely a continuous function but a smooth one. Then the kernel can be expanded as a uniformly convergent series of, say, Lagrange polynomials f_n:

$$K(s, t) = \sum k_{j,m}\, f_j(s)\, f_m(t),$$

where the coefficients $k_{j,m}$ are given by the usual formula for orthonormal expansions:

$$k_{j,m} = \iint K(s, t)\, f_j(s)\, f_m(t)\, ds\, dt.$$

We use now definition (9) of trace with the preceding orthonormal basis f_n:

$$(\mathbf{K} f_n, f_n) = \int \left(\int K(s,t) f_n(t)\, dt \right) f_n(s)\, ds = k_{n,n}$$

according to the formula above for the coefficients $k_{j,m}$. Summing with respect to n gives

$$\operatorname{tr} \mathbf{K} = \sum k_{n,n}.$$

On the other hand, setting $s = t$ in the series for K gives

$$K(s,s) = \sum k_{j,m}\, f_j(s)\, f_m(s).$$

Integrating with respect to s and using the orthogonality of the f_j gives

$$\int K(s,s)\, ds = \sum k_{m,m},$$

identical with the expression derive above for the trace of $\mathbf{K}$.

To handle integral operator whose kernel is merely continuous, we approximate them with operators with smooth kernels. We need

Theorem 13. *An integral operator with a smooth kernel is in trace class.*

Proof. If $\mathbf{K}$ has a smooth kernel, so does $\mathbf{K}^*$, and so does $\mathbf{K}^*\mathbf{K}$. We will estimate the nth eigenvalue λ_n of $\mathbf{K}^*\mathbf{K} = \mathbf{L}$. Since $\mathbf{L}$ is symmetric, we can apply Courant's principle described in chapter 28:

$$\lambda_n = \min_{S_{n-1}} \max_{u \perp S_{n-1}} \frac{(\mathbf{L}u, u)}{(u, u)}.$$

It follows that for any given subspace S_{n-1} of dimension $n - 1$,

$$\lambda_n \leq \max_{u \perp S_{n-2}} \frac{(\mathbf{L}u, u)}{(u, u)}. \tag{33}$$

We choose S_{n-1} to consist of all polynomials of degree $< n - 1$. Then for $u \perp S_{n-1}$,

$$(\mathbf{L}u, u) = \iint L(s,t)\, u(s)\, u(t)\, ds\, dt = \iint [L(s,t) - P_n(s,t)]\, u(s)\, u(t)\, ds\, dt, \tag{34}$$

where P_n is any function of form

$$P_n(s,t) = \sum_0^{n-2} a_j(s)\, t^j + b_j(t)\, s^j.$$

According to results in approximation theory, every smooth function $L(s,t)$ can be well approximated by such functions in the L^2-norm:

$$\iint |L - P_n|^2 \, ds \, dt \leq \text{const } n^{-b},$$

where the exponent b is proportional to the number of continuous derivatives possessed by L. So, by the Schwarz inequality,

$$\left[\iint (L - P_n) u(s) u(t) \, ds \, dt \right]^2 \leq \iint (L - P_n)^2 \, ds \, dt \int u^2(s) u^2(t) ds \, dt$$

$$\leq \text{const.} \, n^{-b}$$

for all u orthogonal to S_{n-1} and L^2 norm equal to 1. It follows from (33) and (34) that $\lambda_n \leq \text{const } n^{-b/2}$. Since $\mathbf{L} = \mathbf{K}^* \mathbf{K}$, $\lambda_n = s_n^2(\mathbf{K})$, and so we have the estimate

$$s_n(\mathbf{K}) \leq \text{const.} \, n^{-b/4}.$$

Clearly, for $b > 4$ the series $\sum s_n(\mathbf{K})$ converges.   □

To approximate $\mathbf{K}(s, t)$ by smooth kernels, we employ mollifiers. Let $p(s)$ be a nonnegative C^∞ function of compact support, and $\int p \, ds = 1$. We define $p_n(s) = n \, p(ns)$, and the mollifying operator $\mathbf{M}_n$ as convolution with p_n:

$$(\mathbf{M}_n u)(s) = \int p_n(s - r) u(r) \, dr.$$

Define $\mathbf{K}_n$ as $\mathbf{M}_n \mathbf{K} \mathbf{M}_n$. $\mathbf{K}_n$ is an integral operator whose kernel is the convolution

$$K_n(s, t) = \iint p_n(s - r) K(r, x) p_n(x - t) \, dr \, dx;$$

$K_n(s, t)$ is a C^∞ function that tends uniformly to $K(s, t)$ as $n \to \infty$. According to theorem 2, the trace class operators form a two-sided ideals in the ring of all bounded operators; since K in theorem 12 is of trace class, so is $\mathbf{K}_n = \mathbf{M}_n \mathbf{K} \mathbf{M}_n$. According to what we have already shown,

$$\text{tr } \mathbf{K}_n = \int K_n(s, s) \, ds.$$

An n tends to ∞, the right side tends to $\int K(s, s) \, ds$. To complete the proof, all we have to show is that the left side tends to tr $\mathbf{K}$. We leave this as

Exercise 7. Show that $\lim \text{tr } \mathbf{K}_n = \text{tr } \mathbf{K}$. (Hint: Prove it for $\mathbf{K}$ degenerate, and then approximate $\mathbf{K}$ in trace norm by a sequence of degenerate operators.)   □

We encountered earlier, in chapter 22, the operator of integration:

$$(\mathbf{V}u)(s) = \int_0^s u(t) \, dt.$$

We showed there that V is a compact operator mapping $C[0, 1]$ into $C[0, 1]$. It is equally true that V is a compact mapping of $L^2[0, 1]$ into $L^2[0, 1]$.

Exercise 8. Show that V maps the unit ball in $L^2[0, 1]$ into a compact subset of $C[0, 1]$.

Note that V is an integral operator, with the discontinuous kernel

$$K(s, t) = \begin{cases} 1 & \text{for } t < s \\ 0 & \text{for } t > s. \end{cases}$$

We showed in chapter 22 that V has no eigenfunctions in $C[0, 1]$. Since V maps $L^2[0, 1]$ into $C[0, 1]$, it follows that V has no eigenfunctions in $L^2[0, 1]$ either. We show now that V is *not* of trace class, by computing its trace with respect to the trigonometric base: $f_n(t) = \cos(2\pi n t)$, $g_n = \sin(2\pi n t)$. By calculus, for $n \neq 0$,

$$V f_n = \frac{g_n}{2\pi n}, \quad V g_n = \frac{(1 - f_n)}{2\pi n},$$

while $(V f_0)(s) = s$. Again, by calculus

$$(V f_0, f_0) = \tfrac{1}{2}, \quad (V f_n, f_n) = 0 \qquad \text{for } n \neq 0, \ (V g_n, g_n) = 0.$$

So the trace of V with respect to the trigonometric base is $\tfrac{1}{2}$, contradicting Lidskii's lemma. □

Exercise 9. Calculate the singular values of V and show that $\sum s_j(V)$ diverges. (Hint: The inverse of V^*V is a differential operator.)

We remind the reader that we showed in chapter 24, theorem 5, that an integral operator whose kernel is Hölder continuous, with Hölder exponent $> \tfrac{1}{2}$, satisfies the integral form of the trace formula.

We close this section by asking—and answering—the following question: Given an integral operator K, how can we decide if it has any nonzero eigenvalues? If the operator is known to be of trace class, we can find its trace as the integral of its kernel along the diagonal; if this is nonzero, then by the trace formula, K has a nonzero eigenvalue. If tr $K = 0$, no conclusion can be drawn. We can then look at K^2, an integral operator whose kernel can be computed from that of K. It is of trace class, and its trace can be computed by integration. If tr $K^2 \neq 0$, then K has a nonzero eigenvalue; otherwise, we inspect the trace of K^3, and so on. What if this process never ends?

Theorem 14. *Let K be an integral operator with a continuous kernel, of trace class. Suppose* tr $K^n = 0$ *for all positive integers n; then K has no nonzero eigenvalues.*

Proof. Denote the nonzero eigenvalues of K by κ_j. The eigenvalues of K^n are κ_j^n, so by the trace formula

$$\operatorname{tr}\mathbf{K}^n = \sum_j \kappa_j^n.\tag{35}$$

It follows from inequality (20) that $\sum |\kappa_j| < |\mathbf{K}|_{\mathrm{tr}}$. We build the entire analytic function

$$F(z) = \sum \left(e^{\kappa_j z} - 1\right).\tag{36}$$

Since $|e^w - 1| < e|w|$ for $|w| < 1$, the series (36) converges for all z. The Taylor coefficients of F at $z = 0$ can be computed by differentiating (36) termwise:

$$F(0) = 0, \quad F^{(n)}(0) = \sum \kappa_j^n.$$

It follows from the assumption that $\operatorname{tr}\mathbf{K}^n = 0$, and so by (35) that all the Taylor coefficients of F are zero, and therefore $F(z)$ itself is zero for all z. We claim that then all κ_j are zero. Suppose not; let $\kappa_1, \ldots, \kappa_j$ be those of largest absolute value. Choose z so that $\kappa_1 z$ is real and positive, and let $|z| \to \infty$. Clearly, the first term in (36) dominates all others, and $F(z) \simeq m\, e^{|\kappa_1||z|}$ where m is the multiplicity of κ_1. This contradicts $F(z) \equiv 0$. □

30.6 THE POISSON SUMMATION FORMULA

It this section we study integral operators of convolution form. Let f be any continuous function on the circle S^1. $\mathbf{T}_f = \mathbf{T}$ is defined as convolution with f:

$$(\mathbf{T}u)(s) = \int_{S^1} f(s - t)\, u(t)\, dt/2\pi.\tag{37}$$

As we saw in chapter 29, the eigenfunctions of $\mathbf{T}$ are the exponentials $e_n(t) = e^{int}$:

$$\mathbf{T}e_n = \int f(s - t)\, e^{int}\, dt/2\pi = \int f(r)\, e^{-inr}\, dr/2\pi\; e^{ins} = a_n e^{ins}.\tag{38}$$

So the eigenvalues are the Fourier coefficients a_n of f.

The kernel of the integral operator $\mathbf{T}$ equals $f(0)$ at every point on the diagonal; so if $\mathbf{T}$ were a trace class operator, $\operatorname{tr}\mathbf{T} = f(0) = \sum a_n$, by the trace formula. This is the same as saying that the Fourier series of f at $s = 0$ converges to $f(0)$, true for sufficiently smooth functions but *not* for all continuous functions; see chapter 11, section 11.2. *This shows that not all integral operators with a continuous kernel are of trace class.*

Consider functions g defined on the whole real line, smooth and decreasing rapidly as $|s|$ tends to infinity. Define now the operator $\mathbf{T}$ as convolution:

$$(\mathbf{T}u)(s) = \int_{\mathbb{R}} g(s - t)\, u(t)\, dt/2\pi,\tag{39}$$

regarded as a mapping of $L^2(S^1)$ into $L^2(S^1)$. We can put this in the form (37) by chopping up $\mathbb{R}$ as the union of intervals $[2\pi m, 2\pi(m+1)]$ of length 2π:

$$(\mathbf{T}u)(s) = \int_{S^1} \sum g(s - t + 2\pi m)\, u(t)\, dt/2\pi. \qquad (35')$$

The eigenvalues of $\mathbf{T}$ are given by formula (38), which can be rewritten as follows:

$$a_n = \int_{S^1} \sum g(r + 2\pi m)\, e^{-inr}\, dr/2\pi = \int_{\mathbb{R}} g(r)\, e^{-inr}\, dr/2\pi = \tilde{g}(n)/2\pi,$$

where $\tilde{g}$ is the Fourier transform of g. The kernel of the integral operator $\mathbf{T}$ equals $\sum g(2\pi m)$ at every point on the diagonal. Therefore the trace formula asserts that

$$\sum g(2\pi m) = 2\pi \sum \tilde{g}(n).$$

This is the classical Poisson summation formula.

The scope of this argument can be enlarged beyond this simple case. The additive group of the reals can be replaced by other, not necessarily commutative, groups, and the group of integer multiples of 2π by other discrete subgroups. The celebrated *Selberg trace formula* is a far-reaching generalization of the Poisson summation formula.

30.7 HOW TO EXPRESS THE INDEX OF AN OPERATOR AS A DIFFERENCE OF TRACES

We recall from chapter 27 the notion of the index of a bounded mapping $\mathbf{F}$ of a Banach space U into another Banach space V. Denote by N the nullspace, by R the range of $\mathbf{F}$; assume that the dimension of N and the codimension of R are both finite. Their difference is defined as the index of $\mathbf{F}$:

$$\text{ind } \mathbf{F} = \dim N - \text{codim } R. \qquad (40)$$

According to theorem 1 of chapter 27, an operator $\mathbf{F} : U \to V$ has an index iff it has a pseudoinverse $\mathbf{G} : V \to U$, such that

$$\mathbf{GF} = \mathbf{I} - \mathbf{T}, \quad \mathbf{FG} = \mathbf{I} - \mathbf{S}, \qquad (41)$$

where $\mathbf{T} : U \to U$ and $\mathbf{S} : V \to V$ are compact maps. In this section we study the case when U and V are Hilbert spaces, and $\mathbf{T}$ and $\mathbf{S}$ are not merely compact but are of trace class:

Theorem 15. *Let U and V be a pair of Hilbert spaces, $\mathbf{F} : U \to V$ and $\mathbf{G} : V \to U$ bounded operators that are pseudoinverses of each other in the sense of (41), where $\mathbf{T} : U \to U$ and $\mathbf{S} : V \to V$ are trace class operators. Then*

$$\text{ind } \mathbf{F} = \text{tr } \mathbf{T} - \text{tr } \mathbf{S}. \qquad (42)$$

Proof. Multiply the first relation in (41) by $\mathbf{F}$ on the left, the second relation by $\mathbf{F}$ on the right, and subtract one from the other:

$$\mathbf{FT} = \mathbf{SF}. \tag{43}$$

Decompose orthogonally U and V as follows:

$$U = N \oplus Z, \quad V = R \oplus W.$$

Define $\mathbf{P}$ as the orthogonal projection of U onto Z. Since the orthogonal complement of Z is the nullspace N of $\mathbf{F}$, it follows that $\mathbf{FP} = \mathbf{F}$. Setting this into (43) gives

$$\mathbf{FPT} = \mathbf{SF}. \tag{44}$$

Note that $\mathbf{PT}$ maps $Z \to Z$, $\mathbf{S}$ maps $R \to R$, and $\mathbf{F}$ is an invertible map of $Z \to R$. We claim that

$$\operatorname{tr} \mathbf{PT}/Z = \operatorname{tr} \mathbf{S}/R. \tag{45}$$

where $\operatorname{tr} \mathbf{PT}/Z$ means the trace of $\mathbf{PT}$ restricted to the invariant subspace Z, and so on.

Proof. Choose any unitary map $\mathbf{M}$ of R onto Z. Multiply (44) on the left by $\mathbf{M}$:

$$\mathbf{MFPT} = \mathbf{MSF} = \mathbf{MSM}^{-1}\,\mathbf{MF}.$$

Multiply this by $(\mathbf{MF})^{-1}$ on the right:

$$(\mathbf{MF})(\mathbf{PT})(\mathbf{MF})^{-1} = \left(\mathbf{MSM}^{-1}\right).$$

All operator in parentheses map $Z \to Z$. So by the commutative property of trace, theorem 4 (iv),

$$\operatorname{tr} \mathbf{PT}/Z = \operatorname{tr} \mathbf{MSM}^{-1}/Z = \operatorname{tr} \mathbf{S}/R. \qquad \square$$

We express now the trace of $\mathbf{T}$ over U in terms of the trace of $\mathbf{PT}$ over Z. Build an orthonormal basis for U consisting of a basis $\{n_j\}$ in N and a basis $\{z_j\}$ in Z. Then

$$\operatorname{tr} \mathbf{T} = \sum (\mathbf{T}n_j, n_j) + \sum (\mathbf{T}z_j, z_j).$$

Since $\mathbf{P}z_j = z_j$, we can rewrite the second sum on the right as

$$\sum (\mathbf{T}z_j, \mathbf{P}z_j) = \sum (\mathbf{PT}z_j, z_j) = \operatorname{tr} \mathbf{PT}/Z.$$

On the other hand, $\mathbf{F} = 0$ on N, so it follows from the first relation of (41) that $\mathbf{T} = \mathbf{I}$ on N. Therefore

$$\sum (\mathbf{T} n_j, n_j) = \sum (n_j, n_j) = \dim N.$$

Putting together the last three relations gives

$$\operatorname{tr} \mathbf{T} = \dim N + \operatorname{tr} \mathbf{PT}/Z. \tag{46}$$

Similarly, we express the trace of S over V in terms of the trace of S over R. We build an orthonormal base for V consisting of a basis $\{w_j\}$ for W and a basis r_j for R. Then

$$\operatorname{tr} \mathbf{S} = \sum (\mathbf{S} w_j, w_j) + \sum (\mathbf{S} r_j, r_j).$$

We identify the second sum on the right as

$$\sum (\mathbf{S} r_j, r_j) = \operatorname{tr} \mathbf{S}/R.$$

It follows from the second relation in (41) that the range of $\mathbf{I} - \mathbf{S}$ lies in R, and therefore is orthogonal to W. In particular, $\big((\mathbf{I} - \mathbf{S}) w_j, w_j\big) = 0$, so

$$\sum (\mathbf{S} w_j, w_j) = \sum (w_j, w_j) = \dim W = \operatorname{codim} R.$$

Putting together the last three relations gives

$$\operatorname{tr} \mathbf{S} = \operatorname{codim} R + \operatorname{tr} \mathbf{S}/R. \tag{46'}$$

Subtract (46') from (46); since we have shown in (45) that $\operatorname{tr} \mathbf{PT}/Z = \operatorname{tr} \mathbf{S}/R$, we obtain the trace formula (42) for the index. □

Even when $\mathbf{G}$ is too crude a pseudoinverse for $\mathbf{F}$ to make $\mathbf{S}$ and $\mathbf{T}$ of trace class, it could happen that for some positive integer n, $\mathbf{S}^n$ and $\mathbf{T}^n$ are trace class:

Corollary 15'. *The spaces U, V, the operator $\mathbf{F}$, $\mathbf{G}$ are as in theorem 15, and $\mathbf{S}^n$ and $\mathbf{T}^n$ of trace class, n some positive integer. Then*

$$\operatorname{ind} \mathbf{F} = \operatorname{tr} \mathbf{T}^n - \operatorname{tr} \mathbf{S}^n. \tag{47}$$

Proof. We replace the pseudoinverse $\mathbf{G}$ by $\mathbf{G}_n = (\sum_0^{n-1} \mathbf{T}^j) \mathbf{G}$. Then

$$\mathbf{G}_n \mathbf{F} = \left(\sum_0^{n-1} \mathbf{T}^j \right) \mathbf{GF} = \left(\sum_0^{n-1} \mathbf{T}^j \right) (\mathbf{I} - \mathbf{T}) = \mathbf{I} - \mathbf{T}^n, \tag{48}$$

where we have used the first relation in (41). Using both relations in (41) we deduce that $\mathbf{TG} = \mathbf{GS}$. Using this relation repeatedly, we get

$$\mathbf{FG}_n = \mathbf{F}\left(\sum_0^{n-1}\mathbf{T}^j\right)\mathbf{G} = \mathbf{FG}\left(\sum_0^{n-1}\mathbf{S}^j\right) = (\mathbf{I} - \mathbf{S})\left(\sum_0^{n-1}\mathbf{S}^j\right) = \mathbf{I} - \mathbf{S}^n. \quad (48')$$

Now we apply theorem 15.   □

If formula (47) holds for one value of n, it holds for all larger ones. This seems peculiar, until we do

Exercise 10. Show that if $\mathbf{S}$ and $\mathbf{T}$ are related as in (41), then each eigenvalue $\neq 1$ of $\mathbf{T}$ is an eigenvalue of $\mathbf{S}$, with the same multiplicity.

Theorem 15 and its corollary can be very useful in calculating the index of operators; see Gilkey.

30.8 THE HILBERT-SCHMIDT CLASS

The last exercise of this chapter summarizes the main properties of the Hilbert-Schmidt (HS) class of operator in Hilbert space.

Exercise 11. A bounded linear operator $\mathbf{K}$ mapping a Hilbert space H into itself belongs to the HS class if for some orthonormal basis $\{e_j\}$ of H,

$$\sum \|\mathbf{K}\,e_k\|^2 < \infty. \quad (49)$$

(a) Show that if $\mathbf{K}$ satisfies (49) for one orthonormal basis, then it satisfies (49) for every orthonormal basis, and the sum in (49) is independent of the basis. The square root of this sum is called the HS-norm, denoted as $\|\mathbf{K}\|_{HS}$.

(b) Show that $\|\mathbf{K}\| \leq \|\mathbf{K}\|_{HS}$.

(c) Show that if $\mathbf{K}$ is HS, so is its adjoint $\mathbf{K}^*$, and $\|\mathbf{K}^*\|_{HS} = \|\mathbf{K}\|_{HS}$.

(d) Show that the HS operators form a complete normed space in the HS-norm.

(e) Show that if $\mathbf{K}$ is HS and $\mathbf{B}$ is any bounded operator, then $\mathbf{BK}$ and $\mathbf{KB}$ are HS, and $\|\mathbf{BK}\|_{HS}$, $\|\mathbf{KB}\|_{HS}$ are both $\leq \|\mathbf{B}\|\,\|\mathbf{K}\|_{HS}$.

(f) Show that $\mathbf{K}$ is HS iff $\sum s_j^2(\mathbf{K}) \leq \infty$.

(g) Show that every HS operator is compact.

(h) Show that every trace class operator is HS.

(i) Show that the product of two HS operators $\mathbf{K}$ and that $\mathbf{H}$ is in trace class, and $\|\mathbf{KH}\|_{\mathrm{tr}} \leq \|\mathbf{K}\|_{HS}\,\|\mathbf{H}\|_{HS}$.

(j) Show that every trace class operator can be written as a product of two HS operators.

30.9 DETERMINANT AND TRACE FOR OPERATOR IN BANACH SPACES

The earliest developments of a determinant theory for operators in a Banach space are due to Lezanski in 1953, Grothedieck in 1956, and Sikorski in 1961. The earliest derivation of the trace formula for a class of operators in a Banach space is due to König; a systematic approach has been developed by Pietsch in a series of publications culminating in his monograph.

Yet another, still broader, systematic approach is presented in the recent excellent monograph of Gohberg, Goldberg, and Krupnik.

BIBLIOGRAPHY

Dunford, N. and Schwartz, J. T. *Linear Operators: Part II, Spectral Theory*. Interscience-Wiley, New York, 1963; see esp. ch. XI, sec. 6.

Gilkey, P. B. *Invariance Theory, the Heat Equation and the Atiyah-Singer Index Theorem*, 2nd. ed. CRC Press, Boca Raton, FL, 1995.

Gohberg, I. C., Goldberg, S., and Kaashoek, M. A. *Classes of Linear Operators*. Vol. 1. Birkhäuser, Boston, 1990.

Gohberg, I. C., Goldberg, S., and Krupnik, N. Traces and determinants of linear operators. *Operator Theory Adv. and Appl.*, **116** (2000).

Gohberg, I. C. and Krein, M. G. Introduction to the theory of linear nonself-adjoint operators. *Nauka*, Moscow (1965); *AMS Trans. Math. Monogr.*, **18** (1969).

Grothendieck, A. La théory de Fredholm. *Bull. Soc. Math.*, France, **84** (1956): 319–384.

Johnson, W. B., König, H., Maurey, B., and Retherford, J. R. Eigenvalues of p-summing and l_p-type operators in Banach spaces. *J. Funct. Anal.*, **32** (1979): 353–380.

König, H. s-numbers, eigenvalues and the trace theorem in Banach spaces. *Studia Math.*, **67** (1980): 157–171.

Lax, P. D. The existence of eigenvalues of integral operators. *Indiana U. Math. J.*, **42** (1993): 889–891.

Leiterer, H. and Pietsch, A. An elementary proof of Lidskii's trace theorem. *Wiss. Ztsch. Friedrich Schiller Univ. Jena, Math.—Nat. R.*, **31** (1982): 587–594.

Lezanski, T. The Fredholm theory of linear equations in Banach spaces. *Studia Math.* **13** (1953): 244–276.

Lidskii, V. B. Nonself-adjoint operators with trace. *Dokl. Akad. Nauk SSR*, **125** (1959): 485–487; *AMS Trans.*, **47** (1961): 43–46.

Mercer, T. Functions of positive and negative type and their connection with the theory of integral equations. *Trans. London Phil. Soc.* (A), **209** (1909): 415–446.

Pietsch, A. *Eigenvalues and s-Numbers*. Cambridge Studies in Advanced Math, **13**. Cambridge University Press, Cambridge, 1987.

Retherford, J. R. *Compact Operators and the Trace Theorem*. London Math. Soc. Student Text, **27**. Cambridge University Press, Cambridge, 1993.

Selberg, A. Harmonic analysis and discontinuous groups in weakly symmetric Riemannian spaces, with applications to Dirichlet series. *J. Indian Math. Soc.*, **20** (1956): 121–129.

Sikorski, R. The determinant theory in Banach spaces. *Colloq. Math.*, **8** (1961): 141–198.

Weyl, H. Inequalities between the two kinds of eigenvalues of a linear transformations. *Proc. Nat. Acad. Sc.*, **35** (1949): 408–411.

31

SPECTRAL THEORY OF SYMMETRIC, NORMAL, AND UNITARY OPERATORS

In this chapter we study operators $\mathbf{M}$ that map a complex Hilbert space H into itself, that are bounded, and that are *symmetric* in the sense that $\mathbf{M}^* = \mathbf{M}$. According to the definition of adjoint this means that for all x and y in H

$$(\mathbf{M}x, y) = (x, \mathbf{M}y). \tag{1}$$

Exercise 1. Show that

(a) The inverse of an invertible symmetric operator is symmetric.
(b) The product of commuting symmetric operators is symmetric.
(c) The set of symmetric operators is closed in the weak topology for operators.

In chapter 28 we saw that every *compact* symmetric operator has a complete set of orthonormal eigenvectors. In this chapter we generalize this result to include symmetric operators that are bounded but not compact. To show how to do this, we reformulate the spectral resolution of compact operators.

Denote by $\{e_n\}$ the eigenvectors of the compact operator $\mathbf{A}$. Every vector x in the Hilbert space H can be expanded in a Fourier series, and so can $\mathbf{A}x$:

$$x = \Sigma a_n e_n, \qquad \mathbf{A}x = \Sigma \lambda_n a_n e_n. \tag{2}$$

Denote by $\mathbf{E}_n$ projection onto the eigenspace with eigenvalue λ_n. Then (2) can be rewritten as

$$x = \Sigma \mathbf{E}_n x, \qquad \mathbf{A}x = \Sigma \lambda_n \mathbf{E}_n x. \tag{2'}$$

We rewrite the sums in (2) as integrals by introducing the *projection-valued measure* $\mathbf{E}(S)$ as follows: for any Borel set S of $\mathbb{R}$,

$$E(S) = \sum_{\lambda_n \in S} E_n.$$

The support of the measure E is the spectrum of A. Using the measure E defined above, we can rewrite $(2')$ in the form

$$x = \int dE(\lambda)x, \qquad Ax = \int \lambda dE(\lambda)x. \qquad (3)$$

Our objective in this chapter is to obtain a *spectral resolution* of form (3) for arbitrary bounded symmetric operators M. The projection-valued measure E that enters the resolution is, of course, no longer a pure point measure.

The following result is as basic as is simple:

Theorem 1. *For B bounded symmetric, (Bx, y) is a bounded, skew symmetric form, linear in x, skew linear y.*

Conversely, let $b(x, y)$ be a skew symmetric form

$$b(y, x) = \overline{b(x, y)}, \qquad (4)$$

linear in x, and bounded:

$$|b(x, y)| \leq c\|x\|\|y\|. \qquad (5)$$

Then b can be expressed as

$$b(x, y) = (x, By) \qquad (6)$$

where B is a bounded symmetric operator, and

$$\|B\| \leq c. \qquad (7)$$

Proof. The direct part is a consequence of the symmetry of B, the Schwarz inequality, and the boundedness of B. To deduce the converse, we fix y and regard $b(x, y)$ as a bounded linear functional of x, with bound $c\|y\|$. According to the Riesz-Frechet representation theorem, we can write this functional as a scalar product with some vector w:

$$b(x, y) = (x, w), \qquad (8)$$

where w is uniquely determined by y; setting $x = w$ in (5) shows that $\|w\| \leq c\|y\|$. Since the left and right sides of (8) depend skew linearly on y and w, respectively, it follows that w is a linear function of y:

$$w = By.$$

This proves (6) and (7). The symmetry of $\mathbf{B}$ is a consequence of skew symmetry of b:

$$(x, \mathbf{B}y) = b(x, y) = \overline{b(y, x)} = \overline{(y, \mathbf{B}x)} = (\mathbf{B}x, y).$$

Note that $(x, \mathbf{B}x) = (\mathbf{B}x, x)$ is real for all x. $\qquad\qquad\qquad\qquad\square$

31.1 THE SPECTRUM OF SYMMETRIC OPERATORS

Theorem 2. *The spectrum of a bounded, symmetric operator* $\mathbf{M}$ *on a Hilbert space is real.*

Proof. We have to show that every nonreal $\lambda = \alpha + i\beta$, $\beta \neq 0$, lies in the resolvent set. Define the function B as follows:

$$B(x, y) = (x, (\mathbf{M} - \lambda)y).$$

This function has all three properties listed as the hypotheses of theorem 6 of chapter 6:

(i) B is linear in x, skew linear in y.
(ii) B is bounded; for by the Schwarz inequality

$$|B(x, y)| \leq \|x\| \|(\mathbf{M} - \lambda)y\| \leq \|x\| \|y\| (\|\mathbf{M}\| + |\lambda|).$$

(iii) $B(y, y)$ is bounded from below,

$$B(y, y) = (y, (\mathbf{M} - \lambda)y) = (y, \mathbf{M}y) - \alpha(y, y) - i\beta(y, y).$$

The first two terms on the right are real, the third imaginary; so

$$|\mathbf{B}(y, y)| \geq |Im\, B(y, y)| = |\beta| \|y\|^2.$$

We appeal now to the Lax-Milgram lemma, theorem 6 in chapter 6, which asserts that every linear function $\ell(x)$ can be represented as $B(x, y)$ for some y, uniquely determined by ℓ. Take $\ell(x) = (x, z)$; there is a uniquely determined y such that $B(x, y) = (x, z)$ for all x. Using the definition of B as $(x, (\mathbf{M} - \lambda)y)$, we conclude that $(\mathbf{M} - \lambda)y = z$. Since z is arbitrary, this shows that $(\mathbf{M} - \lambda)$ is invertible. Therefore λ belongs to the resolvent set of $\mathbf{M}$. $\qquad\qquad\qquad\square$

NOTE. Theorem 15 of chapter 19, stated and proved in the context of Gelfand's theory of commutative B^* algebras, implies theorem 2.

Theorem 3. *The spectral radius of a bounded, symmetric operator* $\mathbf{M}$ *is equal to its norm:*

$$|\sigma(\mathbf{M})| = \|\mathbf{M}\|. \qquad\qquad\qquad (9)$$

Proof. Using the symmetry of $\mathbf{M}$, the Schwarz inequality, and the definition of the norm of $\mathbf{M}^2$, we derive for any x,

$$\|\mathbf{M}x\|^2 = (\mathbf{M}x, \mathbf{M}x) = (x, \mathbf{M}^2 x) \leq \|x\|\|\mathbf{M}^2 x\| \leq \|x\|^2 \|\mathbf{M}^2\|.$$

It follows that $\|\mathbf{M}\|^2 \leq \|\mathbf{M}^2\|$. Repeating this argument k times, we deduce that

$$\|\mathbf{M}\|^n \leq \|\mathbf{M}^n\|, \qquad n = 2^k.$$

Since norm is submultiplicative, the opposite inequality holds also. Therefore, $\|\mathbf{M}\|^n = \|\mathbf{M}^n\|$. Taking the nth root, and using formula (12′) in chapter 17 for the spectral radius, gives

$$|\sigma(\mathbf{M})| = \lim \|\mathbf{M}^n\|^{1/n} = \|\mathbf{M}\|. \qquad \square$$

Theorem 4. *The spectrum of a bounded symmetric operator* $\mathbf{M}$ *lies in the closed interval* $[a,b]$ *on the real axis, where*

$$a = \inf_{\|x\|=1} (x, \mathbf{M}x), \quad b = \sup_{\|x\|=1} (x, \mathbf{M}x). \tag{10}$$

The endpoints of this interval belongs to the spectrum of $\mathbf{M}$.

Proof. Let λ be real and less than a. From the definition (10) of a it follows that for every x in H,

$$(x, (\mathbf{M} - \lambda)x) = (x, \mathbf{M}x) - \lambda(x, x) \geq (a - \lambda)\|x\|^2.$$

It follows that $(x, (\mathbf{M} - \lambda)x)$ yields a norm equivalent to $\|x\|$. Therefore every linear function $\ell(x) = (x, z)$ can be represented uniquely as the associated scalar product $(x, (\mathbf{M} - \lambda)y)$. Since this holds for all x, $(\mathbf{M} - \lambda)y = z$. Since z is an arbitrary element of H, this proves that $\mathbf{M} - \lambda$ is invertible, and so λ does not belong to the spectrum of $\mathbf{M}$. We can deal similarly with $\lambda > b$.

To show that a and b belong to the spectrum, we observe that for $\|x\| = 1$, $|(x, \mathbf{M}x)| \leq \|x\|\|\mathbf{M}x\| \leq \|\mathbf{M}\|$. Therefore by definition (10) of a and b,

$$|a| \leq \|\mathbf{M}\|, \quad |b| \leq \|\mathbf{M}\|. \tag{11}$$

On the other hand, since the spectrum lies in the interval $[a, b]$,

$$|\sigma(\mathbf{M})| \leq \max |a|, |b|. \tag{11′}$$

According to (9), $|\sigma(\mathbf{M})| = \|\mathbf{M}\|$; comparing (11) and (11′), we see that this can be only if $\max |a|, |b| = |\sigma(\mathbf{M})|$. In particular, if $b > |a|$, then b lies in the spectrum of $\mathbf{M}$, and if $|a| > b$, then a lies in the spectrum of $\mathbf{M}$. Replacing $\mathbf{M}$ by $\mathbf{M} + c\mathbf{I}$, with c any constant, we add c to the spectrum of $\mathbf{M}$, as well as to a and b. Applying the results above to the operator $\mathbf{M} + c\mathbf{I}$, and choosing c judiciously we conclude that both a and b belong to the spectrum of $\mathbf{M}$. $\qquad \square$

Theorem 5. *Let* $\mathbf{M}$ *and* $\mathbf{N}$ *denote a pair of symmetric operators. Then*

$$\text{dist}(\sigma(\mathbf{M}), \sigma(\mathbf{N})) \leq \|\mathbf{M} - \mathbf{N}\|, \tag{12}$$

where the distance of the two closed point-sets $\sigma(\mathbf{M})$ *and* $\sigma(\mathbf{N})$ *is defined as the larger of the two quantities*

$$\max_{\nu \text{ in } \sigma(\mathbf{N})} \min_{\mu \text{ in } \sigma(\mathbf{M})} |\nu - \mu|, \qquad \max_{\mu \text{ in } \sigma(\mathbf{M})} \min_{\nu \text{ in } \sigma(\mathbf{N})} |\nu - \mu|. \tag{13}$$

Proof. Denote $\|\mathbf{M} - \mathbf{N}\|$ by d. Suppose that one of the quantities in (13), say the first, is $> d$. Then for some ν in $\sigma(\mathbf{N})$,

$$\min_{\mu \text{ in } \sigma(\mathbf{M})} |\mu - \nu| > d. \tag{14}$$

Such a ν belongs to the resolvent set of $\mathbf{M}$, and so $\mathbf{M} - \nu\mathbf{I}$ is invertible. According to the spectral mapping theorem

$$\sigma((\mathbf{M} - \nu\mathbf{I})^{-1}) = (\sigma(\mathbf{M}) - \nu)^{-1}.$$

It follows from this and (14) that

$$|\sigma(\mathbf{M} - \nu\mathbf{I})^{-1}| < d^{-1}. \tag{15}$$

Since $(\mathbf{M} - \nu\mathbf{I})^{-1}$ is a symmetric operator, its spectral radius equals, according to theorem 3, its norm. So it follows from (15) that $\|(\mathbf{M} - \nu\mathbf{I})^{-1}\| < d^{-1}$.

Next we decompose

$$\mathbf{N} - \nu\mathbf{I} = \mathbf{M} - \nu\mathbf{I} + \mathbf{N} - \mathbf{M} = (\mathbf{M} - \nu\mathbf{I})(\mathbf{I} + (\mathbf{M} - \nu\mathbf{I})^{-1}(\mathbf{N} - \mathbf{M})).$$

The second factor on the right is of the form $\mathbf{I} + \mathbf{K}$, $\mathbf{K} = (\mathbf{M} - \nu\mathbf{I})^{-1}(\mathbf{N} - \mathbf{M})$. Using the estimate above for $(\mathbf{M} - \nu\mathbf{I})^{-1}$ and that $\|\mathbf{N} - \mathbf{M}\| = d$, we conclude that

$$\|\mathbf{K}\| \leq \|(\mathbf{M} - \nu\mathbf{I})^{-1}\| \|\mathbf{N} - \mathbf{M}\| < d^{-1}d = 1,$$

It follows that the second factor $\mathbf{I} + \mathbf{K}$ is invertible by the geometric series. The first factor $(\mathbf{M} - \nu\mathbf{I})$, too, is invertible, so is their product $\mathbf{N} - \nu\mathbf{I}$. But this contradicts ν being in the spectrum of $\mathbf{N}$. $\qquad\square$

31.2 FUNCTIONAL CALCULUS FOR SYMMETRIC OPERATORS

In chapter 28 on compact symmetric operators we first constructed a spectral resolution, that is, a complete orthonormal set of eigenvectors, and then used these eigenvectors to define $f(\mathbf{M})$ for any bounded function f. For general symmetric operators we proceed in the reverse order: first we build a functional calculus $f(\mathbf{M})$ for all real-valued functions f continuous on the spectrum of $\mathbf{M}$, and then we use this to construct a spectral resolution.

Let q be a polynomial with *real* coefficients:

$$q(\lambda) = a_n\lambda^n + \cdots + a_0.$$

Then, if $\mathbf{M}$ is symmetric, so is

$$q(\mathbf{M}) = a_n\mathbf{M}^n + \cdots + a_0\mathbf{I}.$$

According to the spectral mapping theorem, theorem 5 of chapter 17,

$$\sigma(q(\mathbf{M})) = q(\sigma(\mathbf{M})). \tag{16}$$

Combining this with formula (9) of theorem 3, we deduce that

$$\|q(\mathbf{M})\| = \max_{\lambda\in\sigma(\mathbf{M})} |q(\lambda)|. \tag{16'}$$

Let $f(\lambda)$ be any continuous real-valued function on $\sigma(\mathbf{M})$, the spectrum of $\mathbf{M}$. We can approximate $f(\lambda)$ uniformly by polynomials on $\sigma(\mathbf{M})$, for we can extend f continuously to an interval containing $\sigma(\mathbf{M})$. Now according to the Weierstrass approximation theorem, f can be approximated uniformly on this interval by polynomials. So there is a sequence $\{q_n\}$, such that

$$\lim_{n\to\infty} \max_{\lambda\in\sigma(\mathbf{M})} |f(\lambda) - q_n(\lambda)| = 0.$$

It follows that $\{q_n\}$ is a Cauchy sequence:

$$\lim_{m,n\to\infty} \max |q_n(\lambda) - q_m(\lambda)| = 0.$$

It follows then from (16') that

$$\lim_{n,m\to\infty} \|q_n(\mathbf{M}) - q_m(\mathbf{M})\| = 0.$$

Since the bounded operators are complete, $\lim_{n\to\infty} q_n(\mathbf{M})$ exists. We denote this limit by $f(\mathbf{M})$. Its properties are summarized in

Theorem 6. $f \to f(\mathbf{M})$ *is an isometric isomorphism:*

(i) $(f + g)(\mathbf{M}) = f(\mathbf{M}) + g(\mathbf{M})$, $(fg)(\mathbf{M}) = f(\mathbf{M})g(\mathbf{M})$.

(ii) $\|f(\mathbf{M})\| = \max_{\sigma(\mathbf{M})} |f(\lambda)|$.

(iii) $f(\mathbf{M})$ *is symmetric and* $\sigma(f(\mathbf{M})) = f(\sigma(\mathbf{M}))$.

Proof. Property (i) holds when f and g are polynomials; therefore they hold for uniform limits of polynomials.

(ii) Since $f(\mathbf{M})$ is the uniform limit of $q_n(\mathbf{M})$, $\|f(\mathbf{M})\| = \lim_{n\to\infty} \|q_n(\mathbf{M})\|$. Since $f(\lambda)$ is the uniform limit of $q_n(\lambda)$,

$$\max_{\sigma(\mathbf{M})} |f(\lambda)| = \lim_{n\to\infty} \max_{\sigma(\mathbf{M})} |q_n(\lambda)|.$$

These two together, combined with (16′), give (ii).

(iii) The adjoint of a limit of maps is the limit of their adjoints. Since each $p(\mathbf{M})$ is symmetric, it follows that so is $f(\mathbf{M})$. Since $f(\mathbf{M})$ is the uniform limit of $q_n(\mathbf{M})$, it follows from (12) of theorem 5 that $\sigma(f(\mathbf{M}))$ is the limit of $\sigma(q_n(\mathbf{M}))$. Thus (iii) follows from (16). This completes the proof of theorem 6.   □

Recall from chapter 18 that a symmetric operator $\mathbf{M}$ is called *positive* if

$$(\mathbf{M}x, x) \geq 0 \qquad \text{for all } x \text{ in } H.$$

Theorem 7. *A bounded symmetric operator is positive if and only if its spectrum contains only nonnegative numbers:*

$$\sigma(\mathbf{M}) \geq 0.$$

Proof. (i) Suppose that $\sigma(\mathbf{M}) \geq 0$. The function $f(\lambda) = \sqrt{\lambda}$ is continuous for $\lambda \geq o$, which contains $\sigma(\mathbf{M})$. The functional calculus defines $\sqrt{\mathbf{M}} = \mathbf{N}$ as a symmetric operator, which satisfies $\mathbf{N}^2 = \mathbf{M}$. Therefore

$$(\mathbf{M}x, x) = (\mathbf{N}^2 x, x) = (\mathbf{N}x, \mathbf{N}x) \geq 0,$$

which shows that $\mathbf{M}$ is positive.

(ii) Conversely, suppose that $\mathbf{M}$ is positive. According to theorem 4, the infimum of the spectrum of $\mathbf{M}$ is $a = \inf_{\|x\|=1}(x, \mathbf{M}x)$. Since $\mathbf{M}$ is positive, $a \geq 0$, which shows that $\sigma(\mathbf{M}) \geq 0$.   □

Corollary. *Every positive symmetric operator has a positive symmetric square root.*

Exercise 2. Show that a positive symmetric operator has only one positive square root. How many square roots does it have that are not positive?

In theorem 1 of chapter 30 we showed that every compact operator has a polar decomposition:

Every compact operator $\mathbf{T}$ *can be factored as* $\mathbf{T} = \mathbf{U}\mathbf{A}$, *where* $\mathbf{A}$ *is a positive symmetric operator, and* $\mathbf{U}$ *is an isometry on the range of* $\mathbf{A}$, *and zero on the orthogonal complement of the range of* $\mathbf{A}$.

The only place in the construction of the polar decomposition where we used compactness was in taking the square root of $\mathbf{T}^*\mathbf{T}$. Now that we know how to take the square root of any positive symmetric operator, we can remove compactness and assert:

Every bounded operator in Hilbert space has a polar decomposition.

31.3 SPECTRAL RESOLUTION OF SYMMETRIC OPERATORS

According to the Riesz representation theorem, every bounded linear functional ℓ on the space of continuous functions f defined on the compact space $\sigma(\mathbf{M})$ can be described as the integral of f with respect to a uniquely determined measure $m(S)$ defined on the Borel subsets S of $\sigma(\mathbf{M})$, whose total variation is finite. We use now the functional calculus described in theorem 6 to construct the functionals

$$\ell_{x,y}(f) = (f(\mathbf{M})x, y), \tag{17}$$

defined for every pair of points x, y in H. By the Riesz representation theorem there is a complex measure m, uniquely determined, such that

$$(f(\mathbf{M})x, y) = \int f(\lambda)dm_{x,y}. \tag{18}$$

Since $\ell_{x,y}$ depends on x and y, so does the measure m; the manner of the dependence of m on x and y reflect the manner of dependence of ℓ on x, y. We list them in

Theorem 8. *Let $m_{x,y}$ be the measure on $\sigma(\mathbf{M})$ defined by (18).*

(i) *$m_{x,y}$ depends sesquilinearly on x and y, that is, linearly on x, skew linearly on y.*

(ii) *$m_{x,y}$ is skew symmetric in x, y : $m_{y,x} = \overline{m_{x,y}}$.*

(iii) *Total var $m_{x,y} \leq \|x\| \|y\|$.*

(iv) *The measures $m_{x,x}$ are real and nonnegative.*

Proof. Clearly, by (17), $\ell_{x,y}$ is linear in x, skew linear in y. Since m is uniquely determined by ℓ, so is the measure m representing ℓ:

$$m_{x+z,y} = m_{x,y} + m_{z,y} \tag{19}$$

since both measures represent $\ell_{x+z,y}$. This proves (i).

(ii) Since $f(\mathbf{M})$ is symmetric, the left side of (18) is skew symmetric in x, y; using uniqueness of the representing measure, we conclude that m depends skew symmetrically on x, y.

(iii) According to the Riesz theorem, the total variation of the representing measure equals the norm of the functional ℓ. Using the Schwarz inequality and (16'), we conclude that

$$|\ell_{xy}(f)| = |(f(\mathbf{M})x, y)| \le \|f(\mathbf{M})\| \|x\| \|y\| = |f|_{\max} \|x\| \|y\|.$$

This shows that $|\ell_{xy}| \le \|x\| \|y\|$, so (iii) follows.

(iv) According to part (iii) of theorem 6, for f real the spectrum of $f(\mathbf{M})$ is $f(\sigma(\mathbf{M}))$. So for positive functions f, $f(\mathbf{M})$ is a symmetric operator whose spectrum is positive. Therefore according to theorem 7, $f(\mathbf{M})$ is a positive operator. This shows that the linear functional $\ell_{x,x}(f) = (f(\mathbf{M})x, x)$ is positive; but then so is the measure $m_{x,x}$ representing it. $\square$

According to theorem 8, $m_{x,y}(S)$, S a Borel subset of $\sigma(\mathbf{M})$, is a bounded, skew symmetric sesquilinear functional of x and y. We conclude from theorem 1 that for each S there is a bounded symmetric operator $\mathbf{E}(S)$ such that

$$m_{x,y}(S) = (\mathbf{E}(S)x, y). \tag{20}$$

This family of operators has the following properties:

Theorem 9. *Let $\mathbf{E}(S)$ be the family of maps defined by (20), where $m_{x,y}(S)$ is defined by (18).*

(i) $\mathbf{E}^*(S) = \mathbf{E}(S)$.

(ii) $\|\mathbf{E}(S)\| \le 1$.

(iii) $\mathbf{E}(\emptyset) = 0$, $\mathbf{E}(\sigma(\mathbf{M})) = \mathbf{I}$.

(iv) *If $S \cap T = \emptyset$, $\mathbf{E}(S \cup T) = \mathbf{E}(S) + \mathbf{E}(T)$.*

(v) *Each $\mathbf{E}(S)$ commutes with $\mathbf{M}$.*

(vi) $\mathbf{E}(S \cap T) = \mathbf{E}(S)\mathbf{E}(T)$.

(vii) *Each $\mathbf{E}(S)$ is an orthogonal projection. If S and T are disjoint, the range of $\mathbf{E}(S)$ and $\mathbf{E}(T)$ are orthogonal.*

(viii) *All orthogonal projections $\mathbf{E}(S)$, $\mathbf{E}(T)$ commute.*

Proof. (i) is part of theorem 1. Part (ii) follows from part (iii) of theorem 8.

(iii) Since $m_{x,y}(\emptyset) = 0$, it follows from (18) that $\mathbf{E}(S) = 0$. On the other hand, setting $f(\lambda) \equiv 1$, $f(\mathbf{M}) = \mathbf{I}$ in (18) gives for all x and y,

$$(x, y) = \int_{\sigma(\mathbf{M})} dm_{x,y} = (\mathbf{E}(\sigma(\mathbf{M}))x, y),$$

which means that $\mathbf{E}(\sigma(\mathbf{M})) = \mathbf{I}$.

Part (iv) follows from the additivity of the measure $m_{x,y}$. To show (v), we note that since $\mathbf{M}$ commutes with $f(\mathbf{M})$, and is symmetric,

$$(f(\mathbf{M})\mathbf{M}x, y) = (\mathbf{M}f(\mathbf{M})x, y) = (f(\mathbf{M})x, \mathbf{M}y).$$

The functional of f on the left is represented by the measure $m_{\mathbf{M}x,y}$; the functional on the right by the measure $m_{x,\mathbf{M}y}$. Since the functionals are the same, so are the measures:

$$m_{\mathbf{M}x,y} = m_{x,\mathbf{M}y}.$$

Setting this into (20), and using once more the symmetry of $\mathbf{M}$, gives

$$(\mathbf{E}(S)\mathbf{M}x, y) = (\mathbf{E}(S)x, \mathbf{M}y) = (\mathbf{M}\mathbf{E}(S)x, y).$$

Since this holds for all x and y, $\mathbf{E}(S)\mathbf{M} = \mathbf{M}\mathbf{E}(S)$, as claimed in (v).

We postpone the proof of (vi) to section 31.5.

(vii) Setting $S = T$ in (vi) shows that $\mathbf{E}(S) = \mathbf{E}^2(S)$, namely that $\mathbf{E}(S)$ is an idempotent. The geometric expression of this algebraic fact is that $\mathbf{E}(S)$ is a projection. Since, by part (i), $\mathbf{E}$ is symmetric, $\mathbf{E}(S)$ is an orthogonal projection. It follows from (iii) and (vi) that if S and T are disjoint, the range of $\mathbf{E}(S)$ and $\mathbf{E}(T)$ are orthogonal.

Part (viii) results from interchanging S and T in (vi). □

The family of operators $\mathbf{E}(S)$ is an *orthogonal projection-valued measure*.

Exercise 3. (a) Show that $\mathbf{E}(S)$ is countably additive in the strong topology. (Hint: Use the orthogonality of the ranges of $\mathbf{E}(S)$ and $\mathbf{E}(T)$ when S and T are disjoint.)

(b) Show that $\mathbf{E}(S)$ is *not* countably additive in the norm topology.

We summarize:

Theorem 9′. *H is a Hilbert space* $\mathbf{M} : H \to H$ *a bounded, linear symmetric operator. Then there is a uniquely determined orthogonal projection-valued measure* $\mathbf{E}$ *on the spectrum of* $\mathbf{M}$ *such that* $\mathbf{E}(S \cap T) = \mathbf{E}(S)\mathbf{E}(T)$ *and*

$$f(\mathbf{M}) = \int_{\sigma(\mathbf{M})} f(\lambda)d\mathbf{E} \tag{21}$$

for all continuous functions f *on* $\sigma(\mathbf{M})$. *The integral exists in the norm topology.*

Proof. The meaning of (18) and (20) is that (21) holds in the weak topology. To show that it holds in the norm topology, it suffices to show that the Riemann-Stieltjes integral on the right converges in that topology. This can be done in the standard fashion, combined with the estimate

$$\left\| \sum a_j \mathbf{E}(I_j) \right\| \leq \max |a_j|,$$

where $\cup I_j = \sigma(\mathbf{M})$ is a decomposition of $\sigma(\mathbf{M})$ into a finite number of disjoint pieces I_j. The estimate follows from the orthogonality of the ranges of $\mathbf{E}(I_j)$. The uniqueness of $\mathbf{E}(S)$ follows from the uniqueness of the scalar measures (18). □

Take $f \equiv 1$, and $f(\lambda) \equiv \lambda$:

$$\mathbf{I} = \int_{\sigma(\mathbf{M})} d\mathbf{E}, \quad \mathbf{M} = \int_{\sigma(\mathbf{M})} \lambda \, d\mathbf{E}; \tag{22}$$

(22) is called the *spectral resolution* of $\mathbf{M}$.

31.4 ABSOLUTELY CONTINUOUS, SINGULAR, AND POINT SPECTRA

We give now an important further refinement of the spectral resolution. According to the *Lebesgue decomposition theorem* for measures, any measure on $\mathbb{R}$ can be decomposed as the sum of a *point measure*, supported on a denumerable set, a *singular measure*, supported on a set of Lebesgue measure zero, and an *absolutely continuous measure* with respect to Lebesgue measure.

We apply this to the measures $m_{x,y} = (\mathbf{E}x, y)$:

$$m_{x,y} = m_{x,y}^{(p)} + m_{x,y}^{(s)} + m_{x,y}^{(c)}. \tag{23}$$

From the uniqueness of the Lebesgue decomposition we conclude that all three measures on the right in (23) depend linearly on x, skew linearly on y. For any S these sesquilinear functionals can be represented as

$$m_{x,y}^{(p)}(S) = (\mathbf{E}^{(p)}(S)x, y), \quad m_{x,y}^{(s)}(S) = (\mathbf{E}^{(s)}x, y) \quad \text{and} \quad m_{x,y}^{(c)}(S) = (\mathbf{E}^{(c)}x, y).$$

These bounded, symmetric operators $\mathbf{E}^{(p)}$, $\mathbf{E}^{(s)}$, $\mathbf{E}^{(c)}$ have all properties listed in theorems 9; that is, each family is a projection-valued measure, each orthogonal to the other.

Denote the ranges of $\mathbf{E}^{(p)}(\sigma(\mathbf{M}))$, $\mathbf{E}^{(s)}(\sigma(\mathbf{M}))$, and $\mathbf{E}^{(c)}(\sigma(\mathbf{M}))$ as $H^{(p)}$, $H^{(s)}$, and $H^{(c)}$; they are called the *point, singular,* and *absolutely continuous subspace of H with respect to the operator* $\mathbf{M}$. Clearly,

$$H^{(p)} \oplus H^{(s)} \oplus H^{(c)} = H.$$

31.5 THE SPECTRAL REPRESENTATION
OF SYMMETRIC OPERATORS

Spectral representation is an infinite dimensional analogue of the diagonal form of symmetric matrices.

Theorem 10. *For any vector x in the Hilbert space H, and any continuous function f,*

$$\|f(\mathbf{M})x\|^2 = \int |f(\lambda)|^2 dm_{x,x}, \tag{24}$$

THE SPECTRAL REPRESENTATION OF SYMMETRIC OPERATORS   365

where the measure $m_{x,x}$ is the representing measure appearing in formula (18):

$$(f(\mathbf{M})x, y) = \int f(\lambda)dm_{x,y}. \tag{18}$$

Proof. For real-valued f we use the symmetry of $f(\mathbf{M})$ to get $\|f(\mathbf{M})x\|^2 = (f(\mathbf{M})x, f(\mathbf{M})x) = (f^2(\mathbf{M})x, x) = \int f^2(\lambda)dm_{x,x}$, where in the last steps we used (18) with f^2 in place of f.

For complex-valued function $f = g + ih$, similar manipulations lead to (24). $\square$

For any given x denote by J_x the set of elements z of form $z = f(\mathbf{M})x$, f any complex-valued continous function on $\sigma(\mathbf{M})$. Clearly, J_x is an invariant subspace of H, that is J_x is mapped into itself by $\mathbf{M}$. We say that the vector z in J_x is *represented* by the function f. Relation (24) shows that this representation is an *isometry* when the functions f are normed by the $L^2(m_{x,x})$-norm.

If $z = f(\mathbf{M})x$ is represented by $f(\lambda)$, $\mathbf{M}z$ is represented by $\lambda f(\lambda)$.

Denote by K_x the closure of J_x in the Hilbert space H. It follows from isometry that every element z in K_x can be represented isometrically by a function h in $L^2(m_{x,x})$, and that $\mathbf{M}z$ belongs to K_x, represented by $\lambda h(\lambda)$. Conversely, every function h in $L^2(m_{x,x})$ represents some element of K_x. This is called a *spectral representation* of $\mathbf{M}$ acting on K_x.

We recall from chapter 7, section 7.1, that a measure n is called *absolutely continuous* with respect to another measure m if every set that has m-measure zero has n-measure zero. Two measures are called *equivalent* if each is absolutely continuous with respect to the other. According to the Radon-Nikodym theorem, every measure n absolutely continuous with respect to m can be represented as $dn = gdm$, where g is a nonzero positive function integrable with respect to m.

It follows that if we have a spectral representation of a symmetric operator $\mathbf{M}$ as $L^2(m)$, then we can get a spectral representation of $\mathbf{M}$ as $L^2(n)$, n any measure equivalent with m. If z is represented by h in $L^2(m)$, it is represented by $hg^{-1/2}$ in $L^2(n)$.

The space K_x on which we have constructed a spectral representation is, in general, not all of H but only a closed invariant subspace. We appeal now to a result contained in

Exercise 4. Let $\mathbf{M}$ be a symmetric operator acting on a Hilbert space H, K an invariant subspace of $\mathbf{M}$. Show that the orthogonal complement of K also is invariant under $\mathbf{M}$.

Suppose now that K_x above is a proper subspace H. Choose any vector y orthogonal to K_x; according to exercise 4, so is $\mathbf{M}y$. It follows that for any polynomial $q(\lambda)$, $q(\mathbf{M})y$ is orthogonal to K_x. Denote by K_y the closure of the set of vectors $q(\mathbf{M})y$. Clearly, K_y is a closed invariant subspace of H, represented spectrally by $L^2(m_{y,y})$. Using Zorn's lemma and the above construction, we deduce

Theorem 11. *There exists a family $\{K_j\}$ of closed subspaces of H such that*

(i) The K_j are pairwise orthogonal, and they span H:

$$H = K_1 \oplus K_2 \oplus \dots . \tag{25}$$

(ii) Each K_j is invariant under M and is spectrally represented as $L^2(m_j)$.

Such a collection of $\{K_j\}$ is called *a spectral representation* of M on H.

It is easy to derive from a spectral representation of M its spectral resolution. For any measurable set S, define on each K_j the operator $E(S)$ as follows: in the representation of K_j as $L^2(m_j)$, $E(S)$ is multiplication by $c_S(\lambda)$, the *characteristic function* of S, defined as

$$c_S(\lambda) = \begin{cases} 1 & \text{if } \lambda \text{ in } S \\ 0 & \text{otherwise.} \end{cases}$$

Exercise 5. Verify that $E(S)$ as defined above is a spectral resolution of M, namely that the family $\{E(S)\}$ is a projection-valued measure that has all the properties listed in theorems 9 and 9'.

Although a given operator has many spectral representations, the spectral resolution derived from each is the same, as explained in theorem 9'.

We turn now to the last topic of this section: spectral multiplicity. To simplify the discussion, we take the case that the underlying Hilbert space H is separable. It follows that in any spectral representation (25) the family $\{K_j\}$ is denumerable.

For further simplification we assume that the spectrum of M is absolutely continuous with respect to Lebesgue measure in the sense explained in section 31.4. By the Radon-Nikodym theorem, such a measure is equivalent to Lebesgue measure on a subset S of $\mathbb{R}$. Under this assumption the measures m_j entering the spectral representation of theorem 11 can be taken to be Lebesgue measure over sets S_j that support the measure, where the *support* of m is the union of points λ that have the property that the restriction of m to any open interval containing λ has positive measure.

The following are easily verified: if $\{m_j\}$ is a denumerable collection of measures, the support of Σm_j contains the union of the supports of m_j.

Two measures are *singular* with respect to each other if the intersection of their support has Lebesgue measure zero.

Lemma 12. *Take any two spectral representations (25) and (25') of a bounded, symmetric operator M, with absolutely continuous spectrum, acting on a separable Hilbert space:*

$$H = K_1' \oplus K_2' \oplus \dots, \quad K_j' \leftrightarrow L^2(S_j'). \tag{25'}$$

We claim that up to sets of measure zero

$$\cup S_j = \cup S_j'.$$

Proof. Take any index k; choose x to be that vector in K_k that is represented by the function $\equiv 1$ in S_k. For any continuous function f,

$$(f(\mathbf{M})x, x) = \int_{S_k} f(\lambda)d\lambda. \tag{26}$$

Denote by x_j' the projection of x onto K_j', and denote by $g_j(\lambda)$ the function representing x_j' in $L^2(S_j')$. Then $x = \Sigma x_j'$, and

$$(f(\mathbf{M})x, x) = \Sigma(f(\mathbf{M})x_j', x_j') = \Sigma \int f(\lambda)|g_j|^2 d\lambda = \int f(\lambda)|g|^2 d\lambda, \tag{26'}$$

where $|g|^2 = \Sigma|g_j|^2$.

Both (26) and (26') represent the same bounded linear functional. Therefore the representing measures must be equal:

$$c_{S_k} d\lambda = |g^2|d\lambda.$$

The support of the measure on the left is S_k, and the support of the measure on the right is contained in $\cup S_j'$. Therefore $S_k \subset \cup S_j'$. Reversing the role of (25) and (25'), we obtain lemma 12. $\qquad\square$

Definition. The *spectral multiplicity* of a point λ in a spectral representation (25) $\Sigma c_{s_j}(\lambda)$, is the number of sets S_j to which λ belongs; S_j is the support of the jth measure m_j in the representation (25). The spectral multiplicity of λ can be zero, any natural number, or ∞.

A spectral representation is far from unique, since its construction contains many arbitrary choices. So it is far from clear that the spectral multiplicity function has an invariant meaning, the same for all spectral representations of a given bounded symmetric operator. According to a classical result of E. Hellinger:

Theorem 13. *Let* $\mathbf{M}$ *be a bounded, symmetric operator in a separable Hilbert space* H, *whose spectrum is absolutely continuous. Let* $H = K_1 \oplus K_2 \oplus \cdots$ *be any spectral decomposition of* H *for* $\mathbf{M}$, *such that* $K_j \leftrightarrow L^2(S_j)$ *and the action of* $\mathbf{M}$ *is represented as multiplication by* λ.

Spectral multiplicity, defined as $\Sigma c_{s_j}(\lambda)$, *is the same for all spectral representations of* $\mathbf{M}$.

Proof. We will use two simple operations for rearranging spectral decompositions, *splitting* and *combination*.

Splitting. Suppose that a subspace K of H is represented spectrally by $L^2(m)$, m some measure on $\mathbb{R}$. Split m as the sum $m = \Sigma m_j$ of measures that are pairwise singular with respect to each other. Each $L^2(m_j)$ represents some closed subspace K_j of K; their direct sum $K_1 \oplus K_2 \oplus \cdots$ is a spectral decomposition of K.

Combination. This is the reversal of splitting. Suppose that $\{K_j\}$ is a collection of pairwise orthogonal subspaces of H, each represented spectrally by $L^2(m_j)$. Suppose that the m_j are pairwise singular with respect to each other. Then $K = K_1 \oplus K_2 \oplus \cdots$ is represented spectrally by $L^2(m)$, $m = \Sigma m_j$.

We will use splitting and combination to rearrange any spectral representation

$$H = K_1 \oplus K_2 \oplus \cdots, \quad K_j \leftrightarrow L^2(S_j)$$

into a *standard form*, as follows: Decompose each set S_j as

$$S_j = \cup S_j^k,$$

where S_j^k is that subset of S_j which belongs to exactly k of the sets S_l. Define the set M_1 as

$$M_1 = \cup S_j^1. \tag{27}$$

Clearly, M_1 is the set of points that belong to exactly one set S_j, namely points of spectral multiplicity one in the spectral representation under discussion. Similarly we define M_k as

$$M_k = \cup S_j^k; \tag{27'}$$

Clearly, M_k is the set of points of multiplicity k. M_∞ is the set of points of infinite multiplicity. It follows from the construction that

$$\cup S_j = \cup M_k \cup M_\infty. \tag{28}$$

By splitting each S_j and recombining them into sets M_k as described above, we can split and recombine the spectral representations $K_j \leftrightarrow L^2(S_j)$ into spectral representations by $L^2(M_k)$, $k = 1, 2, \cdots, \infty$. There is exactly one subspace represented by $L^2(M_1)$, call it H_1. There are two orthogonal subspaces represented by $L^2(M_2)$, whose direct sum we call H_2, and so on. The subspaces H_k are orthogonal to each other; so we obtain the spectral decomposition

$$H = H_1 \oplus H_2 \oplus \cdots \oplus H_\infty, \tag{29}$$

where H_k is represented spectrally by k copies of $L^2(M_k)$. We call (29) the *standard form* of the spectral representation (25).

We claim that any other spectral representation (25') has the same standard form. Here is the proof: Denote the standard spectral representation obtained from (25') as

$$H = H_1' \oplus H_2' \oplus \cdots \oplus H_\infty', \tag{29'}$$

where H_k' is represented by $L^2(M_k') \oplus \cdots \oplus L^2(M_k')$. We claim that $M_k' = M_k$ and $H_k' = H_k$. To see why, we shall construct vectors in H_k', $k > 1$ that are orthogonal to H_1. We start with $k = 2$.

Let x be the vector in H_1 that is represented by the function $\equiv 1$ in M_1 in the standard form (29). The vectors $f(\mathbf{M})x$ span H_1. Denote by x_2' the projection of x onto H_2' in the standard spectral representation (29'). Denote by $\{g_1, g_2\}$ the functions in M_2' representing x_2'.

Define the functions h_1 and h_2 in M_2' as follows:

$$h_1 = \begin{cases} 1 & \text{where } g_1 = 0 \\ -\bar{g}_2 & \text{elsewhere,} \end{cases} \qquad h_2 = \begin{cases} 1 & \text{where } g_2 = 0 \\ \bar{g}_1 & \text{elsewhere.} \end{cases}$$

Let y be the vector in H_2' represented by $\{h_1, h_2\}$. We claim that y is orthogonal to all vectors of form $f(\mathbf{M})x$. Clearly, since y belongs to H_2', it is orthogonal to all components of $f(\mathbf{M})x$ in the decomposition (29') but the second component. So, using the definition of h_1 and h_2, we get

$$(y, f(\mathbf{M})x) = (y, f(\mathbf{M})x_2') = (\{h_1, h_2\}, f(\lambda)\{g_1, g_2\})$$
$$= \int (h_1\bar{g}_1 + h_2\bar{g}_2)\overline{f}(\lambda)d\lambda = 0.$$

The vector $f(\mathbf{M})y$ is represented by $\{f(\lambda)h_1, f(\lambda)h_2\}$. Therefore, using the definition of h_1 and h_2, we get

$$\|f(\mathbf{M})y\|^2 = \int_{M_2'} |f|^2(|h_1|^2 + |h_2|^2)d\lambda.$$

The formulas for h_1 and h_2 show that $|h_1|^2 + |h_2|^2$ is positive on M_2', and belongs to $L^1(M_2')$. It follows that the closure of $\{f(\mathbf{M})y\}$ has a spectral representation by $L^2(M_2')$. Since y is orthogonal to H_1, so are the vectors $f(\mathbf{M})y$, and so they all belong to $H_1^\perp$, the orthogonal complement of H_1. The Hilbert space $H_1^\perp$ is invariant under $\mathbf{M}$. The subspaces $H_2, \ldots, H_\infty$ furnish a spectral representation of $\mathbf{M}$ on $H_1^\perp$ on the sets $M_2, \ldots M_\infty$. We appeal now to lemma 12, according to which the point-set M_2' belongs to the union of the sets M_k, $k > 1$ entering the spectral representation (29) of $\mathbf{M}$ over the Hilbert space $H_1^\perp$. Therefore we conclude that M_2' is contained in $M_2 \cup M_3 \cup \cdots \cup M_\infty$.

Using the same argument, we conclude that $M_3', \cdots, M_\infty'$ all are contained in $M_2 \cup M_3 \cup \cdots \cup M_\infty$. Since the sets M_k are pairwise disjoint, it follows that the sets M_k', $k > 1$, are disjoint from M_1. On the other hand, from lemma 12 applied to $\mathbf{M}$ acting on the whole space H, we conclude that

$$\cup M_k = \cup M_k'.$$

Combining this with the disjointness of M_1 and M'_k, $k > 1$, we conclude that M_1 is contained in M'_1. Reversing the role of the two spectral representations, we conclude that M'_1 is contained in M_1; therefore $M_1 = M'_2$.

To show that $H_1 = H'_1$, we argue as in the proof of lemma 12. We leave it to the diligent reader to show that for all k, $M_k = M'_k$ and $H_k = H'_k$.   □

Exercise 6. Let (25) be a spectral representation for **M**, $K_j \leftrightarrow L^2(m_j)$, S_j the support of m_j. Show that the closure of the union of S_j is the spectrum of **M**.

Exercise 7. Give an example to show that the spectrum of **M** may contain a set of positive Lebesgue measure whose multiplicity is zero.

Definition. Two bounded, symmetric operators **M** and **N** acting on a Hilbert space H are called *unitarily equivalent* if there is a unitary map **U**, that is, a one-to-one a norm-preserving map of H onto H, that carries **M** into **N**:

$$\mathbf{N} = \mathbf{UMU}^{-1}.$$

Theorem 14. *Two bounded, symmetric operators **M** and **N** whose spectrum is absolutely continuous are unitarily equivalent if and only if they have the same spectral multiplicities.*

Proof. If **M** and **N** are unitarily equivalent, **U** carries the standard spectral representation of **M** into a standard spectral representation of **N**. Conversely, if the spectral multiplicity set M_k and N_k for the two operators **M** and **N** are the same, then the standard spectral representations for **M** and **N** furnish the unitary map **U**.   □

31.6 SPECTRAL RESOLUTION OF NORMAL OPERATORS

A bounded, linear operator N mapping a Hilbert space into itself is called *normal* if it commutes with its adjoint:

$$\mathbf{N}^*\mathbf{N} = \mathbf{NN}^*, \tag{30}$$

Clearly, every bounded symmetric operator is normal; as we will see in the next section, so is every unitary operator.

The spectral resolution of normal operators is analogous to that of symmetric operators except that the spectrum of a normal operator may well contain complex numbers.

Theorem 15. *Let H be a Hilbert space, $\mathbf{N}: H \to H$ a normal operator. Then there is an orthogonal projection-valued measure **E** on the Borel subsets of the spectrum* *of* N *such that*

$$I = \int_{\sigma(N)} d\mathbf{E}, \quad \mathbf{N} = \int_{\sigma(N)} \lambda d\mathbf{E}. \tag{31}$$

The integrals exist in the norm topology.

   Proof. As in theorem 9, we rely on a functional calculus for normal operators. We will make use of Gelfand's theory of commutative B^* algebras, invented for this purpose; see chapters 18 and 19.

   Let $q(\xi, n)$ be any polynomial in two real variables ξ, η. We rewrite it as a polynomial Q in the complex variable $\zeta = \xi + i\eta$ and its conjugate $\bar{\zeta} = \xi - i\eta$:

$$q(\xi, n) = Q(\zeta, \bar{\zeta}). \tag{32}$$

We define the functional calculus by setting

$$\mathbf{Q} = Q(\mathbf{N}, \mathbf{N}^*), \tag{32'}$$

Since $\mathbf{N}$ and $\mathbf{N}^*$ commute, they commute with $\mathbf{Q}$, and $\mathbf{Q}$ commutes with $\mathbf{Q}^*$. We need the following version of the spectral mapping theorem for normal operators:

Lemma 16. *For* $\mathbf{N}$ *a normal operator, and* $\mathbf{Q}$ *of form (32'), the spectrum of* $\mathbf{Q}$ *is of form*

$$\sigma(\mathbf{Q}) = Q(\lambda, \bar{\lambda}), \qquad \lambda \in \sigma(\mathbf{N}). \tag{33}$$

   Proof. The operators $\mathbf{Q}$ of form (32') constitute a commutative algebra of operators with a unit. Denote by $\mathcal{F}$ their closure in the operator norm; $\mathcal{F}$ is a commutative Banach algebra with a unit, and therefore enjoys all the benefits of Gelfand's theory. We appeal in particular to theorem 14 of chapter 18, according to which the spectrum of any $\mathbf{Q}$ in $\mathcal{F}$ is of the form $p(\mathbf{Q})$, where p is a homomorphism of $\mathcal{F}$ into $\mathbb{C}$. For $\mathbf{Q}$ of form (32'),

$$p(\mathbf{Q}) = Q(p(\mathbf{N}), p(\mathbf{N}^*)). \tag{34}$$

According to theorem 16 of chapter 19,

$$p(\mathbf{N}^*) = \overline{p(\mathbf{N})}.$$

Set this into (34); since the numbers $p(\mathbf{N})$ run through the spectrum of $\mathbf{N}$ as p runs through all homomorphisms, we obtain (33). $\square$

   We appeal next to theorem 17 in chapter 19, that asserts that the norm of a normal operator $\mathbf{Q}$ is equal to its spectral radius:

$$\|\mathbf{Q}\| = |\sigma(\mathbf{Q})|. \tag{35}$$

Combining (35) with (33), we obtain the important result that for $\mathbf{Q}$ of form (32'):

$$\|\mathbf{Q}\| = \max_{\lambda \text{ in } \sigma(N)} |Q(\lambda, \bar{\lambda})|, \tag{36}$$

We can now extend the functional calculus from polynomials to all continuous functions on the spectrum of $\sigma(\mathbf{N})$. Every continuous function f on $\sigma(\mathbf{N})$ can be approximated uniformly by a sequence q of polynomials; the relation (36) guarantees that the corresponding sequence of operators $\mathbf{Q}_n$ converge in norm to an operator in $\mathcal{F}$. The resulting functional calculus has the analogues of all properties listed for the symmetric case in theorem 6.

We can now proceed as in our construction of a spectral resolution for normal operators; see theorems 8 and 9. We leave details to the reader.   □

31.7 SPECTRAL RESOLUTION OF UNITARY OPERATORS

Definition. A unitary operator $\mathbf{U}$ is a linear isometric one-to-one mapping of a Hilbert space onto itself:

$$\|\mathbf{U}x\| = \|x\|. \tag{37}$$

Exercise 8. Show that a unitary operator $\mathbf{U}$ preserves scalar products:

$$(\mathbf{U}x, \mathbf{U}y) = (x, y). \tag{37'}$$

It follows from (37') that

$$(x, \mathbf{U}^*\mathbf{U}y) = (x, y);$$

since this holds for all vectors x and y in H, $\mathbf{U}^*\mathbf{U}$ is the identity: $\mathbf{U}^*\mathbf{U} = \mathbf{I}$. Since, by definition, $\mathbf{U}$ is invertible—it is one-to-one and onto—it follows that $\mathbf{U}^*$ is the inverse of $\mathbf{U}$:

$$\mathbf{U}^* = \mathbf{U}^{-1}. \tag{38}$$

Exercise 9. Show that every operator $\mathbf{U}$ that satisfies (38) is unitary.

Exercise 10. Show that the spectrum of a unitary operator lies on the unit circle.

Exercise 11. Show that if $\mathbf{M}$ is a symmetric operator, and k any real number,

$$\mathbf{U} = (\mathbf{M} + ik\mathbf{I})^{-1}(\mathbf{M} - ik\mathbf{I}) \tag{39}$$

is unitary. How about the converse?

Exercise 12. Combine exercises 10 and 11 to show that the spectrum of a bounded symmetric operator is real.

Since every invertible operator commutes with its inverse, it follows from (38) that U and U^* commute. This shows that every unitary operator U is normal; thus, according to theorem 15, U has a spectral resolution of form (31), only in this case the measure E is supported on the unit circle. Since unitary operators are by far the most important among normal operators, we give here a direct proof of this result.

For any vector x in H, form the following doubly infinite sequence $\{a_n\}$:

$$a_n = (U^n x, x), \qquad n \text{ integer.} \tag{40}$$

We claim that this sequence is positive definite in the sense of chapter 14, section 4,

$$\sum_{n,m} a_{n-m} \varphi_n \overline{\varphi_m} \geq o \tag{41}$$

for any finite set of complex numbers φ_n. To see this, we set (40) into (41) and use $U^{-1} = U^*$:

$$\sum (U^{n-m} x, x) \varphi_n \overline{\varphi_m} = \sum (U^n x, U^m y) \varphi_n \overline{\varphi_m}$$

$$= \left(\sum_n \varphi_n U^n x, \sum_m \varphi_m U^m y \right) = \left\| \sum \varphi_n U^n x \right\|^2 \geq o.$$

Then according to Carathéodory's theorem, theorem 7 of chapter 14, a_n are the Fourier coefficients of a *nonnegative* measure on the unit circle:

$$(U^n x, x) = \int e^{in\theta} dm_x. \tag{42}$$

As indicated, the measure m_x depends on x. Setting $n = 0$ in (42) shows that the total measure equals $\|x\|^2$.

For any pair of vectors x, y, $(U^n x, y)$ can be expressed as a simple linear combination of $(U^n(x \pm y, x \pm y)$ and $(U(x \pm iy), x \pm iy)$. Using (42), we can write

$$(U^n x, y) = \int e^{in\theta} dm_{x,y}, \tag{43}$$

where

$$4m_{x,y} = m_{x+y} - m_{x-y} + im_{x+iy} - im_{x-iy}. \tag{44}$$

Clearly, $m_{x,x} = m_x$.

Theorem 17.

(i) *The measures $m_{x,y}$ depend linearly on x, skew linearly on y.*

(ii) *$m_{x,y}$ is a skew symmetric function of x, y: $m_{y,x} = \overline{m}_{x,y}$.*

(iii) *The total mass of $m_{x,y}$ is $\leq \|x\| \|y\|$.*

Proof. (i) Every measure is uniquely determined by its Fourier coefficients. It follows that since the left side of (43) depends linearly on x, so does the right side.

(ii) is an immediate consequence of the definition (44) of the measure $m_{x,y}$.

(iii) Take any set S on the unit circle; we claim that $m_{x,y}(S)$ is a scalar product in H. By (i), it is linear in x; by (ii), it is a skew symmetric function of x and y. Since $m_{x,x}(S) = m_x(S)$, it is nonnegative. Therefore the Schwarz inequality can be applied:

$$|m_{x,y}(S)| \leq m_x(S)^{1/2} m_y(S)^{1/2}.$$

We saw earlier that the total measure m_x is $\|x\|^2$. Since m_x is nonnegative, $m_x(S) \leq \|x\|^2$, $m_y(S) \leq \|y\|^2$, and so $|m_{x,y}(S)| \leq \|x\|\|y\|$. $\qquad\square$

Note that theorem 17 is a literal analogue of theorem 8.

Let S be any Borel set on the unit circle. It follows from theorem 17 that $m_{x,y}(S)$ is a skew symmetric function of x and y, linear in x, and bounded by $\|x\|\|y\|$. It follows then from theorem 1 that it can be represented as

$$m_{x,y}(S) = (\mathbf{E}(S)x, y), \tag{45}$$

$\mathbf{E}(S)$ a bounded, symmetric operator. We claim that this family of operators has the same properties as those listed in theorem 9. Here, for instance, is a demonstration of property (vi):

(vi) $\mathbf{E}(S \cap T) = \mathbf{E}(S)\mathbf{E}(T)$.

Proof. We set (45) into (43), obtaining

$$(\mathbf{U}^n x, y) = \int e^{in\theta} d(\mathbf{E}x, y). \tag{46}$$

Replace n by $n + k$:

$$(\mathbf{U}^{n+k} x, y) = \int e^{in\theta} e^{ik\theta} d(\mathbf{E}x, y); \tag{47}$$

on the other hand, using (46) with x replaced by $\mathbf{U}^k x$, we can express the left side of (47) as

$$\int e^{in\theta} d(\mathbf{E}\mathbf{U}^k x, y). \tag{48}$$

Two measures on the unit circle that have the same Fourier coefficients are identical; therefore

$$e^{ik\theta} d(\mathbf{E}x, y) = d(\mathbf{E}\mathbf{U}^k x, y).$$

Integrate this over the set S:

$$\int c_s(\theta)e^{ik\theta}d(\mathbf{E}x, y) = (\mathbf{E}(S)\mathbf{U}^k x, y), \tag{49}$$

where c_s is the characteristic function of S. We can, using the symmetry of $\mathbf{E}(S)$, rewrite the right side of (49) as $(\mathbf{U}^k x, \mathbf{E}(S)y)$; using formula (46), with k in place of n, we find that this is equal to

$$\int e^{ik\theta}d(\mathbf{E}x, \mathbf{E}(S)y). \tag{49'}$$

The measures on the left in (49) and (49') have the same Fourier coefficients; therefore they are identical:

$$c_s d(\mathbf{E}x, y) = d(\mathbf{E}x, \mathbf{E}(S)y).$$

Integrate both sides over any other Borel set T:

$$\int c_T c_s d(\mathbf{E}x, y) = (\mathbf{E}(T)x, \mathbf{E}(S)y).$$

Since $c_T c_s = c_{S \cap T}$, the left side is $(\mathbf{E}(S \cap T)x, y)$. Using the symmetry of $\mathbf{E}(S)$, we find that the right side is $(\mathbf{E}(S)\mathbf{E}(T)x, y)$. Since they are equal for all x and y,

$$\mathbf{E}(S \cap T) = \mathbf{E}(S)\mathbf{E}(T),$$

as asserted in (vi).   □

We leave it to the reader to verify the rest of the properties listed in theorem 9. Setting (45) into (43), we obtain

$$(\mathbf{U}^n x, y) = \int e^{in\theta}d(\mathbf{E}x, y),$$

which is the weak version of

$$\mathbf{U}^n = \int e^{in\theta}dE. \tag{50}$$

As in theorem 9, the integral on the right exist in the norm topology. This is the spectral resolution of unitary operators.

HISTORICAL NOTE. The spectral resolution of bounded symmetric operators is due to Hilbert. The spectral representation, and theory of spectral multiplicity, is the work of Ernest Hellinger, (1883–1950), a student of Hilbert. He was professor of mathematics at the University of Frankfurt until his dismissal by the Nazis. In the infamous anti-Jewish pogrom in 1938, dubbed "Krystallnacht," he was taken to the dreaded

Dachau concentration camp. Miraculously, he was released; he found a new home in the United States, teaching mathematics at Northwestern University.

BIBLIOGRAPHY

Gelfand, I. M. Normierte Ringe. *Mat. Sbornik, N.S.* (51), **9** (1941): 3–24.

Halmos, P. *Introduction to Hilbert Space and the Theory of Spectral Multiplicity.* Chelsea Publishing, New York, 1951.

Hellinger, E. Neue Begründung der Theorie quadratischen Formen von unendlichvielen Veränderlichen. *J. Mat.*, **136** (1909): 210–271.

Hilbert, D. Grundzüge einer allgemeinen Theorie der linearen Integralgleichungen. *Nachr. Akad. Wiss. Göttingen, Math.—Phys.*, KI (1906): 157–227.

Riesz, F. and Sz. Nagy, B. *Leçons d'analyse fonctionelle.* Akadémiai Kiadó, Budapest, 1952.

Stone, M. H. *Linear Transformations in Hilbert Space and their Applications to Analysis.* AMS Colloquium Publications, **15**. American Mathematical Society, New York, 1932.

Sz. Nagy, B. *Spectraldarstellung linearer Transformationen des Hilbertschen Raumes.* Ergebnisse der Math., **5**. Springer, Berlin, 1942.

Wintner, A. *Spectraltheorie der unendlichen Matricen.* Hirzel, Leipzig, 1929.

32

SPECTRAL THEORY OF SELF-ADJOINT OPERATORS

In this chapter we present the spectral theory of unbounded self-adjoint operators.

We start with the observation, due to Hellinger and Toeplitz, that an operator $\mathbf{M}$ that is defined everywhere on a Hilbert space H and is its own adjoint,

$$(\mathbf{M}x, y) = (x, \mathbf{M}y), \tag{1}$$

is necessarily bounded.

We show first that $\mathbf{M}$ is a closed operator. Suppose that x_n is a convergent sequence, $x_n \to x$, and that $\mathbf{M}x_n$ converges to some vector u. Setting $x = x_n$ into (1), we get

$$(\mathbf{M}x_n, y) = (x_n, \mathbf{M}y),$$

and in the limit we get

$$(u, y) = (x, \mathbf{M}y).$$

By (1), the right side equals $(\mathbf{M}x, y)$. Since this holds for all y, $u = \mathbf{M}x$. This proves that $\mathbf{M}$ is closed. But then, according to the closed graph theorem, theorem 12 of chapter 15, $\mathbf{M}$ is bounded.

It follows from this observation that unbounded operators that are their own adjoints can be defined only on a subspace of Hilbert space. Here is the precise definition, due to von Neumann:

Definition. Let H be a complex Hilbert space, D a dense subspace of H, and $\mathbf{A}$ a linear operator defined on D. The adjoint $\mathbf{A}^*$ of $\mathbf{A}$ is the operator whose domain D^* consists of all vectors v in H for which there is a vector denoted as $\mathbf{A}^*v$ in H such that

$$(\mathbf{A}u, v) = (u, \mathbf{A}^*v) \tag{2}$$

377

holds for all u in D. Since D is dense, for any given v there can be only one such vector $\mathbf{A}^*v$. Clearly, D^* is a linear subspace of H, and $\mathbf{A}^*$ is a linear operator on D^*. $\mathbf{A}$ is called *self-adjoint* if $D^* = D$ and $\mathbf{A}^* = \mathbf{A}$.

32.1 SPECTRAL RESOLUTION

The main result to which this chapter is devoted is the spectral resolution of self-adjoint operators:

Theorem 1. *Let* $\mathbf{A}$ *be a self-adjoint operator in a Hilbert space H; denote the domain of* $\mathbf{A}$ *by D. There is a spectral resolution for* $\mathbf{A}$, *that is, orthogonal projection-valued measure* $\mathbf{E}$ *defined for all Borel measurable subsets of* $\mathbf{R}$, *with the following properties:*

(i) $\mathbf{E}(\emptyset) = \mathbf{0}, \mathbf{E}(\mathbf{R}) = \mathbf{I}$.

(ii) *For any pair of measurable sets S and T,* $\mathbf{E}(S \cap T) = \mathbf{E}(S)\mathbf{E}(T)$.

(iii) *For every measurable set S,* $\mathbf{E}^*(S) = \mathbf{E}(S)$.

(iv) $\mathbf{E}$ *commutes with* $\mathbf{A}$, *that is, for any measurable set S, $\mathbf{E}(S)$ maps the domain D of* $\mathbf{A}$ *into D, and for all u in D,* $\mathbf{A}\mathbf{E}(S)u = \mathbf{E}(S)\mathbf{A}u$.

(v) *The domain D of* $\mathbf{A}$ *consists of all vectors u for which*

$$\int t^2 d(\mathbf{E}(t)u, u) < \infty \tag{3}$$

and

$$\mathbf{A}u = \int t\, d\mathbf{E}(t)u. \tag{4}$$

There are a number of proofs known for this important result an extension of theorem 9 of chapter 31. The first one ever was given by von Neumann, and we will sketch it in section 32.2. Yet another approach will be indicated in section 32.3. A proof due to Marshall Stone will be outlined in chapter 34 on semigroups. The beautiful proof that we present here in all its details is due to Doob and Koopman.

According to the Herglotz-Riesz theorem, theorem 6 of chapter 11, every analytic function $f(\zeta)$ in the unit disk, $|\zeta| < 1$ whose real part is positive, can be expressed uniquely as

$$f(\zeta) = ic + \int \frac{e^{i\theta} + \zeta}{e^{i\theta} - \zeta} dm(\theta), \tag{5}$$

where m is a nonnegative measure of finite total mass on the unit circle and c is real. We can change the scene from the unit disk to the upper half-plane and obtain the following variant:

Theorem 2. *Every analytic function $g(z)$ in the upper half-plane whose imaginary part is positive can be expressed uniquely in the form*

$$g(z) = a + mz + \int \frac{1 + tz}{t - z} \, ds(t), \tag{5'}$$

where s is a nonnegative measure of finite total mass on $\mathbb{R}$, a is real and m nonnegative.

Proof. The transformation

$$\zeta \to z = i \frac{1 + \zeta}{1 - \zeta} \tag{6}$$

maps conformally the unit disk onto the upper half-plane with the point $\zeta = 1$ going to ∞. The function $f(\zeta)$ and $g(z)$ described above are related to each other by

$$g(z) = if(\zeta).$$

Using the inverse of the transformation (6),

$$\zeta = \frac{z - i}{z + i}, \tag{6'}$$

and representation (5) for f, we can write

$$
\begin{aligned}
g(z) = if(\zeta) &= -c + i \int \frac{e^{i\theta}(z + i) + (z - i)}{e^{i\theta}(z + i) - (z - i)} \, dm(\theta) \\
&= -c + i \int \frac{e^{i\theta/2}(z + i) + e^{-\theta/2}(z - i)}{e^{i\theta/2}(z + i) - e^{-i\theta/2}(z - i)} \, dm \\
&= -c + i \int \frac{z \cos \theta/2 - \sin \theta/2}{i \cos \theta/2 + iz \sin \theta/2} \, dm \\
&= -c + \int \frac{-z \cotan \theta/2 + 1}{- \cotan \theta/2 - z} \, dm = a + mz + \int \frac{tz + 1}{t - z} \, ds(t).
\end{aligned}
$$

In the last step we renamed the variables: $a = -c$, $t = - \cotan \theta/2$, $ds(t) = dm(\theta)$, and $m = m(0)$, the mass located at $\theta = 0$. $\qquad \square$

The next result is a sharpening of theorem 2:

Theorem 3 (Nevanlinna). *Every analytic function $g(z)$ in the upper half-plane* Im $z > 0$ *whose imaginary part is positive and satisfies the growth condition*

$$\lim_{y \to \infty} \sup \, y|g(iy)| < \infty \tag{7}$$

can be expressed uniquely in the form

$$g(z) = \int \frac{dn}{t - z},$$ (8)

where n is a nonnegative measure of finite total mass on $\mathbb{R}$. *Furthermore*

$$n(\mathbb{R}) = \lim_{y \to \infty} y \operatorname{Im} g(iy).$$ (9)

Proof. Since g has positive imaginary part in the upper half plane, it can be written in the form (5′). Set $z = iy$ and state the boundedness condition (8) for the real and the imaginary part separately:

$$\lim_{y \to \infty} \sup y|a - \int \frac{y^2 - 1}{t^2 + y^2} t \, ds| \leq M,$$ (7a)

and

$$\limsup y^2 \left(m + \int \frac{1 + t^2}{t^2 + y^2} \, ds \right) \leq M.$$ (7b)

The integral on the left in (7b) tends to 0 as y tends to ∞; it follows therefore that the nonnegative quantity m must be zero. Thus (7b) asserts that

$$\lim \sup \int \frac{y^2}{t^2 + y^2} (1 + t^2) \, ds(t) \leq M.$$

Taking the limit as $y \to \infty$ of the integral on the left, we conclude that

$$\int (1 + t^2) \, ds(t) \leq M.$$

We define

$$dn(t) = (1 + t^2) \, ds(t);$$ (10)

it follows that n is a nonnegative measure whose total mass is $\leq M$. It follows from (7a) that

$$a = \lim_{y \to \infty} \int \frac{y^2 - 1}{t^2 + y^2} t \, ds(t) = \int t \, ds(t).$$

Setting this value of a into (5′) and setting $m = 0$, we obtain

$$g(z) = \int \left(\frac{1 + tz}{t - z} + t \right) ds(t) = \int \frac{1 + t^2}{t - z} \, ds(t) = \int \frac{dn(t)}{t - z}.$$

Uniqueness of the measure follows from uniqueness of the measure in (5). Relation (9) follows from (8).   □

Given any measure n on $\mathbb{R}$, real or complex, of finite total mass, formula (8) defines a pair of analytic functions g in the upper and lower half-plane; g is called the *Cauchy transform* of the measure n.

Lemma 4. *The Cauchy transform (8) is one-to-one, i.e. a complex measure n of finite mass is uniquely determined by its Cauchy transform.*

Proof. We have to show that if $g(z)$ defined by (8) is zero for all nonreal z then $n \equiv 0$. Replace z in (8) by $\bar{z}$ and take the complex conjugate. We get

$$\overline{g}(\bar{z}) = 0 = \int \frac{d\bar{n}}{t - z}.$$

This shows that if the Cauchy transform of n is zero, so are the Cauchy transform of its real and imaginary part. The real part of n can be decomposed into its positive and negative parts:

$$\operatorname{Re} n = n_+ - n_-,$$

n_+ and n_- nonnegative measures. It follows that n_+ and n_- have the same Cauchy transforms. Therefore by the uniqueness part of theorem 3, $n_+ \equiv n_-$, and so $\operatorname{Re} n = 0$. Analogously $\operatorname{Im} n = 0$. □

Definition. A number z belongs to the resolvent set of $\mathbf{A}$ iff $\mathbf{A} - z\mathbf{I}$ maps D one-to-one onto H.

Theorem 5. *Let H be a complex Hilbert space, $\mathbf{A}$ a self-adjoint operator acting in H. All nonreal complex numbers z belong to the resolvent set $\mathbf{A}$.*

Proof. We show first that the range of $\mathbf{A} - z\mathbf{I}$ is a closed subspace of H. The range consists of all vectors u of the form

$$\mathbf{A}v - zv = u, \qquad v \text{ in } D.$$

Take the scalar product of both sides with v:

$$(\mathbf{A}v, v) - z(v, v) = (u, v).$$

Since $\mathbf{A}$ is symmetric, $(\mathbf{A}v, v)$ is real, so the imaginary part on the left side is $-\operatorname{Im} z \|v\|^2$. Using the Schwarz inequality on the right we deduce that

$$|\operatorname{Im} z| \|v\|^2 \leq \|u\| \|v\|,$$

which implies that

$$\|v\| \leq \frac{1}{|\operatorname{Im} z|} \|u\|. \tag{11}$$

Let u_n be a sequence of vectors in the range of $\mathbf{A} - z\mathbf{I}$ that converges to some vector u:

$$\mathbf{A}v_n - zv_n = u_n.$$

It follows from inequality (11) that $\|v_n - v_m\| \leq 1/|\mathrm{Im}\, z| \|u_n - u_n\|$. Therefore also the v_n converge to some limit v. We claim that this limit v is in D. To see this we take the limit of the above relation as $n \to \infty$. The right side tends to u, and the second term on the left tends to $-zv$. Therefore the first term $\mathbf{A}v_n$ also tends to a limit, call it r:

$$r - zv = u. \tag{12}$$

Now take the scalar product of $\mathbf{A}v_n$ with any vector w in D, and use the self-adjointness of $\mathbf{A}$:

$$(\mathbf{A}v_n, w) = (v_n, \mathbf{A}w).$$

Take the limit as $n \to \infty$:

$$(r, w) = (v, \mathbf{A}w).$$

By definition of self-adjointness, this shows that v belongs to the domain D of $\mathbf{A}$, and that $\mathbf{A}v = r$. Combined with (12), this shows that u belongs to the range of $\mathbf{A} - z\mathbf{I}$, so the range is closed.

If the range of $\mathbf{A} - z\mathbf{I}$ were not all of H, there would be a nonzero vector k in H orthogonal to it:

$$(\mathbf{A}v - zv, k) = (\mathbf{A}v, k) - (v, \bar{z}k) = 0$$

for all v in D. By definition of self-adjointness, it follows that k belongs to D, and $Ak = \bar{z}k$. But then $(k, Ak) = z(k, k)$ is not real, contrary to the symmetry of $\mathbf{A}$.

This completes the proof that $\mathbf{A} - z\mathbf{I}$ maps D onto H. That it is one-to-one follows as above, for otherwise some k in D would be mapped into 0 by $\mathbf{A} - z\mathbf{I}$, contrary to (11). $\qquad\qquad\qquad\qquad\qquad\qquad\qquad\qquad\qquad\qquad\qquad\qquad\qquad\qquad \Box$

We denote the resolvent of $\mathbf{A}$ by

$$\mathbf{R}(z) = (\mathbf{A} - z\mathbf{I})^{-1}.$$

It follows from (11) that for z nonreal

$$\|\mathbf{R}(z)\| \leq |\mathrm{Im}\, z|^{-1}. \tag{13}$$

Corollary 1. $\mathbf{R}(z)$ *is an analytic function of z on the resolvent set of* $\mathbf{A}$.

Proof. Choose any vector u in H, and denote $\mathbf{R}(z)u$ as $v(z)$. By definition of $\mathbf{R}$,

$$(\mathbf{A} - z)v(z) = u.$$

Similarly

$$(\mathbf{A} - (z + h))v(z + h) = u.$$

Subtract these two equations and divide by h,

$$(\mathbf{A} - z)\frac{v(z + h) - v(z)}{h} = v(z + h),$$

which is the same as

$$\frac{v(z + h) - v(z)}{h} = \mathbf{R}(z)v(z + h) = \mathbf{R}(z)\mathbf{R}(z + h)u.$$

Using the estimate (13), we conclude that v depends Lipschitz continuously on z. Letting $h \to 0$, we deduce that $v(z)$ is differentiable in the complex plane, so $\mathbf{R}(z)$ is holomorphic in the strong topology. $\square$

We claim that the adjoint of $\mathbf{R}(z)$ is $\mathbf{R}(\bar{z})$. To see this, choose any two vectors u and w in H. Denoting $\mathbf{R}(z)u = v$, $(\mathbf{A} - z)v = u$, we have

$$(u, \mathbf{R}(\bar{z})w) = ((\mathbf{A} - z)v, \mathbf{R}(\bar{z})w) = (v, (\mathbf{A} - \bar{z})\mathbf{R}(\bar{z})w) = (v, w) = (\mathbf{R}(z)u, w). \quad \square$$

For any u in H we define the complex valued function $g(z)$ for z nonreal as follows:

$$g(z) = (\mathbf{R}(z)u, u). \tag{14}$$

To indicate its dependence on u, we will write $g(z) = g_u(z)$ when necessary.

Lemma 6. *For any u in H, g defined by (14) has these properties:*

   (i) *g is an analytic function of z in the upper half-plane $\operatorname{Im} z > 0$, and its imaginary part there is nonnegative.*
   (ii) *$y|g_u(iy)| \le \|u\|^2$.*
   (iii) *$\lim_{y \to \infty} y \operatorname{Im} g_u(iy) = \|u\|^2$.*
   (iv) *$g(\bar{z}) = \overline{g(z)}$.*

Proof. The analyticity of g follows the analytic character of $\mathbf{R}(z)$. Denote $\mathbf{R}(z)u$ by v; then

$$u = (\mathbf{A} - z)v = \mathbf{A}v - zv. \tag{15}$$

Take the scalar product of (15) with v; since by (14) $g = (v, u)$, we get

$$g(z) = (v, u) = (v, \mathbf{A}v - zv) = (v, \mathbf{A}v) - \bar{z}(v, v). \tag{16}$$

Since $\mathbf{A}$ is self-adjoint $(v, \mathbf{A}v) = (\mathbf{A}v, v) = \overline{(v, \mathbf{A}v)}$ is real, so we deduce from (16) that

$$\operatorname{Im} g(z) = y(v, v), \qquad y = \operatorname{Im} z. \tag{17}$$

This proves that the imaginary part of g is positive in the upper half plane, as claimed in (i).

(ii) By the Schwarz inequality,

$$|g_u(z)| = |(\mathbf{R}(z)u, u)| \le \|\mathbf{R}(z)u\|\|u\| \le \|\mathbf{R}(z)\|\|u\|^2;$$

using inequality (13) for $\|\mathbf{R}(z)\|$, we obtain (ii).

(iii) Take the scalar product of (15) with u; since by (14) $g_u = (v, u)$, we get

$$\|u\|^2 = (\mathbf{A}v, u) - zg_u(z).$$

Set $z = iy$ and take the real part of this relation:

$$\|u\|^2 = \operatorname{Re}(\mathbf{A}v, u) + y\operatorname{Im} g_u(iy). \tag{18}$$

To complete the proof of (iii), we have to show that the first term on the right in (18) tends to zero as y tends to ∞. To see this, we first estimate this term by the Schwarz inequality:

$$|\operatorname{Re}(\mathbf{A}v, u)\| \le |(\mathbf{A}v, u)| \le \|\mathbf{A}v\|\|u\|.$$

To see that $\|\mathbf{A}v\|$ tends to zero, we write

$$\mathbf{A}v = \mathbf{A}\mathbf{R}(z)u = (\mathbf{I} + z\mathbf{R}(z))u.$$

Setting $z = iy$ and using inequality (13), we get the estimate

$$\|\mathbf{A}\mathbf{R}(iy)\| \le 1 + y\|\mathbf{R}(iy)\| \le 2. \tag{19}$$

This shows that the operators $\mathbf{A}\mathbf{R}(iy)$ are uniformly bounded. Clearly, it suffices to show that $\mathbf{A}\mathbf{R}(iy)u$ tends to zero for a set of u dense in H. D is such a set, for then by (13)

$$\|\mathbf{A}\mathbf{R}(iy)u\| = \|\mathbf{R}(iy)\mathbf{A}u\| \le \|\mathbf{R}(iy)\|\|\mathbf{A}u\| \le \|\mathbf{A}u\|/y.$$

(iv) Since $\mathbf{R}^*(z) = \mathbf{R}(\bar{z})$, we have

$$g(\bar{z}) = (\mathbf{R}(\bar{z})u, u) = (\mathbf{R}^*(z)u, u) = (u, \mathbf{R}(z)u) = \overline{(\mathbf{R}(z)u, u)} = \overline{g(z)}. \qquad \square$$

Lemma 6 shows that for any u in H the function $g(z)$ defined by (14) satisfies the hypotheses of theorem 3. Therefore we conclude that for $\operatorname{Im} z$ positive $g(z)$ can be represented in form (8):

$$(\mathbf{R}(z)u, u) = \int \frac{dn(t)}{t - z}. \tag{20}$$

The nonnegative measure n depends on the vector u, which we indicate as $n = n_u$. It follows from part (iv) that the representation (20) holds in the lower half-plane as well.

We deduce from relations (9) and part (iii) of lemma 6 that

$$n_u(\mathbf{R}) = \|u\|^2. \tag{21}$$

For any pair of vectors u and v we can express $(\mathbf{R}(z)u, v)$ as a linear combination of $(\mathbf{R}(z)(u \pm v), (u \pm iv))$ and $(\mathbf{R}(z)(u \pm iv), (u \pm iv))$. This leads to an integral representation of $(\mathbf{R}(z)u, v)$ as a Cauchy transform:

$$(\mathbf{R}(z)u, v) = \int \frac{1}{t - z} \, dn_{u,v}. \tag{22}$$

The measure $n_{u,v}$ has a simple expression in terms of $n_{u \pm v}$ and $n_{u \pm iv}$; see formula (44) in chapter 31, section 31.7.

The properties of the measures $n_{u,v}$ are summarized in

Lemma 7.

    (i) $n_{u,u} = n_u$.
    (ii) $n_{u,v}$ depends linearly on u, skew-linearly on v.
    (iii) $n_{u,v}$ is a skew-symmetric function of u and v: $n_{v,u=\bar{n}_{u,v}}$.
    (iv) The total variation of $n_{u,v}$ is $\leq \|u\| \|v\|$.

The proof is identical to the one given for lemma 7 in chapter 31, section 31.7; it is based on the simple explicit expression for $n_{u,v}$. $\qquad\qquad\square$

We appeal now to theorem 1 in chapter 31. A bounded, skew symmetric, skew bilinear functional $b(u, v)$ in a Hilbert space H can be represented uniquely as

$$b(u, v) = (\mathbf{E}u, v),$$

where $\mathbf{E}$ is a bounded symmetric operator acting on H. Lemma 7 says that for any set $S, n(S)_{u,v}$ is such a functional; therefore there exist bounded, symmetric operators $\mathbf{E}(S)$ such that

$$n_{u,v}(S) = (\mathbf{E}(S)u, v), \tag{23}$$

Setting this into (21), we obtain

$$(\mathbf{R}(z)u, v) = \int \frac{1}{t - z} d(\mathbf{E}u, v). \tag{24}$$

We claim that the operators $\mathbf{E}$ defined by (23) furnish the spectral resolution for $\mathbf{A}$, meaning that they have the properties proposed in theorem 1 at the beginning of this chapter.

Proof. (i) Since by (23), $n(\emptyset)_{u,v} = 0$ for all u, v, so is $(\mathbf{E}(\emptyset)u, v)$; this makes $\mathbf{E}(\emptyset) = 0$. On the other hand, combining (21) and (23), we get that for all vectors u,

$$(\mathbf{E}(\mathbf{R})u, u) = n_u(\mathbf{R}) = \|u\|^2 = (u, u).$$

It is easy to deduce from this that $\mathbf{E}(\mathbf{R})$ is the identity.

(ii) We show first that the operators $\mathbf{R}(z)$ and $\mathbf{E}$ commute. To see this, we start with the fact that for arbitrary nonreal complex numbers z and w the operators $\mathbf{R}(z)$ and $\mathbf{R}(w)$ commute. Therefore for any pair of vectors u and v in H,

$$(\mathbf{R}(w)\mathbf{R}(z)u, v) = (\mathbf{R}(z)\mathbf{R}(w)u, v). \tag{25}$$

We use the adjoint of $\mathbf{R}(w)$ to rewrite the left side of (25). Employing then the representation (24) with $\mathbf{R}^*(w)v$ in place of v yields

$$(\mathbf{R}(z)u, \mathbf{R}^*(w)v) = \int \frac{1}{t - z} d(\mathbf{E}u, \mathbf{R}^*(w)v) = \int \frac{1}{t - z} d(\mathbf{R}(w)\mathbf{E}u, v). \tag{26}$$

We rewrite the right side of (25) by employing the representation (24) with $\mathbf{R}(w)u$ in place of u. We get

$$\int \frac{1}{t - z} d(\mathbf{E}\mathbf{R}(w)u, v). \tag{26'}$$

The Cauchy transforms (26) and (26') are identical functions of z. Therefore, by lemma 4, the representing measures are identical:

$$(\mathbf{R}(w)\mathbf{E}(S)u, v) = (\mathbf{E}(S)\mathbf{R}(w)u, v)$$

for every measurable set S. Since this holds for arbitrary u and v, it follows that

$$\mathbf{R}(w)\mathbf{E}(S) = \mathbf{E}(S)\mathbf{R}(w)$$

for arbitrary set S and any nonreal complex number w.

To show that $\mathbf{E}(S \cap T) = \mathbf{E}(S)\mathbf{E}(T)$, we use the resolvent identity

$$\mathbf{R}(z)\mathbf{R}(w) = \frac{\mathbf{R}(z) - \mathbf{R}(w)}{z - w}$$

to rewrite (25); using (24) twice for both z and w yields for the right side of (25),

$$\frac{1}{z - w} \int \left(\frac{1}{t - z} - \frac{1}{t - w} \right) d(\mathbf{E}u, v) = \int \left(\frac{1}{t - z} \right) \left(\frac{1}{t - w} \right) d(\mathbf{E}u, v), \tag{27}$$

We compare now (26) and (27); appealing once more to lemma 4, we conclude that the measures appearing in these two formulas are the same. Therefore for any measurable set S,

$$(\mathbf{R}(w)\mathbf{E}(S)u, v) = \int \frac{c_S(t)}{t - w} d(\mathbf{E}u, v), \tag{28}$$

where $c_S(t)$ is the characteristic function of the set S. We use now formula (24), with w in place of z and $E(S)u$ in place of u, to rewrite the left side of (28) as

$$\int \frac{1}{t-w} d(EE(S)u, v). \tag{28'}$$

We compare (28') with the right side of (28); we appeal once more to lemma 4 to conclude that the measures appearing in these formulas are identical: for any measurable set T,

$$(E(T)E(S)u, v) = \int_T c_S(t) d(Eu, v) = (E(S \cap T)u, v).$$

Since this holds for arbitrary vectors u and v, it follows that $E(T)E(S) = E(S \cap T)$, as asserted in (ii).

(iii) The symmetry of E follows from the skew symmetry of $n_{u,v}(S)$.

(iv) We have already shown that E commutes with $R = R(z)$. To show that it commutes with A, let v be any vector in the domain of A; then v can be written in the form $v = Ru$, u in H. Using the identity $AR = I + zR$, we get

$$EAv = EARu = Eu + zERu.$$

Similarly, since E and R commute, we can use the same identity to write

$$AEv = AERu = AR(Eu) = Eu + zREu = Eu + zERu.$$

Comparing the last two identities we conclude that $EAv = AEv$, and that E maps D into D.

(v) Suppose that v belongs to the domain D of A. Then, since $R(z)$ maps H onto D, we can write v as $v = R(z)u$. It is convenient to choose $z = i$. Let T denote any measurable set. Using previously established properties of E and R and the resolvent identity, we can derive the following string of identities:

$$(E(T)v, v) = (E(T)Ru, Ru) = (R^*RE(T)u, u) = (R(-i)R(i)E(T)u, u)$$

$$= \frac{1}{2i}([R(i) - R(-i)]E(T)u, u)$$

$$= \frac{1}{2i} \int \left[\frac{1}{t-i} - \frac{1}{t+i} \right] d(EE(T)u, u)$$

$$= \int \frac{1}{1+t^2} d(EE(T)u, u) = \int_T \frac{1}{1+t^2} d(Eu, u).$$

This proves that

$$d(Ev, v) = \frac{1}{1+t^2} d(Eu, u), \qquad v = R(z)u,$$

and therefore

$$(1 + t^2)d(\mathbf{E}v, v) = d(\mathbf{E}u, u).$$

It follows that as asserted in (3), for v in D,

$$\int t^2 d(\mathbf{E}v, v) < \infty. \tag{3}$$

Next we show that for v in D the Riemann integral

$$\int t d\mathbf{E}v \tag{4}$$

with respect to the vector-valued $\mathbf{E}v$ measure converges. To see this, consider any finite interval S of $\mathbb{R}$, and any decomposition of $S = \cup S_j$ into disjoint subintervals, each of length less than 1. A Riemann sum corresponding to this decomposition is

$$\sum t_j(\mathbf{E}(S_j)v = \sum t_j v_j, \qquad t_j \epsilon S_j, \tag{4'}$$

where v_j abbreviates $\mathbf{E}(S_j)v$. It follows from property (ii) that the vectors v_j are pairwise orthogonal, so the norm square of the Riemann sum (4') is

$$\sum t_j^2 \|v_j\|^2.$$

Since this is a Riemann sum for the integral (3), it is bounded uniformly for all decompositions. The convergence of the integral (4) follows.

To determine the value of this integral, we take any vector w in H. Let T be any measurable set. Using previously established properties of $\mathbf{E}$ and $\mathbf{R}$, we can derive the following string of identities:

$$(\mathbf{E}(T)v, w) = (\mathbf{E}(T)\mathbf{R}u, w) = (\mathbf{R}\mathbf{E}(T)u, w)$$
$$= \int \frac{1}{t - z} d(\mathbf{E}\mathbf{E}(T)u, w) = \int_T \frac{1}{t - z} d(\mathbf{E}u, w).$$

This proves that

$$d(\mathbf{E}v, w) = \frac{1}{t - z}d(\mathbf{E}u, w),$$

and therefore

$$(t - z)d(\mathbf{E}v, w) = d(\mathbf{E}u, w).$$

Setting $u = (\mathbf{A} - z)v$ on the right above gives

$$t d(\mathbf{E}v, w) = d(\mathbf{E}\mathbf{A}v, w).$$

Integrating this relation over $\mathbb{R}$, we deduce that

$$\int t\, d(\mathbf{E}v, w) = (\mathbf{A}v, w).$$

The left side is scalar product of the integral (4) with w; since w is arbitrary, it follows that

$$\int t\, d\mathbf{E}v = \mathbf{A}v.$$

This completes the proof of theorem 1. $\qquad\qquad\qquad\qquad\qquad\qquad\square$

Corollary. *If x belongs to the domain of $\mathbf{A}^n$, then for every v in H,*

$$(\mathbf{A}^k x, v) = \int t^k d(\mathbf{E}x, v), \qquad k \le m. \tag{29}$$

Exercise 1. Prove relation (29).

We sketch now two other approaches to constructing a spectral resolution of a self-adjoint operator.

32.2 SPECTRAL RESOLUTION USING THE CAYLEY TRANSFORM

We present now von Neumann's original approach employing the Cayley transform of $\mathbf{A}$:

$$\mathbf{U} = (\mathbf{A} - i)(\mathbf{A} + i)^{-1}. \tag{30}$$

We urge the reader to recall from chapter 31, section 31.7, the notion of a unitary operator.

Theorem 8. *The operator $\mathbf{U}$ is unitary, that is, a norm preserving mapping of H onto H.*

Proof. Since the operators $\mathbf{A} \pm i$ map $D(\mathbf{A})$ one-to-one onto H, $\mathbf{U}$ maps H onto itself. We claim that $\mathbf{U}$ is norm preserving. To see this, let u be any vector in H; denote by v and w the vectors

$$v = (\mathbf{A} + i)^{-1} u, \qquad w = \mathbf{U}u.$$

Then

$$(\mathbf{A} + i)v = u, \qquad (\mathbf{A} - i)v = w.$$

Taking scalar products and using the symmetry of **A**, we get

$$\|u\|^2 = ((\mathbf{A}+i)v, (\mathbf{A}+i)v) = \|\mathbf{A}v\|^2 + \|v\|^2 + i[(v, \mathbf{A}v) - (\mathbf{A}v, v)] = \|\mathbf{A}v\|^2 + \|v\|^2,$$

and similarly

$$\|w\|^2 = ((\mathbf{A} - i)v, (\mathbf{A} - i)v) = \|\mathbf{A}v\|^2 + \|v\|^2.$$

This proves that **U** is norm preserving.   □

Then von Neumann appeals to the spectral resolution of the unitary operator **U** in terms of a projection-valued measure on the unit circle; pulled back to the real axis by $(t - i)/(t + i) = e^{i\theta}$, this furnishes the spectral resolution of **A**.

32.3 A FUNCTIONAL CALCULUS FOR SELF-ADJOINT OPERATORS

The resolvent set of **A** consists of all numbers z for which $\mathbf{A} - z\mathbf{I}$ maps D one-to-one onto H. The spectrum of **A** is the complement of the resolvent set. According to theorem 5, the spectrum of **A** lies on the real axis.

Exercise 1. Shows that the spectrum of an unbounded self-adjoint operator is a closed, unbounded set on the real axis.

The *extended spectrum* of an unbounded self-adjoint operator is its spectrum compactified by adjoining ∞.

In this section we will define $f(\mathbf{A})$ for every f that is continuous and real valued on the extended spectrum of the self-adjoint operator **A**.

With start with a few observations about powers of a self-adjoint operator. For any natural number k, the domain of $\mathbf{A}^k$, denoted as $D(\mathbf{A}^k)$, consists of all vectors x for which $x, \mathbf{A}x, \ldots, \mathbf{A}^{k-1}x$ lie in D, the domain of the operator **A**.

Exercise 2. Show that $\mathbf{A}^k$ is a self-adjoint operator.

Lemma 9. *For any self-adjoint operator* **A** *acting on a Hilbert space H,*

(i) $\mathbf{A}^2 + \mathbf{I}$ *has a bounded inverse.*

(ii) $(\mathbf{A}^2 + \mathbf{I})^{-n}$ *maps H into the domain of* $\mathbf{A}^{2n}$.

Proof. (i) according to theorem 5, $\mathbf{A} + i\mathbf{I}$ and $\mathbf{A} - i\mathbf{I}$ both have bounded inverses. Therefore so does their product

$$\mathbf{A}^2 + \mathbf{I} = (\mathbf{A} + i\mathbf{I})(\mathbf{A} - i\mathbf{I}).$$

(ii) We show first that both $(\mathbf{A} + i\mathbf{I})^{-1}$ and $(\mathbf{A} - i\mathbf{I})^{-1}$ map $D(\mathbf{A}^{k-1})$ into $D(\mathbf{A}^k)$. To see this, let y denote any vector in $D(\mathbf{A}^{k-1})$; by theorem 5, $(\mathbf{A} + i\mathbf{I})^{-1}y = x$

belongs to D. For $k = 1$, this is what we want to know; for $k > 1$ we write $y = (A + iI)x$ and rearrange it as

$$Ax = y - ix, \tag{31}$$

which shows that x belongs to $D(A^2)$. For $k = 2$, this is what we want to know; for $k > 2$, let A act on (31), and use (31) on the right:

$$A^2x = Ay - iAx = Ay - iy - x.$$

This shows that x belongs to $D(A^3)$, and so on, until we place x in $D(A^k)$. Ditto for $(A - iI)^{-1}$.

We write now

$$(A^2 + I)^{-n} = [(A - iI)^{-1}(A + iI)^{-1}]^n. \tag{32}$$

Since the k factor $(A \pm iI)^{-1}$ maps $D(A^{k-1})$ into $D(A^k)$, the $2n$ factors on the right in (32) map H into $D(A^n)$. $\qquad\square$

Exercise 3. Show that $(A^2 + I)^{-n}$ maps H onto the domain of A^{2n}.

Let $q(\lambda)$ be any polynomial of degree $\leq 2n$. It follows from lemma 9 that $q(A)(A^2 + I)^{-n}$ is an everywhere defined, bounded operator.

Exercise 4. Show that if the coefficients of q are real, $q(A)(A^2 + I)^{-n}$ is symmetric. (Hint: First show that the domain of A^{2n} is dense in H.)

The following version of the spectral mapping theorem holds:

Lemma 10. *Let q denote a polynomial with real coefficients of degree $\leq 2n$ and abbreviate $q(\lambda)(\lambda^2+1)^{-1}$ as $r(\lambda)$. The spectrum of $r(A)$ consists of all real numbers σ of form $\sigma = r(\lambda)$, λ in the extended spectrum of A.*

Proof. Write

$$r(\lambda) - \sigma = [q(\lambda) - \sigma(\lambda^2 + 1)^n](\lambda^2 + 1)^{-n}.$$

The numerator has real and complex zeros, in conjugate pairs. Factor it as

$$r(\lambda) - \sigma = \prod_1^k \left((\lambda - \rho_j)^2 + \mu_j^2\right) \prod_1^{2(n-k)} (\lambda - \lambda_\ell)(\lambda^2 + 1)^{-n}. \tag{33}$$

Then

$$r(A) - \sigma I = \prod_1^k \left((A - \rho_j I)^2 + \mu_j^2 I\right) \prod_1^{2(n-k)} (A - \lambda_\ell I)(A^2 + I)^{-n}.$$

The partial product

$$\prod_1^k (\mathbf{A} - \rho_j \mathbf{I})^2 + \mu_j^2 \mathbf{I})(\mathbf{A}^2 + \mathbf{I})^{-k}$$

is a one-to-one map of H onto H. The remaining factor is a one-to-one map of H onto H iff all the λ_ℓ belong to the resolvent set of $\mathbf{A}$, and if the total number of zeros of the numerator is $2n$. Since it follows from (33) that $r(\lambda_\ell) = \sigma$, the two conditions above for the invertibility of $r(\mathbf{A}) - \sigma I$ can be stated so: $r(\lambda) \neq \sigma$ for any real λ in the spectrum of $\mathbf{A}$, and $r(\infty) \neq \sigma$. $\square$

According to lemma 9 and exercise 4, for the rational functions $r(\lambda)$ considered above, $r(\mathbf{A})$ is a bounded, symmetric operator. We appeal now to theorem 3 of chapter 31: $\|r(\mathbf{A})\|$ equals the spectral radius of $r(\mathbf{A})$. According to lemma 10 the spectrum of $r(\mathbf{A})$ is the range of $r(\lambda)$ on the extended spectrum $\sigma(\mathbf{A})$ of A. Therefore

$$\|r(\mathbf{A})\| = \max_{\lambda \text{ in } \sigma(\mathbf{A})} |r(\lambda)|. \tag{34}$$

The rational functions r described above form an algebra over the reals. We claim that they separate points of the extended real line. Clearly, if λ_1 and λ_2 have the same sign, $(\lambda^2 + 1)^{-1}$ has different values at λ_1 and λ_2; ditto if λ_1 or λ_2 is ∞. If λ_1 and λ_2 have opposite signs, $\lambda(\lambda^2 + 1)^{-1}$ separates them. Furthermore the constant function belongs to the algebra. We appeal now to the

Stone-Weierstrass Theorem. *An algebra of real valued functions on a compact Hausdorff space that separates points, and contains the constant functions, is dense in the space of all continuous functions, normed by the maximum norm.*

For a proof see section 3 of chapter 13.

It follows that the rational functions $r(\lambda)$ defined above are dense in the space of all continuous functions f on the extended spectrum of $\mathbf{A}$ compactified by the addition of the point ∞. That is, every such continuous function can be approximated uniformly by a sequence of rational functions:

$$\lim r_k = f.$$

It follows that $\{r_k\}$ is a Cauchy sequence:

$$\max_{\lambda \text{ in } \sigma(\mathbf{A})} |r_k(\lambda) - r_\ell(\lambda)| \to 0.$$

We apply now (34) and conclude that $\|r_k(\mathbf{A}) - r_\ell(\mathbf{A})\|$ tends to zero as k, ℓ tend to ∞. The norm limit of the sequence of operator $r_k(\mathbf{A})$ is defined as $f(\mathbf{A})$.

It is easy to verify that the functional calculus we have just defined has all the properties listed in theorem 5 of chapter 31. This functional calculus can be used, just as it was done in chapter 31, section 31.3, to construct a spectral resolution of the operator **A**. A spectral representation can be built along the lines of chapter 31, section 31.5.

NOTES. The first unbounded operators for which a spectral theory has been developed were differential operators, ordinary and partial. These will be discussed, lightly, in chapter 33. The first general spectral theory for unbounded integral operators was developed by Carleman, in the context of singular integral operators.

   In the bibliography we list the early contributions to the spectral theory of self-adjoint operators.

BIBLIOGRAPHY

Carleman, T. *Sur les equations integrales singulières à noyou réel symmétrique.* Almquist and Wiksells, Uppsala, 1923.

Doob, J. L. and Koopman, B. O. On analytic functions with positive imaginary parts. *Bull. AMS,* **40** (1934): 601–606.

Hellinger, E. and Toeplitz, O. Grundlagen einer Theorie der Unendlichen Matricen. *Math. Ann.,* **69** (1910): 289–330.

Lengyel, B. A. and Stone, M. H. Elementary proof of the spectral theorem. *An. Math.,* **37** (1936): 853–864.

Lorch, E. R. Functions of self-adjoint transformations in Hilber space. *Acta. Sc. Math. Szeged,* **7** (1934): 136–146.

Nevalinna, R. Asymptotische Entwickelungen beschränkter Funktionen und das Stieltjessche Moment-problem. *Ann. Acad. Sci. Fennicae,* A **18** (1922).

Riesz, F. Über die linearen Transformationen des komplexen Hilbertschen Raumes, *Acta Sci. Math. Szeged,* **5** (1930): 23–54.

Riesz, F. and Lorch, E. R. The integral representation of unbounded self-adjoint transformations in Hilbert space. *Trans. AMS,* **39** (1936): 331–340.

Stone, M. H. *Linear transformations in Hilbert space, and their application to analysis.* AMS Coll. Publ., **15**. American Mathemaical Society, New York, 1932.

Sz.-Nagy, B. *Spectraldarstelling linearer Transformationen des Hilbertschen Raumes.* Ergebnisse der Math., **5**. Springer, Berlin, 1942.

von Neumann, J. Allgemeine Eigenwerttheorie Hermitescher Functionaloperatoren. *Math. An.,* **102** (1929): 49–131.

33

EXAMPLES OF SELF-ADJOINT OPERATORS

The definition of self-adjointness demands that the domain of such an operator be specified with the greatest precision. This is possible in some cases, but for most partial differential operators with variable coefficients defined on domains and subject to various boundary conditions, it is not possible—and not useful—to give an exact description of their domain. Instead, such operators are defined by some suitable process of extension. In the first part of this chapter we describe such processes.

33.1 THE EXTENSION OF UNBOUNDED SYMMETRIC OPERATORS

Definition. An operator C is called an *extension* of operator B if the domain of C contains the domain of B and $Cu = Bu$ on their common domain.

We are given a linear operator B mapping a dense subspace $D(B)$ of a Hilbert space H into H that is symmetric:

$$(Bu, v) = (u, Bv) \tag{1}$$

for all u and v in $D(B)$. We pose the following questions:

   (i) Is it possible to extend B to a self-adjoint operator?

   (ii) In how many ways?

   (iii) By what process?

We recall the notion of a *closed operator* as one whose graph in $H \times H$ is closed. We spell it out:

Definition. An operator C mapping a dense subspace $D(C)$ of H into H is *closed* if for every sequence $\{u_n\}$ in $D(C)$ that converges to a limit u in H and for which $\{Cu_n\}$ converges to a limit w in H, u belongs to the domain of C and $Cu = w$.

We describe now how to extend minimally any densely defined symmetric operator **B** to a closed symmetric operator. This minimal extension is called the closure of **B** and is denoted as $\overline{\mathbf{B}}$. Take any sequence $\{u_n\}$ of vectors in $D(\mathbf{B})$ that converges to a limit u, and for which $\{\mathbf{B}u_n\}$ converges to a limit w. Set $u = u_n$ in (1):

$$(\mathbf{B}u_n, v) = (u_n, \mathbf{B}v),$$

and let $n \to \infty$. We obtain that

$$(w, v) = (u, \mathbf{B}v) \qquad (1')$$

for all v in $D(\mathbf{B})$. Since $D(\mathbf{B})$ is dense in H, it follows from (1') that the vector w is uniquely determined by the vector u.

Definition. The *closure* $\overline{\mathbf{B}}$ of **B** is defined by setting $\overline{\mathbf{B}}u = w$ for all u, w satisfying (1') for all v in $D(\mathbf{B})$.

Exercise 1. Show that if a closed operator **C** is an extension of a densely defined symmetric operator **B**, then **C** is an extension of $\overline{\mathbf{B}}$ as well.

Theorem 1. *Let **B** be a densely defined symmetric operator, $\overline{\mathbf{B}}$ its closure.*

(i) $\overline{\mathbf{B}}$ *is closed.*

(ii) $\overline{\mathbf{B}}$ *is symmetric.*

(iii) *For any nonreal complex number z, $\overline{\mathbf{B}} - z$ maps $D(\overline{\mathbf{B}})$ one-to-one onto a closed subspace of H.*

Proof. (i) It follows by an easy, tedious argument from the construction of $\overline{\mathbf{B}}$ that $\overline{\mathbf{B}}$ is a closed operator.

(ii) To show that $\overline{\mathbf{B}}$ is symmetric, take any v in $D(\overline{\mathbf{B}})$ and any sequence $\{v_n\}$ in $D(\mathbf{B})$ that converges to v, and for which $\mathbf{B}v_n$ converges to $\overline{\mathbf{B}}v$. Set $v = v_n$ in (1'), and let $n \to \infty$. We obtain

$$\left(\overline{\mathbf{B}}u, v\right) = \left(u, \overline{\mathbf{B}}v\right),$$

the symmetry of $\overline{\mathbf{B}}$.

(iii) Let z be any complex number, u any vector in the domain of $\overline{\mathbf{B}}$, and denote by f the vector

$$\left(\overline{\mathbf{B}} - z\right)u = f.$$

Take the scalar product with u:

$$\left(\overline{\mathbf{B}}u, u\right) - z(u, u) = (f, u).$$

Since $\overline{\mathbf{B}}$ is symmetric, the first term on the left is real; since the imaginary parts of the two sides are equal,

$$|\operatorname{Im} z| \, \|u\|^2 = |\operatorname{Im}(f, u)| \le \|f\| \, \|u\|.$$

It follows from this inequality that

$$\|u\| \le \frac{1}{|\operatorname{Im} z|} \|f\|; \tag{2}$$

this implies that the operator $\mathbf{C} - z$ is one-to-one.

Let $\{f_n\}$ be a sequence of vectors in the range of $\overline{\mathbf{B}} - z$ that converges to f in H:

$$(\overline{\mathbf{B}} - z) \, u_n = f_n. \tag{2'}$$

It follows from (2) that $\{u_n\}$ converges to some limit u in H. But then, by (2'), so does $\overline{\mathbf{B}}u_n$. Since $\overline{\mathbf{B}}$ is a closed operator, u belongs to the domain of $\overline{\mathbf{B}}$ and $\overline{\mathbf{B}}u = f + zu$. □

Here are a few useful corollaries:

Corollary 1. *A self-adjoint operator* $\mathbf{A}$ *is closed.*

Proof. Since by part (ii) of theorem 1, $\overline{\mathbf{A}}$ is symmetric, the domain of $\overline{\mathbf{A}}$ is contained in the domain of $\mathbf{A}^*$. Since $\mathbf{A}^*$ has the same domain as $\mathbf{A}$, the corollary follows. □

Combining corollary 1 with Exercise 1, we deduce

Corollary 1'. *When a self-adjoint operator* $\mathbf{A}$ *is an extension of a densely defined symmetric operator* $\mathbf{B}$, *then* $\mathbf{A}$ *is an extension of* $\overline{\mathbf{B}}$ *as well.*

The following result is basic for identifying self-adjoint operators.

Theorem 2. *A symmetric operator* $\mathbf{A}$ *is self-adjoint iff all nonreal complex numbers* z *belong to its resolvent set.*

Proof. That every nonreal complex z belong to the resolvent set of a self-adjoint operator has been demonstrated in theorem 5 of chapter 32. To show the converse, we take the case that for some z nonreal, both z and $\bar{z}$ belong to the resolvent set of $\mathbf{A}$. We show first that $(\mathbf{A} - z)^{-1}$ is the adjoint of $(\mathbf{A} - \bar{z})^{-1}$. The meaning of this is that for any pair of vector f and g in H,

$$\left((\mathbf{A} - z)^{-1} f, g\right) = \left(f, (\mathbf{A} - \bar{z})^{-1} g\right). \tag{3}$$

To see why this is so, we abbreviate

$$(\mathbf{A} - z)^{-1} f = x, \quad (\mathbf{A} - \bar{z})^{-1} g = y; \tag{4}$$

then we can rewrite (3) as

$$(x, (\mathbf{A} - \bar{z})y) = ((\mathbf{A} - z) x, y). \tag{4'}$$

Since $\mathbf{A}$ is symmetric, this is valid for all x and y in the domain of $\mathbf{A}$. Since $\mathbf{A} - z$ and $\mathbf{A} - \bar{z}$ map $D(\mathbf{A})$ onto H, it follows that (3) holds for all f and g in H.

We are now ready to prove that $\mathbf{A}$ is self-adjoint. What we have to show is that if v belongs to the domain of $\mathbf{A}^*$, then v belongs to the domain of $\mathbf{A}$ and $\mathbf{A}^* v = \mathbf{A}v$. Now v belongs to the domain of $\mathbf{A}^*$, with $\mathbf{A}^* v = w$, if for all x in $D(\mathbf{A})$,

$$(\mathbf{A}x, v) = (x, w). \tag{5}$$

Subtracting $z(x, v)$ from both sides yields

$$((\mathbf{A} - z) x, v) = (x, w - \bar{z}v).$$

Using the abbreviation (4) and relation (3), with $g = w - \bar{z}v$, we can rewrite this as

$$(f, v) = \left((\mathbf{A} - z)^{-1} f, w - \bar{z}v\right) = \left(f, (\mathbf{A} - \bar{z})^{-1}(w - \bar{z}v)\right). \tag{5'}$$

Since (5) holds for all x in $D(\mathbf{A})$, (5)$'$ holds for all f in H; it follows therefore that

$$v = (\mathbf{A} - \bar{z})^{-1} (w - \bar{z}v).$$

Since the range of $(\mathbf{A} - \bar{z})^{-1}$ is $D(\mathbf{A})$, it follows that v belongs to $D(\mathbf{A})$: acting on both sides by $\mathbf{A} - \bar{z}$ shows that $\mathbf{A}v = w$. Since w was defined above as $\mathbf{A}^* v$, $\mathbf{A}^* v = \mathbf{A}v$. ☐

NOTE. In proving the converse, we used only the assumption that some nonreal z and $\bar{z}$ belong to the resolvent set of $\mathbf{A}$.

33.2 EXAMPLES OF THE EXTENSION OF SYMMETRIC OPERATORS; DEFICIENCY INDICES

We turn now to some examples that illustrate the notion of closure of a symmetric operator, and the possibilities of self-adjoint extensions.

Example 1.

Definition. Denote by H the Hilbert space $L^2(\mathbb{R})$, and define the operator $\mathbf{B}$ as $i(d/dx)$ acting on the domain $D(\mathbf{B}) = C_0^1$ consisting of all once differentiable functions on $\mathbb{R}$ with compact support.

Proposition. **B** *is symmetric, and its closure* **B** *is self-adjoint.*

Proof. Integration by parts shows that **B** is symmetric. Let z be any complex number. The range of **B** $-\, z$ consists of all functions f of form

$$i\frac{d}{dx}u - zu = f, \qquad u \in C_0^1. \tag{6}$$

Multiply by e^{izx}:

$$i\frac{d}{dx}\left(e^{izx}u\right) = e^{izx}f. \tag{6'}$$

Integrate over $\mathbb{R}$; since u has compact support, we get

$$0 = \int_{-\infty}^{\infty} e^{izx}f\,dx, \tag{7}$$

a condition satisfied by every function f in the range of **B** $-\, z$. Conversely, every C_o function f on $\mathbb{R}$ that satisfies (7) belongs to the range of **B** $-\, z$. To see this, define u by

$$u(x) = -i \int_{-\infty}^{x} e^{iz(y-x)}f(y)\,dy. \tag{8}$$

Clearly, u has continuous derivative, and if f is zero outside a compact interval S, it follows from (7) and (8) that u too is zero outside S.

The function e^{izx} is not square integrable on $\mathbb{R}$; therefore the set of continuous functions f of compact support that satisfy condition (7) is a *dense* subset of $L^2(\mathbb{R})$. According to part (iii) of theorem 1, for z nonreal the range of $\overline{\textbf{B}} - z$ is closed; so the range is all of H. Since $\overline{\textbf{B}} - z$ is one-to-one, z belongs to the resolvent set of $\overline{\textbf{B}}$. According to theorem 2, it follows that $\overline{\textbf{B}}$ is self-adjoint.

A symmetric operator whose closure is self-adjoint is called *essentially self-adjoint.*   □

Example 2.

Definition. Denote by H the Hilbert space $L^2(\mathbb{R}_+)$, and define the operator **B** as $i(d/dx)$ acting on the domain $D(\textbf{B}) = C_0^1$ consisting of all once differentiable functions whose support is a compact subset of $(0, \infty)$.

Proposition. **B** *is symmetric, but its closure* $\overline{\textbf{B}}$ *is not self-adjoint. Furthermore* **B** *has no self-adjoint extension whatsoever.*

Proof. Symmetry of **B** follows by integration by parts, since all functions in the domain of **B** are zero near 0 and ∞. Arguing as in example 1, we conclude that a continuous function of compact support on $\mathbb{R}_+$ belongs to the range of $\mathbf{B} - z$ iff

$$0 = \int_0^\infty e^{izx} f \, dx. \tag{9}$$

For $\operatorname{Im} z < 0$, the function e^{izx} is not square integrable on $\mathbb{R}_+$; therefore the set of C_0 functions f satisfying (9) is dense in $L^2(\mathbb{R}_+) = H$. It follows from theorem 2 that the range of $\overline{\mathbf{B}} - z$ is all of H.

It is otherwise when $\operatorname{Im} z > 0$, for then the function e^{izx} is square integrable on $\mathbb{R}_+$, and therefore the range of $\overline{\mathbf{B}} - z$ consists of all f in H that satisfy the orthogonality condition (9). It follows that $\overline{\mathbf{B}}$ is not self-adjoint.

To see that $\mathbf{B}$ has no self-adjoint extension, we note that according to corollary 1, such an extension $\mathbf{A}$ would have to be an extension of $\overline{\mathbf{B}}$ as well. Let v be a function in the domain of $\mathbf{A}$ that does not belong to the domain of $\overline{\mathbf{B}}$. Choose any z with $\operatorname{Im} z < 0$; since $\overline{\mathbf{B}} - z$ maps $D(\overline{\mathbf{B}})$ onto H, there is function u in $D(\overline{\mathbf{B}})$ such that

$$\left(\overline{\mathbf{B}} - z\right) u = (\mathbf{A} - z)v.$$

Since $\mathbf{A}$ is an extension of $\overline{\mathbf{B}}$,

$$(\mathbf{A} - z)(v - u) = 0.$$

This is impossible unless $v - u = 0$, for $\mathbf{A}$ is symmetric, and so, according to theorem 1, $\mathbf{A} - z$ has no nonzero nullvector for complex z. □

Example 3.

Definition. Take $H = L^2(0, 1)$ $\mathbf{B} = i(d/dx)$ acting on functions u in $D(\mathbf{B}) = C_0^1$ consisting of continuously differentiable functions on $[0,1]$ that vanish at $x = 0$ and 1.

Proposition. $\mathbf{B}$ *is symmetric, but its closure* $\overline{\mathbf{B}}$ *is not self-adjoint. However,* $\mathbf{B}$ *has self-adjoint extensions.*

Proof. Symmetry of $\mathbf{B}$ follows by integration by parts. Arguing as before, we see that the range of $\overline{\mathbf{B}} - z$, $\operatorname{Im} z \neq 0$, consists of all L^2 functions f that satisfy the orthogonality condition

$$\int_0^1 e^{izx} f \, dx = 0.$$

According to theorem 2, $\overline{\mathbf{B}}$ is not self-adjoint.

We construct now some self-adjoint extensions of $\mathbf{B}$.

Definition. Let α be any complex number $\neq 1$ but of absolute value $1 : |\alpha| = 1$. Define $\mathbf{A}_\alpha$ to be the operator $i(d/dx)$ acting on all C^1 functions that satisfy the boundary condition

$$u(1) = \alpha u(0).$$

Clearly, A_α is an extension of B.

Exercise 2. Show that A_α is symmetric.

We show now that the closure of A_α is self-adjoint. Take any complex z, $\operatorname{Im} z \neq 0$; we claim that every continuous function f belongs to the range of $A_\alpha - z$. To show this, we integrate $(6')$ from 0 to x. Denoting by c the value of u at zero, we get

$$u(x) = c - i \int_0^x e^{z(y-x)} f(y)\,dy.$$

In particular,

$$u(1) = c - i \int_0^1 e^{iz(y-x)} f(y)\,dy.$$

Setting $u(1) = \alpha c$ gives an equation for c that, for $\alpha \neq 1$, has a unique solution.

Since the continuous functions are dense in L^2, it follows that the range of $A_\alpha - z$ is dense. Therefore the range of $\overline{A}_\alpha - z$ is all of H. By theorem 2, $\overline{A}_\alpha$ is self-adjoint.
$\qquad\qquad\qquad\qquad\qquad\qquad\qquad\qquad\qquad\qquad\qquad\qquad\qquad\qquad\qquad\square$

These examples illustrate well the extension problem for symmetric operators. The general result is due to von Neumann.

Theorem 3. *Let C be a densely defined, closed symmetric operator in a Hilbert space H. According to theorem 1, for $\operatorname{Im} z \neq 0$ the range of $C - z$ is a closed subspace of H.*

    (i) The codimension of the range of $C - z$ is the same for all z with $\operatorname{Im} z > 0$. Similarly the codimensions are the same for all z with $\operatorname{Im} z < 0$. These codimensions, denoted as n_+ and n_-, are called the deficiency indices of the operator C.

    (ii) C has a self-adjoint extension iff $n_+ = n_-$.

Proof. We sketch the proof of (ii). Form the Cayley transform of C:

$$V = (C - i)(C + i)^{-1}. \tag{10}$$

V maps the range of $C+i$ onto the range of $C-i$. As shown in the proof of theorem 8 in chapter 32, V is an isometry. Clearly, an isometry can be extended to a unitary operator U iff the codimension of its domain and range are equal. Suppose that $n_+ = n_-$; then V can be so extended. Call this extension U. We claim that the inverse

Cayley transform of $\mathbf{U}$,

$$\mathbf{A} = i(\mathbf{I} + \mathbf{U})(\mathbf{I} - \mathbf{U})^{-1}, \tag{10'}$$

defined on the range of $\mathbf{I} - \mathbf{U}$, is a self-adjoint extension of $\mathbf{C}$. First we have to show that $\mathbf{U}$ has a Cayley inverse, namely that $\mathbf{I} - \mathbf{U}$ has no nullvector. To see that $\mathbf{I} - \mathbf{U}$ annihilates only the zero vector, suppose that $(\mathbf{I} - \mathbf{U})n = 0$. By adjointness, for any vector y,

$$0 = ((\mathbf{I} - \mathbf{U})n, y) = (n, (\mathbf{I} - \mathbf{U}^*)y);$$

it follows that n is orthogonal to the range of $\mathbf{I} - \mathbf{U}^*$. Since $\mathbf{U}$ is unitary, this equals the range of $(\mathbf{I} - \mathbf{U}^*)\mathbf{U} = \mathbf{U} - \mathbf{U}^*\mathbf{U} = \mathbf{U} - \mathbf{I}$.

From formula (10) we get that $\mathbf{V} - \mathbf{I} = -2i(\mathbf{C} + i\mathbf{I})^{-1}$. According to theorem 1, the range of $(\mathbf{C} + i\mathbf{I})^{-1}$ is the domain of $\mathbf{C}$, a dense subspace of H. Since $\mathbf{U}$ is an extension of $\mathbf{V}$, the range of $\mathbf{I} - \mathbf{U}$ contains the range of $\mathbf{I} - \mathbf{V}$, and so it too is dense in H. This proves $n = 0$, and so $\mathbf{I} - \mathbf{U}$ is invertible. Since according to (10') the domain of $\mathbf{A}$ is the range of $\mathbf{I} - \mathbf{A}$, it follows that $\mathbf{A}$ is densely defined.

Next we show that $\mathbf{A}$ is symmetric. Let u and v be a pair of vectors in the domain of $\mathbf{A}$. By definition (10') of $\mathbf{A}$,

$$(\mathbf{A}u, v) = i\left((\mathbf{I} + \mathbf{U})(\mathbf{I} - \mathbf{U})^{-1}u, v\right).$$

By adjointness, this equals

$$i\left(u, (\mathbf{I} - \mathbf{U}^*)^{-1}(\mathbf{I} + \mathbf{U}^*)v\right).$$

Since $\mathbf{U}$ is unitary, $\mathbf{U}^* = \mathbf{U}^{-1}$; so the above can be rewritten as

$$i\left(u, (\mathbf{I} - \mathbf{U}^{-1})^{-1}(\mathbf{I} + \mathbf{U}^{-1})v\right) = i\left(u, (\mathbf{U} - \mathbf{I})^{-1}\mathbf{U}(\mathbf{I} + \mathbf{U}^{-1})v\right)$$

$$= i\left(u, (\mathbf{U} - \mathbf{I})^{-1}(\mathbf{U} + \mathbf{I})v\right) = (u, \mathbf{A}v).$$

In the last step we used the fact that in formula (10') defining $\mathbf{A}$, the two factors commute.

To show that $\mathbf{A}$ is not only symmetric but self-adjoint we use theorem 2 and verify that every nonreal complex number z belongs to the resolvent set of $\mathbf{A}$. Using definition (10') of $\mathbf{A}$, we write

$$\mathbf{A} - z\mathbf{I} = i\left(\mathbf{I} + \mathbf{U} + iz(\mathbf{I} - \mathbf{U})\right)(\mathbf{I} - \mathbf{U})^{-1} = i\left((1 + iz)\mathbf{I} + (1 - iz)\mathbf{U}\right)(\mathbf{I} - \mathbf{U})^{-1}.$$

We saw in chapter 31, section 31.7, that the spectrum of a unitary operator lies on the unit circle. Since for nonreal z, $(1 + iz)/(1 - iz)$ does not lie on the unit circle, the first factor on the right above maps H onto H; the second factor maps the domain of $\mathbf{A}$ onto H. This shows that $\mathbf{A} - z\mathbf{I}$ maps the domain of $\mathbf{A}$ one-to-one onto H. $\square$

REMARK. The deficiency indices are nonnegative integers or ∞. We assume that H is separable, so there is only one kind of ∞.

Exercise 3. Prove part (i) of theorem 3.

Exercise 4. What are the deficiency indices of Examples 1, 2, and 3?

Theorem 3 has this important

Corollary. *Let K be a Hilbert space over the reals, and* **B** *a densely defined symmetric operator on K. Such a* **B** *has a self-adjoint extension to the complexification of K.*

Proof. The complexification of K is $H = K + iK$; there is a natural extension of **B** to H. There is a natural complex conjugation in H: $\overline{u + iv} = u - iv$, u and v in K. Conjugation commutes with the action of **B**.

Denote by **C** the closure of **B**; **C** too commutes with complex conjugation. It follows that the range of $\mathbf{C} - z\mathbf{I}$ is the complex conjugate of the range of $\mathbf{C} - \bar{z}\mathbf{I}$. It follows that the codimension of the range of $\mathbf{C} - z\mathbf{I}$ equals the codimensions of the range of $\mathbf{C} - \bar{z}\mathbf{I}$. This proves that the deficiency indices of **C** are equal; so according to theorem 3, **C** has a self-adjoint extension.   □

33.3 THE FRIEDRICHS EXTENSION

In this section we describe an enormously useful method, due to Friedrichs, of constructing a self-adjoint extension of a large class of symmetric operator, such as Schroedinger operators. The way this extension is carried out imposes certain boundary conditions automatically.

Definition. A symmetric operator **L** defined on a dense subspace D of a Hilbert space H is semibounded (from below) if

$$c\|u\|^2 \leq (u, \mathbf{L}u) \tag{11}$$

for some constant c and all u in D.

In what follows we will take the constant c to be 1; this can be accomplished by augmenting **L** by a sufficiently large multiple of the identity. We define on D a new scalar product, denoted by $(v, u)_\mathbf{L}$, as follows:

$$(v, u)_\mathbf{L} = (v, \mathbf{L}u). \tag{12}$$

The symmetry of the operator **L** guarantees that $(v, u)_\mathbf{L}$ is skew symmetric, and semiboundedness, with $c = 1$, shows that $(u, u)_\mathbf{L}$ is positive. We define the **L**-norm as

$$\|u\|_{\mathbf{L}} = (u, u)_{\mathbf{L}}^{1/2}.$$

It follows from (11) with $c = 1$ that the **L**-norm of every u in D is bigger than its original norm:

$$\|u\| \leq \|u\|_{\mathbf{L}}. \tag{11'}$$

The subspace D is, in general, not complete in the **L**-norm; we can complete it. Denote its completion by $H_{\mathbf{L}}$; it consists of equivalence classes of Cauchy sequences in the **L**-norm. By (11'), a Cauchy sequence in the **L**-norm is also a Cauchy sequence in the norm of H. Since H is complete, such a Cauchy sequence has a limit in H, this defines a natural mapping of $H_{\mathbf{L}}$ into H.

Lemma. *The natural mapping of $H_{\mathbf{L}}$ into H is one-to-one.*

Proof. Let $\{u_n\}$ be a Cauchy sequence in the **L**-norm of vectors in D that tends to $u^{\mathbf{L}}$ in $H_{\mathbf{L}}$. As noted above, $\{u_n\}$ is a Cauchy sequence also in the original norm; denote its limit in H by u. By definition of the **L**-scalar product, for every v in D,

$$(u_n, v)_{\mathbf{L}} = (u_n, \mathbf{L}v).$$

The limit of this relation as n tends to ∞ is

$$\left(u^{\mathbf{L}}, v\right)_{\mathbf{L}} = (u, \mathbf{L}v).$$

This shows that the **L**-scalar product of $u^{\mathbf{L}}$ with any v in D is uniquely determined by u. Since D is a dense subspace of $H_{\mathbf{L}}$, $u^{\mathbf{L}}$ is completely determined by u. $\square$

In view of the lemma, the natural map $H_{\mathbf{L}} \to H$ is an *embedding* of $H_{\mathbf{L}}$ into H. We will regard $H_{\mathbf{L}}$ as a subspace of H; note that D is contained in $H_{\mathbf{L}}$. We will define now the Friedrichs extension of **L**, to be denoted as $\mathbf{L}^F$, as follows: Take any vector g in H, and for every v in H define

$$\ell(v) = (v, g); \tag{13}$$

ℓ is a *bounded* linear functional of v:

$$|\ell(v)| \leq \|v\| \|g\|. \tag{14}$$

It follows from (11') that for all v in $H_{\mathbf{L}}$,

$$|\ell(v)| \leq \|v\|_{\mathbf{L}} \|g\|; \tag{14'}$$

EXAMPLES OF SELF-ADJOINT OPERATORS

this shows that $\ell(v)$ is a bounded linear functional on $H_{\mathbf{L}}$. According to the Riesz-Frechet representation theorem, we can write for all v in $H_{\mathbf{L}}$,

$$\ell(v) = (v, w)_{\mathbf{L}}, \tag{13'}$$

w some vector in $H_{\mathbf{L}}$. We denote the set of all such vectors w as D^F.

It follows from (13') that for all v in $H_{\mathbf{L}}$, the value of $\ell(v)$ is determined by w. Comparing this with (13), we conclude that g is determined by w, meaning that g is a function of w. Since ℓ depends linearly on g, it follows that g is a linear function of w. We denote this function as $\mathbf{L}^F$:

$$\mathbf{L}^F w = g, \qquad w \text{ in } D^F. \tag{15}$$

Combining the two representations, (13) and (13') of $\ell(v)$, and using the definitions (15), we get

$$(v, w)_{\mathbf{L}} = (v, \mathbf{L}^F w) \tag{16}$$

for all w in D^F and all v in $H_{\mathbf{L}}$.

Take g in (13) to be $g = \mathbf{L}u$, u some vector in D. Then for all v in D,

$$\ell(v) = (v, g) = (v, \mathbf{L}u) = (v, u)_{\mathbf{L}}.$$

Comparing this with (13'), we conclude that $u = w$ and $\mathbf{L}^F w = \mathbf{L}u$. In words, D is a subspace of D^F, and $\mathbf{L}^F$ is an extension of $\mathbf{L}$.

Theorem 4. $\mathbf{L}^F$ *is a self-adjoint extension of* $\mathbf{L}$.

Proof. We show first that $\mathbf{L}^F$ is symmetric. Restrict in (16) the vector v to D^F; interchanging w and v in (16) gives

$$(w, v)_{\mathbf{L}} = (w, \mathbf{L}^F v).$$

Since both scalar products are skew-symmetric, we deduce from this that

$$(v, w)_{\mathbf{L}} = (\mathbf{L}^F v, w).$$

Comparing this with (16), we conclude that $\mathbf{L}^F$ is symmetric.

The vector g in the definition (13) of the functional ℓ is an arbitrary vector in H; the vector w in (15) is uniquely determined by g. This shows that $\mathbf{L}^F$ is an invertible operator, namely that it maps its domain D^F one-to-one onto H.

Denote the inverse of $\mathbf{L}^F$ by $\mathbf{M}$. Since $\mathbf{L}^F$ is symmetric, so is $\mathbf{M}$. According to the Hellinger-Toeplitz result described at the beginning of chapter 32, the symmetric operator $\mathbf{M}$ is bounded. According to theorem 2 of chapter 31, every nonreal complex number z belongs to the resolvent set of such an operator $\mathbf{M}$. The formula

$$z^{-1}\mathbf{I} - \mathbf{M}^{-1} = \mathbf{M}^{-1}(\mathbf{M} - z\mathbf{I})z^{-1} \tag{17}$$

shows that then z^{-1} belongs to the resolvent set of $\mathbf{M}^{-1}$. According to theorem 2, this implies that $\mathbf{M}^{-1} = \mathbf{L}^F$ is self-adjoint.   □

Exercise 5. Show that the inverse of a symmetric operator is symmetric.

We give now some examples of semibounded operator and their Friedrichs extension.

Example 4. $H = L^2(0, 1), \mathbf{L} = -(d^2/dx^2) + q, q$ some continuous function on $[0, 1]$. The domain of $\mathbf{L}$ is $C_0^2(0, 1)$. Since every u in $D(\mathbf{L})$ is zero at the endpoints, integration by parts gives

$$\|u\|_L^2 = (u, \mathbf{L}u) = \int \left(u_x^2 + qu^2\right) dx.$$

Clearly, inequality (11) is satisfied, with $c = \min q$. Let us assume that $c = \min q = 1$.

Proposition. *Every function in H_L is continuous on the closed interval $[0, 1]$ and vanishes at the endpoint.*

Proof. For every u in C_0^2, we deduce using the Schwarz inequality that for every a, b in $[0, 1]$ that

$$|u(b) - u(a)| = \left|\int_a^b u_x dx\right| \le \sqrt{b-a} \left(\int_a^b u_x^2 dx\right)^{1/2} \le \sqrt{b-a}\|u\|_L. \quad (18)$$

It follows that a Cauchy sequence in the $\mathbf{L}$-norm converges uniformly, and that the limit u in H_L satisfies (18) and is zero at the endpoints.   □

Since $D(\mathbf{L}^F)$ is contained in H_L, this shows that the extension process imposes zero Dirichlet boundary conditions on functions in the domain of $\mathbf{L}^F$.

Exercise 6. Show that the closure of the operator $\mathbf{L}$ in example 4 is not self-adjoint.

Exercise 7. $H = L^2(0, 1), \mathbf{L} = -(d/dx)p(d/dx) + q$, where p is a positive function in C^1, q in C, the domain of $\mathbf{L}$ is $C^2(0, 1)$. Show that every u in the domain of $\mathbf{L}^F$ is continuous in $[0, 1]$ and is zero at the endpoints.

Example 5. G a bounded domain in the x, y plane, $H = L^2(G)$, the space of square integrable functions in G, $\mathbf{L} = -\Delta = -(\partial_x^2 + \partial_y^2)$, the domain of $\mathbf{L}$ is the space $C_0^2(G)$.

Proposition. *When G has a smooth boundary, every function in the domain of $\mathbf{L}^F$ vanishes on the boundary of G in the following sense:*

$$\int_{C_n} u^2 \, ds$$

tends to zero as the curves C_n approach the curve boundary G, and the tangents of C_n approach the tangent of C.

Exercise 8. Prove this proposition. (Hint: Show that $\|u\|_L^2 = \int (u_x^2 + u_y^2) \, dx \, dy$.)

Exercise 9. Show that the operator L in example 5 has deficiency indices ∞, ∞.

33.4 THE RELLICH PERTURBATION THEOREM

In this section we present a result of Rellich that says, roughly speaking, that if we add to a self-adjoint operator A a symmetric operator T that is not too large compared to A then the sum $A + T$ is self-adjoint. Here is the precise result:

Theorem 5. *Let A denote a self-adjoint operator acting in a Hilbert space H, with domain $D(A)$. Let T be a symmetric operator in H whose domain includes the domain of A and which is smaller than A in this sense: there exist numbers a and b, $b < 1$, such that for all u in $D(A)$,*

$$\|Tu\|^2 \le a^2 \|u\|^2 + b^2 \|Au\|^2. \tag{19}$$

Then $A + T$ defined on $D(A)$ is self-adjoint.

Proof. We show first that the operator $A + T$ is closed. As a first step we take the square root of inequality (19) and deduce

$$\|Tu\| \le a\|u\| + b\|Au\|. \tag{19'}$$

For any u in $D(A)$,

$$Au = (A + T)u - Tu.$$

So, by the triangle inequality combined with (19'), we get

$$\|Au\| \le \|(A + T)u\| + \|Tu\| \le \|(A + T)u\| + a\|u\| + b\|Au\|.$$

Since $b < 1$, it follows that

$$\|Au\| \le (1 - b)^{-1}\|(A + T)u\| + (1 - b)^{-1}a\|u\|.$$

It follows from this inequality that any vector in the domain of the closure of $A + T$ belongs to the domain of the closure of A. Since A is self-adjoint, it is closed; it follows that so is $A + T$.

We will show now that ic and $-ic$ belong to the resolvent set of $\mathbf{A} + \mathbf{T}$ when $c > a$, where a is the constant appearing in inequality (19). Since $\mathbf{A} + \mathbf{T}$ is a closed operator, we can appeal to theorem 1 and conclude that $\mathbf{A} + \mathbf{T} + ic$ maps $D(\mathbf{A})$ one-to-one onto a closed subspace of H. We claim that this subspace is all of H; for it not, there would be a vector v in H perpendicular to the range of $\mathbf{A} + \mathbf{T} + ic$:

$$((\mathbf{A} + \mathbf{T} + ic)u, v) = 0 \qquad (20)$$

for all u in $D(\mathbf{A})$. Since $\mathbf{A}$ is self-adjoint, the range of $\mathbf{A} + ic$ is all of H so that there is w in $D(\mathbf{A})$ that is mapped into v:

$$(\mathbf{A} + ic)w = v.$$

Setting this into (20) and choosing $u = w$ gives

$$(\mathbf{A}w + icw, \mathbf{A}w + ics) + (\mathbf{T}w, \mathbf{A}w + ics) = 0.$$

Estimating the second term by the Schwarz inequality yields after an algebraic manipulation

$$\|\mathbf{A}w + icw\|^2 \leq \|\mathbf{T}w\|^2.$$

Since $\mathbf{A}$ is symmetric, the left side of equal $\|\mathbf{A}w\|^2 + c^2\|w\|^2$; the right side is bounded by (19):

$$\|\mathbf{A}w\|^2 + c^2\|w\|^2 \leq a^2\|w\|^2 + b^2\|\mathbf{A}w\|^2.$$

Since $b < 1$ and $a < c$, we conclude that $w = 0$; this makes $v = 0$, and shows that the range of $\mathbf{A} + \mathbf{T} + ic$ is all of H. Ditto for the range of $\mathbf{A} + \mathbf{T} - ic$. According to theorem 2, this implies that $\mathbf{A} + \mathbf{T}$ is self-adjoint. $\qquad\square$

Corollary 2. *Let* $\mathbf{B}$ *denote an essentially self-adjoint operator, that is, one whose closure* $\overline{\mathbf{A}}$ *is self-adjoint. Let* $\mathbf{T}$ *denote a symmetric operator whose domain includes the domain of* $\mathbf{B}$ *and that satisfies inequality (19). Then the domain of* $\overline{\mathbf{T}}$ *includes the domain of* $\overline{\mathbf{B}}$ *and* $\overline{\mathbf{B}} + \overline{\mathbf{T}}$ *is self-adjoint on* $D(\overline{\mathbf{B}})$.

Exercise 10. Prove corollary 5.

Example 6. H is $\mathbf{L}^2(\mathbb{R})$, $\mathbf{B} = -d^2/dx^2$, $D(\mathbf{B}) = C_0^2(\mathbb{R})$, the space of twice differentiable functions of compact support. $\mathbf{T}$ is multiplication by some real-valued function q in class $\mathbf{L}^2$; $D(\mathbf{T})$ consist of all $C_0(\mathbb{R})$ functions.

Proposition.

 (i) $\mathbf{B}$ *is essentially self-adjoint.*
 (ii) $\mathbf{T}$ *is bounded by* $\mathbf{B}$ *in the sense of inequality (19).*

Proof. Part (i) is proved by showing that for nonreal z the range of $\mathbf{B} - z\mathbf{I}$ is a dense subset of H. This can be verified following the analysis given in example 1 in section 33.2. To show part (ii), we start with inequality (18) derived in the discussion of example 4:

$$|u(b) - u(a)| \leq \sqrt{b - a}\|u_x\|. \tag{18}$$

given any point a in $\mathbb{R}$, there is a point b in the interval $[a - \frac{1}{2}, a + \frac{1}{2}]$ such that

$$|u(b)| \leq \left(\int_{a-(1/2)}^{a+(1/2)} u(x)\,dx\right)^{(1/2)} \leq \|u\|.$$

Then we deduce from (18) that for any a,

$$|u(a)| \leq \|u\| + \sqrt{\tfrac{1}{2}}\|u_x\|.$$

Taking the sup over all a, we conclude that

$$|u|_{L^\infty} \leq \|u\| + \sqrt{\tfrac{1}{2}}\|u_x\|.$$

Squaring this inequality gives

$$|u|_{L^\infty}^2 \leq 2\|u\|^2 + \|u_x\|^2. \tag{21}$$

Integration by parts gives

$$(\mathbf{B}u, u) = -\int u_{xx}\bar{u}\,dx = \int |u_x|^2\,dx = \|u_x\|^2.$$

Applying the Schwarz inequality on the left followed by the arithmetic-geometric mean inequality gives

$$\|u_x\|^2 \leq \|u\|^2\|\mathbf{B}u\|^2 \leq \frac{1}{2\epsilon}\|u\|^2 + \frac{\epsilon}{2}\|\mathbf{B}u\|^2 \tag{22}$$

for any positive ϵ. Putting together the two inequalities (21) and (22) gives

$$|u|_{L^\infty}^2 \leq \left(2 + \frac{1}{2\epsilon}\right)\|u\|^2 + \frac{\epsilon}{2}\|\mathbf{B}u\|^2. \tag{23}$$

We turn now to $\mathbf{T}$:

$$\|\mathbf{T}u\|^2 = \int q^2 u^2\,dx \leq \int q^2\,dx\,|u|_{L^\infty}^2 = Q|u|_{L^\infty}^2,$$

where Q abbreviates $\int q^2 dx$. Using the estimate (23) on the right side, we get

$$\|\mathbf{T}u\|^2 \leq Q\left(2 + \frac{1}{2\epsilon}\right)\|u\|^2 + Q\frac{\epsilon}{2}\|\mathbf{B}u\|^2. \tag{24}$$

Taking ϵ small enough, the coefficient of $\|\mathbf{B}u\|^2$ in (24) can be made < 1; this shows that inequality (19) is satisfied for all u in $D(\mathbf{B})$. We appeal now to corollary 5 to conclude that $\mathbf{B} + \mathbf{T}$ is self-adjoint. $\square$

Exercise 11. Prove that $\mathbf{B}$ is essentially self-adjoint.

Example 7. H is $L^2(\mathbb{R})$, $\mathbf{B} = -d^2/dx^2$, and $D(\mathbf{B})$ consist of twice differentiable functions of compact support that are zero at the origin. $\mathbf{T}$ is multiplication by a real-valued function q subject to these restrictions:

(i) $|q(x)| \leq \frac{c}{|x|^p}$ for $|x| \leq |1|$; $p < 1$.

(ii) q is L^2 outside of the interval $[-1, 1]$.

Proposition.

(i) $\mathbf{B}$ is essentially self-adjoint.

(ii) $\mathbf{T}$ is bounded by $\mathbf{B}$ in the sense of inequality (19).

Proof. For part (i), see the remark in the previous example. For part (ii), we need another inequality in addition to (23). It too is based on inequality (18). We set $a = 0$ and use the fact that $u(0) = 0$. Squaring, we get

$$|u(b)|^2 \leq |b| \|u_x\|^2. \tag{25}$$

Next we break up the integral for $\|\mathbf{T}u\|^2$ into two parts:

$$\|\mathbf{T}u\|^2 = \int_{\mathbb{R}} q^2 u^2 \, dx = \int_I + \int_{\mathbb{R}-I}, \tag{26}$$

where I is the interval $[-1, 1]$. On I we use the bound (i) for q and the estimate (25) for u:

$$\int_I q^2 u^2 dx \leq c^2 \int_I \frac{|x|}{|x|^{2p}} dx \|u_x\|^2 = \frac{c^2}{1-p} \|u_x\|^2.$$

We use (22) to estimate $\|u_x\|^2$ on the right side:

$$\int_I q^2 u^2 \leq \frac{c^2}{1-p}\left(\frac{1}{2\epsilon}\|u\|^2 + \frac{\epsilon}{2}\|\mathbf{B}u\|^2\right). \tag{27}$$

We estimate the right side of (26), using (27) for the first term, and (24) for the second term:

$$\|Tu\|^2 \le 2Q\|u\|^2 + \left(\frac{c^2}{1-p} + Q\right)\left(\frac{1}{2\epsilon}\|u\|^2 + \frac{\epsilon}{2}\|Bu\|^2\right). \qquad (28)$$

Since ϵ is arbitrary, we can take it so small that the coefficient of $\|Bu\|^2$ in (28) is less than 1. So an inequality of form (19) is satisfied for all u in $D(\mathbf{B})$. By corollary 5, $\overline{B} + \overline{T}$ is self-adjoint. $\qquad\square$

Exercise 12. Carry out the details of the proof that $\mathbf{B}$ is essentially self-adjoint.

33.5 THE MOMENT PROBLEM

In chapter 14, section 14.7, we have formulated the

Hamburger Moment Problem. What sequences $a_0, a_1, \ldots$ of real numbers can be represented as moments of a mass distribution on $\mathbb{R}$:

$$a_n = \int_{\mathbb{R}} t^n dm, \qquad (29)$$

where m is a nonnegative measure whose support is larger than a finite set of points.

As observed in chapter 14, every sequence $a_0, a_1, \ldots$ of form (29) is *Hankel positive*, which means that the quadratic form

$$Q = \Sigma a_{n+k}\xi_n\xi_k \qquad (30)$$

is positive for every nonzero choice of real numbers $\xi_1, \ldots, \xi_N$, N arbitrary. This is evident from the formula

$$\Sigma a_{n+k}\xi_n\xi_k = \int \Sigma t^{n+k}\xi_n\xi_k dm = \int \left(\Sigma t^n\xi_n\right)^2 dm. \qquad (30')$$

Clearly, the right side of (30') is nonnegative. Furthermore, when the support of the measure m is not finite, it is not confined to the zeros of the polynomial $\Sigma\xi_n t^n$; this shows that the right side of (30') is positive. Note that it follows that a_0 is positive.

Hans Hamburger has shown that this necessary condition for the a_n to be represented in form (29) is also sufficient:

Theorem 6 (Hamburger). *Let $\{a_n\}$ be a sequence of real numbers such that the quadratic form (30) is positive. The the a_n are the moments of a nonnegative measure on the real axis $\mathbb{R}$; that is, they can be represented in the form (29).*

Proof. Denote by D the linear space of all finite sequences of real numbers:

$$x = (\xi_0, \xi_1, \ldots, \xi_N, 0, \ldots).$$

Define the scalar product (x, y) on D by the quadratic form Q in (30):

$$(x, y) = Q(x, y) = \Sigma a_{n+k}\xi_n\eta_k. \tag{31}$$

Denote the completion of D with respect to the norm $\|x\| = Q^{1/2}(x, x)$ by K, and denote by $H = K + iK$ the complexification of K.

Define on D the operator $\mathbf{R}$ as *right shift*, that is,

$$\mathbf{R}x = (0, \xi_0, \xi_1, \ldots). \tag{32}$$

Denote by e the unit vector in D:

$$e = (1, 0, 0, \ldots).$$

Then

$$\mathbf{R}^n e = (0, \ldots, 0, 1, 0 \ldots),$$

and it follows from the definition (31) of the scalar product that

$$(e, \mathbf{R}^n e) = a_n. \tag{33}$$

$\mathbf{R}$ is symmetric, for we can write, denoting $n - 1$ as ℓ,

$$(\mathbf{R}x, y) = \Sigma a_{n+k-1}\xi_n\eta_k = \Sigma a_{\ell+k+1}\xi_\ell\eta_k. \tag{34}$$

The sum on the right is, clearly, a symmetric function of x and y.

Since $\mathbf{R}$ is a symmetric operator acting on D, a dense subspace of the real Hilbert space K, it has, according to the corollary of theorem 3, a self-adjoint extension to the complexification of H of K. Denote by $\mathbf{E}$ the projection-valued measure that gives the spectral resolution of such an extension of $\mathbf{R}$. According to theorem 1 of chapter 32,

$$(e, \mathbf{R}^n e) = \int t^n d(\mathbf{E}e, e);$$

combining this with (33), we get the sought-after representation (29), with $m = (\mathbf{E}e, e)$. $\qquad\qquad\square$

NOTE. It took Hamburger 150 pages to prove his theorem. Using the theory of self-adjoint operators, it takes less than a page.

A related problem is the *Stieljes moment problem* used to represent a sequence of real numbers as the moments of a mass distribution on the *positive* real axis $\mathbb{R}_+$:

$$a_n = \int_{\mathbb{R}_+} t^n \, dm. \tag{35}$$

The positivity of the quadratic form Q defined in (30) is certainly a condition necessary for a representation of form (35). So is the positivity of

$$\Sigma a_{n+k+1}\xi_n\xi_k. \tag{36}$$

This is evident from the formula

$$\Sigma a_{n+k+1}\xi_n\xi_k = \int_{\mathbb{R}_+} t^{n+k+1}\xi_n\xi_k \, dm = \int_{\mathbb{R}_+} t(\Sigma t^n\xi_n)^2 \, dm.$$

The positivity of (36) can be expressed by setting $y = x$ in (34) and writing

$$(\mathbf{R}x, x) \geq 0.$$

This inequality can be expressed by saying that $\mathbf{R}$ is a positive operator on D. Then $\mathbf{R}$ remains positive on the complexification of D, and so Friedrichs' procedure can be used to extend $\mathbf{R}$ to H as a positive self-adjoint operator. As before, we use the spectral projections $\mathbf{E}$ of the Friedrichs extension of $\mathbf{R}$ to write

$$a_n = (e, \mathbf{R}^n e) = \int t^n d(\mathbf{E}e, e).$$

The support of the spectral measure $\mathbf{E}$ lies on the spectrum of the extension of $\mathbf{R}$. Since this is a positive operator, its spectrum lies on the positive axis $\mathbb{R}_+$; see theorem 7 in chapter 31. It follows that the measure $m = (\mathbf{E}e, e)$ is supported on the positive axis $\mathbb{R}_+$. This completes the proof of the following proposition:

Theorem 7 (Stieltjes). *Let $\{a_n\}$ be a sequence of real numbers such that the quadratic forms (30) and (36) are positive. Then the a_n are the moments of a nonnegative measure on the positive axis $\mathbb{R}_+$; that is, they can be represented in the form (35).*

What about uniqueness of the representing measure? The following is a simple example of a sequence that can be represented as the moments of two distinct measures:

Let f be a real-valued even C_0^∞ function on $\mathbb{R}$ that vanishes to ∞ order at the origin; that is, all its derivatives are zero there. Denote its Fourier transform by g. Since f is even, g is real valued, and since f is C_0^∞, $g(t)$ tends to zero faster than any negative power of t as t tends to ∞. By Fourier inversion,

$$\int g(t)e^{ist} \, dt = f(s).$$

Differentiate n times both sides with respect to s and set $s = 0$:

$$i^n \int g(t)t^n = \frac{d^n f}{ds^n}\Big|_{s=0} = 0 \,,$$

meaning that all moments of g are zero. Writing g as the difference of its positive and negative part, $g = g_+ - g_-$, we conclude that the nonnegative functions g_+ and g_- have the same moments. Since g is not $\equiv 0$, g_+ and g_+ are distinct, a case of nonuniqueness for the Hamburger moment problem.

It is almost as easy to give examples of moment problems whose solutions are unique, for instance, when the sequence of moments is bounded,

$$|a_n| \leq \text{const.}$$

An example of such a sequence is the moments of Lebesgue measure on $[0, 1]$, and zero everywhere else. Here

$$\int_0^1 t^n \, dt = \frac{1}{n+1}$$

is a bounded sequence.

Let r be an arbitrary positive number; multiply (29) by $r^n/n!$ and sum from 0 to N:

$$\sum_0^N a_n \frac{r^n}{n!} = \int \sum_0^N \frac{(rt)^n}{n!} \, dm.$$

Since the a_n are bounded, the left side is $\leq \text{const.} \ e^r$. Since the integrand on the right is positive and tends to e^{rt}, uniformly on any compact interval, as N tends to ∞, it follows that e^{rt} is integrable with respect to the measure m, and that

$$\int e^{rt} \, dm = \sum_0^\infty a_n \frac{r^n}{n!}.$$

It follows that this relation holds for any complex r, in particular, for $r = is$:

$$\int e^{ist} \, dm = \Sigma a_n \frac{(is)^n}{n!}.$$

This shows that the Fourier transform of m is uniquely determined by its moments a_n. Since a finite measure is uniquely determined by its Fourier transform, it follows that the moments $\{a_n\}$ uniquely determine the measure m.

The story is similar for the Stieltjes moment problem. Stieltjes already has given an example of a sequence $\{a_n\}$ that is the sequence of moments of two distinct measures on $\mathbb{R}_+$, and the other hand sequences $\{a_n\}$ that are the moment of a single measure on $\mathbb{R}_+$.

It is an interesting problem—and at any rate a classical one—to characterize those moment problems that have a unique solution.

Theorem 8.

*(i) The moment problem (29) has a unique solution iff the operator **R** is essentially self-adjoint, that is, has a unique self-adjoint extension.*

*(ii) The moment problem (34) has a unique solution iff the operator **R** has a unique nonnegative self-adjoint extension.*

Neither part of the theorem is obvious. Even if **R** has two distinct self-adjoint extensions R_1 and R_2, it is not clear that the measures (E_1e, e) and (E_2e, e) are distinct. On the other hand, when **R** is not essentially self-adjoint, there are solutions m of the moment problem that are *not* of the form (Ee, e), where **E** is the spectral resolution of some self-adjoint extension of **R**.

For a proof of theorem 8, and for a review of the literature of the moment problem, we refer to the article by Barry Simon; see also Henry Landau's article in the AMS Symposium volume edited by him, as well as the books by Akhiezer, and Shohat and Tamarkin.

HISTORICAL NOTES. Stieltjes introduced the integral named after him in connection with his work on the moment problem.

The theory of self-adjoint operator was created by von Neumann to fashion a framework for quantum mechanics. The operators in Schrödinger's theory that are associated with atoms are partial differential operators whose coefficients are singular at certain points; these singularities correspond to the unbounded growth of the force between two electrons that approach each other. To define such differential operators as self-adjoint ones is not a trivial task. Examples 5 and 6 presented in section 33.4 allow some singularities in the potential q, but the ones occurring in quantum mechanics are more singular still. I recall in the summer of 1951 the excitement and elation of von Neumann when he learned that Kato has proved the self-adjointness of the Schrödinger operator associated with the helium atom.

And what do the physicists think of these matters? In the 1960s Friedrichs met Heisenberg, and used the occasion to express to him the deep gratitude of the community of mathematicians for having created quantum mechanics, which gave birth to the beautiful theory of operators in Hilbert space. Heisenberg allowed that this was so; Friedrichs then added that the mathematicians have, in some measure, returned the favor. Heisenberg looked noncommittal, so Friedrichs pointed out that it was a mathematician, von Neumann, who clarified the difference between a self-adjoint operator and one that is merely symmetric. "What's the difference," said Heisenberg.

BIBLIOGRAPHY

Akhiezer, N. I. *The Classical Moment Problem.* Hafner, New York, 1965.

Friedrichs, K. O. Spektraltheorie halbbeschränhter Operatoren. *Math. An.,* **109** (1934): 465–487, 685–713.

Hamburger, H. Über eine Erweiterung des Stieltjesschen Moment Problems. *Math. An.,* **81** (1920): 235–319, **82** (1921): 120–164, 168–187.

Kato, T. Fundamental properties of Hamiltonian operators of Schrödinger type. *Trans. AMS.* **70** (1951): 195–211.

Kato, T. On the existence of solutions of the helium wave equation. *Trans. AMS,* **70** (1951): 212–218.

Kato, T. *Perturbation Theory for Linear Operators.* Die Grundlehren der Math. Wiss. in Einzeldarstellung, **132**. Springer, Berlin, 1966.

Landau, H. J., ed. *Moments in Mathematics. Proc. Symp. Appl. Math.*, **37**. American Mathematical Society. Providence, RI, 1987.

Reed, M. and Simon, B. *Methods of Modern Mathematical Physics: Vol. 1, Functional Analysis.* Academic Press, New York, 1972.

Rellich, F. Störungstheorie der Spektralzerlegung. *Math. An.,* **116** (1939): 555–570.

Shohat, J. A. and Tamarkin, J. D. *The Problem of Moments.* AMS Surveys **1**. American Mathematical Society, New York, 1943.

Simon, B. The classical moment problem as a self-adjoint finite difference operator. *Adv. Math.,* **137** (1998): 82–203.

Stieltjes, T. *Recherches sur les fractions continue.* Ann. Fac. Sc. Univ. Toulouse, **8** (1894–95): J1–J22; **9** A5–A47.

Stone, M. H. *Linear Transformations in Hilbert Space and Their Applications to Analysis.* AMS Coll. Publ., **15**. American Mathematical Society, New York, 1932.

von Neumann, J. Allgemeine eigenwertheorie Hermitescher Funktionaloperatoren. *Math. An.,* **102** (1929): 49–131.

von Neumann, J. *Mathematische Grundlagen der Quantenmechanic.* Die Grundlehren der Math. Wiss. in Einzeldarstellung, **37**. Springer, Berlin, 1932.

34

SEMIGROUPS OF OPERATORS

The natural source of semigroups of operators are partial differential equations describing evolution in time, and flows generated by dynamical systems. In this chapter we present an abstraction of these concrete situations, a point of view initiated by Hille. In the next two chapters we will present illustrations and applications of the theory. For a detailed treatment of this subject, we recommend Hille-Phillips, Yosida, and Goldstein's excellent monograph.

Definition. A *one-parameter semigroup* of operators over a complex Banach space X is a family of bounded linear operators $\mathbf{Z}(t)$, $t \geq 0$, each mapping $X \to X$, with the following properties:

$$\mathbf{Z}(t + s) = \mathbf{Z}(t)\mathbf{Z}(s) \qquad \text{for all } t, s \geq 0; \mathbf{Z}(0) = \mathbf{I}. \tag{1}$$

Equation (1) is the multiplicative property of exponential functions. We show next that under an additional continuity property, equation (1) characterizes exponential functions.

Theorem 1.

(i) *Let* $\mathbf{G}\colon X \to X$ *be a bounded linear map. Define* $\mathbf{Z}(t)$ *to be*

$$\mathbf{Z}(t) = e^{t\mathbf{G}}, \qquad t \geq 0, \tag{2}$$

where the exponential of the operator is defined by the power series

$$e^{t\mathbf{G}} = \sum_{0}^{\infty} \frac{t^n \mathbf{G}^n}{n!}. \tag{3}$$

Then $\mathbf{Z}(t)$ *is a one-parameter semigroup of operators, continuous in the norm topology for operators.*

(ii) *Conversely, let* $\mathbf{Z}(t)$: $X \rightarrow X$ *be a one-parameter semigroup of operators continuous in the norm topology at* $t = 0$:

$$\lim_{t \to 0} |\mathbf{Z}(t) - \mathbf{I}| = 0. \tag{4}$$

Then $\mathbf{Z}(t)$ *is of form* (2), $\mathbf{G}$ *some bounded linear map* $X \rightarrow X$.

NOTE. Formula (2) defines $\mathbf{Z}(t)$ also for $t < 0$; these operators form a group.

Proof. (i) is a special case of the functional calculus for operators; see chapter 17, theorem 4 (ii). So is part (ii) of theorem 1 above. Suppose that $\mathbf{Z}(t)$ is a one-parameter semigroup uniformly continuous at $t = 0$. The logarithm function

$$\log(1 + \zeta) = \zeta - \frac{\zeta^2}{2} + \cdots$$

is analytic in the unit disk around 1. Therefore for any operator $\mathbf{Z}$ that differs from $\mathbf{I}$ by an operator of norm < 1 we can define

$$\log \mathbf{Z} = \log(\mathbf{I} + \mathbf{Z} - \mathbf{I}) = \mathbf{Z} - \mathbf{I} - \frac{(\mathbf{Z} - \mathbf{I})^2}{2} + \cdots. \tag{5}$$

Exercise 1. Denote by $\mathbf{Z}$ and $\mathbf{W}$ two operators mapping X into X that commute. Suppose that $\|\mathbf{Z} - \mathbf{I}\| < \frac{1}{3}$ and $\|\mathbf{W} - \mathbf{I}\| < \frac{1}{3}$. Show that $\|\mathbf{ZW} - \mathbf{I}\| < 1$, and prove that

$$\log \mathbf{ZW} = \log \mathbf{Z} + \log \mathbf{W},$$

where the logarithms are defined by formula (5).

By (4), there is an $a > 0$ such that $|\mathbf{Z}(t) - \mathbf{I}| < \frac{1}{3}$ for $t < a$. We define $\mathbf{L}(t) = \log \mathbf{Z}(t)$, $t < a$, by formula (5). The multiplicative property (1) implies that $\mathbf{Z}(t)$ and $\mathbf{Z}(s)$ commute; therefore we deduce from (1) and (5) that

$$\mathbf{L}(t + s) = \mathbf{L}(t) + \mathbf{L}(s), \qquad t + s < a.$$

From this we deduce that for all rational $t < a$, $t^{-1}\mathbf{L}(t)$ is independent of t; denote this operator by $\mathbf{G}$:

$$\mathbf{L}(t) = t\mathbf{G}, \qquad t \text{ rational, } < a. \tag{6}$$

The multiplicative property (1), that is, $\mathbf{Z}(t + h) - \mathbf{Z}(t) = \mathbf{Z}(t)[\mathbf{Z}(h) - \mathbf{I}]$, and continuity at $t = 0$ show that $\mathbf{Z}(t)$ is continuous for all t in the uniform topology. By (5), the same is true for $\mathbf{L}(t)$ where $t < a$. From this we deduce that (6) holds not just for t rational but all $t < a$. Exponentiating gives (2) for $t < a$. By (1), it holds for all t. $\qquad \square$

34.1 STRONGLY CONTINUOUS ONE-PARAMETER SEMIGROUPS

The most interesting semigroups are not of the form (2) but are associated with differential equations, of which a typical example is the heat equation,

$$u_t - \partial_x^2 u = 0,$$

u being a periodic function of x. A solution of the heat equation is uniquely determined by specifying its initial values, namely its value at $t = 0$. The initial value can be specified as an arbitrary continuous function; the absolute value of the corresponding solution at any later time does not exceed the maximum absolute value of the prescribed initial values. Denote by $\mathbf{Z}(t)$ the operator relating the initial values $u(x, 0)$ of solutions to $u(x, t)$. Clearly, these operators form a one-parameter semigroup in the sense of (1). Yet they are not of the form (2); for if they were, they could be extended to negative values of t to form a group of operators. But it is well known that such extension of solutions of the heat equation backward in time is not possible in general.

The heat equation semigroup $\mathbf{Z}(t)$ is not uniformly continuous at $t = 0$; yet it retains a less stringent kind of continuity that expresses the fact that each solution u is a continuous function of t:

Definition. A one-parameter semigroup $\mathbf{Z}(t)$ is *strongly* continuous at $t = 0$ if

$$\text{s}- \lim_{t \to 0} \mathbf{Z}(t)x = x \tag{7}$$

for all x in X.

Theorem 2. *Denote by $\mathbf{Z}(t)$ a one-parameter semigroup of operators that is strongly continuous at $t = 0$.*

(i) *There exist constants b and k such that $\mathbf{Z}(t)$ is bounded in norm by*

$$|\mathbf{Z}(t)| \le b e^{kt}. \tag{8}$$

(ii) *$\mathbf{Z}(t)x$ is a strongly continuous function of t for every x in X.*

Proof. (i) We claim that $|\mathbf{Z}(t)|$ is uniformly bounded in some neighborhood of $t = 0$. To see this, suppose, to the contrary, that there is a sequence $t_j \to 0$ such that $|\mathbf{Z}(t_j)| \to \infty$. By the principle of uniform boundedness (see chapter 14, theorem 7) $\mathbf{Z}(t_j)x$ could not converge to x for all x in X. This violates strong continuity at $t = 0$; therefore there exists an $a > 0$, $b > 0$ such that $|\mathbf{Z}(t)| \le b$ for $t \le a$.

Any t can be decomposed as $t = na + r$, $0 \le r < a$. By the semigroup property, $\mathbf{Z}(t) = \mathbf{Z}^n(a) \mathbf{Z}(r)$. So

$$|\mathbf{Z}(t)| \le |\mathbf{Z}(a)|^n |\mathbf{Z}(r)| \le b^{n+1} \le b e^{kt},$$

where $k = \frac{1}{a} \log b$. This proves (8).

(ii) For any pair of positive numbers $s < t$ we can, by the semigroup property, write

$$\mathbf{Z}(t)x - \mathbf{Z}(s)x = \mathbf{Z}(s)\big[\mathbf{Z}(t-s)x - x\big].$$

Combining (7) and (8) strong continuity follows. □

Next we show, speaking loosely, that a strongly continuous semigroup can be interpreted as an exponential function of an *unbounded*, not everywhere defined, operator. We start with a few facts and notions about unbounded operators:

Definition. Let D be a dense linear subspace of a Banach space X, $\mathbf{G}$ a linear operator mapping D into X. We call the operator $\mathbf{G}$ *closed* if whenever $\{x_n\}$ is a sequence of vectors such that $x_n \to x$ and $\mathbf{G}x_n \to y$, then x lies in D, and $\mathbf{G}x = y$. D is called the *domain* of $\mathbf{G}$, and is denoted as $D(\mathbf{G})$.

According to the closed graph theorem, a closed linear operator defined at every point of a Banach space X is bounded. The operators we encounter in this chapter are defined only on a dense subspace of X and are *unbounded*.

Definition. Let $\mathbf{G}$ be a closed operator with domain $D(\mathbf{G})$; a complex number ζ belongs to the *resolvent set* of $\mathbf{G}$, denoted as $\rho(\mathbf{G})$, if $\zeta\mathbf{I} - \mathbf{G}$ maps $D(\mathbf{G})$ one-to-one onto X. The *spectrum* of $\mathbf{G}$, denoted as $\sigma(\mathbf{G})$, is the complement of its resolvent set.

Suppose that ζ is in the resolvent set of $\mathbf{G}$; then $\zeta\mathbf{I} - \mathbf{G}$ is invertible. Its inverse is called the *resolvent* of $\mathbf{G}$ and is denoted as

$$\mathbf{R}(\zeta) = (\zeta\mathbf{I} - \mathbf{G})^{-1};$$

it maps X onto $D(\mathbf{G})$. Since $\mathbf{G}$ is closed, so is $\mathbf{R}(\zeta)$; it follows then from the closed graph theorem, theorem 12 of chapter 15, that the resolvent $\mathbf{R}(\zeta)$ is a *bounded operator*.

Exercise 2. Suppose that the resolvent set of $\mathbf{G}$ is not empty, and that ζ belongs to $\rho(\mathbf{G})$. Show that a complex number γ belongs to the spectrum of $\mathbf{G}$ iff $(\zeta - \gamma)^{-1}$ belongs to the spectrum of $\mathbf{R}(\zeta)$. We can express this symbolically as

$$\sigma(\mathbf{R}(\zeta)) = (\zeta - \sigma(\mathbf{G}))^{-1}. \tag{9}$$

One can think of (9) as an instance of the spectral mapping theorem for unbounded operators.

Exercise 3. Deduce from (9) that the spectrum of $\mathbf{G}$ is a closed set in the complex plane.

Next we define the transpose of an unbounded operator:

Definition. Let $\mathbf{G}$ be a densely defined, closed linear operator in a Banach space X. We define its *transpose* $\mathbf{G}'$ by the relation

$$(\mathbf{G}x, \ell) = (x, \mathbf{G}'\ell). \tag{10}$$

The meaning of (10) is this: The domain of $\mathbf{G}'$ consists of those linear functionals ℓ for which the left side of (10) is a *bounded* linear functional of x, defined on $D(\mathbf{G})$. Since $D(\mathbf{G})$ is dense in X, this bounded linear functional can be extended uniquely from $D(\mathbf{G})$ to all of X; this extension is denoted by $\mathbf{G}'\ell$, its domain by $D(\mathbf{G}')$.

Exercise 4. Show that the transpose of a densely defined linear operator is closed.

The difficulty with the preceding definition is that the domain of $\mathbf{G}'$ is an elusive thing. The following result is useful for pinning down the domain of $\mathbf{G}'$.

Theorem 3. *Let X be a reflexive Banach space, $\mathbf{G}$ a densely defined, closed linear operator mapping $D(\mathbf{G})$ into X, whose resolvent set is not empty. Then its transpose $\mathbf{G}'$ is a densely defined, closed linear operator mapping $D(\mathbf{G}')$ into X', and $\rho(\mathbf{G}') = \rho(\mathbf{G})$.*

Proof. Let ζ be a complex number in the resolvent set of $\mathbf{G}$; subtract $\zeta(x, \ell)$ from both sides of (10):

$$((\mathbf{G} - \zeta\mathbf{I})x, \ell) = (x, (\mathbf{G}' - \zeta\mathbf{I})\ell). \tag{10'}$$

Denote the resolvent of $\mathbf{G}$ by $\mathbf{R}(\zeta) = -(\mathbf{G}-\zeta\mathbf{I})^{-1}$ and define $\mathbf{R}'(\zeta)$ as the transpose of $\mathbf{R}(\zeta)$:

$$(y, \mathbf{R}'(\zeta)m) = (\mathbf{R}(\zeta)y, m). \tag{10''}$$

We claim that $\mathbf{R}'(\zeta)$ is the resolvent of $\mathbf{G}'$. We show first that $\mathbf{R}'(\zeta)$ is one-to-one; for otherwise there would be a nonzero m such that $\mathbf{R}'(\zeta)m = 0$. Setting this into (10''), we conclude that the range of $\mathbf{R}(\zeta)$ is annihilated by m; since the range of $\mathbf{R}(\zeta)$ is $D(\mathbf{G})$, assumed to be dense, this is a contradiction. Now define $\mathbf{R}'^{-1}(\zeta)$ to be $-\mathbf{G}' + \zeta\mathbf{I}'$ and set $\mathbf{R}(\zeta)y = x$ and $\mathbf{R}'(\zeta)m = \ell$ into (10''). Thus we obtain (10').

The domain of $\mathbf{G}'$ as defined above is the range of $\mathbf{R}'(\zeta)$. It remains to be shown that $\mathbf{G}'$ cannot be extended any further. To see this, we note that the range of $\mathbf{G}' - \zeta\mathbf{I}'$ as defined above is all of X'. Therefore, if we extend $\mathbf{G}'$ further, $\mathbf{G}' - \zeta\mathbf{I}'$ would annihilate some nonzero ℓ. Setting this into (10'), we would conclude that ℓ annihilates the range of $\mathbf{G} - \zeta\mathbf{I}$, a contradiction since ζ belongs to the resolvent of $\mathbf{G}$.

The last task is to show that the domain of $\mathbf{G}'$, identified above as the range of $\mathbf{R}'(\zeta)$, is dense in X. If it were not, then there would be a y in X that annihilates the range of $\mathbf{R}'(\zeta)$ (it is at this point that we use the reflexivity of X). Setting this into (10''), we conclude that $\mathbf{R}(\zeta)y = 0$, contrary to the fact that ζ belongs to the resolvent set of $\mathbf{G}$. $\square$

Exercise 5. Show that in the Hilbert space setting the conclusion of theorem 3 has to be modified as follows: $\varrho(\mathbf{G}^*) = \overline{\varrho(\mathbf{G})}$.

Definition. Let $\mathbf{Z}(t)$ be a strongly continuous one-parameter semigroup of operators: $X \rightarrow X$, meaning that (1), (2), and (9) are satisfied. Its *infinitesimal generator* $\mathbf{G}$ is defined by

$$\mathbf{G}x = s - \lim_{h \to 0} \frac{\mathbf{Z}(h)x - x}{h}; \tag{11}$$

the *domain* of $\mathbf{G}$, denoted as $D(\mathbf{G})$, consists of all x for which the strong limit (11) exists.

Theorem 4. *Let* $\mathbf{Z}(t)$ *be a strongly continuous one-parameter semigroup,* $\mathbf{G}$ *its infinitesimal generator.*

  (i) $\mathbf{G}$ *commutes with* $\mathbf{Z}(t)$, *in the sense that if x belongs to* $D(\mathbf{G})$, *so does* $\mathbf{Z}(t)x$, *and*

$$\mathbf{GZ}(t)x = \mathbf{Z}(t)\mathbf{G}x. \tag{12}$$

  (ii) *The domain of* $\mathbf{G}$ *is dense.*

  (iii) *The domain of* $\mathbf{G}^n$, *n any natural number, is dense.*

  (iv) $\mathbf{G}$ *is a closed operator.*

  (v) *All complex numbers ζ whose real part is $> k$ belong to the resolvent set of* $\mathbf{G}$, *where k is the constant appearing in inequality (8). The resolvent of* $\mathbf{G}$ *is the Laplace transform of* $\mathbf{Z}$.

Proof. (i) Using (1), we can factor the difference quotient in two ways:

$$\frac{\mathbf{Z}(t + h) - \mathbf{Z}(t)}{h}x = \mathbf{Z}(t)\frac{\mathbf{Z}(h) - I}{h}x = \frac{\mathbf{Z}(h) - I}{h}\mathbf{Z}(t)x. \tag{13}$$

When x belongs to $D(\mathbf{G})$, the middle term converges as $h \to 0$ to $\mathbf{Z}(t)\mathbf{G}x$. Therefore the terms on the right and left converge also, and we deduce from (13) that

$$\frac{d}{dt}\mathbf{Z}(t)x = \mathbf{Z}(t)\mathbf{G}x = \mathbf{GZ}(t)x \tag{14}$$

for every x in $D(\mathbf{G})$. This proves part (i).

  (ii) We claim that an integrated form of (14),

$$\mathbf{Z}(t)x - x = \mathbf{G}\int_0^t \mathbf{Z}(s)x \, ds, \tag{15}$$

is valid for all x in X. Since $\mathbf{Z}(s)$ is strongly continuous, the integrand on the right in (15) is a continuous function of s; therefore the integral can be defined as the limit of

Riemann sums. To prove (15), we want to evaluate the action of $\mathbf{G}$ on this integral. Letting $\mathbf{Z}(h)$ act on the integrand, and using the semigroup property, we get

$$\frac{\mathbf{Z}(h) - \mathbf{I}}{h} \int_0^t \mathbf{Z}(s)x\,ds = \frac{1}{h} \int_0^t \big[\mathbf{Z}(s+h)x - \mathbf{Z}(s)x\big]\,dx$$

$$= \frac{1}{h} \int_t^{t+h} \mathbf{Z}(s)x\,ds - \frac{1}{h} \int_0^h \mathbf{Z}(s)x\,ds.$$

Since $\mathbf{Z}(s)x$ is strongly continuous, the terms on the right converge to the left side of (15); this proves that for any x in X, $\int_0^t \mathbf{Z}(s)x\,ds$ belongs to $D(\mathbf{G})$, and that (15) holds.

It follows from the strong continuity of $\mathbf{Z}$ that for any x in X,

$$\lim_{t \to 0} \frac{1}{t} \int_0^t \mathbf{Z}(s)x\,ds = x;$$

this proves that $D(\mathbf{G})$ is dense in X.

(iii) We argue similarly about the domain of higher powers of $\mathbf{G}$. Denote by ϕ any infinitely differentiable function on $\mathbb{R}$ supported on $[0, 1]$. For any x in X we define

$$x_\phi = \int \phi(s)\,\mathbf{Z}(s)x\,ds.$$

The same argument as above shows that x_ϕ belongs to the domain of $\mathbf{G}$, and that

$$\mathbf{G}x_\phi = -\int \phi'(s)\,\mathbf{Z}(s)x\,ds.$$

Clearly, x_ϕ belongs to the domain of all $\mathbf{G}^n$. We choose now a sequence of ϕ_j that are nonnegative, satisfy $\int \phi_j\,ds = 1$, and whose support tends to zero. Appealing once more to strong continuity, we conclude that x_{ϕ_j} tends to x. This proves that $D(\mathbf{G}^n)$ is dense in X.

(iv) We claim that the following integrated form of (14) is valid for all x in $D(\mathbf{G})$:

$$\mathbf{Z}(t)x - x = \int_0^t \mathbf{Z}(s)\mathbf{G}x\,ds, \tag{15'}$$

where the integral on the right is a Riemann integral. For proof we appeal to the basic theorem of calculus: if two functions whose value lies in a Banach space, and that have continuous strong derivatives, have the same derivative and are equal at $t = 0$, then they are equal for all t.

Exercise 6. Prove the basic theorem of calculus for vector-valued functions.

We apply the fundamental theorem to the functions on the two sides of (15'). Both are 0 for $t = 0$. By (14), the derivative of the function on the left is $\mathbf{Z}(t)\mathbf{G}x$. The derivative of the indefinite integral on the right is also $\mathbf{Z}(t)\mathbf{G}x$; this proves (15').

Let $\{x_n\}$ be a sequence of elements in the domain of $\mathbf{G}$ such that $x_n \to x$, $\mathbf{G}x_n \to y$. We claim that x lies in $D(\mathbf{G})$, and that $\mathbf{G}x = y$. Take x to be x_n in (15'),

$$\mathbf{Z}(t)x_n - x_n = \int_0^t \mathbf{Z}(s)\mathbf{G}x_n \, ds,$$

and let $n \to \infty$. Both sides converge, and their limits are equal:

$$\mathbf{Z}(t)x - x = \int_0^t \mathbf{Z}(s)y \, ds.$$

Divide by t, and let $t \to 0$. The right side tends to y; this shows that x belongs to $D(\mathbf{G})$ and that $\mathbf{G}x = y$.

(v) The Laplace transform of $\mathbf{Z}$ is defined as

$$\mathbf{L}(\zeta)x = \int_0^\infty e^{-\zeta s} \mathbf{Z}(s)x \, ds, \tag{16}$$

where the integral on the right is defined as the limit of the Riemann integral from 0 to T as $T \to \infty$. Since, by (8), $\mathbf{Z}$ grows at most exponentially, $|\mathbf{Z}(s)| \leq b\,e^{ks}$, it follows that the integral on the right of (16) converges when Re $\zeta > k$, and that

$$|\mathbf{L}(\zeta)x| \leq \int_0^\infty b\,e^{(k-\mathrm{Re}\,\zeta)s} |x|\, ds = \frac{b}{\mathrm{Re}\,\zeta - k}|x|.$$

This shows that $\mathbf{L}(\zeta)$ is a bounded operator and that

$$|\mathbf{L}(\zeta)| \leq \frac{b}{\mathrm{Re}\,\zeta - k}. \tag{17}$$

We claim that $\mathbf{L}(\zeta) = \mathbf{R}(\zeta)$, that is, the inverse of $\zeta\mathbf{I} - \mathbf{G}$. To prove this, we look at the modified semigroup $e^{-\zeta t}\mathbf{Z}(t)$. It is easily verified that this is also a strongly continuous semigroup, and that its infinitesimal generator is $\mathbf{G} - \zeta\mathbf{I}$, where $\mathbf{G}$ denotes the generator of the original semigroup. We apply (15) to the modified semigroup:

$$e^{-\zeta t}\mathbf{Z}(t)x - x = (\mathbf{G} - \zeta I)\int_0^t e^{-\zeta s}\mathbf{Z}(s)x \, ds.$$

Suppose that Re $\zeta > k$; as $t \to \infty$, the left side tends to $-x$. The integral on the right tends to $\mathbf{L}(\zeta)x$; since $\mathbf{G}$ is a closed operator, we conclude that

$$x = (\zeta\mathbf{I} - \mathbf{G})\,\mathbf{L}(\zeta)x.$$

This shows that $\mathbf{L}(\zeta)$ is a right inverse of $(\zeta\mathbf{I} - \mathbf{G})$. To deduce from this that $\mathbf{L}(\zeta)$ is the inverse of $\zeta - \mathbf{G}$, we use (15') in place of (15). $\qquad\square$

As we will see in chapters 35 and 36, there are plenty of semigroups of operators that are strongly, but not uniformly, continuous; however replacing strong with weak continuity doesn't add much.

Theorem 5. *Let X denote a Banach space, $\mathbf{Z}(t)$: $X \rightarrow X$ a one-parameter family of bounded linear operators that is weakly continuous at $t = 0$:*

$$\lim_{t \to 0} (\mathbf{Z}(t)x, \ell) = (x, \ell)$$

for all x in X and ℓ in X'. Then $\mathbf{Z}(t)$ is strongly continuous.

Proof. This is a somewhat surprising result, for weak continuity demands much less than strong continuity. The proof, which we omit here, is rather tricky; see Hille-Phillips or Goldstein. □

34.2 THE GENERATION OF SEMIGROUPS

Theorem 6. *A strongly continuous semigroup of operators is uniquely determined by its infinitesimal generator.*

Proof. Suppose that $\mathbf{Z}(t)$ and $\mathbf{W}(t)$ have the same generator $\mathbf{G}$. Take x in $D(\mathbf{G})$; using the commutation rules (14) and (12), we get

$$\frac{d}{dt}\mathbf{W}(t)\mathbf{Z}(s - t)x = \mathbf{W}(t)\mathbf{G}\mathbf{Z}(s - t)x - \mathbf{W}(t)\mathbf{G}\mathbf{Z}(s - t)x = 0 \qquad (18)$$

for all x in $D(\mathbf{G})$. Then, by the fundamental theorem of calculus, we conclude that $\mathbf{W}(s)x = \mathbf{Z}(s)x$. Since $D(\mathbf{G})$ is dense, this holds for all x in X. □

We show now how to reconstruct $\mathbf{Z}(t)$ from $\mathbf{G}$ in the case where $\mathbf{Z}(t)$ is a *contraction*, that is,

$$|\mathbf{Z}(t)| \leq 1 \qquad \text{for all } t \geq 0. \qquad (19)$$

Theorem 7.

(i) *The infinitesimal generator $\mathbf{G}$ of a strongly continuous semigroup of contractions has every positive, real λ in its resolvent set, and*

$$|\mathbf{R}(\lambda)| = |(\lambda\mathbf{I} - \mathbf{G})^{-1}| \leq \frac{1}{\lambda}. \qquad (20)$$

(ii) *Conversely, every densely defined unbounded operator $\mathbf{G}$ whose resolvent set includes the positive reals, and whose resolvent is bounded by (20), is the generator of a strongly continuous semigroup of contractions.*

This remarkable result is called the Hille-Yosida theorem, after its discoverers.

Proof. Part (i) is just a restatement of inequality (17), with $b = 1$, $k = 0$.

We give Yosida's proof of part (ii). It is based on approximating $\mathbf{G}$ by $\mathbf{G}_n = n\mathbf{GR}(n)$ and letting $n \to \infty$. The identity

$$\mathbf{G}_n = n^2\mathbf{R}(n) - n\mathbf{I} \tag{21}$$

shows that $\mathbf{G}_n$ is a bounded operator. We approximate $\mathbf{Z}(t)$ by $\mathbf{Z}_n(t) = e^{t\mathbf{G}_n}$ where the exponential is defined as an infinite series. We show first: if (20) holds, then for all x in X,

$$\lim_{n \to \infty} n\mathbf{R}(n)x = x. \tag{22}$$

We use the identity

$$n\mathbf{R}(n) - \mathbf{I} = \mathbf{R}(n)\mathbf{G}$$

and inequality (20) to deduce that for x in $D(\mathbf{G})$,

$$|n\mathbf{R}(n)x - x| = |\mathbf{R}(n)\mathbf{G}x| \leq \frac{1}{n}|\mathbf{G}x|.$$

This proves (22) for x in $D(\mathbf{G})$. Since, by (20), the operator $n\mathbf{R}(n)$ has norm ≤ 1 for all n, and since $D(\mathbf{G})$ is dense in X, it follows that (22) holds for all x in X.

Next we show that for all x in $D(\mathbf{G})$,

$$\lim_{n \to \infty} \mathbf{G}_n x = \mathbf{G}x. \tag{23}$$

By definition of $\mathbf{G}_n$, for all x in $D(\mathbf{G})$, $\mathbf{G}_n x = n\mathbf{GR}(n)x = n\mathbf{R}(n)\mathbf{G}x$, so (23) follows from (22).

By definition of $\mathbf{Z}_n$, we obtain, using formula (21) for $\mathbf{G}_n$, that

$$\mathbf{Z}_n(t) = e^{t\mathbf{G}_n} = e^{-nt}e^{n^2\mathbf{R}(n)}t = e^{-nt}\sum_{0}^{\infty}\frac{(n^2t)^m}{m!}\mathbf{R}^m(n).$$

Using (20), we deduce that each $\mathbf{Z}_n(t)$ is a contraction:

$$|\mathbf{Z}_n(t)| \leq e^{-nt}\sum\frac{(n^2t)^m}{m!}\left(\frac{1}{n}\right)^m = e^{-nt}e^{nt} = 1. \tag{24}$$

To estimate the difference of $\mathbf{Z}_n$ and $\mathbf{Z}_k$, we use the fact that $\mathbf{G}_k$ and $\mathbf{G}_n$ commute with $\mathbf{Z}_n$ and $\mathbf{Z}_k$:

$$\frac{d}{dt}\mathbf{Z}_n(s-t)\mathbf{Z}_k(t)x = \mathbf{Z}_n(s-t)\mathbf{Z}_k(t)[\mathbf{G}_k - \mathbf{G}_n]x.$$

Using (24), we get that the norm of the right side is $\leq |\mathbf{G}_n x - \mathbf{G}_k x|$; integrating this inequality with respect to t from 0 to s, we deduce that

$$|\mathbf{Z}_n(s)x - \mathbf{Z}_k(s)x| \leq s|\mathbf{G}_n x - \mathbf{G}_k x|. \tag{25}$$

Combining (23) and (25), we deduce that for all x in $D(G)$, the limit

$$\lim_{n \to \infty} \mathbf{Z}_n(s)x = \mathbf{Z}(s)x \tag{26}$$

exists, uniformly on bounded sets of s. It follows then from the uniform bounded-ness of $|\mathbf{Z}_n|$ that (26) exists for all x in X. It follows directly that since $\mathbf{Z}_n(s)$ is a semigroup, so is $\mathbf{Z}(s)$. Since the convergence in (26) is uniform for $s < S$, strong continuity of $\mathbf{Z}_n$ implies strong continuity of $\mathbf{Z}$. Since by (24) each $\mathbf{Z}_n$ is a contraction, so is their strong limit $\mathbf{Z}$.

It remains to show that the generator of $\mathbf{Z}$ is $\mathbf{G}$. Apply $(15')$ to $\mathbf{Z}_n$:

$$\mathbf{Z}_n(t)x - x = \int_0^t \mathbf{Z}_n(s)\mathbf{G}_n x \, ds.$$

Suppose that x lies in $D(G)$; taking the limit $n \to \infty$, we get, using (23), that

$$\mathbf{Z}(t)x - x = \int_0^t \mathbf{Z}(s)\mathbf{G} x \, ds.$$

Denote by $\mathbf{H}$ the generator of $\mathbf{Z}(t)$. Dividing the above by t and letting $t \to 0$, we conclude that $D(\mathbf{H})$ includes $D(\mathbf{G})$ and that $\mathbf{H} = \mathbf{G}$ on $D(\mathbf{G})$. In other words, $\mathbf{H}$ is an *extension* of $\mathbf{G}$. However, since, by theorem 4, $\lambda > 0$ belongs to the resolvent set of both $\mathbf{G}$ and $\mathbf{H}$, $\mathbf{H}$ cannot be a proper extension of $\mathbf{G}$. $\square$

Another proof due to Hille is based on replacing the differential equation (14) by the backward difference equation

$$\frac{\mathbf{W}(t)x - \mathbf{W}(t - h)x}{h} = \mathbf{G}\mathbf{W}(t)x.$$

Solve this for $\mathbf{W}(t)$:

$$\mathbf{W}(t) = (\mathbf{I} - h\mathbf{G})^{-1}\,\mathbf{W}(t - h).$$

Set $h = t/n$, and set $t = h, 2h, \ldots$; we obtain recursively

$$\mathbf{W}(t) = \left(\mathbf{I} - \frac{t}{n}\mathbf{G}\right)^{-n}. \tag{27}$$

We denote these operators as $\mathbf{Z}_n$. We claim:

(i) Each $\mathbf{Z}_n(t)$ is a contraction.

(ii) $\mathbf{Z}_n$ converges strongly to a semigroup whose generator is $\mathbf{G}$.

(i) is an immediate consequence of (20). We omit the verification of (ii); see example 1 in section 34.3.

34.3 THE APPROXIMATION OF SEMIGROUPS

The result described in this section is of considerable importance in analyzing the convergence of discrete approximation to the solution of partial differential equations. We will first state it in the language of semigroups. We start with a strongly continuous one-parameter semigroup of operators $\mathbf{Z}(t)$, as in the previous sections, with infinitesimal generator $\mathbf{G}$. We discretise time into integer multiples of a small unit h, imagined to tend to zero eventually, and define recursively discrete approximations $u^{(n)}$ to $\mathbf{Z}(nh)x$ as follows:

$$u^{(n+1)} = \mathbf{C}_h\, u^{(n)}, \qquad u^{(0)} = x, \tag{28}$$

$\mathbf{C}_h$ a bounded operator depending on h. The function $u(t) = \mathbf{Z}(t)x$ satisfies the differential equation

$$\frac{\partial}{\partial t} u = \mathbf{G}x. \tag{29}$$

Therefore $(u^{n+1} - u^n)/h$ ought to be an approximation to $\mathbf{G}u$. Computing this difference quotient from (28), we are led to the following condition for the operator $\mathbf{C}_h$, called *consistency*:

$$\lim_{h \to 0} \left| \left(\frac{\mathbf{C}_h - I}{h} - \mathbf{G} \right) u(t) \right| = 0 \tag{30}$$

for $u(t) = \mathbf{Z}(t)x$, uniformly for all $0 \le t \le 1$, for a set of x that is dense in the domain D of $\mathbf{G}$.

The recursion (28) can be solved inductively to yield

$$u^{(n)} = \mathbf{C}_h^n x \tag{31}$$

as an approximation to $\mathbf{Z}(nh)x$. We impose now a second condition, called *stability*, that requires the approximations to depend boundedly on the initial vector x:

$$|\mathbf{C}_h^n| \le \text{const.} \qquad \text{for } nh \le 1 \tag{32}$$

and for some constant independent of n or h.

The following result is called *Lax's equivalence theorem*:

Theorem 8. *Let $\mathbf{Z}(t)$ be a strongly continuous one-parameter semigroup acting on a Banach space X. Let (28) be a discrete approximation to $\mathbf{Z}(t)$, satisfying the consistency condition (30). Then $u^{(n)} = \mathbf{C}_h^n x$ tends to $\mathbf{Z}(t)x$ as nh tends to t, for all x in X, iff the scheme is stable in the sense of condition (32).*

Proof. The necessity of condition (32) follows from the principle of uniform boundedness; see theorem 7 of chapter 15. To prove sufficiency, we use the following version of a high school algebra identity, valid for noncommuting operators

A and **B**:

$$\mathbf{A}^n - \mathbf{B}^n = \mathbf{A}^{n-1}(\mathbf{A} - \mathbf{B}) + \mathbf{A}^{n-2}(\mathbf{A} - \mathbf{B})\mathbf{B} + \cdots + (\mathbf{A} - \mathbf{B})\mathbf{B}^{n-1}.$$

We take $\mathbf{A} = \mathbf{C}_h$, $\mathbf{B} = \mathbf{Z}(h)$, and let both sides act on a vector x:

$$\mathbf{C}_h^n x - \mathbf{Z}(nh)x = \sum_0^{n-1} \mathbf{C}_h^{n-j-1}(\mathbf{C}_h - \mathbf{Z}(h))\mathbf{Z}(jh)x$$

$$= \sum_0^{n-1} \mathbf{C}_h^{n-j-1}(\mathbf{C}_h - \mathbf{Z}(h))u(jh); \tag{33}$$

here we have used the semigroup property, and the notation $u(t) = \mathbf{Z}(t)x$.

By the definition of infinitesimal generator, for every x in the domain of $\mathbf{G}$,

$$\mathbf{Z}(h)x = x + h\mathbf{G}x + s_h,$$

where s_h is a vector whose norm $|s_h|$ is $o(h)$. So, for any t,

$$\mathbf{Z}(h)u(t) = \mathbf{Z}(h)\mathbf{Z}(t)x = \mathbf{Z}(t)\mathbf{Z}(h)x$$
$$= \mathbf{Z}(t)[x + h\mathbf{G}x + S_h] = u(t) + h\mathbf{G}u(t) + \mathbf{Z}(t)s_h.$$

We use this expression of $\mathbf{Z}(h)u$ to write

$$(\mathbf{C}_h - \mathbf{Z}(h))u = (\mathbf{C}_h - \mathbf{I} - h\mathbf{G})u + \mathbf{Z}(t)s_h.$$

Setting this into (33) gives

$$\mathbf{C}_h^n x - \mathbf{Z}(nh)x = \sum \mathbf{C}_h^{n-j-1}[\mathbf{C}_h - \mathbf{I} - h\mathbf{G}]u(jh) + \sum \mathbf{C}_h^{n-j-1}\mathbf{Z}(jh)s_h. \tag{33'}$$

Using the consistency condition (30), the estimate $|s_h| = o(h)$, the stability condition (32), and the boundedness of $\|Z(t)\|$, we conclude that for $t \leq 1$ the right side of (33') is bounded in norm by $\sum_0^{n-1} o(h) = no(h)$, which for $nh \leq 1$ tends to zero as h tends to zero. This proves that $\mathbf{C}_h^n x$ tends to $\mathbf{Z}(t)x$, as nh tends to t and h tends to zero, for all x for which (30) holds and for $t \leq 1$. The set of such x is dense in X; since the operators $\mathbf{C}_h^n$ and $\mathbf{Z}(t)$ are uniformly bounded, the conclusion holds for all x. The extension to $t > 1$ is obvious. $\square$

The equivalence theorem is the framework of finite difference and other discrete approximations for solving time-dependent partial differential equations; see Lax and Richtmyer. There $\mathbf{C}_h$ is a difference operator in the space variables, G a partial differential operator; consistency is verified, in the class of smooth functions, by using Taylor's theorem.

The equivalence theorem is valid beyond the setting of semigroups; it hold equally for approximations of form (28) to solutions of differential equations $u_t = \mathbf{G}u$, where $\mathbf{G}$ is a linear operator that depends on t.

The literature of this field is enormous (e.g., see Richtmyer and Morton).

Here are some applications of theorem 8 in an abstract framework.

Example 1. Suppose that $\mathbf{Z}(t)$ is a semigroup of contractions, and let us choose for $\mathbf{C}_h$ the backward difference operator (27) employed by Hille in the proof sketched at the end of the previous section:

$$\mathbf{C}_h = (\mathbf{I} - h\mathbf{G})^{-1}. \tag{34}$$

We claim that this choice of $\mathbf{C}_h$ leads to a consistent and stable difference scheme. Consistency follows from this string of algebraic identities:

$$\frac{\mathbf{C}_h - \mathbf{I}}{h} - \mathbf{G} = (\mathbf{I} - h\mathbf{G})^{-1}\frac{(\mathbf{I} - (\mathbf{I} - h\mathbf{G}))}{h} - \mathbf{G} = (\mathbf{I} - h\mathbf{G})^{-1}\mathbf{G} - \mathbf{G}$$

$$= (\mathbf{I} - h\mathbf{G})^{-1}[\mathbf{G} - (\mathbf{I} - h\mathbf{G})\mathbf{G}] = h(\mathbf{I} - h\mathbf{G})^{-1}\mathbf{G}^2$$

$$= (h^{-1}\mathbf{I} - \mathbf{G})^{-1}\mathbf{G}^2.$$

Clearly, if x belongs to the domain of $\mathbf{G}^2$, then it follows from estimate (20) that the consistency condition (30) is satisfied. Since we have shown in theorem 4 that the domain of $\mathbf{G}^2$ is dense, consistency follows.

Exercise 7. Prove, using (20), that the scheme (34) is stable.

Theorem 8 is only one of several approximation theorems for semigroups. Other versions are due to Trotter, Kato, and Chernoff (see sections 7 and 8 of chapter 1 of Goldstein, and Strang). A useful consequence of Trotter's theorem is *Trotter's product formula.*

Suppose that $\mathbf{G}$ *and* $\mathbf{H}$ *are generators of strongly continuous semigroups* $\mathbf{T}$ *and* $\mathbf{S}$ *of contractions, and that the closure of* $\mathbf{G} + \mathbf{H}$ *also generates such a semigroup* $\mathbf{Z}$. *Then as* $nh \to t, h \to 0$,

$$\lim [\mathbf{T}(h)\mathbf{S}(h)]^n x = \mathbf{Z}(t)x \tag{35}$$

for all x in X.

The significance of this result is that many infinitesimal generations of physical processes have a natural decomposition as a sum. Also in many problems it is advantageous to compute approximately $\mathbf{T}(h)$ and $\mathbf{S}(h)$ by entirely different methods. Strang has pointed out that

$$\left[\mathbf{T}\left(\frac{h}{s}\right)\mathbf{S}(h)\mathbf{T}\left(\frac{h}{2}\right)\right]^n x = \mathbf{T}\left(\frac{h}{2}\right)[\mathbf{S}(h)\mathbf{T}(h)]^{n-1}\mathbf{S}(h)\mathbf{T}\left(\frac{h}{2}\right)x$$

is a much better approximation to $\mathbf{Z}(nh)x$ than $[\mathbf{T}(h)\mathbf{S}(h)]^n x$.

Exercise 8. Why?

It is natural to ask if it is possible in theorem 8 to *dispense* with the hypothesis that $\mathbf{G}$ is the generator of a strongly continuous semigroup, and instead *deduce* this from the existence of a *stable* difference scheme (28) that is *consistent* with $\mathbf{G}$. In the context of partial differential equations, this would amount to proving the existence of solutions of partial differential equations by approximating them with solutions of difference equations. We offer only a weak abstract result; for simplicity we take X to be a Hilbert space.

Let G be a densely defined closed operator whose adjoint also is densely defined. A *weak solution* of the equation

$$\frac{\partial}{\partial t}u - Gu = 0, \quad u(0) = x, \tag{36}$$

is defined as follows:

Let $w(t)$ be any continuously differentiable vector-valued function of t whose values lie in the domain of G^*, and $G^*w(t)$ is a continuous function of t. Form the scalar product of $w(t)$ with (36) and integrate with respect to t from 0 to 1:

$$\int_0^1 \left(w, \frac{\partial}{\partial t}u - Gu\right) dt = 0$$

Integrate by parts, and use the adjointness of G and G^*. Restrict w to vanish for $t \geq 1$; we get

$$\int \left(\frac{\partial}{\partial t}w + G^*w, u\right) dt + (w(0), x) = 0. \tag{37}$$

A function $w(t)$ satisfying all conditions specified above is called an *admissible test function*.

Definition. An integrable vector-valued function $u(t)$ that satisfies (37) for all admissible test functions w is called a *weak solution* of (36).

We describe now briefly how approximations of form (28) as can be used to construct weak solutions. We rewrite (28) as

$$\frac{u^{(n+1)} - u^{(n)}}{h} = G_h u^{(n)}, \quad u^{(o)} = x, \tag{38}$$

where

$$G_h = \frac{C_h - I}{h}.$$

We replace the consistency condition (30) by its dual:

$$G_h^* w \rightarrow G^* w \tag{39}$$

for all w in the domain of G^*.

The solution of (38) is given by formula (31): $u^{(n)} = C_h^n x$. We retain the stability condition (32).

We introduce the Hilbert space H as the completion in the H-norm of continuous vector-valued functions $w(t)$; the H-norm is defined as

$$\|w\|_H^2 = \int_0^1 \|w(t)\|^2 dt.$$

Given a solution of the difference equation (38), we extend it as a function of t by setting

$$u_h(t) = u^{(n)} \quad \text{for } nh \leq t < (n+1)h.$$

Clearly, the H-norm of u_h is

$$\|u_h\|_H^2 = h \sum_0^N \|u^{(n)}\|^2, \qquad Nh = 1.$$

The stability condition implies that $\|u^{(n)}\| \leq$ const. for all n, $nh \leq 1$; it follows that $\|u_h\|_H \leq$ const. for all h. We appeal now to the weak sequential compactness of bounded sets in Hilbert space (see chapter 10) and conclude that we can select a subsequence of $h \to 0$ so that $\{u_h\}$ converges weakly to some limit u in H. We claim

Theorem 9. *The weak limit u is a weak solution of (36).*

Proof. To show this, take any admissible test function $w(t)$, and take the scalar product of $w^{(n)} = w(nh)$ with (38). Multiply by h and sum with respect to n:

$$h \sum_0^N \left(w^{(n)}, \frac{u^{(n+1)} - u^{(n)}}{h} \right) - \left(w^{(n)}, G_h u^{(n)} \right) = 0.$$

Sum by parts, and use the adjointness of G_h and G_h^*. Since $w(t) = 0$ for $t \geq 1$, we get

$$h \sum_1^N \left(\frac{w^{(n)} - w^{(n-1)}}{h} + G_h^* w^{(n)}, u^{(n)} \right) + (w(0), x) = 0. \qquad (40)$$

Now replace in (40) the function $w(t)$ by $w(t - s)$, and integrate over the interval $0 \leq s \leq h$. The result can be written as

$$\int \left(\frac{w(t) - w(t - h)}{h} + G_h^* w(t), u \right) dt + (w(0), x). \qquad (40')$$

Now let h tend to zero along the previously selected subsequence. Since admissible test functions are differentiable, and $G_h^* w(t)$ tends to $G^* w(t)$ uniformly for all $t \leq 1$,

(40') tends to

$$\int \left(\frac{\partial}{\partial t} w + \mathbf{G}^* w, u \right) dt + (w(0), x) = 0.$$

This proves that u is a weak solution of (36).   □

REMARK 1. The preceding argument shows that the weak limit lies in H; that is, it is square integrable. It is easy to show that it is in fact bounded. Take any interval $[a, b]$ in $[0, 1]$; according to the stability condition, $\|u^{(n)}\|^2 \leq$ const. for all n. It follows that

$$\int_a^b \|u_h(t)\|^2 dt = h \sum_{a < nh < b} \|u^{(n)}\|^2 \leq (b - a)\text{const.}$$

The L^2-norm is lower semicontinuous under weak convergence, so

$$\int_a^b \|u(t)\|^2 dt \leq \liminf \int_a^b \|u_h\|^2 dt \leq (b - a)\text{const.}$$

This proves that $\|u(t)\| \leq$ const. for almost all t.

REMARK 2. The consistency condition (39) can be relaxed by requiring (39) hold only on some dense linear subspace W of the domain of $\mathbf{G}^*$. This requires a corresponding change in what test functions are admissible.

REMARK 3. A famous theorem of Friedrichs shows that for first-order partial differential operators, a weak solution u is a *strong solution*, that is, an L^2 limit of genuine solutions. I don't know of any comparable abstract result.

34.4 PERTURBATION OF SEMIGROUPS

Rellich has shown (see theorem 5 in chapter 33) that if one adds to a self-adjoint operator $\mathbf{A}$ another symmetric operator not too large compared to $\mathbf{A}$, then the sum also is self-adjoint. A similar result holds for generators of contraction semigroups. Before we state it, we explain a reformulation due to Lumer and Phillips of the Hille-Yosida condition on the generators of contraction semigroups, For simplicity we state it only for semigroups acting on a Hilbert space.

Lemma 10 (Lumer-Phillips). *Let $\mathbf{G}$ denote a densely defined operator on a Hilbert space H whose resolvent set includes $\mathbb{R}_+$. Then inequality (20), necessary and sufficient for $\mathbf{G}$ to generate a semigroup of contractions, is equivalent to*

$$\text{Re}\,(x, \mathbf{G}x) \leq 0 \tag{20'}$$

for all x in the domain of $\mathbf{G}$.

An operator satisfying (20') is called *dissipative*.

Proof. Condition (20) asserts that for all u in H and all $\lambda > 0$,

$$\|(\lambda \mathbf{I} - \mathbf{G})^{-1} u\|^2 \leq \frac{1}{\lambda^2} \|u\|^2.$$

Denote $(\lambda \mathbf{I} - \mathbf{G})^{-1} u$ as x. Then the inequality above can be rewritten as

$$(x, x) \leq \frac{1}{\lambda^2} (\lambda x - \mathbf{G} x, \lambda x - \mathbf{G} x).$$

Expanding the right side, cancelling (x, x) on both sides, and multiplying by λ, rearranging terms, we get

$$(x, \mathbf{G} x) + (\mathbf{G} x, x) \leq \frac{1}{\lambda} \|\mathbf{G} x\|^2.$$

Letting λ tend to ∞ we obtain (20').

The converse can be shown by running the proof backward.   □

The following result is due to Trotter:

Theorem 11. *Suppose that* $\mathbf{G}$ *is the generator of a semigroup of contractions on a Hilbert space* H. *Let* $\mathbf{H}$ *be an operator with the following properties:*

 (i) *The domain of* $\mathbf{H}$ *includes the domain of* $\mathbf{G}$.

 (ii) $\mathbf{H}$ *is dissipative.*

 (iii) *There exist numbers* a *and* b, $a < 1$, *such that*

$$\|\mathbf{H} x\| \leq a \|\mathbf{G} x\| + b \|x\| \tag{41}$$

for all x *in the domain of* $\mathbf{G}$. *Then* $\mathbf{G} + \mathbf{H}$, *defined on the domain of* $\mathbf{G}$, *generates a semigroup of contractions.*

Proof. Since $\mathbf{G}$ is the generator of a strongly continuous semigroup, it is a closed operator. We claim that so is $\mathbf{G} + \mathbf{H}$. To see this, let x_n be a convergent sequence such that $(\mathbf{G} + \mathbf{H}) x_n = y_n$ also converges. We write $\mathbf{G} x_n = y_n - \mathbf{H} x_n$ and form the difference

$$\mathbf{G}(x_n - x_m) = y_n - y_m - \mathbf{H}(x_n - x_m).$$

Using inequality (41) on the right, we conclude that $\mathbf{G} x_n$ converges, and consequently also $\mathbf{H} x_n$ converges. Since $\mathbf{G}$ is closed, $\mathbf{G} x_n \to \mathbf{G} x$ where $x_n \to x$. So x belongs to the domain of $\mathbf{G}$. That $\mathbf{H} x_n$ converges to $\mathbf{H} x$ follows from (41).

We claim that λ sufficiently large positive belongs to the resolvent set of $\mathbf{G} + \mathbf{H}$. First we show that for every x in the domain of $\mathbf{G}$,

$$\|x\| \leq \frac{1}{\lambda} \|(\lambda \mathbf{I} - \mathbf{G} + \mathbf{H})x\|. \tag{42}$$

According to the Lumer-Phillips lemma, $\mathbf{G}$, being the generator of a semigroup of contractions, is dissipative. $\mathbf{H}$ is dissipative by hypothesis; therefore so is their sum. Inequality (42) follows from the converse part of the Lumer-Phillips lemma.

If follows from (42) that the range of $\mathbf{G} + \mathbf{H} - \lambda \mathbf{I}$ is closed. To show that it is the whole space we argue indirectly; if it were not, there would a nonzero vector v perpendicular to the range: for all x in $D(\mathbf{G})$,

$$((\mathbf{G} + \mathbf{H} - \lambda \mathbf{I})x, v) = 0. \tag{43}$$

Since $\mathbf{G}$ generates a semigroup of contractions, $\mathbf{G} - \lambda \mathbf{I}$ is invertible; so there is an x in the domain of $\mathbf{G}$ such that

$$(\mathbf{G} - \lambda \mathbf{I})x = v. \tag{44}$$

Setting this into (43), we get

$$\|v\|^2 + (\mathbf{H}x, v) = 0.$$

Using the Schwarz inequality to estimate the second term, we get

$$\|v\| \leq \|\mathbf{H}x\|. \tag{45}$$

Using (44) to express v, and inequality (41) to estimate the right side of (45), we get

$$\|\mathbf{G}x - \lambda x\| \leq a\|\mathbf{G}x\| + b\|x\|.$$

Square both sides; using the fact that $\mathbf{G}$ is dissipative, we get

$$\|\mathbf{G}x\|^2 + \lambda^2 \|x\|^2 \leq a^2 \|\mathbf{G}x\|^2 + 2ab\|\mathbf{G}x\|\|x\| + b^2 \|x\|^2. \tag{46}$$

Since $a < 1$, for λ large enough we conclude that $\|x\| = 0$. That makes $x = 0$ and $v = (\mathbf{G} - \lambda \mathbf{I})x = 0$; this proves that the range of $\mathbf{G} + \mathbf{H} - \lambda \mathbf{I}$ is the whole space.

This, combined with the Lumer-Phillips lemma, shows that $\mathbf{G} + \mathbf{H}$ generates a semigroup of contractions.   □

A statement and proof of Trotter's perturbation theorem in a Banach space setting is given in Goldstein.

34.5 THE SPECTRAL THEORY OF SEMIGROUPS

We saw at the beginning of this chapter that when the generator $\mathbf{G}$ is a bounded operator, the semigroup is the exponential of $\mathbf{G}$:

$$\mathbf{Z}(t) = e^{t\mathbf{G}}, \tag{47}$$

by the spectral mapping theorem, theorem 4 of chapter 18,

$$\sigma(\mathbf{Z}(t)) = e^{t\sigma(\mathbf{G})}. \tag{48}$$

When $\mathbf{G}$ is unbounded, (47) holds only in a symbolic sense. The question is: does (48) hold?

It is easy to show that if γ is an eigenvalue of $\mathbf{G}$, then $e^{t\gamma}$ is an eigenvalue of $\mathbf{Z}(t)$. To see this, let u be a corresponding eigenvector: $\mathbf{G}u = \gamma u$. Then

$$\frac{d}{dt}e^{-\gamma t}\mathbf{Z}(t)u = \mathbf{Z}(t)(\mathbf{G} - \gamma\mathbf{I})u = 0,$$

which means that $e^{-\gamma t}\mathbf{Z}(t)$ is independent of t. Since its value at $t = 0$ is u, it follows that $e^{-\gamma t}\mathbf{Z}(t)u = u$ for all t. This shows that $e^{\gamma t}$ is an eigenvalue of $\mathbf{Z}(t)$.

Ralph Phillips has shown that the same conclusion holds for any γ in the spectrum of $\mathbf{G}$:

Theorem 12. $\mathbf{Z}(t)$ *is a strongly continuous one-parameter semigroup of operators,* $\mathbf{G}$ *its generator. Then*

$$\sigma(\mathbf{Z}(t)) \supset e^{t\sigma(\mathbf{G})}. \tag{48$'$}$$

Proof. The operators $\mathbf{Z}(t)$ and $\mathbf{R}(\zeta)$ commute for any ζ in the resolvent set of $\mathbf{G}$. Adjoin to this collection of operators their resolvents, and denote by $\mathcal{A}$ the closure in the uniform topology of the algebra generated by these operators. $\mathcal{A}$ is a Banach algebra, and the spectrum of $\mathbf{Z}$ and $\mathbf{R}$ as operators mapping $X \to X$ is the same as their spectrum as elements of the algebra $\mathcal{A}$.

We define the one-parameter family of operators $\mathbf{V}(t)$ by

$$\mathbf{V}(t) = \mathbf{R}(\zeta)\mathbf{Z}(t), \qquad \zeta > k; \tag{49}$$

ζ lies in the resolvent of $\mathbf{G}$. We claim that $\mathbf{V}(t)$ depends continuously on t in the *uniform* topology. To see this, substitute the formula (16) for $\mathbf{R}$ into (49):

$$\begin{aligned}
\mathbf{V}(t)x &= \int_0^\infty \mathbf{Z}(s)e^{-\zeta s}\,\mathbf{Z}(t)x\,ds \\
&= \int_0^\infty \mathbf{Z}(s+t)xe^{-\zeta s}\,ds = e^{\zeta t}\int_t^\infty \mathbf{Z}(r)xe^{-\zeta r}\,dr;
\end{aligned}$$

continuity in the uniform topology follows. Since $\mathbf{Z}$ is a one-parameter semigroup that commutes with $\mathbf{R}$, it follows from (49) that

$$\mathbf{R}(\zeta)\mathbf{V}(t+s) = \mathbf{V}(t)\,\mathbf{V}(s). \tag{50}$$

By (9), γ belongs to the spectrum of $\mathbf{G}$ iff $(\zeta - \gamma)^{-1}$ belongs to the spectrum of $\mathbf{R}(\zeta)$:

$$\sigma(\mathbf{R}(\zeta)) = (\zeta - \sigma(\mathbf{G}))^{-1}. \tag{51}$$

According to Gelfand's theory (see theorem 14 of chapter 18) the spectrum of $\mathbf{R}$ is $\sigma(\mathbf{R}) = \{p(\mathbf{R})\}$ as p ranges over all homomorphisms p of $\mathcal{A}$ into $\mathbb{C}$. Combining this with (51), we conclude that for every γ in $\sigma(\mathbf{G})$ there is a $p : \mathcal{A} \to \mathbb{C}$ such that

$$p(\mathbf{R}(\zeta)) = (\zeta - \gamma)^{-1}. \tag{51'}$$

Let p act on (50); since p is multiplicative, we get

$$p(\mathbf{R})p(\mathbf{V}(t + s)) = p(\mathbf{V}(t))p(\mathbf{V}(s)). \tag{52}$$

It follows from (51') that $p(\mathbf{R}) \neq 0$. We define

$$m(t) = (p\mathbf{V}(t))/p(\mathbf{R}) \tag{53}$$

and rewrite (52) as

$$m(t + s) = m(t)m(s). \tag{54}$$

We have shown above that $\mathbf{V}(t)$ is a continuous function of t in the uniform topology; the homomorphisms are continuous in the uniform topology. Combining these, we conclude that $p(\mathbf{V}(t))$, and therefore $m(t)$, is a continuous function of $\mathbb{R} \to \mathbb{C}$. As is well known, all continuous solutions of (54) are of the form

$$m(t) = e^{\kappa t}. \tag{55}$$

Apply p to (49) to get $p(\mathbf{V}(t)) = p(\mathbf{R})p(\mathbf{Z}(t))$; combining this with (53) and (55) gives

$$p(\mathbf{Z}(t)) = e^{\kappa t}. \tag{55'}$$

Now multiply (16) by $\mathbf{R}$:

$$\mathbf{R}^2 x = \int_0^\infty e^{-\zeta s} \mathbf{R}\,\mathbf{Z}(s)\,x\,ds.$$

As shown above, $\mathbf{R}\mathbf{Z}(s)$ is continuous in the *uniform* topology; therefore the integral above exists in the uniform topology

$$\mathbf{R}^2 = \int_0^\infty e^{-\zeta s} \mathbf{R}\,\mathbf{Z}(s)\,ds.$$

Apply p to both sides and use (55'):

$$p(\mathbf{R})^2 = \int_0^\infty e^{-\zeta s} p(\mathbf{R}) p(\mathbf{Z}(s))\,ds = p(\mathbf{R}) \int_0^\infty e^{-\zeta s} e^{\kappa s}\,ds = \frac{p(\mathbf{R})}{\zeta - \kappa}.$$

Dividing by $p(\mathbf{R})$, we get $p(\mathbf{R}) = (\zeta - \kappa)^{-1}$. Comparing this with (51'), we conclude that $\kappa = \gamma$. Setting this into (55') gives

$$p(\mathbf{Z}(t)) = e^{\gamma t}. \tag{56}$$

According to Gelfand's theory, theorem 14 of chapter 19, $p(\mathbf{Z}(t))$ is in the spectrum of $\mathbf{Z}(t)$. Since γ is any point in the spectrum of $\mathbf{G}$, we conclude from (56) that $\sigma(\mathbf{Z}(t))$ contains $e^{t\sigma(\mathbf{G})}$, as asserted in (48'). □

The inclusion in (48') is in some cases proper. Phillips has, however, shown

Theorem 12'. *Let $\mathbf{Z}(t)$ denote a strongly continuous one-parameter semigroup of operators, $\mathbf{G}$ its infinitesimal generator. Suppose that for some $T > 0$, $\mathbf{Z}(t)$ is uniformly continuous for $t \geq T$. Then every nonzero point in the spectrum of $\mathbf{Z}(t)$ is of the form $\sigma(\mathbf{Z}(t)) = e^{t\sigma(\mathbf{G})}$.*

Proof. The spectrum of $\mathbf{Z}(t)$ is the range of the homomorphisms $p : \mathcal{A} \to \mathbb{C}$. Applying p to $\mathbf{Z}(s + t) = \mathbf{Z}(s)\mathbf{Z}(t)$ and abbreviating $p(\mathbf{Z}(t))$ as $n(t)$, we get

$$n(s + t) = n(s)n(t).$$

Since $\mathbf{Z}(t)$ is assumed uniformly continuous for $t \geq T$, it follows from the definition that $n(t)$ is continuous for $t \geq T$. The only such solutions of this functional equation $\not\equiv 0$ are the exponentials: $n(t) = e^{\nu t}$. The rest of the proof proceeds as that of theorem 12. □

Theorem 13. *Let $\mathbf{Z}(t)$ be a strongly continuous one-parameter semigroup of operators, $\mathbf{G}$ its infinitesimal generator. Suppose that for some $T > 0$, $\mathbf{Z}(T)$ is compact. Then*

(i) *the nonzero part of the spectrum of $\mathbf{Z}(t)$ is of the form $\sigma(\mathbf{Z}(t)) = e^{t\sigma(\mathbf{G})}$.*

(ii) *the spectrum of $\mathbf{G}$ consists of discrete points $\{\gamma_j\}$, $\operatorname{Re}\gamma_1 \geq \operatorname{Re}\gamma_2 \geq \ldots, \operatorname{Re}\gamma_j$ $\to -\infty$.*

(iii) *for every vector x, $\mathbf{Z}(t)x$ has an asymptotic expansion for large t of the form*

$$\mathbf{Z}(t)x \sim \Sigma e^{\gamma_j t} p_j(t), \tag{57}$$

where p_j are polynomials in t whose coefficients are generalized eigenvectors of $\mathbf{G}$.

Exercise 9. Prove theorem 13.

We turn now to the question of the transpose of a semigroup and of its generator.

Theorem 14. *Let* X *be a reflexive Banach space,* $\mathbf{Z}(t) : X \to X$ *a strongly continuous one-parameter semigroup of operators. Then its transpose* $\mathbf{Z}'(t) : X' \to X'$ *is likewise a strongly continuous one-parameter semigroup of operators, generated by the transpose of the generator of* $\mathbf{Z}(t)$.

Proof. By definition of the transpose,

$$\big(\mathbf{Z}(t)x, \ell\big) = (x, \mathbf{Z}'(t)\ell) \tag{58}$$

for all x in X and all ℓ in X'. From this and the reflexivity of X we deduce that $\mathbf{Z}'(t)$ is weakly sequentially continuous. But then, by theorem 5, $\mathbf{Z}'(t)$ is strongly continuous.

Denote the generator of $\mathbf{Z}'(t)$ by $\mathbf{H}$, that of $\mathbf{Z}(t)$ by $\mathbf{G}$. Choose x to be in $D(\mathbf{G})$, ℓ in $D(\mathbf{H})$. Differentiate (58) with respect to t and set $t = 0$:

$$(\mathbf{G}x, \ell) = (x, \mathbf{H}\ell).$$

Comparing this with the definition of $\mathbf{G}'$, we conclude that $\mathbf{G}'$ is an extension of $\mathbf{H}$. By theorem 4, all $\lambda > k$ belong to the resolvent set of $\mathbf{H}$; the resolvent set of $\mathbf{G}'$, by theorem 3 the same as that of $\mathbf{G}$, also contains these points. But then $\mathbf{G}'$ cannot be a proper extension of $\mathbf{H}$, and thus $\mathbf{G}'$ is the generator of $\mathbf{Z}'(t)$.   □

BIBLIOGRAPHY

Chernoff, P. R. Note on product formulas for operator semigroups. *J. Func. Anal.*, **2** (1968): 238–242.

Friedrichs, K. O. The identity of the weak and strong extension of differential operators. *Trans. AMS*, **55** (1944): 132–151.

Goldstein, J. A. *Semigroups of Linear Operators and Applications*. Oxford University Press, Oxford, 1985.

Hille, E. and Phillips, R. S. *Functional Analysis and Semigroups*. AMS Coll. Publ., **31**. American Mathematical Society, New York, 1957.

Kato, T. On the Trotter-Lie product formula. *Proc. Japan Acad.*, **50** (1974): 694–698.

Lax, P. D. and Richtmyer, R. D. Survey of the stability of linear finite difference equations. *CPAM*, **9** (1956): 267–293.

Lumer, G. and Phillips, R. S. Dissipative operators in a Banach space. *Pac. J. Math.*, **11** (1961): 679–698.

Phillips, R. S. Spectral theory for semigroups of linear operators. *Trans. AMS*, **74** (1951): 393–415.

Richtmyer, R. D. and Morton, K. W. *Difference Methods for Initial Value Problems*, 2nd ed. Interscience, New York, 1967.

Strang, G. Approximation of semigroups and the consistency of difference schemes. *Proc. AMS*, **20** (1969): 1–7.

Trotter, H. F. On the product of semigroups of operators. *Proc. AMS*, **10** (1959): 545–551.

Trotter, H. F. Approximation of semi-groups of operators. *Pac. J. Math.*, **8** (1958): 887–919.

Yosida, K. On the differentiability and the representation of one-parameter semigroups of linear operators. *J. Math. Soc. Jap.*, **1** (1948): 15–21.

Yosida, K. *Functional Analysis*. Springer Verlag, 1965.

35

GROUPS OF UNITARY
OPERATORS

The mathematical landscape is full of groups of unitary operators. The ones we will consider in this chapter, strongly continuous one-parameter groups $U(t)$, $-\infty < t < \infty$, come mostly from three sources: processes where energy is conserved, such as those governed by wave equations of all sorts; processes where probability is preserved, for instance, ones governed by Schrödinger equations; and Hamiltonian and other measure-preserving flows.

35.1 STONE'S THEOREM

Theorem 1. *Let **A** be a self-adjoint operator acting on a Hilbert space H.*

 (i) *There exists a strongly continuous group **U**(t) of unitary operators whose infinitesimal generator is i**A**.*

 (ii) *Conversely, every strongly continuous group of unitary operators is generated by i**A**, **A** some self-adjoint operator.*

Proof. (i) We saw in theorem 5 of chapter 32 that every nonreal complex number z belongs to the resolvent set of any self-adjoint operator **A**, and [see (13) in chapter 32] that the resolvent is bounded by

$$\|\mathbf{R}(z)\| \leq |\operatorname{Im} z|^{-1}.$$

It follows from this that both i**A** and $-i$**A** satisfy the hypothesis of the Hille-Yosida theorem, theorem 7 in chapter 34, so both i**A** and $-i$**A** generate strongly continuous semigroups of contraction; denote these by **U**(t) and **V**(t). We claim that **V**(t) and **U**(t) are inverses. To see this, take any x in the domain of **A** and form

$$\mathbf{U}(t)\mathbf{V}(t)x.$$

This is a differentiable function of t, and its derivative is zero:

$$\mathbf{U}(t)\mathbf{A}\mathbf{V}(t)x - \mathbf{U}(t)\mathbf{A}\mathbf{V}(t)x = 0.$$

This proves that $\mathbf{U}(t)\mathbf{V}(t)x$ is independent of t. Since it is x at $t = 0$, it is x for all t. In other words, $\mathbf{U}(t)\mathbf{V}(t)$ is the identity on the domain of $\mathbf{A}$. Since the domain of $\mathbf{A}$ is dense in the Hilbert space H on which $\mathbf{A}$ acts, $\mathbf{U}(t)\mathbf{V}(t)$ is, by continuity, the identity on all of H. Reversing $\mathbf{U}$ and $\mathbf{V}$ shows that they are indeed inverses of each other.

According to the Hille-Yosida theorem, both $\mathbf{U}(t)$ and $\mathbf{V}(t)$ are contractions. On the other hand, their product is $\mathbf{I}$; this can only be if both are norm preserving. Since they are invertible, they are unitary; see section 31.7 for the basics of unitary operators.

Define $\mathbf{U}(t)$ for t negative as $\mathbf{V}(-t)$. Clearly, $\mathbf{U}(s + t) = \mathbf{U}(s)\mathbf{U}(t)$ is satisfied for all real t and s, and $d\mathbf{U}(t)x/dt = \mathbf{A}\mathbf{U}(t)x$ for all x in the domain of $\mathbf{A}$.

(ii) We turn now to proving the converse proposition. Let $\mathbf{U}(t)$, $-\infty < t < \infty$, be a strongly continuous group of unitary operators. Then $\mathbf{U}(t)$ and $\mathbf{V}(t) = \mathbf{U}(-t)$ are strongly continuous semigroups of contractions; their generators are negatives of each other, $\mathbf{G}$ and $-\mathbf{G}$. It follows from theorem 7 in chapter 34 applied to $\mathbf{G}$ and $-\mathbf{G}$ that all nonzero real numbers belong to the resolvent set of $\mathbf{G}$. Since $\mathbf{U}(t)$ is unitary,

$$\|\mathbf{U}(t)x\|^2 = (\mathbf{U}(t)x, \mathbf{U}(t)x) = \|x\|^2.$$

Choose x in the domain of $\mathbf{G}$; differentiating and setting $t = 0$, we get

$$(\mathbf{G}x, x) + (x, \mathbf{G}x) = 0.$$

Replacing x by $x + y$, we deduce that the real part of

$$(\mathbf{G}x, y) + (x, \mathbf{G}y) = 0. \tag{1}$$

Replacing y by iy shows that (1) holds for all x and y in the domain of $\mathbf{G}$. Equation (1) says that $\mathbf{G}$ is antisymmetric. It follows that $\mathbf{G}^*$ is an extension of $-\mathbf{G}$. According to the Hilbert space version of theorem 3 in chapter 34 (see exercise 5), the resolvent set of $\mathbf{G}^*$ is the complex conjugate of the resolvent set of $\mathbf{G}$. We have shown above that all nonzero real numbers belong to the resolvent set of $\mathbf{G}$; therefore they belong to the resolvent set of $\mathbf{G}^*$. Since they also belong to the resolvent set of $-\mathbf{G}$, the one cannot be an extension of the other. Therefore $\mathbf{G}^* = -\mathbf{G}$. □

The spectral resolution of a self-adjoint transformation $\mathbf{A}$ can be thought of as defining the functions $c_S(\mathbf{A})$, where c_S is the characteristic function of the Borel set S. The three constructions we gave in chapter 33 for the spectral resolution of $\mathbf{A}$ all started with a more limited functional calculus. Section 33.1 was based on the resolvent $(\mathbf{A} - z\mathbf{I})^{-1}$, z nonreal; in section 33.2 we used the Cayley transform $(\mathbf{A} - i\mathbf{I})(\mathbf{A} + i\mathbf{I})^{-1}$; section 33.3 was based on all $f(\mathbf{A})$, f continuous on $\mathbb{R} \cup \infty$. Stone's theorem can be interpreted as defining the exponential functions $e^{it\mathbf{A}}$; we sketch how to build out of this functional calculus the spectral resolution of $\mathbf{A}$.

Lemma 2. *Let* $\mathbf{U}(t)$ *be a strongly continuous one-parameter group of unitary operators acting on a Hilbert space H. Let u be any vector in H; then the function*

$$a(t) = (\mathbf{U}(t)u, u) \tag{2}$$

is positive definite in the sense of Bochner; see chapter 14, section 14.4. That is, $a(t)$ is skew symmetric, continuous, and

$$\sum a(t_j - t_k)\phi_j\bar{\phi}_k \geq 0$$

for all choice of $t_1, \ldots, t_N$ on $\mathbb{R}$, and all complex numbers $\phi_1, \ldots, \phi_N$.

Proof. By the definition (30), the group property of $U(t)$, and the fact that $\mathbf{U}(-t) = \mathbf{U}(t)^{-1} = \mathbf{U}^*(t)$, we have

$$\sum a(t_j - t_k)\phi_j\bar{\phi}_k = \sum (\mathbf{U}(t_j - t_k)u, u)\phi_j\bar{\phi}_k = \sum (\mathbf{U}(t_k)^{-1}\mathbf{U}(t_j)u, u)\phi_j\bar{\phi}_k$$

$$= \sum (\mathbf{U}(t_j)u, \mathbf{U}(t_k)u)\phi_j\bar{\phi} = \left(\sum \phi_j\mathbf{U}(t_j)u, \sum \phi_k\mathbf{U}(t_k)u\right)$$

$$= \left\|\sum \phi_j\mathbf{U}(t_j)u\right\|^2,$$

clearly a nonnegative quantity.   □

According to Bochner's theorem, theorem 8 in chapter 14, a positive definite function is the Fourier transform of a nonnegative measure; so the function (2) can be represented as

$$(\mathbf{U}(t)u, u) = \int e^{it\lambda}\, dm(\lambda). \tag{3}$$

Setting $t = 0$, we obtain $\|u\|^2 = m(\mathbf{R})$. The measure m depends on the vector u; it is the Fourier transform of $(\mathbf{U}(t)u, u)$ and is therefore uniquely determined by u.

The right side of (2) is a quadratic function of u; we associate with it a skew bilinear function

$$(\mathbf{U}(t)u, v) = \int e^{it\lambda}\, dm(\lambda; u, v), \tag{4}$$

where the measures $m(u, v)$ are formed out of the measures $m(u)$ by polarization.

Lemma 3. *The measure $m(u, v)$ has the following properties:*

   (i) $m(u, v) = m(u)$.
  (ii) $m(v, u) = \overline{m(u, v)}$.
 (iii) $m(u, v)$ depends linearly on u, skew linearly on v.
  (iv) $|m(S, u, v)| \leq \|u\| \|v\|$ for any Borel set S.

Proof. The proof is the same as of theorem 17 in chapter 31.   □

We appeal now to theorem 1 of chapter 31 on bounded sesquilinear forms to conclude that for any Borel set S,

$$m(S, u, v) = (\mathbf{E}(S)u, v),$$

where $\mathbf{E}(S)$ is a bounded, symmetric operator, $\|\mathbf{E}\| \leq 1$. So we can rewrite (5) as

$$(\mathbf{U}(t)u, v) = \int e^{it\lambda} \, d(\mathbf{E}u, v). \tag{5}$$

We claim that $\mathbf{E}$ is the resolution of the identity for the operator $\mathbf{A}$, meaning that it has properties stated in theorem 1 of chapter 32. The proof is analogous to the arguments used in chapter 31, section 31.7, on the spectral resolution of a single unitary operator. We leave it to the reader to complete the details. □

35.2 ERGODIC THEORY

Strongly continuous groups of unitary operators serve as a good setting for the mean ergodic theorem. We present the abstract theorem in this section, the connection with ergodicity of dynamical systems in the next section.

Theorem 4 (von Neumann). *Let* $\mathbf{U}(t)$ *be a strongly continuous one-parameter group of unitary operators mapping a Hilbert space* H *onto itself.*

(i) *Denote by* F *the set of all vectors* f *in* H *that remain fixed under the action of the group:* $\mathbf{U}(t)f = f$ *for all* t. *Then* F *is a closed linear subspace of* H.

(ii) *Denote by* $\mathbf{M}(t)$ *the averaging operator*

$$\mathbf{M}(t)g = \frac{1}{t} \int_0^t \mathbf{U}(s)g \, ds. \tag{6}$$

Then as $t \to \infty$, $\mathbf{M}(t)$ *converges strongly, meaning that for every* g *in* H,

$$s - \lim_{t \to \infty} \mathbf{M}(t)g = \mathbf{P}g$$

exists, and the strong limit $\mathbf{P}$ *is the orthogonal projection onto* F.

Proof. The original proof of von Neumann relied on the spectral resolution of unbounded self-adjoint operators. The simple proof presented here is due to Eberhardt Hopf.

That F is a closed linear subspace is obvious, for the nullspace of the continuous operator $\mathbf{U}(t) - \mathbf{I}$ is closed and linear and so is their intersection F. For part (ii) we need first

Lemma 5. *Let* $\mathbf{U}$ *be a unitary operator,* E *the nullspace of* $\mathbf{U} - \mathbf{I}$, *and* R *its range. We claim that* E *is the orthogonal complement of* R.

Proof. For any two vectors g, h in H,

$$((U - I)g, h) = (g, (U^* - I)h).$$

Using $U^*U = I$, see section 31.7, we can rewrite this as

$$((U - I)g, h) = (g, (I - U)U^*h),$$

which shows that if g is orthogonal to the range R, the right side is zero for all h. Then so is the left side; therefore g lies in the nullspace E. The converse can be proved conversely.   □

We take $U = U(r)$, r any real number. According to the orthogonal decomposition theorem applied to the closure $\overline{R}$ of the range of $U - I$, every g in H can be decomposed as

$$g = e + z, \tag{7}$$

where z belongs to $\overline{R}$ and e is orthogonal to R. According to lemma 5, e belongs to the nullspace E of $U(r) - I$. We claim that $M(t)z$ tends to zero as t tends to ∞. To see this approximate z arbitrarily closely by a vector z_ϵ in R:

$$\|z - z_\epsilon\| < \epsilon, \quad z_\epsilon = (U(r) - I)h.$$

From the definition (6) of M and because $\|U(s)\| = 1$, it follows that $\|M(t)\| \leq 1$. So for all t,

$$\|M(t)z - M(t)z_\epsilon\| = \|M(t)(z - z_\epsilon)\| \leq \epsilon.$$

Using the definition (6) of M and that $U(s)U(r) = U(s + r)$, we write

$$M(t)z_\epsilon = M(t)(U(r) - I)h = M(t)U(r)h - M(t)h$$

$$= \frac{1}{t} \int_0^t U(s)U(r)h \, ds - \frac{1}{t} \int_0^t U(s)h \, ds$$

$$= \frac{1}{t} \int_t^{t+r} U(s)h \, ds - \frac{1}{t} \int_0^r U(s)h \, ds.$$

Each term on the right side is in norm less than $r\|h\|/t$; this shows that $\|M(t)z_\epsilon\|$ tends to zero as $t \to \infty$. Since $M(t)z_\epsilon$ differs by less than ϵ from $M(t)z$, and ϵ is arbitrary, it follows that $\|M(t)z\|$ tends to zero as $t \to \infty$.

Next we show that for e in the nullspace E of $U(r) - I$, $M(t)e$ tends strongly to a limit as $t \to \infty$, and that this limit belongs to E. $U(t + r)e = U(t)U(r)e = U(t)e$ shows that $U(t)e$ is a periodic function of t with period r; therefore, writing $t \bmod r$ as $t = nr + q$, $0 \leq q < r$, n some natural number, we have

$$M(t)e = \frac{1}{t} \int_0^t U(s)e \, ds = \frac{n}{t} \int_0^r U(s)e \, ds + \frac{1}{t} \int_0^q U(s)e \, ds.$$

Clearly, as $t \to \infty$, the second term on the right tends to zero, and the first term tends to

$$\frac{1}{r} \int_0^r \mathbf{U}(s)e \, ds. \tag{8}$$

Since $\mathbf{U}(s)$ commutes with $\mathbf{U}(r) - \mathbf{I}$, it maps its nullspace E into E. Therefore (8) belongs to E.

According to (7), every g in H can be decomposed as $e + z$; it follows that the limit $\mathbf{M}(t)g$ as $t \to \infty$ exists for every g in H, and that the limit belongs to E, the set of vectors fixed under $\mathbf{U}(r)$. Since r is arbitrary, it follows that the limit belongs to F, the set of vectors fixed under the action of every operator of the group.

The operators $\mathbf{U}(s)$ act as the identity on F; we claim that they map the orthogonal complement of F into the orthogonal complement of F. To see this, suppose that w is orthogonal to F; the relation

$$(\mathbf{U}(s)w, f) = (w, \mathbf{U}^*(s)f) = (w, \mathbf{U}(-s)f) = (w, f) = 0$$

shows that then $\mathbf{U}(s)w$ is orthogonal to F. Since the operators $\mathbf{M}(t)$ defined in (6) are averages of the operators $\mathbf{U}(s)$, they also act as the identity on F and map the orthogonal complement of F into itself. The same is true of the strong limit of $\mathbf{M}(t)$. Since we have shown that this strong limit maps H into F, it follows that the strong limit maps the orthogonal complement of F into 0. This shows the $s - \lim_{t \to \infty} \mathbf{M}(t)$ is orthogonal projection onto F. $\qquad\square$

35.3 THE KOOPMAN GROUP

Let M be an open, compact differentiable manifold, with some prescribed volume element V. We wish to study volume-preserving flows along vector fields D, namely solutions of the differential equation

$$\frac{dx}{dt} = D(x). \tag{9}$$

Among such flows are those along Hamiltonian vector fields. Denote by $x(y; t)$ the position at time t of that solution of (9) whose value at time zero is y; the mappings $y \to x(y; t)$ are volume preserving. Since M is compact, its volume is finite. The vector field D is independent of t; therefore

$$x(y(z; s), t) = x(z; s + t). \tag{9'}$$

Bernard Koopman has associated with every such flow a one-parameter group of unitary operators, acting on the Hilbert space of square integrable functions g on M:

$$(\mathbf{U}(t)g)(y) = g(x(y; t)). \tag{10}$$

It follows from the volume-preserving character of the flow that the operators $U(t)$ defined above preserve the L^2-norm; in fact they preserve all the L^p-norms, $1 \leq p \leq \infty$.

What does the mean ergodic theorem, theorem 4, say about these flows; that is, what can we say here about the space F of functions that remain fixed under the operators (10)? Obviously every constant function remains fixed; are there any others? Let f be such a function; we may take f to be real. Let c be any real number, the set S_c of points y in M where $f(y) < c$ is then *invariant* under the flow; that is, if y belongs to S_c, so do all points $x(y; t)$. If f is nonconstant, there is a value of c for which the invariant set S_c is nontrivial in the sense that neither S_c nor its complement in M has measure zero. Conversely, if there is a nontrivial set S invariant under the flow, its characteristic function

$$f(y) = \begin{cases} 1 & \text{if } y \text{ in S} \\ 0 & \text{if not} \end{cases}$$

remains fixed under the group (10). So we have shown:

Only the constant functions on M are fixed under all operators of the Koopman group associated with a given volume preserving flow on M iff M has no nontrivial measurable subset that is invariant under the flow.

A flow that has no nontrivial invariant subsets is called *metrically transitive*.

Suppose that there are no nontrivial invariant sets under the flow (9). Then F consists of constants, and according to the mean ergodic theorem, for every function g in $L^2(M)$, $M(t)g$ tends to the projection of g into the space of constants. What is that projection? Clearly, it is the *mean value* of g over M with respect to the prescribed volume:

$$Pg = \frac{1}{\text{Vol(M)}} \int_M g(x) \, dV. \tag{11}$$

The projection (11) is called the *space average* of the function g; the limits of the averaging operators M defined by (6) are called *time averages* of g. So, loosely speaking, the mean ergodic theorem applied to the Koopman group asserts that the space average and the time average of an arbitrary L^2 function are equal, provided that the flow (9) in question has no nontrivial invariant subset.

In statistical mechanics the manifold M is phase space, the vector field derived from the Hamiltonian of N interacting particles, $N \sim 10^{23}$. The time average is interpreted as the measured value of a function g of N variables; the time of measurement is large on the scale relevant in theorem 4. The significance of the ergodic theorem, first proposed in a rudimentary form by Ludwig Boltzmann, is that instead of having to solve a differential equation of form (9), involving about 10^{23} unknown functions, we merely have to evaluate an integral (11) over a 10^{23}-dimensional manifold.

Jack Schwartz has pointed out that the functions g whose measured values have thermodynamic significance are highly symmetric functions of their 10^{23} variables. The equality of the time and space average of such highly special functions might well be due to reasons other than the ergodic theorem.

We remark that in general, it is very hard to decide which flows have nontrivial invariant subsets and which don't. An amusing example is given in Lax.

35.4 THE WAVE EQUATION

The classical wave equation is

$$u_{tt} - \Delta u = 0,$$

Δ being the Laplace operator:

$$\Delta = \partial_x^2 + \partial_y^2 + \partial_z^2;$$

here subscripts denote partial derivatives. Let u be a solution of the wave equation in full space-time $\mathbb{R}^3 \times R$, that tends to zero, together with its first derivatives, sufficiently rapidly as $x^2 + y^2 + z^2 \to \infty$. For such solutions the *law of conservation of energy* may be derived as follows: multiply the wave equation by u_t, and integrate over $\mathbb{R}^3$:

$$\int_{\mathbb{R}^3} u_t (u_{tt} - \Delta u) \, dV = 0.$$

Integration by parts changes this to

$$\int (u_t u_{tt} + u_{xt} u_x + \ldots + u_{zt} u_z) \, dV = 0.$$

The integral is the t-derivative of

$$E(t) = \frac{1}{2} \int (u_t^2 + u_x^2 + u_y^2 + u_z^2) \, dV;$$

so we conclude that $E(t)$ is independent of t. Such a quantity is called a *constant of motion*, or a conserved quantity.

$E(t)$ is called *energy* ; it is a quadratic functional of u and u_t. Its square root is the *energy norm*.

The initial data $u(o)$ and $u_t(o)$ completely determine the solution for all time. If any solutions have the same initial data, their difference, also a solution of the wave equation, has zero initial data, and so zero initial energy. By the conservation of energy, the difference has zero energy at all times, and so zero data at all times.

We denote by H the completion in the energy norm of space of all initial data $\{u(o), u_t(o)\}$ that have finite energy. H is a Hilbert space.

Denote by $\mathbf{U}(t)$ the solution operators, that is, the operator that maps initial data into data at time t:

$$\mathbf{U}(t) : \{u(o), u_t(o)\} \to \{u(t), u_t(t)\}.$$

It is not hard to show (e.g., by taking the spatial Fourier transform) that the initial value problem, that is, the problem of finding a solution in full space-time of the wave equation with prescribed initial data $u(o)$ and $u_t(o)$, can be solved if the initial data are sufficiently smooth and have compact support. So the operators $\mathbf{U}(t)$ are well defined on such data. Since these data form a dense subspace of all data H, we can, by continuity, define the operator $\mathbf{U}(t)$ on all of H. It is not hard to show that *the extended operators form a strongly continuous group of unitary operators.* Strong continuity can be verified for smooth solutions with smooth initial data, and then extended by continuity to all initial data with finite energy. The group property $\mathbf{U}(s+r) = \mathbf{U}(s)\mathbf{U}(r)$ expresses the fact that if $u(x, y, z, t)$ is a solution of the wave equation, so is $u(x, y, z, t - s)$. Unitarity is merely a restatement of the conservation of energy, plus the fact that each operator $\mathbf{U}(t)$ is invertible; the inversion is accomplished by solving the initial value problem backward.

Exercise 1. What is the infinitesimal generator of the group formed by the solution operator of the wave equation?

We describe now an important extension: the study of solutions of the wave equation for all time but not in the whole three-dimensional space, only in the exterior of some obstacle B. On the obstacle all solutions are required to satisfy the boundary condition: $u = 0$ on the boundary of B.

The conservation of energy is derived the same way as before. This time integration by parts produces the boundary term

$$\int_{\partial B} u_t u_n dS, \qquad u_n = \text{normal derivative of } u,$$

which is zero for all u that vanish on the boundary. Note that we could also have chosen to impose the boundary condition $u_n = 0$, to make the boundary term vanish. In either case the laws of conservation of energy follows.

We then proceed as before to construct the solution operators $\mathbf{U}(t)$, which form a strongly continuous group of unitary operators in the energy norm. The only new complication is that it is harder to prove the existence for all times of solutions of the wave equation with prescribed smooth, initial data, and satisfying the boundary condition. But this is only a technical difficulty, which should not be allowed to obscure the simple underlying structure of solutions with finite energy in the exterior of an obstacle.

35.5 TRANSLATION REPRESENTATION

The spectral representation of a self-adjoint operator $\mathbf{A}$ acting in a Hilbert space H is described in chapter 32. When $\mathbf{A}$ has spectrum of multiplicity 1, there exists a nonnegative measure m on $\mathbb{R}$ and a unitary mapping between H and $L^2(\mathbb{R}, m)$:

$$H \longleftrightarrow L^2(\mathbb{R}, m)$$

such that the action of A is represented as multiplication by the variable $\lambda \in \mathbb{R}$. When the spectrum of A is multiple, the spectral representation is a unitary mapping between H and a possibly infinite Cartesian product of L^2 spaces on $\mathbb{R}$ equipped with various nonnegative measures:

$$H \longleftrightarrow \Pi L^2(\mathbb{R}, m_j).$$

The action of A is represented by multiplication by $\lambda \in \mathbb{R}$ in each component.

According to Stone's theorem, i times a self-adjoint operator A generates a strongly continuous one-parameter group of unitary operator $U(t)$. We denote these operators symbolically as $U(t) = \exp i At$.

Theorem 6. *In a spectral representation for a self-adjoint operator A, the action of $U(t) = \exp i At$ is represented as multiplication by $\exp i \lambda t$.*

Proof. The proof of Stone's theorem is based on the Hille-Yosida theorem, theorem 7 of chapter 34. Yosida's proof of this theorem constructs $\exp i At$ as the strong limit

$$\exp i As = s - \lim_{n \to \infty} e^{i G_n t}, \qquad G_n = n^2 R(n) - n I. \tag{12}$$

In the course of deriving the spectral representation, chapter 31, section 31.5, we have shown that the action of $R(n) = (n - iA)^{-1}$ is represented as multiplication by $(n - i\lambda)^{-1}$. It follows that the action of the right side of (12) is represented as multiplication by $e^{i n \lambda t/(n - i\lambda)}$.

Let u be any vector in H, represented by the functions $\{k_j(\lambda)\}$. Then $e^{i G_n t} u$ is represented by $\{e^{i n \lambda t/(n - i\lambda)} k_j(\lambda)\}$; the $L^2(m_j)$ limit as $n \to \infty$ of these functions is $\{e^{i\lambda t} k_j(\lambda)\}$. □

We consider now the case where the spectrum is *absolutely continuous and of uniform multiplicity on* $\mathbb{R}$. That means that all the measures m_j in (18) are absolutely continuous with respect to Lebesgue measure and their support is all of $\mathbb{R}$. As remarked at the end of chapter 31, in this case the measures entering the representation can be taken as Lebesgue measure on all of $\mathbb{R}$, and (18) can be rewritten as

$$H \longleftrightarrow \Pi L^2(\mathbb{R}).$$

It is convenient to put together a Cartesian product of $L^2(\mathbb{R})$ spaces as a single space $L^2(N, \mathbb{R})$ consisting of L^2 functions whose values lie in an auxiliary Hilbert space N; the dimension of N equals the number of components in the Cartesian product, possibly ∞. Note that the dimension of N is the multiplicity of the spectrum of A. So we can write the spectral representation as

$$H \longleftrightarrow L^2(N, \mathbb{R}). \tag{13}$$

By taking a Fourier inverse, we can obtain from (13) another representation

$$H \longleftrightarrow L^2(N, \mathbb{R}). \tag{13'}$$

That is, if the vector u in H is represented in (13) by the function $f(\lambda)$, then (13')
assigns to u the representation

$$k(x) = \frac{1}{\sqrt{2\pi}} \int f(\lambda) e^{-ix\lambda} \, d\lambda.$$

It follows from theorem 6 that $(\exp i\mathbf{A}t)u$ is represented in (13) by $\exp(i\lambda t) f(\lambda)$,
and so in (13') by $k(x - t)$. For this reason (13') is called a *translation representation* of H for the unitary group generated by $i\mathbf{A}$. Conversely, from a translation
representation we can construct a spectral representation by Fourier inversion.

We turn now to a geometrical characterization of translation representations due
to Sinai. Let H be a Hilbert space, $\mathbf{U}(t)$ a strongly continuous one-parameter group
of unitary operators on H that has a translation representation. Denote by F the
subspace of H consisting of vectors represented by functions supported on $\mathbb{R}_-$:

$$F \longleftrightarrow L^2(N, \mathbb{R}_-). \tag{14}$$

Clearly, $\mathbf{U}(r)F$ consists of vectors represented by functions supported on $(-\infty, r)$.
It follows then that the one-parameter family $\mathbf{U}(r)F$ increases as r does, going from
$\{0\}$ to H as r goes from $-\infty$ to ∞. We express this more precisely thus:

$$\mathbf{U}(r)F \subset F \qquad \text{for } r < 0, \tag{15a}$$

$$\cap \, \mathbf{U}(r)F = \{0\}, \tag{15b}$$

$$\overline{\cup \mathbf{U}(r)F} = H. \tag{15c}$$

Theorem 7. *Conversely, let H be a Hilbert space, $\mathbf{U}(t)$ a strongly continuous one-
parameter group of unitary operators mapping H into H. Let F be a closed subspace of H, and suppose that all three conditions (15) are satisfied. Then H has a
translation representation (13') for $\mathbf{U}(t)$ where F is given by (14).*

Exercise 2. Show that $\mathbf{U}(s)F \subset \mathbf{U}(t)F$ when $s < t$.

Sinai deduced this result from von Neumann's theorem about the Heisenberg
commutation relation. Phillips and the author gave an independent proof and showed
how to deduce from it the result of von Neumann; we will present that in the next
section. We give here the proof due to Phillips and Lax; it is a little technical, but
pretty, at least in a parent's eyes.

Proof. We will deduce theorem 7 from another representation theorem:

Theorem 8. *Let K be a Hilbert space, $\mathbf{Z}(t)$ a strongly continuous one-parameter
semigroup of contractions mapping K into K. Assume furthermore that $\mathbf{Z}(t)$ tends*

to 0 *strongly as t tends to* ∞:

$$\lim_{s \to \infty} \mathbf{Z}(t)k = 0 \tag{16}$$

for all k in K. Then K can be unitarily represented as a closed subspace of
$L^2(N, \mathbb{R}_-)$, *N some auxiliary Hilbert space, so that the action of* $\mathbf{Z}(t)$ *is translation to the right by t, restricted to* $\mathbb{R}_-$.

Proof. Denote the generator of the semigroup $\mathbf{Z}$ by $\mathbf{G}$, $D(\mathbf{G})$ the domain of $\mathbf{G}$. We want to define first a representation for vectors g in $D(\mathbf{G})$, and then extend this representation by continuity to all of K. To any g in $D(\mathbf{G})$, we assign the vector-valued function $\gamma(s)$ as

$$\gamma(s) = \mathbf{Z}(-s)g, \qquad s \le 0. \tag{17}$$

Thus the function γ is defined on $\mathbb{R}_-$ and its values lie in D. We define now a new norm in D, denoted as $\|g\|_N$, so that the L^2 norm of γ is equal to $\|g\|$. That is, we require that for all g in $D(\mathbf{G})$,

$$\|g\|^2 = \int_{-\infty}^{0} \|\gamma(s)\|_N^2 \, ds = \int_{0}^{\infty} \|\mathbf{Z}(t)g\|_N^2 \, dt. \tag{18}$$

Since $\mathbf{Z}$ maps D into D, this must hold for g replaced by $\mathbf{Z}(h)g$,

$$\|\mathbf{Z}(h)g\|^2 = \int_{0}^{\infty} \|\mathbf{Z}(s+h)g\|_N^2 = \int_{h}^{\infty} \|\mathbf{Z}(s)g\|_N^2 \, ds. \tag{18'}$$

Differentiate both sides of (18') and set $h = 0$; we get

$$(\mathbf{G}g, g) + (g, \mathbf{G}g) = -\|g\|_N^2, \tag{19}$$

We take this as the *definition of* $\|g\|_N$. Note that since we assumed that the $\mathbf{Z}(t)$ are contractions, the left side of (18') is a nonincreasing function of h, and therefore the left side of (19) is nonpositive. This shows that our new norm $\|g\|_N$ is nonnegative. We define the auxiliary Hilbert space N as the completion of D in the norm $\|g\|_N$, modulo the vectors of norm zero.

We verify now that with this definition of N and the N-norm, the representation (17) is an isometry, namely that (18) holds for all g in D. Since $\mathbf{Z}(t)g$ belongs to D, we can use definition (17) and (19); writing $s = -t$, we get

$$\|\gamma(s)\|_N^2 = \|\mathbf{Z}(t)g\|_N^2 = -2\mathrm{Re}(\mathbf{G}\mathbf{Z}(t)g, \mathbf{Z}(t)g) = -\frac{d}{dt}\|\mathbf{Z}(t)g\|^2. \tag{19'}$$

We integrate this with respect to t from 0 to r, and using the hypothesis that $\|\mathbf{Z}(r)g\|$ tends to 0 as r tends to ∞, we obtain (18).

Since D is dense in K, we can extend by continuity the representation (17) as an isometry of all K. Clearly, in this representation $\mathbf{Z}(t)$ acts as translation followed by restriction to $\mathbb{R}_-$. $\square$

We return now to theorem 7. Denote by $\mathbf{P}$ projection onto the subspace F and define the operators $\mathbf{Z}(t)$ by

$$\mathbf{Z}(t) = \mathbf{PU}(t), \qquad t \geq 0. \tag{20}$$

Lemma 9. *Suppose that* $\mathbf{U}$ *has the properties postulated in (15). Then* $\mathbf{Z}(t)$ *defined by (20) form a strongly continuous one-parameter semigroup of contractions mapping* $F \to F$, *and* $\mathbf{Z}(t)$ *tends strongly to* 0 *as* $t \to \infty$.

Proof. Since $\mathbf{U}(t)$ is strongly continuous, so is $\mathbf{Z}(t)$. Since $\mathbf{U}(t)$ and $\mathbf{P}$ are contractions, so is $\mathbf{Z}(t)$. To show that $\mathbf{Z}(t)$ is a semigroup, take any vector f in F; by definition of $\mathbf{Z}$,

$$\mathbf{Z}(r)\mathbf{Z}(s)f = \mathbf{PU}(r)\mathbf{PU}(s)f = \mathbf{PU}(r)[\mathbf{U}(s)f + p]$$
$$= \mathbf{PU}(r+s)f + \mathbf{PU}(r)p = \mathbf{Z}(r+s)f + \mathbf{PU}(r)p; \tag{21}$$

here p denotes some vector orthogonal to F. We claim that also $\mathbf{U}(r)p$ is orthogonal to F for $r > 0$. To see this, take any vector f in F; since $\mathbf{U}(r)$ is unitary, $\mathbf{U}^*(r) = \mathbf{U}^{-1}(r) = \mathbf{U}(-r)$:

$$(f,\ \mathbf{U}(r)p) = \big(\mathbf{U}^*(r)f,\ p\big) = (\mathbf{U}(-r)f,\ p).$$

According to assumption (15a), $\mathbf{U}(-r)f$ belongs to F; since p is orthogonal to F, the last term on the right is zero. Since $\mathbf{P}$ maps vectors orthogonal to F into 0, it follows that the last term on the right in (21) is zero; this shows that the $\mathbf{Z}(t)$ form a semigroup.

To prove that $\mathbf{Z}(t)$ tends strongly to zero as s tends to ∞, we first show that the set of vectors of form

$$\mathbf{U}(t)F^\perp, \qquad t < 0, \tag{22}$$

where $F^\perp$ denotes the orthogonal complement of F in H, is *dense* in H. We argue indirectly and suppose not; then there would be a nonzero vector v in H orthogonal to all vectors $\mathbf{U}(t)F^\perp$. The unitary operator $\mathbf{U}(t)$ maps orthogonal complements into orthogonal complements. The orthogonal complement of $\mathbf{U}(t)F^\perp$ is $\mathbf{U}(t)F$; so it follows that v belongs to $\mathbf{U}(t)F$. This holds for all $t < 0$, but it contradicts (15b).

Given any f in F and any $\epsilon > 0$, we can, according to the above, find g in $F^\perp$ and $r < 0$ such that $\|f - \mathbf{U}(r)g\| < \epsilon$. Therefore, writing $s = -r$, we have $\|\mathbf{U}(s)f - g\| < \epsilon$. The projection $\mathbf{P}$ maps g into zero. Therefore applying the projection $\mathbf{P}$, we conclude that $\|\mathbf{Z}(s)f\| = \|\mathbf{PU}(s)f\| < \epsilon$. Since the operators $\mathbf{Z}$ are contractions, $\|\mathbf{Z}(t)f\| < \epsilon$ for all $t > s$. $\square$

It follows from lemma 9 that theorem 8 is applicable to the semigroup $\mathbf{Z}(t) = \mathbf{PU}(t)$ defined on F. There is an isometric translation representation that assigns to each f in F a function ϕ in $L^2(N, \mathbb{R}_-)$:

$$f \longleftrightarrow \phi(s) \tag{23}$$

so that for any $t > 0$,

$$\mathbf{PU}(t)f \longleftrightarrow c(t)\phi(s - t), \tag{23'}$$

where c is the characteristic function of $\mathbb{R}_-$. The representation is isometric:

$$\|f\|^2 = \int_{-\infty}^{0} \|\phi(s)\|_N^2 \, ds, \tag{24}$$

$$\|\mathbf{PU}(t)f\|^2 = \int_{-\infty}^{-t} \|\phi(s)\|_N^2 \, ds. \tag{25}$$

We extend the representation to all vectors in H of form $\mathbf{U}(t)f$, f in F, by setting

$$\mathbf{U}(t)f \longleftrightarrow \phi(s - t). \tag{26}$$

It follows from (24) and the isometry of $\mathbf{U}(t)$ that this assignment is isometric. We claim that (26) is consistent with (23'); that is, if $\mathbf{U}(t)f$ belongs to F, then the right sides of (23') and (26) are equal. Clearly, this is so iff the function ϕ representing f is zero for $-t < s < 0$. That this condition is satisfied can be seen by comparing (24) and (25). The left sides are equal; therefore so are the right sides, which can be only if $\phi = 0$ in $(-t, 0)$.

According to (15c), vectors of form $\mathbf{U}(r)f$, f in F, are dense in H. Therefore the representation (26) can be extended by continuity to all u in H:

$$u \longleftrightarrow k(x).$$

This representation is isometric:

$$\|u\|^2 = \int_{\infty}^{\infty} \|k(x)\|_N^2 \, dx.$$

transmutes the action of $\mathbf{U}$ into translation,

$$\mathbf{U}(t)u \longleftrightarrow k(x - t),$$

and the action of $\mathbf{P}$ into truncation,

$$\mathbf{P}u \longleftrightarrow c(s)k(s).$$

It is not hard to show that the range of this representation is all of $L^2(N, \mathbb{R})$; see Lax-Phillips, 1981. $\qquad \square$

The conditions under which a unitary group has a translation representation may appear somewhat special. Nevertheless, there are natural, interesting, and nontrivial examples coming from wave propagation as described in section 35.4. There the underlying Hilbert space H is the set of all initial data $\{u(x, 0), u_t(x, 0)\}$ defined in $\mathbb{R}^3$ with the *energy norm*

$$\|\{u(x), u_t(x)\}\|_E^2 = \int \left(\sum u_x^2 + u_t^2\right) dx.$$

The group $U(t)$ is the group of solution operators for the wave equation

$$u_{tt} - \Delta u = 0,$$
$$U(t) : \{u(x, 0), u_t(x, 0)\} \longrightarrow \{u(x, t), u_t(x, t)\}.$$

The unitary character of $U(t)$ expresses the *conservation of energy* and the *reversibility of time*.

The role of the distinguished subspace F is taken by the so-called *incoming initial data*. These are initial data of incoming solutions $u(x, t)$, which are zero inside the *backward light cone* :

$$u(x, t) = 0 \qquad \text{for } |x| < -t.$$

It is far from obvious that there are any incoming solutions at all. We show now how to construct them by relying on *Huygens's principle* for wave propagation the three-dimensional space. Conceptually this principle says that solutions of the wave equation propagate information with speed equal to 1. Technically this means that the value of a solution $u(y, s)$, (y, s) in $\mathbb{R}^3 \times \mathbb{R}$, uniquely determined by the initial data $\{u(0), u_t(0)\}$, depends only on the values of $u(0)$ and $u_t(0)$ and their space derivatives, at the intersection of the light cone with the initial plane, that is, the points x satisfying $|x - y| = |s|$.

Let $f = \{f_1, f_2\}$ be data that are zero for $|x| > S$. We claim that for $T > S$, $U(-T)f$ is incoming.

Exercise 3. Show, using Huygens's principle, that $U(-T)f$ is incoming.

We show now that the space F of incoming data has all three properties listed in (15a) to (15c):

(i) Let u denote an incoming solution, f its initial data. The solution whose initial data are $U(r)f$ is $u^{(r)}(x, t) = u(x, t + r)$. Since u is incoming,

$$u(x, t) = 0 \qquad \text{for } |x| < -t.$$

Therefore

$$u^{(r)}(x, t) = 0 \qquad \text{for } |x| < -(t + r).$$

When $r < 0$, u_r is incoming; this shows that $U(r)F \subset F$ for $r < 0$.

(ii) The relation above shows that $u^{(r)}(x,0) = 0$ for $|x| < -r$ and, after differentiating with respect to t, that $u_t^{(r)}(x,0) = 0$ for $\|x\| < -r$. It follows that the intersection of $\mathbf{U}(r)F$ contains only the zero data.

(iii) Let u be any of the incoming solutions constructed above, f its initial data. It follows from the construction that $\mathbf{U}(T)f$ can be taken as arbitrary data supported in $|x| < T$. Clearly, the union of $\mathbf{U}(T)F$ as $T \to \infty$ is dense in H.

Huygens's principle is valid in any space of odd dimension, and so is the analysis of incoming data. One could deduce the properties of incoming data from explicit formulas for solutions of the wave equation in $\mathbb{R}^n \times \mathbb{R}$, but our derivation is more illuminating.

Far more interesting is the case of wave propagation in the exterior of an obstacles, discussed in section 35.4. Suppose that the obstacle is contained inside a ball around the origin of radius R. We define an incoming solution $u(x,t)$ as vanishing inside the cone

$$u(x,t) = 0 \qquad \text{for } |x| < -t + R.$$

Note that such an incoming solution satisfies both boundary conditions discussed in section 35.4. Define again F as the initial data of incoming solutions. Properties (15a) and (15b) can be immediately verified as before, but property (15c) lies much deeper (see chapter V of the book by Lax and Phillips). We will return to this example in chapter 36, section 36.5.

Another very interesting example is furnished by the automorphic wave equation in hyperbolic space. We will take it up in chapter 37.

35.6 THE HEISENBERG COMMUTATION RELATION

In quantum mechanics the *state* of a physical system is a *unit vector* u $\|u\| = 1$, in a Hilbert space H over $\mathbb{C}$ associated with the physical system. Each *observable* is identified with a *self-adjoint operator*, constructed according to the so-called rules of quantization.

Definition. The *expected value* of an observable $\mathbf{A}$ in state u is defined to be $(u, \mathbf{A}u)$. Here we assume that u belongs to the domain of $\mathbf{A}$.

The term "expected value" implies an uncertainty in the measurement of the observable in state u:

Definition. The *uncertainty* in the measurement of an observable $\mathbf{A}$ in state u is the square root of the expected value in the state u of $(\mathbf{A} - a\mathbf{I})^2$, where a is the expected value of $\mathbf{A}$. We denote this uncertainty by $\Delta(\mathbf{A}, u)$:

$$\Delta^2(\mathbf{A}, u) = \left(u, (\mathbf{A} - a)^2 u \right) = \|\mathbf{A}u - au\|^2$$

$$= \|\mathbf{A}u\|^2 - 2a(u, \mathbf{A}u) + a^2 = \|\mathbf{A}u\|^2 - a^2. \tag{27}$$

The third formula shows that the uncertainty $\Delta(\mathbf{A}, u)$ equals zero only if $\mathbf{A}u - au = 0$, namely if u is an eigenstate of $\mathbf{A}$.

Let $\mathbf{A}$ and $\mathbf{B}$ denote a pair of observables. According to (27), both can be measured with absolute certainty in the same state u iff u is an eigenvector of both $\mathbf{A}$ and $\mathbf{B}$. Pairs of commuting operators have common eigenvectors, but in general, this is not to be expected.

Suppose that the pair of self-adjoint operators $\mathbf{A}$ and $\mathbf{B}$ satisfy the *Heisenberg commutation relation*

$$\mathbf{AB} - \mathbf{BA} = i\mathbf{I}. \tag{28}$$

$\mathbf{A}$ and $\mathbf{B}$ have no common eigenvector, for such a vector would be mapped into zero by the left side of (28). This shows that such observables $\mathbf{A}$ and $\mathbf{B}$ cannot both be measured with absolute certainty. Heisenberg has made this uncertainty quantitative:

Theorem 10. *Suppose a pair of self-adjoint operators $\mathbf{A}$ and $\mathbf{B}$ satisfy the Heisenberg commutation relation (28). Then in any state u that belongs to the domain of both $\mathbf{A}$ and $\mathbf{B}$ the uncertainties in the measurement of $\mathbf{A}$ and $\mathbf{B}$ satisfy the inequality*

$$\Delta(A, u)\Delta(B, u) \geq \tfrac{1}{2}. \tag{29}$$

(29) is called the Heisenberg uncertainty principle.

Proof. We start by reformulating the commutation relation. Let (28) act on a vector u, and form the scalar product with u. Operating formally, using the symmetry of the operators $\mathbf{A}$ and $\mathbf{B}$, we get

$$(\mathbf{B}u, \mathbf{A}u) - (\mathbf{A}u, \mathbf{B}u) = i\|u\|^2. \tag{28'}$$

We require (28′) to hold for all vectors in the domain of both $\mathbf{A}$ and $\mathbf{B}$. This is a kind of weak interpretation of (28). Let t denote any real number. According to the Schwarz inequality, for any unit vector u that belongs to the domain of both $\mathbf{A}$ and $\mathbf{B}$,

$$|(u, \mathbf{A}u + it\mathbf{B}u)|^2 \leq \|\mathbf{A}u + it\mathbf{B}u\|^2. \tag{30}$$

Denote the expected value of $\mathbf{A}$ and $\mathbf{B}$ by a and b, respectively; then inequality (30) can be written as

$$a^2 + b^2 t^2 \leq \|\mathbf{A}u\|^2 + i\,[(\mathbf{B}u, \mathbf{A}u) - (\mathbf{A}u, \mathbf{B}u)]t + \|\mathbf{B}u\|^2 t^2. \tag{30'}$$

Using (28′), we see that the middle term on the right in (30′) equals $-t$. Using this and the notation introduced in (27), we can by subtracting the left side in (30′) from the right side rewrite (30′) as

$$0 \le \left(\|\mathbf{A}u\|^2 - a^2 \right) - t + \left(\|\mathbf{B}u\|^2 - b^2 \right) t^2 = \Delta^2(\mathbf{A}, u) - t + \Delta^2(\mathbf{B}, u)t^2.$$

In words, the quadratic polynomial on the right is nonnegative for all real t. Therefore its discriminant is nonpositive:

$$1 - 4\Delta^2(\mathbf{A}, u)\Delta^2(\mathbf{B}, u) \le 0;$$

this is inequality (29). $\qquad\square$

What pair of operators satisfy the commutation relations (28)? An elegant argument of Wielandt shows that no *bounded* operators do. To see this, we deduce by induction from (28) that for all natural numbers n,

$$i\, n\mathbf{B}^{n-1} = \mathbf{A}\mathbf{B}^n - \mathbf{B}^n\mathbf{A}. \tag{28''}$$

Taking the norm of both sides and using the triangle and product inequalities on the right gives

$$n\|\mathbf{B}^{n-1}\| \le 2\|\mathbf{A}\|\,\|\mathbf{B}^n\| \le 2\|\mathbf{A}\|\,\|\mathbf{B}\|\,\|\mathbf{B}^{n-1}\|,$$

which implies that $\|\mathbf{B}^{n-1}\| = 0$ for $n > 2\|\mathbf{A}\|\,\|\mathbf{B}\|$; so $\mathbf{B}^{n-1} = \mathbf{O}$. Backward recursion based on (28″) shows that then $\mathbf{B}^k = \mathbf{O}$ for all k.

On the other hand, the operators $\mathbf{A} = i(d/d\mu)$, $\mathbf{B} = \mu$, acting on the Hilbert space $L^2(\mathbb{R})$ of functions $f(\mu)$ do satisfy (28), for

$$i\frac{d}{d\mu}\mu f - \mu i\frac{d}{d\mu}f = if.$$

It was shown by von Neumann that this pair is, except for multiplicity and unitary equivalence, the only pair of operators satisfying the commutation relation. Before stating precisely and proving this, we follow Weyl in reformulating (28). Consider

$$\mathbf{U}(s)\mathbf{B}\mathbf{U}(-s), \tag{31}$$

where $\mathbf{U}(s)$ is the unitary group generated by $i\mathbf{A}$. On the domain of $\mathbf{A}$, $\mathbf{U}(s)$ satisfies

$$\frac{d}{dt}\mathbf{U}(s) = i\mathbf{A}\mathbf{U}(s) = i\mathbf{U}(s)\mathbf{A}.$$

Differentiating (31) formally and using the commutation relation (28), we get

$$\frac{d}{dt}\mathbf{U}(s)\mathbf{B}\mathbf{U}(-s) = i\mathbf{U}(s)\,[\mathbf{A}\mathbf{B} - \mathbf{B}\mathbf{A}]\,\mathbf{U}(-s) = -\mathbf{I}.$$

Integrating this relation gives

$$U(s)\mathbf{B}U(-s) = \mathbf{B} - s\mathbf{I}, \tag{32}$$

called the *Weyl form of the commutation relation*. It is taken to mean that for all real values of s the self-adjoint operators on the two sides of (32) are identical.

Exercise 4. Deduce from (32) that $U(s)$ maps the domain of $\mathbf{B}$ onto itself.

Exercise 5. Denote by $V(t)$ the unitary group generated by $i\mathbf{B}$. Deduce from (32) that for all real s and t,

$$U(s)V(t) = e^{ist}V(t)U(s).$$

(Hint: Differentiate with respect to t.)

The next result is due to von Neumann:

Theorem 11. *Let* $\mathbf{A}$ *and* $\mathbf{B}$ *a pair of self-adjoint operators acting in a Hilbert space* H, $U(t)$ *the unitary group of operators generated by* $i\mathbf{A}$. *Suppose that the Weyl relation (32) is satisfied. Then there is a representation of* H *as* $L^2(N, \mathbb{R})$ *so that*

$$\mathbf{A} = i\frac{d}{d\mu}, \quad \mathbf{B} = \mu.$$

Proof. We remarked earlier that Sinai derived the translation representation theorem from von Neumann's theorem. Since we have given an independent proof of the translation representation theorem, we are entitled to reverse Sinai's proof.

Let $\mathbf{E}(\lambda)$ be the spectral resolution for the self-adjoint operator B:

$$\mathbf{B} = \int \lambda d\mathbf{E}(\lambda). \tag{33}$$

Then

$$U(s)\mathbf{B}U(-s) = \int \lambda d(U(s)\mathbf{E}(\lambda)U(-s)) \tag{34}$$

and

$$\mathbf{B} - s\mathbf{I} = \int (\lambda - s)d\mathbf{E}(\lambda) = \int \lambda d\mathbf{E}(\lambda + s). \tag{34'}$$

The operators on the left of (34) and (34′) are self-adjoint; the integrals on the right give their spectral resolution. Since, according to (32), the two operators are identical, so are their spectral resolutions: for every Borel set T,

$$U(s)\mathbf{E}(T)U(-s) = \mathbf{E}(T + s). \tag{35}$$

Denote by F the range of $\mathbf{E}(\mathbb{R}_-)$. We claim that the group $\mathbf{U}(s)$ and the subspace F behave as indicated in (15). Setting $T = \mathbb{R}_-$ in (35), we conclude that $\mathbf{U}(s)F$ is the range of $\mathbf{E}(\mathbb{R}_- + s)$. According to spectral theory, these form a one-parameter family of subspaces increasing with s, going from $\{0\}$ to H as s goes from $-\infty$ to ∞. These are the properties of F stated in (15a) to (15c).

We appeal now to theorem 7, according to which H has a representation as $L^2(N, \mathbb{R})$ in which $\mathbf{U}(s)$ acts as translation, and $F = L^2(N, \mathbb{R}_-)$. This shows that the generator $i\mathbf{A}$ of $\mathbf{U}$ is represented as $-d/d\mu$. It further follows that the range of $\mathbf{E}(\mathbb{R}_- + s)$ is $\mathbf{U}(s)F = L^2(N, \mathbb{R}_- + s)$. Thus $\mathbf{E}(\mathbb{R}_- + s)$ is multiplication by the characteristic function of $\mathbb{R}_- + s$. Setting this into (33), we see that $\mathbf{B}$ is multiplication by μ. $\qquad\square$

PHILOSOPHICAL–HISTORICAL NOTE. The uncertainty principle is one of those notions of mathematical physics that have profoundly changed philosophical thinking. Other examples are quantum jump, the special theory of relativity, Gödel's incompleteness theorem, and maybe black holes. It has even entered public consciousness. An example occurs in Michael Frayn's play *Copenhagen*, triumphantly presented in London and on Broadway. The play revolves around a visit that Heisenberg payed to Bohr in Copenhagen on September 21, 1941, the high water mark of Germany's conquests. Heisenberg claimed that he came with a vague proposal that scientists on neither side should make an effort to build a nuclear bomb. Bohr's recollection was that Heisenberg came to gather information; he denied that Heisenberg's account had "any basis in the actual events." The playwright suggests that perceiving the same event differently is a manifestation of a kind of uncertainty principle in human communication.

Arnold Kramish, physicist and historian of the nuclear age, has evidence that the visit to Copenhagen was an intelligence gathering mission, no uncertainty about it. It was triggered by an article in 1941 in the Swedish newspaper *Stockholms-Tidningen* describing an effort in the United States to build a new type of bomb out of uranium, with unprecedented explosive power. This article was picked up by Dr. P. K. Schmidt, head of the Press Branch of the German Foreign Office. Schmidt forwarded the report to the physicist Carl von Weizsäcker, son of the German Foreign Minister Ernst von Weizsäcker, who on September 4, 1941, informed the Abwehr, the Intelligence Branch of the German High Command, and Bernhard Rüst, Reichsminister in charge of the ongoing German uranium project, of which Heisenberg and Weizsäcker were leading members. A "cultural" visit of Heisenberg and Weizsäcker to Copenhagen two weeks later was arranged at the highest level.

It is ironical that the report in the Swedish newspaper was premature; the American uranium project did not start until 1942. By an ever greater irony of fate, Heisenberg learned nothing on this trip from Bohr, but Bohr found out that the Germans had an active uranium project. In 1943, when Bohr escaped to America, he warned the leaders of the Manhattan Project of the danger of a possible German nuclear bomb.

BIBLIOGRAPHY

Hopf, E. *Ergodentheorie*. Ergebnisse der Math., **2**, Springer, Berlin, 1937.

Heisenberg, W. *Z. Phys.*, **43** (1927).

Koopman, B. O. Hamiltonian systems and transformations in Hilbert space. *Proc. Nat. Acad. Sci. USA*, **17** (1931): 315–318.

Lax, P. D. The ergodic character of sequences of pedal triangles. *Am. Math. Monthly*, **97** (1990): 377–381.

Lax, P. D. and Phillips, R. S. *Scattering Theory*. Pure and Applied Mathematics, **26**. Academic Press, New York, 1967.

Lax, P. D. and Phillips, R. S. *The Translation Representation Theorem*. Integral Equations and Operator Theory **4**. Birkhäuser, Boston, 1981, pp. 416–421.

von Neumann, J. Die Eindeutigkeit der Schrödingerschen Operatoren. *Math. An.*, **104** (1931).

von Neumann, J. Proof of the quasi-ergodic hypothesis. *Proc. Nat. Acad. Sci. USA*, **18** (1932): 70–82.

Schwartz, J. The pernicious influence of mathematics on science. *Logic, Methodology and Philosophy of Science. Proc. 1960 Int. Congr.* E. Nagel, P. Suppes, and Tarski, eds. Stanford University Press, Stanford, 1962, pp. 356–360.

Sinai, Ja. G. Dynamical systems with countable Lebesgue spectrum. *Izv. Akad. Nauk SSSR*, **25** (1961): 899–924.

Stone, M. Linear transformations in Hilbert space, IV. *Proc. Nat. Acad. Sci. USA*, **15** (1929): 198–200.

Weyl, H. Quantenmechanic und Gruppentheorie. *Z. Phys.*, **46** (1927): 1–46.

Wielandt, H. Über die Unbeschränkheit der Operatoren der Quantenmechanic. *Math. An.*, **121** (1949): 21.

36

EXAMPLES OF STRONGLY CONTINUOUS SEMIGROUPS

36.1 SEMIGROUPS DEFINED BY PARABOLIC EQUATIONS

In chapter 16, section 16.5, we studied solutions of the heat equation

$$u_t = u_{xx} \tag{1}$$

that are defined for all $t \geq 0$ and all x, and that decay sufficiently rapidly as $|x| \to \infty$. We have shown there, see theorem 13, that such solutions are uniquely determined for all t by their initial values. The solution operators $S(t)$ relate initial values of solutions to their values at time t:

$$S(t)u(0) = u(t).$$

The results of section 16.5 can be summarized as follows in semigroup language:

The solution operators $S(t)$ form a strongly continuous semigroup of contraction operators in any of the $L^p(\mathbb{R})$ spaces, $1 \leq p < \infty$, and on the space of continuous functions on $\mathbb{R}$ that are 0 at $\pm\infty$.

It was remarked already in chapter 16 that similar results hold for a far more general class of equations than the heat equation: we may replace in (1) the second space derivative of u by any second-order elliptic operator in any number of space variables E acting on u:

$$u_t = Eu, \tag{2}$$

$$E = \sum a_{ij} \partial_i \partial_j + b_i \partial_i + c, \qquad \partial_i = \frac{\partial}{\partial x_i}, \tag{2'}$$

where (a_{ij}) is a real, uniformly positive definite symmetric matrix whose entries are smooth functions of x. The coefficients b_i and c are also smooth functions of x. The whole space may be replaced by a bounded domain in $\mathbb{R}^n$, on whose boundary u is required to satisfy a single boundary condition, say $u = 0$. The solution operators again form a semigroup. The domain of the infinitesimal generator includes all smooth functions that satisfy the boundary condition, and on such functions the generator acts as the operator $\mathbf{E}$ defined in (2'). A possible approach to proving the existence of solutions of (2) with prescribed initial values is to make the proper extension of the operator $\mathbf{E}$ and then verify that $\mathbf{E}$ thus extended satisfies the hypotheses of the Hille-Yosida theorem.

36.2 SEMIGROUPS DEFINED BY ELLIPTIC EQUATIONS

In the first example we take the Banach space $C(S^m)$, the space of continuous functions u on the m dimensional unit sphere. For each such function u there is a uniquely determined *harmonic* function $h = h(rw)$ in the $(m+1)$-dimensional unit ball which equals u on the boundary of the unit ball:

$$\Delta h = 0, \quad h(\omega) = u(\omega), \quad \omega \text{ in } S^m. \tag{3}$$

We define the semigroup $\mathbf{Z}(t)$ as follows:

$$\mathbf{Z}(t)u = h(e^{-t}\omega), \tag{4}$$

where h is the harmonic function defined by (3).

Theorem 1.

   (i) The operators $\mathbf{Z}(t)$ are contractions in the maximum norm.

  (ii) $\mathbf{Z}(t)$ form a strongly continuous one-parameter semigroup of operators.

 (iii) For $t > 0$ the operator $\mathbf{Z}(t)$ is compact.

Proof. (i) follows from the maximum principle for harmonic functions. (ii) The semigroup property is clear. Since if $h(x)$ is a harmonic function, so is $h(cx)$ for any constant c; therefore $\mathbf{Z}(t)\mathbf{Z}(s)u = h(e^{-t}e^{-s}\omega) = h(e^{-(s+t)}\omega) = \mathbf{Z}(s + t)u$. Strong continuity is a consequence of the continuity of harmonic functions in the unit ball whose boundary values on the unit sphere are continuous.

(iii) To prove the compactness of the operator $\mathbf{Z}(t)$, $t > 0$, we have to show that the image of the unit ball under $\mathbf{Z}(t)$ lies in a compact set. This image consists of the harmonic functions on the sphere $|x| = e^{-t}$ which are bounded by 1 in the unit ball. According to the Arzela-Ascoli criterion (see the beginning of chapter 22) for precompactness in the maximum norm a set of functions has to be equicontinuous. It is a well-known property of harmonic functions that on any compact subset of their domain of definition, their first (or any higher-order) derivatives are bounded by a constant multiple of their maximum on their domain of definition. Since uniformly bounded first derivatives guarantee equicontinuity, it follows that the image of the unit ball in $C(S^m)$ under $\mathbf{Z}(t)$, $t > 0$, is precompact.   ☐

In our second example we replace the maximum norm by the $L^2(S^m)$ norm. The same results hold, as well as an additional property:

Theorem 1′.

(i) *The operators $\mathbf{Z}(t)$ defined by (3) and (4) are contractions in the $L^2(S^m)$ norm.*

(ii) *$\mathbf{Z}(t)$ form a strongly continuous semigroup in the $L^2(S^n)$ norm.*

(iii) *$\mathbf{Z}(t)$ is a compact operator for $t > 0$.*

(iv) *The operators $\mathbf{Z}(t)$ are real symmetric.*

Proof. (i) We state and prove an L^2 analogue of the maximum principle. For simplicity let us take the case $m = 1$. Every harmonic function h defined in the unit disk can be expanded into a Fourier series:

$$h = \Sigma r^n (a_n \cos n\theta + b_n \sin n\theta),$$

where r and θ are polar coordinates. By the Parseval relation

$$\int h^2(r, \theta) d\theta = \pi \Sigma r^{2n} (a_n^2 + b_n^2), \tag{5}$$

clearly an increasing function of r.

This completes the proof of (i). To deduce (ii), we go back to theorem 1, that $\mathbf{Z}(t)u$ is continuous in the max norm for u continuous; it follows that $\mathbf{Z}(t)u$ is continuous in the L^2 norm as well. Since the continuous functions are a dense subspace of L^2, it follows that $\mathbf{Z}(t)u$ is continuous in the L^2 norm for all u in L^2.

(iii) The proof of compactness of $\mathbf{Z}(t)$ for $t > 0$ can be deduced as before, since it is possible to estimate a harmonic function and its derivatives at points of the unit ball with $r < 1$ in terms of r and the square integral of the harmonic function on the unit sphere.

(iv) The symmetry of $\mathbf{Z}$ is expressed by

$$(\mathbf{Z}u, v) = (u, \mathbf{Z}v)$$

In view of the definition (4) of $\mathbf{Z}$, this means that for any pair of harmonic functions h and k in the unit ball, and that any $s < 1$,

$$\int h(s, \omega) k(1, \omega) \, d\omega = \int h(1, \omega) k(s, \omega) \, d\omega. \tag{6}$$

To prove this, we want to deform continuously the left side of (6) into the right side through the one-parameter family:

$$\int h(p, \omega) k(q, \omega) \, d\omega, \quad q = \frac{s}{p}, \quad s \leq p \leq 1. \tag{7}$$

In the open interval $s < p < 1$ the integral in (7) depends differentiably on p; its derivative with respect to p is

$$\int \left[h_r(p)\, k(q) - h(p)\, k_r(q)\, \frac{q}{p} \right] d\omega, \tag{8}$$

where we have used the fact that $dq/dp = -q/p$. Now define the function ℓ by

$$\ell(r, \omega) = k\left(\frac{q}{p} r, \omega \right);$$

ℓ is a harmonic function in the ball of radius p/q, and

$$\ell(p, \omega) = k(q, \omega), \quad \ell_r(p, \omega) = \frac{q}{p}\, k_r(q, \omega).$$

Setting this into (8), we obtain

$$\int [h_r(p)\, \ell(p) - h(p)\ell_r(p)] \, d\omega. \tag{9}$$

We recall now Green's formula:

$$\int_G [\ell \Delta h - h \Delta \ell]\, dx = \int_{\partial G} [\ell h_n - h\ell_n]\, dS. \tag{10}$$

Since h and ℓ are harmonic functions, the left side of (10) is zero; therefore so is the right side. Now take G to be the ball of radius p; the normal derivatives h_n and ℓ_n are then derivatives with respect to the radius r, and $dS = p^m\, d\omega$. The right side of (10) is p^m times (9). This proves that (9) is zero; therefore (8) is independent of p. Letting p tend to s and to 1, we deduce that the two sides of (6) are equal.   □

We show next that the generator $\mathbf{G}$ of $\mathbf{Z}(t)$ defined in theorem 1′ is self-adjoint. To see this, let u and v belong to the domain of $\mathbf{G}$; since $\mathbf{Z}(t)$ is symmetric,

$$(\mathbf{Z}(t)u, v) = (u, \mathbf{Z}(t)v).$$

Differentiate with respect to t, and set $t = 0$:

$$(\mathbf{G}u, v) = (u, \mathbf{G}v),$$

proving that $\mathbf{G}$ is symmetric. We saw in chapter 32 that every complex number with sufficiently large real part belongs to the resolvent set of $\mathbf{G}$. On the other hand, we have shown in chapter 33 that every unbounded symmetric operator whose resolvent includes points in both the upper and lower half of the complex plane is self-adjoint. Combining these two facts, we conclude that $\mathbf{G}$ is self-adjoint.

Exercise 1. Show that if the generator $\mathbf{G}$ of a semigroup $\mathbf{Z}(t)$ is self-adjoint, so is $\mathbf{Z}(t)$. (Hint: Use the functional calculus of self-adjoint operators.)

What is the spectrum of $\mathbf{G}$? Since $\mathbf{Z}(t)$ is compact for $t > O$, its spectrum is pure point spectrum, a discrete set of points accumulating only at zero. It follows from

Phillips' spectral mapping theorem (see chapter 34, section 34.5) that the spectrum also of G is discrete and accumulates only at $-\infty$. It turns out that the spectrum of G can be determined explicitly:

Let γ denote a point of the spectrum of G; since G is self-adjoint, γ is real. By the aforementioned spectral mapping theorem, $e^{\gamma t}$ belongs to the spectrum of $Z(t)$; since $Z(t)$ is a contraction, γ is ≤ 0. Let $e(w)$ be a corresponding eigenfunction, $h(rw)$ the harmonic function whose boundary value is $e(\omega)$. By definition (4), $Z(t)e = h(e^{-t}w) = e^{\gamma t}e(\omega)$. Setting $e^{-t} = r$, we get that

$$h(rw) = r^{-\gamma}e(\omega). \tag{11}$$

It follows from (11) that γ is a nonpositive integer; otherwise, the function $r^{-\gamma}e(\omega)$ would have only a finite number of derivatives at $r = 0$, contrary to the fact that harmonic functions in the unit ball are infinitely differentiable (even analytic) at the origin.

Theorem 2. *The spectrum of the generator G of the semigroup $Z(t)$, defined in theorem 1', consists of all integers ≤ 0.*

Proof. What remains to show is that every nonnegative integer n belongs to the spectrum of G. Consider the space P of homogeneous polynomials $p(x_0, \cdots, x_m)$ of degree n; denote its dimension by d_n. The Laplace operator maps such a p into the space of homogeneous polynomial of degree $n - 2$. By linear algebra, a subspace of P of dimension at least $d_n - d_{n-2}$ is mapped into zero; that subspace consists of harmonic polynomials. The boundary value $e(\omega) = p(\omega)$ of such a polynomial satisfies

$$Z(t)e = p(e^{-t}\omega) = e^{-nt}e(\omega),$$

and so is an eigenfunction of $Z(t)$. Differentiation with respect to t shows that e is an eigenfunction of G with eigenvalue $-n$. $\qquad\square$

36.3 EXPONENTIAL DECAY OF SEMIGROUPS

In chapter 34, section 34.1, we saw that a strongly continuous one-parameter semigroup grows at most exponentially as $t \to \infty$. In this section we investigate how fast such a semigroup can decay.

Suppose that the infinitesimal generator G, acting on a Hilbert space H, is self-adjoint and bounded from above. Using the spectral resolution of G we can express the semigroup as

$$Z(t)f = \int e^{st} \, d\mathbf{E}f.$$

It follows from this formula that for $f \neq 0$, $\mathbf{Z}(t)$ decreases at some exponential rate as t tends to ∞. The main result of this section (see Lax) includes a perturbation of this result.

Theorem 3. *Let H be a Hilbert space, $\mathbf{G}$ the generator of a strongly continuous semigroup. Assume that there is an infinite sequence $\{\xi_n\}$ of real numbers, tending to $-\infty$, such that the resolvent of $\mathbf{G}$ is uniformly bounded by some constant d^{-1} on all lines $\operatorname{Re} \lambda = \xi_n$ in the complex plane:*

$$\|(\mathbf{G} - \lambda I)^{-1}\| \leq d^{-1}, \qquad \operatorname{Re} \lambda = \xi_n. \tag{12}$$

Let $u(t)$, $0 \leq t$, be a vector-valued function whose values lie in the domain of $\mathbf{G}$. Assume that $\mathbf{G}u(t)$ is a strongly continuous function of t, and that u is strongly differentiable, and its derivative u_t is strongly continuous. We require $u(t)$ to satisfy for all t the inequality

$$\|u_t - \mathbf{G}u\| \leq k\|u\|, \qquad 0 < t, \tag{13}$$

where k is some constant less than d. Then, unless identically zero, $u(t)$ decays no faster than exponentially in the L^2 sense, that is there is a positive number b such that

$$\int_0^\infty \|u(t)\|^2 e^{bt} = \infty. \tag{14}$$

Proof. The proof is based on an inequality contained in

Lemma 4. *Let $\mathbf{G}$ be the generator of a strongly continuous semigroup, whose resolvent is bounded by d^{-1} for $\operatorname{Re} \lambda = \xi$. Let $u(t)$, $-\infty < t < \infty$, be a vector-valued function whose values lie in the domain of $\mathbf{G}$. We assume that $\mathbf{G}u(t)$ is a strongly continuous function of t, and that $u(t)$ has a strongly continuous first derivative with respect to t. Then*

$$d^2 \int_{\mathbb{R}} \|u(t)\|^2 e^{-2\xi t}\, dt \leq \int_{\mathbb{R}} \|\mathbf{G}u - u_t\|^2 e^{-2\xi t}\, dt, \tag{15}$$

provided that the integral on the left is finite.

Proof. Define $v(t) = e^{-\xi t} u(t)$; then $(\mathbf{G}u - u_t)e^{-\xi t} = (\mathbf{G} - \xi)v - v_t$. Its Fourier transform is $(\mathbf{G} - \xi - i\tau)\hat{v}(\tau)$, where $\hat{v}$ is the Fourier transform of v. Since, by assumption, $\|(\mathbf{G} - \lambda)^{-1}\| \leq d^{-1}$ for $\operatorname{Re} \lambda = \xi$,

$$d^2\|\hat{v}(\tau)\|^2 \leq \|(\mathbf{G} - \xi - i\tau)\hat{v}(\tau)\|^2. \tag{16}$$

According to Parseval's theorem, the Fourier transform preserves the L^2-norm. Integrating (16) we obtain (15), provided that v is in L^2. $\square$

Exercise 2. Prove Parseval's theorem for functions whose values lie in a Hilbert space.

We turn now to theorem 3. First we extend $u(t)$ for negative values of t, such as by setting $u(t) = u(0)a(t)$ for $t < o$, where $a(t)$ is a smooth function of compact support, $a(0) = 1$. We apply inequality (15) to u thus extended, with $\xi = \xi_n$ where (12) is satisfied. On the right, we apply for $t > 0$ inequality (13), and on the left, we drop the integral over $\mathbb{R}_-$. We get

$$d^2 \int_{\mathbb{R}_+} \|u(t)\|^2 e^{-2\xi_n t}\, dt \leq K_n + k^2 \int_{\mathbb{R}_+} \|u(t)\|^2 e^{-2\xi_n t}\, dt, \tag{17}$$

where K_n is the integral of $\|Gu - u_t\|^2 e^{-2\xi_n t}$ over $\mathbb{R}_-$, and therefore tends to zero as n tends to ∞. Since k is less than d, we deduce that

$$\int_{\mathbb{R}_+} \|u(t)\|^2 e^{-2\xi_n t}\, dt \leq K_n/(d^2 - k^2), \tag{17'}$$

provided that the integral on the left is finite. If, contrary to (14), the left side is finite for all ξ_n, it follows from (17') that $u(t) \equiv 0$ for all $t > 0$. ☐

Note that in the proof we only required the resolvent of $\mathbf{G} - \lambda$ to be bounded as in (12), and not that it be defined everywhere.

Note that for $\mathbf{G}$ self-adjoint, condition (12) is equivalent to this:
The intervals $(\xi_n - d, \xi_n + d)$ are free of the spectrum of $\mathbf{G}$.
So as a corollary of theorem 3 we get

Theorem 3'. *Let* $\mathbf{G}$ *be a self-adjoint operator, bounded from above, whose spectrum has infinitely many gaps of width* $2d$. *Let* $u(t)$ *be a vector-valued function as in theorem 3, satisfying inequality (13); then, unless identically zero,* $u(t)$ *decays no faster than exponentially, in the sense of (14).*

The restriction $k < d$ is sharp; an example to the contrary when $k < d$ is given in Lax.

Here is an application of theorem 3'. Let $\mathbf{L}$ be a partial differential operator of the form $\mathbf{L} = \Delta - c$, where Δ is the Laplace operator and c is the operator of multiplying by the function $c(x)$, assumed to be smooth and positive. The L^2 analogue of the maximum principle (5) holds for solutions of $\mathbf{L}w = 0$:

Lemma 5. *Let* w *be a solution of* $\mathbf{L}w = 0$, $\mathbf{L} = \Delta - c$, c *positve. Then*

$$I(r) = \int w^2(r, \omega)\, d\omega \tag{18}$$

is an increasing function of r.

Proof. Differentiate (18) with respect to r; we get

$$\frac{d}{dr} I(r) = 2 \int w_r w \, d\omega. \tag{18'}$$

Next we use Green's theorem to transform the following integral over the ball B_r of radius r:

$$0 = \int_{B_r} w L w \, dx = \int_{B_r} (w \Delta w - cw^2) \, dx = \int_{S_r} w w_r \, d\omega - \int_{B_r} (w_x^2 + cw^2) \, dx.$$

Since c is a positive function, the positivity of (18') follows, and so does the lemma.
□

We designate by H the Hilbert space $L^2(S)$, S, the unit sphere. We take $\mathbf{G}$ to be the generator of the semigroup $\mathbf{Z}(t)$ discussed in theorem 1' of section 36.2. Recall that $\mathbf{Z}(t)$ is defined by equation (4):

$$\mathbf{Z}(t)u = h(e^{-t}\omega), \tag{4}$$

where h is the harmonic function defined in (3):

$$\Delta h = 0, h(\omega) = u(\omega), \qquad \omega \text{ in } S. \tag{3}$$

Differentiating (4) at $t = 0$ gives

$$\mathbf{G}u = -h_r. \tag{19}$$

Let w be a solution of $\mathbf{L}w = 0$ in an open set containing the unit ball. Define $u(t)$ to be $w(e^{-t}\omega)$. Then at $t = 0$

$$u_t = -w_r. \tag{20}$$

We now show that this function u satisfies inequality (13) of theorem 3, with $k = k(t)$ a function that tends to zero as t tends to ∞.

Let v be any smooth function on S; define $p(x)$ to be the harmonic function in the unit ball whose value on S is v:

$$\Delta p = 0, \quad p(\omega) = v(\omega). \tag{21}$$

Multiply (21) by h and integrate over the unit ball. Since h and p are both harmonic functions, Green's theorem applied on the unit ball B gives

$$0 = \int_S (p_r h - p h_r) \, d\omega. \tag{22}$$

A similar application of Green's theorem gives

$$\int_B p \Delta w - (\Delta p)w = \int_S (p w_r - p_r w) \, d\omega.$$

Using the fact that $\Delta p = 0$ and $\Delta w = cw$, we get

$$\int_B pcw = \int (pw_r - p_r w)\,d\omega.$$

Adding this to (22) gives

$$\int p(w_r - h_r) + p_r(h - w)\,d\omega = \int_B pcw\,dx. \tag{23}$$

At $t = 0$, $u(\omega) = w(\omega)$, and by (3), $h(\omega) = w(\omega)$, According to (21), $p(w) = v(\omega)$. Furthermore, according to (19), $h_r = -Gu$, and by (20), $w_r = -u_t$. Setting these into (23) gives

$$(v, Gu - u_t) = \int_B pcw\,dx, \tag{23'}$$

where $(,)$ is the scalar product in H. By the Schwarz inequality the right side of (23′) is less than

$$c_{\max}\left(\int_B p^2\,dx \int w^2\,dx\right)^{1/2}. \tag{24}$$

It was shown in section 36.2 that for a harmonic function p, $\int p^2(r\omega)\,d\omega$ is an increasing function of r. It follows that

$$\int_B p^2(x)\,dx = \int_0 \int_S p^2(r\omega)\,d\omega r^m\,dr \leq \frac{1}{m+1}\int_S p^2(\omega)\,d\omega = \frac{1}{m+1}\|v\|^2,$$

where $\|\,\|$ is the norm in H. Similarly, using lemma 5, we can estimate

$$\int_B w^2\,dx \leq \frac{1}{m+1}\|u\|^2.$$

These estimates show that (24) is bounded by $c_{\max} 1/(m+1)\|v\|\|u\|$. So we deduce from (23′) that

$$|(v, Gu - u_t)| \leq \frac{1}{m+1}c_{\max}\|v\|\|u\|. \tag{25}$$

According to corollary 1′ of chapter 6, the norm of every element f of the real Hilbert space H can be characterized as follows:

$$\|f\| = \sup(v, f),$$

where v ranges over a set of unit vector, $\|v\| = 1$, that are dense in the unit sphere of H. So we conclude from (25) that for $t = 0$,

$$\|Gu - u_t\| \le \frac{1}{m+1} c_{\max} \|u\|. \tag{26}$$

This is inequality (13), with $k = 1/(m+1)c_{\max}$.

Recall that $u(t)$ in H is defined as the value on the unit sphere of $w(e^t x)$. The function $w(t) = w(e^{-t} x)$ satisfies the differential equation

$$\Delta w(t) = e^{-2t}(\Delta w)(e^{-t}x) = e^{-2t} c(e^{-t}x) w(t).$$

Therefore applying (26) to $u(t)$, we get

$$\|Gu - u_t\| \le \frac{e^{-2t}}{m+1} c_{\max} \|u\|. \tag{26'}$$

This is inequality (13), with $k(t) = e^{-2t} c_{\max}/(m+1)$.

According to theorem 2 the spectrum of G lies on the negative integers; therefore it contains infinitely many gaps of length 1. The factor on the right in inequality (26') tends to zero as t tends to ∞. Therefore we can apply theorem 3' to $u(t) = w(e^{-t}\omega)$ and conclude, as in (14), that unless $w \equiv 0$, there is a constant b such that

$$\int_0^\infty \|u(t)\|^2 e^{bt}\, dt = \infty. \tag{14}$$

Denote $e^{-t} = r$, and $x = r\omega$. Since $dt = -r\,dr$, and $dx = r^m\, d\omega dr$, we can rewrite (14) as follows:

$$\int w^2(x)|x|^{-b-m+1}\, dx = \infty. \tag{14'}$$

We can restate our conclusion (14') as follows:

Theorem 6. *A solution w of a partial differential equation $\Delta w + cw = 0$ cannot have a zero of infinite order unless $w \equiv 0$.*

NOTE. It is well known (to specialists in partial differential equations) that solutions of linear elliptic equations with *analytic* coefficients are themselves analytic. Such solutions cannot have zeros of infinite order. So the result derived above is novel only in the case when the function $c(x)$ is not analytic. The first such theorem was obtained by Carleman in two space variables, and by Müller for any m; a very general result is due to Calderon.

36.4 THE LAX-PHILLIPS SEMIGROUP

In chapter 35, section 35.5, we introduced the concept of a translation representation of a group of unitary operators $U(t)$ acting on a Hilbert space H. A key role there was played by a subspace F, which we will here call an *incoming subspace* and denote as F_-. The incoming subspace, F_-, is assumed to have the properties enumerated in equation (15) in chapter 35:

$$U(r)F_- \subset F_- \qquad \text{for } r < 0, \tag{27a}$$

$$\cap U(r)F_- = \{0\}, \tag{27b}$$

$$\overline{\cup U(r)F_-} = H. \tag{27c}$$

We also require the existence of an *outgoing subspace* F_+, satisfying analogous conditions:

$$U(r)F_+ \subset F_+ \qquad \text{for } r > 0, \tag{28a}$$

$$\cap U(r)F_+ = \{0\}, \tag{28b}$$

$$\overline{\cup U(r)F_+} = H. \tag{28c}$$

Furthermore we require F_- and F_+ to be orthogonal to each other.

Theorem 7. *Let* $U(t)$ *be a strongly continuous one-parameter group of unitary operators acting on a Hilbert space* H, F_-, *and* F_+ *a pair of incoming and outgoing subspaces in the sense of properties (27) and (28), orthogonal to each other. Denote by* P_- *orthogonal projection onto the orthogonal complement of* F_-, *and by* P_+ *orthogonal projection onto the complement of* F_+. *Denote by* K *the orthogonal complement in* H *of* $F_- \oplus F_+$. *Then*

$$Z(t) = P_+ U(t) P_-, \qquad t \geq 0, \tag{29}$$

is a strongly continuous semigroup of contractions on K *that tends strongly to zero as* t *tends to* ∞.

Proof. Clearly, each $Z(t)$ is a contraction. To show that they map K into K, we have to demonstrate that for k, $Z(t)k$ is orthogonal to both F_+ and F_-. Since P_+ is projection onto the orthogonal complement of F_+, it follows from (29) that the range of $Z(t)$ is orthogonal to F_+.

Since P_- is the identity on K, $Z(t)k = P_+ U(t)k$. We claim that for $t \geq 0$, $U(t)k$ is orthogonal to F_-. To see this, take any f_- in F_- and write

$$(U(t)k, f_-) = (k, U^*(t)f_-) = (k, U(-t)f_-). \tag{30}$$

According to (27), for $t \geq 0$, $U(-t)$ maps F_- into F_-, and therefore that scalar product on the right in (30) is zero. Since $P_+ U(t)k$ differs from $U(t)k$ by a vector in F_+, assumed to be orthogonal to F_-, it follows that $P_+ U(t)k$, too, is orthogonal to F_-.

Next we show that $Z(t)$ form a semigroup. According to properties (28) for $t \geq 0$, $U(t)$ maps F_+ into F_+. Since P_+ removes the F_+ component, it follows that

$$P_+ U(t) P_+ = P_+ U(t), \qquad t \geq 0.$$

Thus we can write for k in K that

$$\mathbf{Z}(t)\mathbf{Z}(s)k = \mathbf{P}_+\mathbf{U}(t)\mathbf{P}_+\mathbf{U}(s)k = \mathbf{P}_+\mathbf{U}(t)\mathbf{U}(s)k = \mathbf{P}_+\mathbf{U}(t+s)k = \mathbf{Z}(t+s)k.$$

To show that $\mathbf{Z}(t)$ tends strongly to zero, we use property (28c) that the union of $\mathbf{U}(t)F_+$ is dense in H. Therefore given any k in K, and any $\varepsilon > 0$, there is a vector f_+ in F_+, and r such that $\mathbf{U}(r)f_+$ differs by less than ε from k. Since $\mathbf{P}_+\mathbf{U}(t)$ is a contraction,

$$\|\mathbf{P}_+\mathbf{U}(t)(k - \mathbf{U}(r)f_+)\| < \varepsilon.$$

But by (28c) for $t+r > 0$, $\mathbf{P}_+\mathbf{U}(t+r)f_+$ is zero; therefore for $t > -r$, $\|\mathbf{P}_+\mathbf{U}(t)k\| = \|\mathbf{Z}(t)k\| < \varepsilon$. $\square$

In the next section we will present an example that lends substance and interest to the abstract theory described in this chapter. An equally interesting example will be given in chapter 37, section 37.9.

36.5 THE WAVE EQUATION IN THE EXTERIOR OF AN OBSTACLE

Let B be a smoothly bounded domain, the obstacle, in $\mathbb{R}^3$, contained in the ball of radius R, R some positive number. At the end of section 35.4 of the preceding chapter we have looked at solutions of the wave equation

$$u_{tt} - \Delta u = 0,$$

defined for all x in the exterior of B and for all times, and which are zero on the boundary of B. We have shown there that such solutions conserve energy, that is,

$$E = \frac{1}{2}\int (u_t^2 + u_x^2)\,dx \tag{31}$$

is independent of time, where the integration is over all points x in the exterior of B. The square root of energy is the energy norm; we denote by H the completion in the energy norm of all initial data $\{u(0), u_t(0)\}$ that are smooth, of compact support in the exterior of B, and are zero on the boundary of B. We denote the energy norm of the initial data $f = \{f_1, f_2\}$ as $\|f\|_E$.

As already remarked in section 35.4, using techniques in the theory of partial differential equations one can show that there exist for all times, positive and negative, solutions of the wave equation in the exterior of B with prescribed smooth initial data with finite energy that are zero on the boundary of B. Conservation of energy shows that these solutions are uniquely determined by their initial data. So we may speak of the operator $\mathbf{U}(t)$ that maps initial data into data at time t:

$$\mathbf{U}(t) : \{u(o), u_t(o)\} \rightarrow \{u(t), u_t(t)\}.$$

We can extend, by continuity, these operators $U(t)$ to all of H. The operators thus extended form a strongly continuous one-parameter group of unitary operators.

We recall now from the end of section 35.5 the notion of an *incoming solution* $u(x, t)$ of the wave equation in the exterior of an obstacle as one that is zero in the *backward cone*:

$$u(x, t) = 0 \quad \text{for } |x| < -t + R; t \leq 0.$$

The closure in the energy norm of initial data of such solutions is defined as the *incoming subspace* F_- for our group of unitary operators. Properties (27a) and (27b) in are easily verified. Property (27c) lies much deeper. A full proof is given in Lax and Phillips; here we will merely point out its close connection with *local energy decay*. We shall show that for every bounded subset G of the exterior of the obstacle, and for every f in H,

$$\lim_{t \to -\infty} \|U(t)f\|_{E,G} = 0. \tag{32}$$

where $\| \ \|_{E,G}$ denotes the local energy norm:

$$\|h\|_{E,G}^2 = \frac{1}{2} \int_G \left(|h_{1,x}|^2 + h_2^2 \right) dx.$$

Property (27c) implies that given any $\varepsilon > 0$ there exist a number T and g in F_- such that

$$\|f - U(T)g\| < \varepsilon.$$

By definition of F_-, for $T + t$ negative $U(T + t)g$ is zero in the ball $|x| < R - T + t$. For t large enough negative this ball contains G. Since $U(t)$ preserves energy, $\|U(t)f - U(t + T)g\|_E = \|f - U(T)g\|_E < \varepsilon$; it follows that for t large enough negative $\|U(t)f\|_{E,G} < \varepsilon$. This proves (32). $\qquad\square$

The importance of local energy decay is that also the inverse implication holds. In fact property (27c) follows from a weaker form of energy decay:

$$\lim_{t \to \infty} \inf \|U(t)f\|_{E,G} = 0. \tag{32'}$$

For a derivation of property (27c) from (32), and a proof of (32) we refer to chapter V of Lax and Phillips. We remark that the key step in the derivation of (32) is to show that the generator of the group $U(t)$ has no point spectrum.

In a similar fashion we define *outgoing solutions* v as those that are zero in the *forward cone*:

$$v(x, t) = 0 \quad \text{for } |x| < t + R, t \geq 0.$$

The closure in the energy norm of the initial data of outgoing solutions forms the *outgoing subspace* F_+. As before, properties (28a) and (28b) are easily verified, while property (28c) is hard work.

Lemma 8. *F_- and F_+ are orthogonal to each other.*

Proof. We mimic the construction of incoming data as described at the end of section 35.5, chapter 35: choose any number T greater than R. It follows from Huygens's principle that a solution $u(x, t)$ *in free space* $\mathbb{R}^3 \times \mathbb{R}$ whose data at time T are zero for $|x| > T - R$ is incoming for $t \leq 0$.

It follows from the conservation of energy that if u and v are two solutions of the wave equation in free space for $0 \leq t \leq T$, then the energy scalar product

$$(\{u(t), u_t(t)\}, \{v(t), v_t(t)\})_E$$

of two solutions of the wave equation is independent of t. In particular

$$(\{u(o), u_t(o)\}, \{v(o), v_t(o)\})_E = (\{u(T), u_t(T)\}, \{v(T), v_t(T)\})_E. \qquad (33)$$

For u constructed as above, and v any outgoing solution, the right side of (33) is zero, since the functions $u(T), u_t(T)$ are supported inside the ball of radius $T - R$, whereas an outgoing solution v is zero there.

The argument above makes plausible but doesn't quite prove lemma 8, since we haven't shown that the incoming solutions u constructed above with the aid of Huygens's principle are dense among all incoming solutions. I assure the reader that a straightforward proof of lemma 8 can be found in Lax and Phillips.   □

The use of Huygens's principle to show the orthogonality of incoming and outgoing data is quite natural, for in two space dimensions; where Huygens's principle fails, incoming and outgoing data are not orthogonal.

The semigroup $\mathbf{Z}(t)$ is a natural tool for studying the interaction of waves with the obstacle. Incoming waves have not yet interacted with the obstacle, and outgoing waves never will. Their removal by the projections $\mathbf{P}_-$ and $\mathbf{P}_+$ clears the way for the study of waves that interact all the time with the obstacle.

There is an intimate relation between the shape of the obstacle and the spectrum of the operators $\mathbf{Z}(t)$ and their infinitesmal generator $\mathbf{G}$. We start with a preliminary result:

Theorem 9. *Let k be any positive number. The operator $(k\mathbf{I} - \mathbf{G})^{-1}\mathbf{Z}(2R)$ is compact.*

For proof we refer to chapter V of the Lax and Phillips book. A fairly immediate consequence of theorem 9 is

Corollary 9′. *$\mathbf{G}$ has a pure point spectrum accumulating only at ∞.*

Exercise 3. Deduce corollary 9' from theorem 9. (Hint: Use the spectral theory of semigroups described in chapter 34, section 34.5.)

The link between the geometry of the obstacle and the spectrum of $\mathbf{Z}(t)$ is the *geometrical optics* description of wave propogation. The relevant geometric property is how long the obstacle can retain a ray, defined as a path consisting of straight line segment, reflected at the boundary of the obstacle according to the classical laws of reflection. Denote by $\ell(B)$ the supremum of the length of reflected rays that are contained within the ball of radius R.

Theorem 10.

(i) If $\ell(B) < \infty$, then $\mathbf{Z}(t)$ is compact for t large enough.
(ii) If $\ell(B) = \infty$, $\|\mathbf{Z}(t)\| = 1$ for all t.

The proof of this theorem is too technical to be included here; a few remarks will have to suffice. Lax and Phillips have pointed out that part (i) would follow from a generalized Huygens principle, which states, roughly, that in the exterior of an obstacle the sharp part of signals propagates along rays, including rays reflected from the obstacle. This generalized Huygens principle was proved be Melrose and Taylor.

Combining (i) with local energy decay, we conclude that all eigenvalues of $\mathbf{Z}(t)$ are of the form $e^{\gamma_0 t}$, $0 > \mathrm{Re}\,\gamma_j \geq \cdots \to -\infty$. It follows that $\|\mathbf{Z}(t)\|$ decays exponentially as $t \to \infty$. Lax, Morawetz, and Phillips have shown that for star-shaped obstacles (for which $\ell(B) < 2R$), $\|\mathbf{Z}(t)\|_E \leq e^{-\alpha t}$, $\alpha > 0$. Their proof is based on the nonstandard energy estimates of Morawetz, and does not rely on the generalized Huygens principle.

Part (ii) is based on the notion that the wave equation has solutions whose energy is contained in an arbitrarily preassigned neighborhood of any given reflected ray. Such solutions were constructed by Jim Ralston.

When $\ell(B) = \infty$, the real parts of the eigenvalues of the generator $\mathbf{G}$ do not tend to $-\infty$. Interesting information about their location was obtained by Ikawa.

BIBLIOGRAPHY

Calderon, A. P. Uniqueness in the Cauchy problem for partial differential equations. *Am. J. Math.*, **80** (1958): 16–36.

Carleman, T. Sur un problem d'unicité pour les systems d'equations aux dériées partielles a deux variables indépendentes. *Arkiv. Math.*, 26B, **17** (1939): 1–9.

Ikawa, M. Decay of solutions of the wave equation in the exterior of two convex obstacles. *Osaka J. Math.* **19** (1982): 459–509.

Lax, P. D. A stability theorem for solutions of abstract differential equations, and its application to the study of local behavior of solutions of elliptic equations. *CPAM*, **9** (1956): 747–766.

Lax, P. D. and Phillips, R. S. *Scattering Theory*, Academic Press, New York, 1967.

Lax, P. D., Morawetz, C. S., and Phillips, R. S. Exponential decay of solutions of the wave equation in the exterior of a star-shaped obstacle. *CPAM*, **16** (1963): 477–486.

Melrose, R. Singularities and energy decay in acoustic scattering. *Duke Math. J.*, **46** (1979): 43–59.

Morawetz, C. S. The decay of solutions of the exterior initial–boundary value problem for the wave equation. *CPAM*, **14** (1961): 561–568.

Müller, C. On the behaviour of the solutions of the differential equation $\Delta u = F(x, u)$ in the neighborhood of a point. *CPAM*, **7** (1954): 505–515.

Ralston, J. Solution of the wave equation with localized energy. *CPAM*, **22** (1969): 807–823.

Taylor, M. Propagation, reflection and diffraction of singularities of solutions to wave equations. *Bull. AMS*, **84** (1978): 589–611.

37

SCATTERING THEORY

Two souls dwell in the bosom of scattering theory. One is mathematical, and handles the unitary equivalence of operators with continuous spectra. The other is in physics, and deals with such notions as quasi-stationary states, cross sections, and what is observable in quantum mechanics.

37.1 PERTURBATION THEORY

Perturbation theory was developed by Lord Rayleigh, and by Erwin Schrödinger as a means of solving problems in physics, classical and quantum mechanical. A rigorous theory was set in place by Franz Rellich. In this section we describe the simplest results; an extensive discussion can be found in Kato's magisterial book.

Theorem 1. *Let* $\mathbf{A}$ *be a self-adjoint operator in a Hilbert space* H, *and* α *an isolated eigenvalue of* $\mathbf{A}$ *of multiplicity one. Let* $\mathbf{D}$ *be a bounded symmetric operator,* $\|\mathbf{D}\| \leq 1$. *Then for* ϵ *small enough* $\mathbf{A} + \epsilon\mathbf{D}$ *has an isolated eigenvalue of multiplicity one in* $[\alpha - \epsilon, \alpha + \epsilon]$.

Proof. We need

Lemma 2. *Let* $\mathbf{P}$ *and* $\mathbf{Q}$ *be orthogonal projections in a Hilbert space whose difference has norm less than* 1:

$$\|\mathbf{P} - \mathbf{Q}\| < 1. \tag{1}$$

Then the range of $\mathbf{P}$ *and* $\mathbf{Q}$ *have the same dimension.*

Proof. If the dimension of the range of $\mathbf{P}$ were greater than that of $\mathbf{Q}$, then there would be a nonzero vector u in the range of $\mathbf{P}$ that is orthogonal to the range of $\mathbf{Q}$. For such u, $\mathbf{P}u = u$, $\mathbf{Q}u = 0$, so $\|(\mathbf{P} - \mathbf{Q})u\| = \|u\|$, contrary to (1). $\qquad\square$

Let C be a circle centered at the isolated eigenvalue α of $\mathbf{A}$, with radius so small that C lies in the resolvent set of $\mathbf{A}$. Let the only point of the spectrum of $\mathbf{A}$ contained inside the circle C be α. Define the mapping $\mathbf{P}$ by

$$\mathbf{P} = \oint (\zeta - \mathbf{A})^{-1} \, d\zeta. \tag{2}$$

Clearly, $\mathbf{P}$ is the projection $\mathbf{E}(\alpha)$, where $\mathbf{E}$ is the orthogonal projection-valued measure entering the spectral resolution of $\mathbf{A}$. By assumption, the range of $\mathbf{P}$ is one dimensional.

For ϵ small enough, $\zeta - (\mathbf{A} + \epsilon \mathbf{D})$ is invertible for all ζ on C, and its inverse differs little from $(\zeta - \mathbf{A})^{-1}$. So we may define $\mathbf{P}_\epsilon$ by

$$\mathbf{P}_\epsilon = \oint (\zeta - \mathbf{A} - \epsilon \mathbf{D})^{-1} \, d\zeta; \tag{2'}$$

$\mathbf{P}_\epsilon$ is also an orthogonal projection, and $\|\mathbf{P} - \mathbf{P}_\epsilon\|$ tends to zero as ϵ tends to zero. It follows therefore from lemma 2 that for ε small the range of $\mathbf{P}_\epsilon$ is one dimensional; every vector in the range is an eigenvector of $\mathbf{A} + \epsilon \mathbf{D}$. The corresponding eigenvalue tends to α as ϵ tends to zero. $\square$

Exercise 1. Prove that this eigenvalue of $\mathbf{A} + \epsilon \mathbf{D}$ differs from α by $\leq \epsilon$.

Corollary. *It follows from formula $(2')$ that the eigenvector and eigenvalue of $\mathbf{A} + \epsilon \mathbf{D}$ depends analytically on ϵ.*

Exercise 2. Denote the eigenvalue of $\mathbf{A} + \epsilon \mathbf{D}$ by $\alpha(\epsilon)$, the corresponding eigenvector of unit norm by $u(\epsilon)$. Show that

$$\frac{d}{d\epsilon} \alpha = (\mathbf{D}u, \, u).$$

How about higher derivatives?

REMARK. Suppose that α is an isolated eigenvalue of $\mathbf{A}$ of multiplicity n. The argument used to prove theorem 1 shows that in this case $\mathbf{A} + \epsilon \mathbf{D}$ has n eigenvalues in $[\alpha - \epsilon, \alpha + \epsilon]$, some of them possibly multiple.

From formula $(2')$ we can conclude that the eigenvalues of $\mathbf{A} + \epsilon \mathbf{D}$ are *algebraic* functions of ϵ, with a possible algebraic singularity at $\epsilon = 0$. Rellich has pointed out that an algebraic function can be expanded in a Puiseux series, a power series in $\epsilon^{1/p}$, p the order of the branch point at $\epsilon = 0$; all branches of this power series represent solutions of the algebraic equation; that is, they are eigenvalues of $\mathbf{A} + \epsilon \mathbf{D}$. But for $p \neq 1$ some of these branches are complex valued for ϵ real; since all the eigenvalues of the self-adjoint operator $\mathbf{A} + \epsilon \mathbf{D}$ are real, p must be 1. So all eigenvalues of $\mathbf{A} + \epsilon \mathbf{D}$ are analytic functions of ϵ. $\square$

We turn now to the other extreme:

Theorem 3. *Let* A *be a self-adjoint operator in a Hilbert space* H, α *an essential point in its spectrum, that is, for every interval* I *containing* α, $E(I)$ *has infinite-dimensional range. Let* C *be any symmetric compact operator; then* α *belongs to the essential spectrum of* $A + C$.

Proof. If not, for some interval I containing α, the range of $E_C(I)$ would be finite dimensional, where E_C is the spectral resolution of $A + C$. Adding another compact operator to $A + C$ would eliminate the finite number of eigenvalues in I altogether. So it suffices to prove that α belongs to the spectrum of $A + C$.

To see this, we show that the norm of the resolvent of $A + C$ at $\alpha + i\eta$ tends to ∞ as η approaches zero. Decompose the compact operator C as

$$C = F + S, \qquad \text{where } F = \sum_1^N \gamma_j(x, f_j) f_j; \quad \|S\| < \eta. \tag{3}$$

For every x in the range of $E(\alpha - \eta, \alpha + \eta)$, $\|(\dot{A} - \alpha - i\eta)x\| < 2\eta\|x\|$. Since the range of $E(\alpha - \eta, \alpha + \eta)$ is infinite dimensional, we can choose x so that it is orthogonal to all the f_j, $j = 1, \ldots, N$. For such an x, $Fx = 0$. Using this, as well as (3), we get

$$\|(A + C - \alpha - i\eta)x\| = \|(A - \alpha - i\eta)x + Sx\| \leq \|(A - \alpha - i\eta)x\| + \|Sx\| < 3\eta\|x\|. \tag{4}$$

This proves that the norm of $(A + C - \alpha - i\eta)^{-1}$ is greater than $1/3\eta$. $\qquad\square$

Theorem 3 is slightly misleading. Suppose that the spectrum of A is absolutely continuous; every point of it belongs to the essential spectrum of A, and therefore to the essential spectrum of $A + C$, C any symmetric compact operator. But, as Weyl observed, the spectrum of $A + C$ need not be continuous; he showed that there exists, for any positive ϵ, a compact operator C, $\|C\| < \epsilon$, such that the spectrum of $A + C$ is pure point spectrum, dense in the formerly continuous spectrum. This result— spectral curdling—was sharpened by von Neumann, who showed that C may be taken not only compact, but of Hilbert-Schmidt class (see chapter 30, section 30.8), with arbitrary small Hilbert-Schmidt norm.

For a symmetric compact operator C, with eigenvalues $\{\gamma_j\}$, the Hilbert-Schmidt norm is $(\sum \gamma_j^2)^{1/2}$. More generally one can define the p-cross norm as $(\sum |\gamma_j|^p)^{1/p}$; see Shatten. Kuroda has further sharpened von Neumann's result by showing that C may be taken to have arbitrarily small p-cross norm, $p > 1$.

For $p = 1$, the p-cross norm is the trace norm; see chapter 30, section 30.2. For C of trace class the story changes dramatically: Marvin Rosenblum has shown, in his dissertation, that perturbing a self-adjoint operator by adding a symmetric operator of trace class leaves the continuous spectrum continuous. The next two sections are devoted to this result; we follow Kato's formulation.

37.2 THE WAVE OPERATORS

The scene is a separable Hilbert space H, and a pair of self-adjoint operators, $\mathbf{A}$ and $\mathbf{B}$. As we observed in chapter 35, $\mathbf{A}$ and $\mathbf{B}$ generate one-parameter unitary groups, which we denote as $e^{it\mathbf{A}}$ and $e^{it\mathbf{B}}$. Define the one-parameter family of unitary operators $\mathbf{W}(t)$ by

$$\mathbf{W}(t) = e^{it\mathbf{B}} e^{-it\mathbf{A}}. \tag{5}$$

It is the strong limits of $\mathbf{W}(t)$ as t tends to $\pm\infty$ that occupy center stage:

$$\mathbf{W}_+ = s - \lim_{t \to \infty} \mathbf{W}(t), \quad \mathbf{W}_- = s - \lim_{t \to -\infty} \mathbf{W}(t), \tag{6}$$

provided that they exist. $\mathbf{W}_+$ and $\mathbf{W}_-$ are called the *wave operators*; when we need to underline their dependence on $\mathbf{A}$ and $\mathbf{B}$, we will denote them as $\mathbf{W}_\pm(\mathbf{B}, \mathbf{A})$.

Exercise 3. Show that

$$\mathbf{W}_\pm (\mathbf{C}, \mathbf{A}) = \mathbf{W}_\pm (\mathbf{C}, \mathbf{B}) \, \mathbf{W}_\pm (\mathbf{B}, \mathbf{A}) \, ,$$

provided that all the wave operators on the right exists.

Since the wave operators are the strong limits of unitary operators, they are isometric:

$$\|\mathbf{W}_\pm u\| = \|u\|. \tag{7}$$

Taking the adjoint of (5) yields

$$\mathbf{W}^*(t) = e^{it\mathbf{A}} e^{-it\mathbf{B}}.$$

Since a strong limit (even weak limit) of the adjoint is the adjoint of the limit of a sequence of operators, we conclude that

$$\mathbf{W}_\pm(\mathbf{A}, \mathbf{B}) = \mathbf{W}^*_\pm(\mathbf{B}, \mathbf{A}), \tag{8}$$

provided that all these wave operators exist.

It follows from definition (5) that for every real s,

$$\mathbf{W}(t + s) = e^{is\mathbf{B}} \mathbf{W}(t) e^{-is\mathbf{A}}.$$

Taking the limit as t tends to ∞, we get

$$\mathbf{W}_+ = e^{is\mathbf{B}} \mathbf{W}_+ e^{-is\mathbf{A}},$$

and similarly for $\mathbf{W}_-$. We can rewrite this as

$$\mathbf{W}_+ e^{is\mathbf{A}} = e^{is\mathbf{B}} \mathbf{W}_+. \tag{9}$$

Form the difference quotients

$$W_+ \frac{e^{isA} - I}{s} v = \frac{e^{isB} - I}{s} W_+ v,$$

where v is a vector in the domain of A. Take the limit $s \to 0$; on the left we get $iW_+ Av$. Therefore the limit on the right exists too, so $W_+ v$ lies in the domain of B and

$$W_+ A = B W_+. \tag{9'}$$

A similar relation holds for W_-.

The wave operator W_+ maps the Hilbert space H onto a subspace K of H. Relation (7) shows that this mapping is unitary. Relation (9) shows that e^{isB} maps K into K; we claim that it maps the orthogonal complement $K^\perp$ of K into $K^\perp$. To see this, denote by z any vector orthogonal to K. The relation $(W_+^* z, h) = (z, W_+ h)$ shows that $K^\perp$ is the nullspace of W_+^*. Using the adjoint of relation (9), with $-s$ in place of s, we have

$$e^{isA} W_+^* = W_+^* e^{isB},$$

which gives $W_+^* e^{isB} z = e^{isA} W_+^* z = 0$. This shows that the subspace K *reduces* the operator B, namely that the operator B restricted to $K \cap D$ and to $K^\perp \cap D$ are self-adjoint mappings on K and on $K^\perp$, respectively. Relation (9') shows that A acting on H and B acting on K are *unitary equivalent*.

Suppose that not only the wave operator $W_+(B, A)$ but also $W_+(A, B)$ exist; then it follows from (8) that $W_+^*(B, A) = W(A, B)$, so by (7) $W_+^*(B, A)$ is an isometry. It follows that its nullspace is trivial, so $K = H$. We summarize:

Theorem 4. *If both wave operators $W_+(B, A)$ and $W_+(A, B)$ exist, then A and B are unitarily equivalent.*

Operators that are unitarily equivalent have the same spectrum. In particular, they have the same point spectrum. Since in scattering theory B is a perturbation of A, it is highly unlikely that a perturbation would leave the eigenvalue of A unchanged; see exercise 2. It is clear that if A has a point eigenvalue, the wave operators $W_\pm(A, B)$ would exist only under the most exceptional circumstances. However, this is easily remedied by restricting A to the absolutely continuous subspace of $H^{(c)}$ of H, as explained in chapter 31, section 31.4.

Exercise 4. In chapter 31, section 31.4, we dealt only with bounded operators; show how to extend these notions to unbounded operators.

Definition. The *generalized wave operator*, also denoted as $W_+(B, A)$ is defined as

$$\lim_{t \to \pm\infty} W(t) P_c = W_\pm, \tag{6'}$$

where $\mathbf{W}(t)$ is defined by (5), and $\mathbf{P}_c$ is the projection of the Hilbert space H onto the subspace $H^{(c)}$ on which $\mathbf{A}$ has an absolutely continuous spectrum.

Exercise 5. Show that if the generalized wave operator $\mathbf{W}_+(\mathbf{B}, \mathbf{A})$ exists, it maps $H^{(c)}$ onto a subspace K of H that reduces the operator $\mathbf{B}$, and that $\mathbf{B}$ has an absolutely continuous spectrum on K.

Exercise 6. Suppose that both generalized wave operators $\mathbf{W}_+(\mathbf{B}, \mathbf{A})$ and $\mathbf{W}_+(\mathbf{A}, \mathbf{B})$ exist; then the absolutely continuous parts of $\mathbf{A}$ and $\mathbf{B}$ are unitarily equivalent.

37.3 EXISTENCE OF THE WAVE OPERATORS

The following result is due to Cook; see also Jauch and Kuroda:

Theorem 5. *Suppose that there is a dense subset J of $H^{(c)}$ such that for all u in J*

(i) $e^{-it\mathbf{A}} u$ *belongs to the intersection* $D(\mathbf{A}) \cap D(\mathbf{B})$ *of the domain of* $\mathbf{A}$ *and* $\mathbf{B}$.
(ii) $(\mathbf{B} - \mathbf{A}) e^{-it\mathbf{A}} u$ *is a continuous function of* t.
(iii) $\|(\mathbf{B} - \mathbf{A}) e^{-it\mathbf{A}} u\|$ *is integrable up to* $+\infty$.

Then the wave operator $\mathbf{W}_+(\mathbf{B}, \mathbf{A})$ exists.
A similar result holds for $\mathbf{W}_-$.

Proof. It follows from (i) that for u in J, $\mathbf{W}(t)u = e^{it\mathbf{B}}e^{-it\mathbf{A}}u$ is differentiable, and that

$$\frac{d}{dt}\mathbf{W}(t)u = ie^{it\mathbf{B}}(\mathbf{B} - \mathbf{A})e^{-it\mathbf{A}}u. \tag{10}$$

By (ii), this derivative is continuous; therefore integration gives

$$\mathbf{W}(b)u - \mathbf{W}(a)u = i\int_a^b e^{it\mathbf{B}}(\mathbf{B} - \mathbf{A})e^{-it\mathbf{A}}u\,dt. \tag{10'}$$

Since $e^{it\mathbf{B}}$ is a unitary operator,

$$\|\mathbf{W}(b)u - \mathbf{W}(a)u\| \leq \int_a^b \|(\mathbf{B} - \mathbf{A})e^{-it\mathbf{A}}u\|\,dt.$$

It follows from (iii) that the right side tends to zero as a, b tends to ∞.
    This shows that the limit of $\mathbf{W}(t)u$ as $t \to +\infty$ exists for all u in J. Since J is dense in H, and since the operators $\mathbf{W}(t)$ are norm preserving, it follows that the strong limit (6) exists.   □

Lemma 6. *Suppose $\mathbf{B}$ differs from $\mathbf{A}$ by a bounded operator:*

$$\mathbf{B} = \mathbf{A} + \mathbf{D}, \quad \|\mathbf{D}\| < \infty,$$

and suppose that the generalized wave operator $\mathbf{W}_+(\mathbf{B}, \mathbf{A})$ *exists. Then for every vector u in* $H^{(c)}$

$$\|\mathbf{W}_+ u - \mathbf{W}(a) u\|^2 = -2 \operatorname{Im} \int_a^\infty \left(e^{it\mathbf{A}} \, \mathbf{W}_+^* \, \mathbf{D} \, e^{-it\mathbf{A}} u, u \right) dt. \qquad (11)$$

Proof. Since $\mathbf{A}$ and $\mathbf{B}$ differ by a bounded operator, formula (10) holds for u in the domain of $\mathbf{A}$, and so does formula $(10')$. Since the domain of $\mathbf{A}$ is dense in H, $(10')$ holds for all u in H. Let b in formula $(10')$ tend to $+\infty$. By hypothesis, for every u in $H^{(c)}$, $\mathbf{W}(b) u$ tends strongly to $\mathbf{W}_+ u$, so we get

$$\mathbf{W}_+ u - \mathbf{W}(a) u = i \int_a^\infty e^{i\mathbf{B}t} \, \mathbf{D} \, e^{-it\mathbf{A}} u \, dt. \qquad (10'')$$

Since both $\mathbf{W}(a)$ and $\mathbf{W}_+$ are isometric,

$$\|\mathbf{W}_+ u - \mathbf{W}(a)u\|^2 = 2\|u\|^2 - 2\operatorname{Re}(\mathbf{W}(a)u, \mathbf{W}_+ u) = 2\operatorname{Re}(\mathbf{W}_+ u - \mathbf{W}(a)u, \mathbf{W}_+ u).$$

Expressing $\mathbf{W}_+ u - \mathbf{W}(a) u$ on the right from $(10)''$, and using the transpose of identity (9), with s replaced by $-t$, gives (11). $\qquad \square$

Theorem 7. *(Rosenblum) Suppose that the self-adjoint operators* $\mathbf{A}$ *and* $\mathbf{B}$ *differ by an operator* $\mathbf{D}$ *of trace class. Then the generalized wave operators* $\mathbf{W}_+(\mathbf{B}, \mathbf{A})$ *and* $\mathbf{W}_+(\mathbf{A}, \mathbf{B})$ *exist.*

As observed in exercise 6, it follows that the absolutely continuous parts of $\mathbf{A}$ and $\mathbf{B}$ are unitarily equivalent.

Proof. We treat first the case when the perturbation $\mathbf{D}$ is of rank one:

$$\mathbf{D}u = c(u, f) f, \quad \|f\| = 1.$$

We denote by K the smallest closed subspace of H that contains f and reduces the operator $\mathbf{A}$. The subspace K can be obtained as the closure of all vectors of the form $b(\mathbf{A}) f$, where b are bounded, continuous functions on $\mathbb{R}$; see chapter 31, section 31.5. Note that $\mathbf{B} = \mathbf{A}$ on the orthogonal complement of K, so for $u \perp K$, $\mathbf{W}(t)$ is the identity, as are the wave operators. It suffices then to prove the existence of the wave operators for $\mathbf{A}$ and $\mathbf{B}$ restricted to K.

Decompose f as

$$f = g + h, \quad g = \mathbf{P}_c f, \quad h = (\mathbf{I} - \mathbf{P}_c)f.$$

The closure of the set of vectors $b(\mathbf{A}) g$ is the subspace $K^{(c)}$ on which $\mathbf{A}$ is absolutely continuous; $\mathbf{A}$ is singular on its orthogonal complement. Since g belongs to $H^{(c)}$, the measure $(\mathbf{E}g, g)$ is absolutely continuous. So $K^{(c)}$ can be represented by $L^2(S)$, S some Borel subset of $\mathbb{R}$, and $\mathbf{A}$ acts as multiplication by λ in this representation.

With $\mathbf{D}u$ defined as $c(u, f) f$, we take for simplicity $c = 1$, and define

$$\mathbf{D}_c\, u = \mathbf{P}_c\, \mathbf{D}\mathbf{P}_c = (\mathbf{P}_c\, u, f)\, \mathbf{P}_c\, f = (u, g)\, g;$$

$g = \mathbf{P}_c\, f$ is represented by a square integrable function supported on S. Thus the existence of the wave operators $\mathbf{W}_+(\mathbf{B}, \mathbf{A})$, $\mathbf{B}$ a rank one perturbation of $\mathbf{A}$, has been reduced to the case where the Hilbert space is $L^2(S)$, $\mathbf{A}$ is multiplication by λ, and $\mathbf{B} = \mathbf{A} + \mathbf{D}$, $\mathbf{D}u = (u, g)\, g$.

We can regard $L^2(S)$ as a closed subspace of $L^2(\mathbb{R}) = H$. We extend $\mathbf{A}$ to H as multiplication by λ, and $\mathbf{D}u$ as $(u, g)\, g$, where g is defined as zero on the complement of S. It suffices to prove the existence of $\mathbf{W}_+$ on the extended space.

We prove the existence of the wave operators first in the case where g is a smooth function, it and its derivatives rapidly decrease as λ tends to $\pm\infty$. We appeal to theorem 5; we choose the dense subset J to consist of all smooth functions $u(x)$ that decrease rapidly together with their derivatives as $\lambda \to \pm\infty$. We verify now the three hypotheses of theorem 5:

(i) The domain of $\mathbf{A}$ and $\mathbf{B}$ consists of functions $u(\lambda)$ for which $\lambda\, u(\lambda)$ is square integrable; hypothesis (i) is clearly satisfied.

(ii)

$$\mathbf{D}_c\, e^{-it\mathbf{A}}\, u = \left(e^{-it\mathbf{A}}\, u, g\right) g = \int e^{-it\mu}\, u(\mu)\, \overline{g}(\mu)\, d\mu\, g(\lambda)$$
$$= \widetilde{u\overline{g}}(-t)\, g(\lambda), \tag{12}$$

where $\sim$ denotes the Fourier transform. Since both u and g decay rapidly at infinity, $u\overline{g}$ is in L^1, and so its Fourier transform is continuous, as required in (ii). Since the derivatives of u and g decay rapidly, $\widetilde{u\overline{g}}$ decays faster than any power of t as $t \to \pm\infty$; therefore requirement (iii) is satisfied. So we conclude that the generalized wave operators $\mathbf{W}_\pm(\mathbf{A}, \mathbf{B})$ exist.

To pass to arbitrary g, we make use of the identity (11). Replacing $\mathbf{D}\, e^{-it\mathbf{A}}\, u$ in this formula by its expression in formula (12), we get

$$\|\mathbf{W}_+ u - \mathbf{W}_{(a)} u\|^2 = -2\,\mathrm{Im} \int_a^\infty \widetilde{u\overline{g}}(t) \left(e^{it\mathbf{A}}\, \mathbf{W}_+^*\, g, u\right) dt. \tag{13}$$

Abbreviate $\mathbf{W}_+^*\, g$ as g^*; then we can rewrite the second factor in the integral above as

$$\left(e^{it\mathbf{A}}\, g^*, u\right) = \int e^{it\lambda}\, g^*(\lambda)\, \overline{u}(\lambda)\, d\lambda = \widetilde{g^*\overline{u}}(t).$$

We estimate the integral on the right in (13) by the Schwarz inequality:

$$\|\mathbf{W}_+ u - \mathbf{W}(a)\, u\|^2 \leq 2 \left(\int_a^\infty |\widetilde{u\overline{g}}|^2\, dt \int_a^\infty |\widetilde{g^*\overline{u}}|^2\, dt\right)^{1/2}.$$

By Parseval's identity the Fourier transform is an isometric mapping times 2π. So the second factor on right side above is less than

$$2\pi \int |g^* u|^2 \, d\lambda \le 2\pi \, |u|_\infty^2 \int |g^*|^2 \, d\lambda,$$

where $|u|_\infty$ is the maximum of $|u(\lambda)|$ on $\mathbb{R}$. Furthermore

$$\int |g^*|^2 \, d\lambda = \|g^*\|^2 = \|\mathbf{W}_+^* \, g\|^2 \le \|g\|^2,$$

since $\mathbf{W}_+^*$, being the adjoint of the isometry $\mathbf{W}_+$, has norm equal to 1. Putting these inequalities together, we get

$$\|\mathbf{W}_+ u - \mathbf{W}(a) \, u\| \le (8\pi)^{1/4} \, |u|_\infty^{1/2} \left(\int_a^\infty |u\widetilde{g}|^2 \right)^{1/4}.$$

A similar inequality holds for a replaced by b; so by the triangle inequality

$$\|\mathbf{W}(b) \, u - \mathbf{W}(a) \, u\| \le (8\pi)^{1/4} \, |u|_\infty^{1/2}$$
$$\cdot \left\{ \left[\int_a^\infty |u\widetilde{g} \, (t)|^2 \, dt \right]^{1/4} + \left[\int_b^\infty |u\widetilde{g} \, (t)|^2 \, dt \right]^{1/4} \right\}. \tag{13'}$$

This inequality was proved under the assumption the u and g are smooth and rapidly decreasing. We show now that it holds for all g that are merely square integrable. To this end we approximate in the L^2-norm any given g in $L^2(\mathbb{R})$ by a sequence of g_n that are smooth and rapidly decreasing at ∞. Inequality (13') holds for g_n, with $\mathbf{W}_n(t) = e^{i\mathbf{B}_n t} e^{-it\mathbf{A}}$, where $\mathbf{B}_n = \mathbf{A} + (\cdot, g_n) g_n$. As $g_n \to g$, the right side of this inequality tends to the right side of (13'); we claim that so does the left side. To show that, we have to demonstrate that $e^{i\mathbf{B}_n t}$ tends to $e^{i\mathbf{B}t}$; this is easily done. $e^{i\mathbf{B}t} u = u(t)$ and $e^{i\mathbf{B}_n t} = u_n(t)$ are the solutions of the differential equations

$$\frac{d}{dt} u - i\mathbf{B}u = 0, \quad \frac{d}{dt} u_n - i\mathbf{B}_n u_n = 0, \quad u(0) = u_n(0) = u,$$

where u belongs to the domain of $\mathbf{B}$. Subtracting the equations, we get

$$\frac{d}{dt}(u - u_n) - i\mathbf{B}(u - u_n) = i(\mathbf{B} - \mathbf{B}_n) u_n.$$

Acting on both sides by $e^{-it\mathbf{B}}$ and integrating, we get

$$u(s) - u_n(s) = i \int_0^s e^{i(s-t)\mathbf{B}} (\mathbf{B} - \mathbf{B}_n) u_n \, dt.$$

Since $e^{i(s-t)\mathbf{B}}$ has norm 1 and $\|u_n(t)\| \le \|u\|$, we deduce that

$$\|u(s) - u_n(s)\| \le s\|\mathbf{B} - \mathbf{B}_n\| \, \|u\|.$$

As g_n tends to g, $\|\mathbf{B} - \mathbf{B}_n\| = \|(\bullet, g) g - (\bullet, g_n) g_n\|$ tends to zero; this completes the proof that $e^{i\mathbf{B}_n t}$ tends to $e^{i\mathbf{B}t}$ in the uniform topology; strong convergence would have been enough. This completes the proof that the inequality (13') is valid for all g in L^2, and all smooth u that decrease rapidly at ∞.

The right side of (13') tends to zero as a and b tend to ∞; therefore so does the left side. This proves the convergence of $\mathbf{W}(t) u$ as t tends to ∞. Since the smooth, rapidly decreasing functions u are dense in $L^2(\mathbb{R})$, and since the operators $\mathbf{W}(t)$ have norm 1, it follows that the limit exists for all u in $L^2(\mathbb{R})$, and therefore the generalized wave operator $\mathbf{W}_+(\mathbf{B}, \mathbf{A})$ exists. Since the role of $\mathbf{A}$ and $\mathbf{B}$ is symmetric, the wave operator $\mathbf{W}_+(\mathbf{A}, \mathbf{B})$ also exists; this shows that the absolutely continuous parts of $\mathbf{A}$ and $\mathbf{B}$ are unitarily equivalent.

We pause for a moment to point out that even when the operator $\mathbf{A}$ has absolutely continuous spectrum, a rank one perturbation can create a point spectrum of $\mathbf{B}$. Take $\mathbf{A}$ to be multiplication by x acting on $L^2(\mathbb{R})$, and $\mathbf{D} = (\bullet, f) f$, f in L^2 and smooth (Lipschitz continuity will do); we are looking for an eigenfunction u of $\mathbf{A} + \mathbf{D}$, with eigenvalue of τ:

$$x u(x) + (u, f) f(x) = \tau u(x).$$

Solve for $u(x)$:

$$u(x) = \frac{(u, f)}{\tau - x} f(x).$$

In order for u to be square integrable, τ has to be a zero of f. Taking the scalar product of both sides with f gives, after canceling the factor (u, f), the relation

$$\int \frac{f^2(x)}{\tau - x} \, dx = 1,$$

a second condition, in addition to $f(\tau) = 0$. If both conditions are satisfied, $\mathbf{A} + \mathbf{D}$ has τ as an eigenvalue.

Exercise 7. Show that f can be so chosen that $\mathbf{A} + \mathbf{D}$ has n eigenvalues, n any natural number. Can $\mathbf{B}$ have infinitely many eigenvalues?

The eigenvalue τ can disappear at the slightest change in the function f; for although the changed f will have a zero near τ, the second integral condition will in general not be satisfied. This phenomenon is called the *instability of the point spectrum embedded in the continuous spectrum*.

We return now to complete the proof of theorem 7.

Let $\mathbf{D}$ be a symmetric operator of trace class. Denote its eigenvalues by d_k, the corresponding normalized eigenvector by f_k. By (20) of chapter 30,

$$\|\mathbf{D}\|_{\text{tr}} \geq \sum |d_k|,$$

and $\mathbf{D}$ itself can be expressed as

$$\mathbf{D} = \sum d_k(\ .\ f_k)\, f_k.$$

Denote by $\mathbf{D}_n$ the finite sum

$$\mathbf{D}_n = \sum_1^n d_k(\ ,\ f_k)\, f_k.$$

Define $\mathbf{B}_n$ as $\mathbf{A} + \mathbf{D}_n$. By what we have already proved, the generalized wave operators $W_+(\mathbf{B}_n, \mathbf{B}_{n-1})$ exist. Therefore (see exercise 3) the wave operator $W_+(\mathbf{B}_n, \mathbf{A})$ exists; we denote it as $\mathbf{W}_{n+}$. We use now formula (11):

$$\|\mathbf{W}_{n+}\, u - \mathbf{W}_n(a)\, u\|^2 = -2\,\mathrm{Im} \int_a^\infty \left(e^{it\mathbf{A}}\, \mathbf{W}_{n+}^*\, \mathbf{D}_n\, e^{-it\mathbf{A}} u,\, u \right) dt.$$

Using the definition of $\mathbf{D}_n$, we can write this as

$$\|\mathbf{W}_{n+}\, u - \mathbf{W}_n(a)\, u\|^2 = -2\,\mathrm{Im} \int_a^\infty \sum_1^n d_k \left(e^{-it\mathbf{A}} u,\, f_k \right) \left(e^{it\mathbf{A}}\, \mathbf{W}_{n+1}^*\, f_k,\, u \right) dt.$$

Using the Schwarz inequality, first with respect to the sum, then the integral, we get

$$\|\mathbf{W}_{n+}\, u - \mathbf{W}_n(a)\, u\| \leq 2 \left[\sum_1^n |d_k| \int_a^\infty \left| \left(e^{-it\mathbf{A}} u,\, f_k \right) \right|^2 dt \right]^{1/2}$$
$$\cdot \left[\sum_1^n |d_k| \int_a^\infty \left| \left(e^{it\mathbf{A}} f_k^*,\, u \right) \right|^2 dt \right]^{1/2}, \tag{14}$$

where f_k^* is an abbreviation for $\mathbf{W}_{n+}^*\, f_k$. As we saw earlier, $\|f_k^*\| \leq \|f\|_k = 1$.

The vector u for which we want to prove the existence of the strong limit $\mathbf{W}(t)\, u$ as $t \to \infty$ belong to the absolutely continuous subspace for the operator $\mathbf{A}$, so $u = \mathbf{P}_c\, u$. Replace u by $\mathbf{P}_c\, u$ on the right in (14); using the fact that $\mathbf{P}_c$ commutes with $e^{it\mathbf{A}}$, we can rewrite the right side of (14) with f_k and f_k^* replaced by $\mathbf{P}_c\, f_k$ and $\mathbf{P}_c\, f_k^*$. Rather than rewriting (14), we just assume that f_k and f_k^* belong to the absolutely continuous subspace of $\mathbf{A}$.

Next we use the spectral resolution of $\mathbf{A}$ to rewrite the terms on the right in (14) as follows:

$$(e^{-it\mathbf{A}} u,\, f) = \int e^{-it\lambda}\, d(\mathbf{E}u,\, f)$$

and

$$(e^{it\mathbf{A}} f^*,\, u) = \int e^{it\lambda}\, d(\mathbf{E}f^*,\, u).$$

Since u, f, and f^* belong to the absolutely continuous subspace of $\mathbf{A}$, we can rewrite the integrals on the right in terms of the Radon-Nikodym derivative as

$$\int e^{-it\lambda}\frac{d}{d\lambda}(\mathbf{E}u, f)\,d\lambda, \qquad \text{etc.} \qquad (15)$$

By the Schwarz inequality,

$$\frac{d}{d\lambda}(\mathbf{E}u, f) \le \left[\frac{d}{d\lambda}(\mathbf{E}u, u)\frac{d}{d\lambda}(\mathbf{E}f, f)\right]^{1/2}.$$

Exercise 8. Prove this inequality.

Suppose that the vector u has the property that

$$\sup_\lambda\frac{d}{d\lambda}(\mathbf{E}u, u) < \infty. \qquad (16)$$

Then, by the Schwarz inequality,

$$\frac{d}{d\lambda}\mathbf{E}(u, f) \le m\left[\frac{d}{d\lambda}(\mathbf{E}f, f)\right]^{1/2},$$

where m is the square root of the sup (16). This shows that $d(\mathbf{E}u, f)/d\lambda$ is square integrable, and so by Parseval's theorem, (15) is square integrable as function of t.

Denote the function (15) as $\eta(t)$, and the corresponding integral with f replaced by f^* as $\eta^*(t)$. Setting these in (14), we get

$$\|\mathbf{W}_{n+}u - \mathbf{W}_n(a)u\|^2 \le 2\left[\sum_1^n |d_k|\int_a^\infty |\eta_k|^2\,dt\right]^{1/2}\left[\sum_1^n |d_k|\int_a^\infty |\eta_k^*|^2\,dt\right]^{1/2}. \tag{14'}$$

By Parseval's formula,

$$\int_a^\infty |\eta_k^*|^2\,dt \le \int_{-\infty}^\infty |\eta_k^*|^2\,dt = 2\pi\int_{\mathbb{R}}\left|\frac{d}{d\lambda}(\mathbf{E}f^*, u)\right|^2\,d\lambda$$

$$\le 2\pi\, m^2\int\frac{d}{d\lambda}(\mathbf{E}f^*, f^*)\,d\lambda = 2\pi\, m^2\|f^*\|^2 \le 2\pi\, m^2.$$

So we can rewrite (14') as

$$\|\mathbf{W}_{n+}u - \mathbf{W}_n(a)u\|^2 \le (8\pi)^{1/2}\left[\sum_1^n |d_k|\int_a^\infty |\eta_k|^2\,dt\right]^{1/2}m\|\mathbf{D}\|_{\text{tr}}^{1/2}; \tag{14''}$$

here we have used the fact that $\sum_1^n |d_k| \le \|\mathbf{D}\|_{\text{tr}}$. Replacing a by b and using the triangle inequality, we get from (14'') that

$\|\mathbf{W}_n(b)\, u - \mathbf{W}_n(a)\, u\| \leq$

$$(8\pi\|\mathbf{D}\|_{\mathrm{tr}})^{1/4}\, m^{1/2}\left(\left[\sum_1^n |d_k| \int_a^\infty |\eta_k|^2\, dt\right]^{1/4} + \left[\sum_1^n |d_k| \int_b^\infty |\eta_k|^2\, dt\right]^{1/4}\right).$$

Now we pass to the limit as n tends to ∞. Since $\mathbf{B}_n = \mathbf{A} + \mathbf{D}_n$ tends uniformly to $\mathbf{B} = \mathbf{A} + \mathbf{D}$, $\mathbf{W}_n(a)$, $\mathbf{W}_n(b)$ tend uniformly to $\mathbf{W}(a)$, $\mathbf{W}(b)$. The right side tends to the infinite sum. In the resulting inequality we let a and b tend to ∞. We claim that the right side tends to zero. This follows from these facts:

(i) $\displaystyle\int_{-\infty}^\infty |\eta_k(t)|^2\, dt \leq 2\pi m^2.$

(ii) $\displaystyle\sum |d_k| < \infty.$

(iii) $\displaystyle\lim_{a\to\infty} \int_a^\infty |\eta_k|^2\, dt = 0.$

This proves the existence of the strong limit $W(b)\, u$ as b tends to ∞.

This conclusion is valid for those vectors u that satisfy (16). The last step of the proof is to verify that the set of such u is dense in $H^{(c)}$. This is not hard; let v be any vector in $H^{(c)}$. For such a v, the measure $(\mathbf{E}v, v)$ is absolutely continuous with respect to Lebesgue measure, so we can write

$$\|v\|^2 = \int d\,(\mathbf{E}v, v) = \int \frac{d}{d\lambda}\,(\mathbf{E}v, v)\, d\lambda.$$

$d(\mathbf{E}v, v)/d\lambda$ is nonnegative and in L^1. Denote by S_m the set of λ where $d(\mathbf{E}v, v)/d\lambda > m$. Set $v_m = (\mathbf{I} - \mathbf{E}(S_m))\, v$; denote by μ_m the measure

$$(\mathbf{E}v_m, v_m).$$

If we denote the measure $(\mathbf{E}v, v)$ by μ, the two measures are related by $d\mu_m = (1 - c_s)d\mu$, where c_s is the characteristic function of the set S_m. It follows that the Radon-Nikodym derivatives are similarly related:

$$\frac{d}{d\lambda}(\mathbf{E}v_m, v_m) = (1 - c_S)\frac{d}{d\lambda}(\mathbf{E}u, u).$$

Since S_m was chosen as the set where $d\mu/d\lambda > m$, it follows that $d\mu_m/d\lambda > m \leq m$ for all λ. The vectors v_m tend to v as m tends to ∞. This completes the proof that the vectors u that satisfy (16) are dense in $H^{(c)}$. Since the domain of the wave operator is a closed subspace of $H^{(c)}$, this completes the proof that the generalized wave operator $\mathbf{W}_+(\mathbf{B}, \mathbf{A})$ exists when $\mathbf{A}$ and $\mathbf{B}$ differ by an operator of trace class. $\qquad\square$

Of course, the wave operator $\mathbf{W}_-(\mathbf{B}, \mathbf{A})$ also exists, and since the $\mathbf{A}$ and $\mathbf{B}$ enter the hypothesis symmetrically, $\mathbf{W}_+(\mathbf{A}, \mathbf{B})$ also exist. Therefore the absolutely continuous parts of $\mathbf{A}$ and $\mathbf{B}$ are unitarily equivalent.

37.4 THE INVARIANCE OF WAVE OPERATORS

The scope of the main result of section 37.3 is greatly extended by the following result: Let $\phi(\lambda)$ be a real valued function with the following properties:

(i) ϕ is piecewise differentiable.

(ii) ϕ' is positive, continuous and of bounded variation locally.

Theorem 8 (Birman-Kato). *Let* **A** *and* **B** *be a pair of self-adjoint operators on a Hilbert space that differ by an operator of trace class. Let ϕ be a function as above. Then the wave operator* $\mathbf{W}_\pm(\phi, (\mathbf{B}), \phi(\mathbf{A}))$ *exist and are independent of ϕ.*

Since $\mathbf{W}_\pm(\phi(\mathbf{A}), \phi(\mathbf{B}))$ *also exist, the absolute continuous parts of $\phi(\mathbf{A})$ and $\phi(\mathbf{B})$ are unitarily equivalent.*

Even when ϕ is not monotonic, the existence of $\mathbf{W}_\pm(\phi(\mathbf{B}), \phi(\mathbf{A}))$ can be shown. Only in this case these wave operators are no longer equal to $\mathbf{W}_\pm(\mathbf{B}, \mathbf{A})$ but are composites of them.

For a proof of these result, see chapter X of Kato's book.

37.5 POTENTIAL SCATTERING

As an example of the abstract theory developed in section 37.3, take the Hilbert space $H = L^2(\mathbb{R}^3)$, $\mathbf{A} = -\Delta$, the negative Laplace operator, and $\mathbf{B} = -\Delta + q$, q a real-valued function. We take the simplest case when the potential q is a square integrable bounded function.

Theorem 9. *For q bounded and square integrable, the wave operators* $\mathbf{W}_\pm(\mathbf{B}, \mathbf{A})$ *exist, where* $\mathbf{A} = -\Delta$, $\mathbf{B} = -\Delta + q$.

Proof. Fourier transformation is the spectral representation for $-\Delta$, and shows that its spectrum is absolutely continuous, fills $\mathbb{R}$, and is of infinite multiplicity. We appeal now to theorem 5 in section 37.3. We take for the dense subset J the linear span of functions of the form $e^{-(x-a)^2/2}$, a in $\mathbb{R}^3$. For u of this form, we can solve the equation $u_t = -i\,\mathbf{A}u = i\,\Delta u$ explicitly by Fourier transform:

$$\widetilde{u}_t = -i\xi^2\,\widetilde{u}, \quad \widetilde{u}(0) = e^{-\xi^2/2 + ia\cdot\xi},$$

so $\widetilde{u}(\xi, t) = e^{-(2it+1)\xi^2/2 + ia\cdot\xi}$. Taking the Fourier inverse gives

$$u(x, t) = e^{-it\mathbf{A}}\,u(x) = (1 + 2it)^{-3/2}\,e^{-(x-a)^2/(2+4it)}.$$

Clearly, this function belong to the common domain of **A** and **B**, as required in (i). Conditions (ii) and (iii) are also fulfilled, since

$$\|(\mathbf{B} - \mathbf{A})e^{-it\mathbf{A}}u\| = \|qu(x, t)\| \leq \|q\| \, |1 + 2it|^{-3/2}$$

is a continuous function of t, integrable over the whole t axis.

It remains to show that the linear span of the functions $e^{-(x-a)^2/2}$ is dense in $L^2(\mathbb{R}^3)$. According to the spanning criterion(see theorem 7 of chapter 6) we have to show that a function f in $L^2(\mathbb{R}^3)$ that is orthogonal to all functions $e^{-(x-a)^2/2}$ is zero. To see this, we rewrite this condition as

$$0 = \int f(y + a) \, e^{-y^2/2} \, dy = \int e^{ia\xi} \, \tilde{f}(\xi) \, e^{-\xi^2/2} \, d\xi;$$

here in the first step we changed the variable of integration to $x - a = y$, and in the second step we use Parseval's relation. The last equation says that the Fourier inverse of $\tilde{f}e^{-\xi^2/2}$ is identically zero. But then so is $\tilde{f}e^{-\xi^2/2}$, and so is $\tilde{f}(\xi)$ and f itself.

□

We conclude from theorem 9 that $-\Delta$ is unitarily equivalent to $\mathbf{B} = -\Delta + q$ acting on an invariant subspace of $\mathbf{B}$. In particular, the continuous spectrum of $-\Delta + q$ contains the whole positive axis, of infinite multiplicity.

The restriction that q be square integrable can be relaxed. For sharper result, see chapter X of Kato's book, and the literature quoted there. However, if q is merely bounded, then with probability 1, $\Delta + q$ has a pure point spectrum. This result, of importance in solid state physics, is called Anderson localization; for example, see Fröhlich and Spencer.

On the other hand, if the potential $q(x)$ tends to zero fast enough as $|x| \to \infty$, one can show that $(\lambda - \mathbf{A})^{-1}$ and $(\lambda - \mathbf{B})^{-1}$ differ by a trace class operator when $\lambda < 0$. Then it follows from theorem 8 that $W_\pm(\mathbf{A}, \mathbf{B})$ and $W_\pm(\mathbf{B}, \mathbf{A})$ exist, and so $-\Delta$ and the absolutely continuous part of $-\Delta + q$ are unitarily equivalent.

37.6 THE SCATTERING OPERATOR

Suppose that $\mathbf{B}$ is a perturbation of $\mathbf{A}$, and that the generalized wave operators $W_\pm(\mathbf{B}, \mathbf{A})$ exist and map the absolutely continuous part of $\mathbf{A}$ onto the absolutely continuous part of $\mathbf{B}$. Then, by (9'),

$$W_+\mathbf{A} = \mathbf{B}W_+ \quad \text{and} \quad W_-\mathbf{A} = \mathbf{B}W_-.$$

From these relations, we deduce that

$$W_-^{-1} W_+\mathbf{A} = W_-^{-1}\mathbf{B}W_+ = \mathbf{A}W_-^{-1}W_+.$$

In words, $W_-^{-1} W_+$ commutes with $\mathbf{A}$. This operator, $W_-^{-1} W_+$, is called the *scattering operator*, and is denoted as $\mathbf{S}$.

The physical significance of the scattering operator is this:

Think of $\mathbf{A}$ and $\mathbf{B}$ as an unperturbed and perturbed Schrödinger operators, where for large distances the perturbation is negligible. The operators discussed in section 37.5 form such a pair, for the potential $q(x)$ tends to zero as x tends to ∞. For large positive times, most of the signal has propagated to large distances, so that the signal $e^{it\mathbf{B}} u$ differs very little from a signal governed by the unperturbed equation, call it $e^{it\mathbf{A}} u_+$. Similarly, for t large, negative $e^{it\mathbf{B}} u$ differs very little from a signal $e^{it\mathbf{A}} u_-$. Letting t tend to $\pm\infty$, we deduce that

$$u_+ = \lim_{t \to \infty} e^{-it\mathbf{A}} e^{it\mathbf{B}} u = \mathbf{W}_-^{-1} u$$

and

$$u_- = \lim_{t \to -\infty} e^{-it\mathbf{A}} e^{it\mathbf{B}} u = \mathbf{W}_+^{-1} u.$$

The operator linking u_- and u_+, $\mathbf{W}_-^{-1} \mathbf{W}_+$, is the *scattering operator*. Thus the scattering operator links the state of the perturbed system in the remote past to its state in the dim future.

This time-dependent picture of the scattering process has been described in 1945 by Møller. A stationary picture has been formulated by John Wheeler in 1937. It was elaborated by Heisenberg in 1943; Heisenberg's motivation was that a physical theory should only deal with observable quantities. The forces acting on electrons surrounding a nucleus cannot be measured; a physical experiment measures only the outcome, taking place at $t = \infty$ on the atomic time scale, and compares it to the setup, the state of the system at $t = -\infty$. The task of scattering theory is to reconstruct the atomic forces from the scattering operator. This is no place to say anything about this fascinating problem, another gift bestowed on mathematics by physics; we refer to the review article of Ludvig Faddeev, and to volume 2 of Reed and Simon.

Since $\mathbf{S}$ commutes with $\mathbf{A}$, the natural description of $\mathbf{S}$ is in a spectral representation of $\mathbf{A}$, where $\mathbf{S}$ acts a multiplication operator. We will elaborate this in a slightly different setting in the next section.

HISTORICAL NOTE. In 1930 Heisenberg lectured at the University of Chicago on the new quantum mechanics. His assistant there was the young American physicist Frank Hoyt, who helped prepare the English lecture notes. During the Second World War Hoyt joined the Manhattan project for building nuclear weapons; one of his assignments was to scrutinize every wartime publication of Heisenberg and see if it could be a by-product of bomb research. Hoyt studied very thoroughly the two papers Heisenberg published in 1943 on scattering theory, and as he told me later, concluded that they had no bearing on nuclear weapons. This may have saved Heisenberg's life, for the OSS, the wartime precurser of the CIA, had been training an agent to assassinate him.

37.7 THE LAX-PHILLIPS SCATTERING THEORY

The setting is the same as in section 4 of chapter 36: a unitary group $U(t)$ acting on a separable Hilbert space H, and a pair of incoming and outgoing subspaces F_- and F_+, orthogonal to each other, each satisfying properties (27a)–(27c), respectively (28a)–(28c)

$$U(t) \text{ map } F_- \text{ into } F_- \text{ for } t < 0, \tag{17a}$$

$$\text{The intersection of } U(t) F_- \text{ is } \{0\}. \tag{17b}$$

$$\text{The union of } U(t) F_- \text{ is dense in } H, \tag{17c}$$

and similarly for F_+, changing the sign of t in (i).

We appeal now to the translation representation theorem 7 of chapter 35, which says that if the above three conditions are satisfied, then H can be represented isometrically as $L^2(N, \mathbb{R})$, so that the action of $U(t)$ is represented by translation to the right by t. Furthermore the incoming subspace F_- is represented by $L^2(N, \mathbb{R}_-)$. Similarly, since F_+ is an outgoing subspace, there is an *outgoing* representation of H as $L^2(N, \mathbb{R})$, where $U(t)$ is represented by translation, and F_+ is represented by $L^2(N, \mathbb{R}_+)$. Since the dimensions of the auxiliary spaces N appearing in the incoming and outgoing representations are equal to the multiplicity of the spectrum of the generator of the group $U(t)$, the two auxiliary spaces can be taken to be the same.

Let u be any vector in H, k_- and k_+ its incoming and outgoing representers. We define S to be the operator relating the two:

$$S k_- = k_+. \tag{18}$$

We call S the *scattering operator* associated with $U(t)$, F_- and F_+.

In chapter II of Lax and Phillips's *Scattering Theory*, we show how to construct an unperturbed group U_0 of unitary operators so that the scattering operator, defined in section 37.6 in terms of the wave operators linking U_0 and U, is the same as the one defined in (18).

Theorem 10. *Let* $U(t)$ *be a unitary group of operators,* F_- *and* F_+ *orthogonal incoming and outgoing subspaces,* S *the scattering operator defined in* (18).

(i) S *is unitary.*

(ii) S *commutes with translation.*

(iii) S *maps* $L^2(N, \mathbb{R}_-)$ *into itself.*

Proof.

(i) Since both k_- and k_+ represent u isometrically, $\|k_-\| = \|u\| = \|k_+\|$, so S is an isometry. Since it maps $L^2(N, \mathbb{R})$ onto $L^2(N, \mathbb{R})$, it is unitary.

(ii) Since $k_-(x - t)$ and $k_+(x - t)$ both represent $U(t) u$, S maps the translate of k_- onto the same translate of k_+.

(iii) Any k_- in $L^2(N, \mathbb{R}_-)$ is the incoming representation of a vector u in F_-. Such a vector u is orthogonal to F_+; therefore its outgoing representation k_+ is orthogonal to the outgoing representation of F_+, which is $L^2(N, \mathbb{R}_+)$. This shows that k_+ belongs to $L^2(N, \mathbb{R}_-)$.

Property (iii) is called *causality*, and can be put in the following words: *The value of k_+ on $\mathbb{R}_+$ depends on the value of k_- only on $\mathbb{R}_+$.*

The adjoint $\mathbf{S}^*$ of $\mathbf{S}$ has analogous properties:

(i) $\mathbf{S}^*$ *is unitary.*

(ii) $\mathbf{S}^*$ *commutes with translation.*

(iii) $\mathbf{S}^*$ *maps* $L^2(N, \mathbb{R}_+)$ *into itself.*

Let u be a vector in H, k_- and k_+ its incoming and outgoing translation representations. The Fourier transforms of k_- and k_+, denoted as f_- and f_+, are the incoming and outgoing *spectral representations* of u. We denote their relation as

$$S f_- = f_+. \tag{19}$$

Theorem 10$'$.

(i$'$) S *is unitary;*

(ii$'$) S *commutes with multiplication by bounded, measurable functions;*

(iii$'$) S *maps the Fourier transform of $L^2(N, \mathbb{R}_-)$ into itself.*

Proof. (i$'$) follows from part (i) of theorem 10, since $\mathbf{S}$ is unitary and so is the Fourier transform.

(ii$'$) The Fourier transform transmutes translation by an amount a into multiplication by $e^{ia\lambda}$. Therefore, by part (ii) of theorem 10, S commutes with multiplication by $e^{ia\lambda}$. Given any bounded, measurable function $b(\lambda)$, we can approximate it by a sequence b_n of the form $b_n(\lambda) = \sum_1^n c_j e^{ia_j \lambda}$, so that $\lim b_n(\lambda) = b(\lambda)$ a.e., and so that the functions b_n are uniformly bounded on $\mathbb{R}$. It follows that for any f in $L^2(N, \mathbb{R})$, $b_n f$ tends to bf, and $b_n S f$ tends to $bS f$, in the $L^2(N, \mathbb{R})$ norm. Since S is a bounded operator, $S b_n f$ tends to $S b f$. Since $S b_n f = b_n S f$, (ii$'$) follows.

(iii$'$) is a restatement of part (iii) of theorem 10 in the spectral representation.   □

Let k be an L^2 function supported on $\mathbb{R}_-$. Then its Fourier transform f can be extended as an analytic function to the lower half $\mathbb{C}_-$ of the complex plane, $\zeta = \lambda + i\eta$, $\eta < 0$, by the formula

$$f(\zeta) = \frac{1}{\sqrt{2\pi}} \int_{-\infty}^0 k(x) e^{i\zeta x} dx. \tag{20}$$

Theorem 11 (Paley-Wiener). *The Fourier transform f of a function k in $L^2(N, \mathbb{R}_-)$ is a vector-valued function in $L^2(N, \mathbb{R})$, that has an analytic extension $f(\zeta)$ into $\mathbb{C}_-$, with the following properties:*

(i) *For fixed $\eta < 0$, $f(\lambda + i\eta)$ is a vector valued L^2 functon of λ. As η tends to $-\infty$, $\|f(\cdot + i\eta)\|$ tends to 0.*

(ii) *As η tends to 0, $f(\cdot + i\eta)$ tends to f in the L^2-norm.*

Conversely, any function f with properties (i) and (ii) is the Fourier transform of an $L^2(N, \mathbb{R}_-)$ function.

A proof for scalar-valued functions, employing nothing more than the Cauchy integral theorem, is presented in chapter 38. An extension to the vector-valued case is, as so often, straightforward. $\qquad\square$

We will denote the Fourier transform of $L^2(N, \mathbb{R}_-)$ as H_-, and the Fourier transform of $L^2(N, \mathbb{R}_+)$ as H_+. H_+ can be characterized as consisting of vector-valued functions in $L^2(N, \mathbb{R})$ that have analytic extension into the upper half-plane C_+, with properties analogous to those enumerated in theorem 11.

Theorem 12. *The operator S defined in (19) can be realized as multiplication by an operator-valued function $\mathcal{M}(\lambda)$, mapping N into N.*

(i) *$\mathcal{M}(\lambda)$ is unitary for almost all λ.*

(ii) *$\mathcal{M}(\lambda)$ is the boundary value of an operator valued function $\mathcal{M}(\zeta)$ in holomorphic C_-.*

(iii) *For each ζ in C_-, $\mathcal{M}(\zeta)$ is a contraction, mapping $N \to N$.*

The function $\mathcal{M}(\zeta)$ is called the *scattering matrix*.

Proof. We want first to tackle (ii) and (iii). Let u be any vector in F_-. According to part (iii′) of theorem 10′, its incoming and outgoing spectral representations f_- and f_+ both belong to H_-. So f_- and f_+ are vector-valued analytic functions in C_-.

We now show that for any ζ in C_- and any f_- in H_-, the value of $f_+(\zeta)$ is determined by the value of $f_-(\zeta)$. To prove this, it suffices to show that if $f_-(\zeta) = 0$, then $f_+(\zeta) = 0$. We factor such an f_- as

$$f_-(\lambda) = \frac{\lambda - \zeta}{\lambda + \zeta} \, g(\lambda).$$

It follows from the Paley-Wiener theorem, theorem 11, that g belongs to H_-. Since by theorem 10′, S commutes with multiplication by functions bounded on $\mathbb{R}$,

$$f_+\lambda = S f_- = S \frac{\lambda - \zeta}{\lambda + \zeta} g = \frac{\lambda - \zeta}{\lambda + \zeta} Sg.$$

Since g belongs to H_-, by theorem 10′ so does Sg; setting $\lambda = \zeta$ in the relation above shows that $f_+(\zeta) = 0$.

$f_+(\zeta)$ is related to $f_-(\zeta)$ by a linear mapping of $N \to N$; we denote it as $\mathcal{M}(\zeta)$:

$$\mathcal{M}(\zeta) f_-(\zeta) = f_+(\zeta). \tag{21}$$

To show that $\mathcal{M}(\zeta)$ is strongly analytic, take $f_-(\lambda) = n/(\lambda - i)$, n any vector in N. Clearly, f_- belongs to H_-; therefore so does f_+. Set this pair in (21):

$$\frac{1}{\zeta - i} \mathcal{M}(\zeta) n = f_+(\zeta);$$

since $f_+(\zeta)$ is analytic in $\mathbb{C}_-$, so is $\mathcal{M}(\zeta) n$.

(iii) Take any ζ in $\mathbb{C}_-$, any vector n in N, and define

$$k_+(x) = \begin{cases} e^{i\zeta x} n & x < 0 \\ 0 & 0 < x. \end{cases} \tag{22}$$

For any positive r,

$$k_+(x - r) = e^{-i\zeta r} e^{i\zeta x} n \qquad \text{for } x < r. \tag{22'}$$

Set (22) into (22'):

$$k_+(x - r) - e^{-i\zeta r} k_+(x) = 0 \qquad \text{for } x < 0. \tag{23}$$

Define

$$k_- = \mathbf{S}^* k_+. \tag{24}$$

Since $\mathbf{S}^*$ commutes with translation, and since $\mathbf{S}^*$ maps $L^2(N, R_+)$ into itself, we deduce from (23) that for all positive r,

$$k_-(x - r) - e^{-i\zeta r} k_-(x) = 0 \qquad \text{for } x < 0. \tag{23'}$$

This implies that

$$k_-(x) = e^{i\zeta x} m \qquad \text{for } x < 0, \tag{22''}$$

where m is some vector in N. We define as before the incoming and outgoing spectral representations as

$$f_- = \mathbf{F} k_-, \qquad f_+ = \mathbf{F} k_+,$$

where $\mathbf{F}$ is the Fourier transformation. Using formulas (22) and (22'') we get

$$f_+(\lambda) = \frac{1}{\sqrt{2\pi}} \int_{-\infty}^{0} e^{i(\lambda+\zeta)x} n \, dx = \frac{-i}{\sqrt{2\pi}} \frac{n}{\lambda + \zeta} \tag{25}$$

and

$$f_-(\lambda) = \frac{-i}{\sqrt{2\pi}} \frac{m}{\lambda + \zeta} + a_+(\lambda). \tag{25'}$$

where a_+ is the Fourier transform of k_- restricted to $\mathbb{R}_+$. Therefore a_+ belongs to H_+.

Let p be any vector in N. Using formula (25), we get

$$\left(f_+, \frac{p}{\lambda + \zeta} \right) = \frac{-i}{\sqrt{2\pi}} \int \frac{(n, p)_N}{(\lambda + \zeta)(\lambda + \bar{\zeta})} \, d\lambda.$$

The calculus of residues gives

$$\left(f_+, \frac{p}{\lambda + \zeta} \right) = -\sqrt{2\pi} \frac{(n, p)_N}{2\mathrm{Im}\, \zeta}. \tag{26}$$

Similarly, using (25'), we get

$$\left(f_-, \frac{p}{\lambda + \zeta} \right) = \frac{-i}{\sqrt{2\pi}} \int \frac{(m, p)_N}{(\lambda + \zeta)(\lambda + \bar{\zeta})} \, d\lambda + \int \frac{(a_+(\lambda), p)_N}{\lambda + \bar{\zeta}} \, d\lambda.$$

Since a_+ belongs to H_+, we can in the second integral on the right shift the path of integration from the real axis to $\lambda + i\kappa$, $\kappa > 0$. Estimating the integral by the Schwarz inequality, we see that it tends to zero as κ tends to ∞. So we obtain

$$\left(f_-, \frac{p}{\lambda + \xi} \right) = -\sqrt{2\pi} \frac{(m, p)_N}{2\mathrm{Im}\, \zeta}. \tag{26'}$$

The Fourier transform of (24) is $f_- = S^* f_+$. We write, using this and (25), that

$$\left(\frac{p}{\lambda + \zeta}, f_- \right) = \left(\frac{p}{\lambda + \zeta}, S^* f_+ \right)$$

$$= \left(S\left(\frac{p}{\lambda + \zeta} \right), f_+ \right) = \frac{i}{\sqrt{2\pi}} \int \left(S\left(\frac{p}{\lambda + \zeta} \right), n \right)_N \frac{d\lambda}{\lambda + \bar{\zeta}}.$$

Since $p/(\lambda + \zeta)$ belongs to H_-, so does $S(p/\lambda + \zeta)$; thus the integrand above is meromorphic in the lower half-plane, with a simple pole at $-\bar{\zeta}$. We shift the path of integration from the real axis to $\lambda + i\kappa$ and let κ tend to $-\infty$. We obtain, using (21), that

$$\left(\frac{p}{\lambda + \xi}, f_- \right) = \sqrt{2\pi} \left(\mathcal{M}(-\bar{\zeta}) p, n \right)_N \frac{1}{2\mathrm{Im}\, \zeta}. \tag{27}$$

Compare this with (26'):

$$(m, p)_N = \left(n, \mathcal{M}(-\bar{\zeta}) p \right)_N = \left(\mathcal{M}^*(-\bar{\zeta}) n, p \right)_N.$$

Since this holds for all p in N,

$$m = \mathcal{M}^*(-\bar{\zeta}) n. \tag{28}$$

We will now estimate the norm of m. Applying the Schwarz inequality to the left side of (26'), we get

$$\frac{\sqrt{2\pi}}{2|\text{Im}\,\zeta|}\,|(m,p)_N| \le \|f_-\|\,\left\|\frac{p}{\lambda+\zeta}\right\|. \tag{29}$$

Using the definition of f_- as $\mathbf{F}k_-$, (24) and that S^* is a unitary operator, we get

$$\|f_-\| = \|k_-\| = \|S^*k_+\| = \|k_+\|. \tag{30}$$

Using the definition (22) of k_+ gives

$$\|k_+\|^2 = \int_0^\infty \left|e^{i\zeta x}\right|^2\,dx\,|n|_N^2 = \frac{1}{2\,|\text{Im}\,\zeta|}\,|n|_N^2. \tag{30'}$$

By calculus,

$$\left\|\frac{p}{\lambda+\zeta}\right\|^2 = \int \frac{d\lambda}{|\lambda+\zeta|^2}\,|p|_N^2 = \frac{\pi}{|\text{Im}\,\zeta|}\,|p|_N^2. \tag{30''}$$

Setting (30), (30'), and (30'') into the right side of (29) gives

$$|(m,p)_N| \le |n|_N\,|p|_N.$$

Since this holds for all p in N, it follows that $|m|_N \le |n|_N$. In light of (28), $|\mathcal{M}^*(-\bar\zeta)|_N \le 1$, which implies that $|\mathcal{M}(-\bar\zeta)|_N \le 1$, as asserted in (iii).

$(\mathcal{M}(\zeta)\,n,p)$ is a bounded analytic function in $\mathbb{C}_-$. According to a basic result of the theory of analytic functions,

$$\lim_{\eta\to 0}\,(\mathcal{M}(\lambda+i\eta)\,n,p)_N \qquad \eta < 0, \tag{31}$$

exists for a.a. real λ. Take for n and p a denumerable dense set of vector in N; since $|\mathcal{M}(\zeta)| \le 1$, it follows that the limit (31) exists for all n and p in N, for almost all λ. Denote the weak limit (31) by $\mathcal{M}(\lambda)$; clearly, $|\mathcal{M}(\lambda)| \le 1$ a.e.

Let n be any vector in N. The function $f_-(\lambda) = n/(\lambda - i)$ belongs to H_-; therefore it is the incoming spectral representation of some function u in H_-. The outgoing spectral representation f_+ of u also belongs to H_-. Set $\zeta = \lambda + i\eta$ in (21):

$$\frac{1}{\lambda+i\eta-i}\,\mathcal{M}(\lambda+i\eta)\,n = f_+(\lambda+i\eta)\,, \qquad \eta < 0. \tag{21'}$$

and let η tend to zero. The right side tends to f_+ in the $L^2(N,\mathbb{R})$ norm; therefore, so does the left side. It is not hard to deduce from this that $\mathcal{M}(\lambda+i\eta)\,n$ tends *strongly* to $\mathcal{M}(\lambda)\,n$, for a.a. real λ.

Since S is an isometry, we deduce from (21') that

$$\|f_-\|^2 = \left\| \frac{n}{\lambda - i} \right\|^2 = |n|_N^2 \int \frac{1}{|\lambda - i|^2} \, d\lambda = \|f_+\|^2 = \int \frac{1}{|\lambda - i|^2} \, |\mathcal{M}(\lambda) \, n|_N^2 \, d\lambda.$$

We noted earlier that $|\mathcal{M}(\lambda)| \leq 1$ for a.a. λ, so it follows from the above that $|\mathcal{M}(\lambda) \, n|_N = |n|_N$ for a.a. λ.

The operator S is multiplication by $\mathcal{M}(\lambda)$; therefore the operator S^* is multiplication by $\mathcal{M}^*(\lambda)$. Since S^* is an isometry, it follows that $\mathcal{M}^*(\lambda)$ is an isometry for a.a. λ. We conclude that $\mathcal{M}(\lambda)$ is unitary for a.a. λ. This completes the proof of theorem 12. $\qquad \Box$

Note that $\mathcal{M}^*(\lambda)$ is the boundary value of the function $\mathcal{M}^*(\bar{\zeta})$, holomorphic for ζ in $\mathbb{C}_+$.

37.8 THE ZEROS OF THE SCATTERING MATRIX

We recall from chapter 36, section 36.4, the Lax-Phillips semigroup

$$Z(t) = P_+ \, U(t) \, P_-,$$

where P_-, P_+ are projections onto the orthogonal complements of F_- and F_+, a pair of incoming and outgoing subspaces, orthogonal to each other. The semigroup $Z(t)$ acts on the space $K = H \ominus F_- \ominus F_+$.

Since the semigroup Z and the scattering matrix have the same ingredients, there is bound to be some relation between them. Here it is; denote the infinitesimal generator of $Z(t)$ by G:

Theorem 13. *A complex number γ, $\mathrm{Re}\,\gamma < 0$, belongs to the point spectrum of G if and only if $\mathcal{M}^*(i\bar{\gamma})$ has a nontrivial nullspace.*

Proof. Let u be an eigenvector of G:

$$Gu = \gamma u, \quad Z(t)u = e^{\gamma t} u. \tag{32}$$

Let k_+ be the outgoing translation representation of u. Since u belongs to K, it is orthogonal to F_+; therefore k_+ is zero on $\mathbb{R}_+$. In the outgoing representation $Z(t)$ acts as translation by t, followed by restriction to the negative axis; so (32) becomes

$$k_+(x - t) = e^{\gamma t} k(x), \quad x < 0, t > 0.$$

It follows that

$$k_+(x) = \begin{cases} e^{-\gamma x} n & \text{for } x < 0 \\ 0 & \text{for } 0 < x, \end{cases}$$

n some vector in N.

The outgoing spectral representation of u is the Fourier transform of k_+:

$$f_+(\lambda) = \mathbf{F}\,k_+ = c\,\frac{n}{i\lambda - \gamma}, \qquad c = \frac{1}{\sqrt{2\pi}}.$$

Since $\mathcal{M}^{-1}(\lambda) = \mathcal{M}^*(\lambda)$ for λ real, the incoming spectral representation is, by (21),

$$f_-(\lambda) = \frac{c}{i\lambda - \gamma}\,\mathcal{M}^*(\lambda)\,n. \tag{33}$$

Since u belongs to K, it is orthogonal to F_-; therefore f_- is orthogonal to H_-. Consequently f_- belongs to H_+, and thus has an analytic extension to $\mathbb{C}_+$. Formula (33) gives a meromorphic extension of f_- to $\mathbb{C}_+$; it is analytic iff the potential pole at $\lambda = -i\,\gamma$ is cancelled by a zero of $\mathcal{M}^*(\bar{\zeta})\,n$ at $\zeta = -i\gamma$:

$$\mathcal{M}^*(i\,\bar{\gamma})\,n = 0.$$

The reverse of this argument gives the converse proposition.   □

The proof presented above gives a little more: the dimension of the nullspace of $\mathbf{G} - \gamma\,\mathbf{I}$ equals the dimension of the nullspace of $\mathcal{M}^*(i\,\bar{\gamma})$. More generally, one can show:

Theorem 13′. *A complex number γ belongs to the resolvent set of* $\mathbf{G}$ *iff* $S(i\,\bar{\gamma})$ *is invertible.*

Proof. For a proof we refer to section 3, chapter III of Lax and Phillips's *Scattering Theory.*

According to theorem 12, the scattering matrix $\mathcal{M}(\lambda)$ can be extended analytically into the lower half-plane $\mathbb{C}_-$. Suppose that $\mathcal{M}(\lambda)$ is continuous in the norm topology along an interval I of the real axis. Then $\mathcal{M}$ can be continued analytically across I by the operator version of the Schwarz reflection principle:

$$\mathcal{M}(\zeta) = \mathcal{M}^*(\bar{\zeta})^{-1} \tag{34}$$

for ζ in $\mathbb{C}_+$ near I.

37.9 THE AUTOMORPHIC WAVE EQUATION

Faddeev and Pavlov have given a beautiful application of the Lax-Phillips scattering theory to automorphic solutions of the wave equation in the hyperbolic plane. The Poincaré model of the hyperbolic plane $\mathbb{H}$ is the upper half-plane (x, y), $y > 0$, equipped with the Riemannian metric

$$ds^2 = \frac{dx^2 + dy^2}{y^2}. \tag{35}$$

The isometries, called *hyperbolic motions,* can be expressed elegantly using the complex variable $z = x + iy$ as

$$z \to \frac{az + b}{cz + d}, \quad a, b, c, d \quad \text{real}, \quad ad - bc = 1. \tag{36}$$

Exercise 9. Show that the metric (35) is invariant under the hyperbolic motions (36).

The group G of hyperbolic motions has many interesting *discrete subgroups* Γ, which have the property that the images of any point under the mapping in Γ accumulate only at ∞. A function $u(x, y)$ is called *automorphic* with respect to the subgroup Γ if $u(\gamma(x, y)) = u(x, y)$ for every motion γ contained in Γ.

A domain P in $\mathbb{H}$ is called a *fundamental domain* for Γ if

(i) every point in $\mathbb{H}$ can be mapped into a point of $\overline{P}$ by some γ in Γ.

(ii) no two points of P are mapped into each other by any γ in Γ.

A boundary point of P will be mapped into another boundary point by some γ in Γ.

Exercise 10. Show that the image of a fundamental domain by any γ in Γ is another fundamental domain.

A fundamental domain is called a *fundamental polygon* if its boundary consists of a finite number of geodesics. The geodesics in the Poincaré model are circles whose center is located on the line $y = 0$, and their limits, the lines $x = \text{const}$

The discrete subgroups that are amenable to scattering theory have fundamental polygons that are unbounded. In this section we will look at the simplest such subgroup, the *modular group* Γ consisting of all hyperbolic motions of form (36), where a, b, c, d are integers. Clearly, these form a subgroup, and it is not hard to show that this subgroup is discrete. A fundamental domain for the modular group is the geodesic triangle T bounded by the geodesic arcs $x = \pm\frac{1}{2}, y > \sqrt{\frac{3}{4}}$, and $x^2 + y^2 = 1, -\frac{1}{2} < x < \frac{1}{2}$.

Exercise 11. Draw a picture of T.

Exercise 12. Look up a proof of the fact that T is a fundamental domain for the modular group.

Exercise 13. Show that the modular group is generated by the two transformations $z \to z + 1$ and $z \to -1/z$.

Exercise 14. Show that the fundamental triangle T has finite hyperbolic area. Calculate its area.

The motion $z \to z + 1$ carries the side $x = -\frac{1}{2}$ of T onto the other side $x = \frac{1}{2}$. The motion $z \to -1/z$ carries the third side $x^2 + y^2 = 1$ onto itself, mapping the point (x, y) to $(-x, y)$. Denote by p and p' pairs of boundary points linked to each other by a mapping in the modular group. A C^1 automorphic function u satisfies the boundary conditions

$$u(p) = u(p'), \qquad u_n(p) = -u_n(p') \tag{37}$$

at corresponding boundary points of T, where u_n denotes the outward normal derivative.

The Laplace-Beltrami operator in the Poincaré model is

$$\Delta_\mathbb{H} = -y^2(\partial_x^2 + \partial_y^2). \tag{38}$$

Exercise 15. Show that $\Delta_\mathbb{H}$ is invariant under the hyperbolic motions (36).

Denote by $(\ ,\ )_T$ the L^2 scalar product over T with respect to the hyperbolic area element:

$$(u, v)_T = \int_T uv \frac{dx\,dy}{y^2}. \tag{39}$$

Let u be a C^2 automorphic function that is zero for y near ∞. Integration by parts yields

$$(\Delta_\mathbb{H} u, u)_T = \int_T (u_x^2 + u_y^2)\,dx\,dy; \tag{40}$$

the boundary term is zero because of the boundary conditions (37).

Formula (40) shows that the operator $\Delta_\mathbb{H}$, defined for all automorphic function is symmetric and nonnegative. Its Friedrichs extension, also denoted as $\Delta_\mathbb{H}$, is a self-adjoint operator. What is its spectrum? It turns out that it is more natural to renormalize $\Delta_\mathbb{H}$ as

$$L = \Delta_\mathbb{H} - \tfrac{1}{4}\mathbf{1}. \tag{41}$$

Theorem 14.

(i) *On the interval $[-\frac{1}{4}, 0]$ the spectrum of $\mathbf{L}$ consists of the single point $-\frac{1}{4}$.*

(ii) *$\mathbf{L}$ has infinitely many positive eigenvalues, accumulating at ∞. The odd eigenfunctions span the space of odd functions in T.*

(iii) *$\mathbf{L}$ has absolutely continuous spectrum of multiplicity 1 on $\mathbb{R}_+$.*

Proof. We will not give a complete proof of part (i), for it would take us too far afield.

(i) Using formula (40) and the definition of **L**, we write

$$(\mathbf{L}u, u)_T = \int_T \int \left(u_x^2 + u_y^2 - \frac{u^2}{4y^2} \right) dx\, dy. \tag{42}$$

Let a be any number > 2. We divide T into two parts, $T = T_a \cup T^a$, T_a denoting the part of T below $y = a$, T^a the part above. Let $u(y)$ be C^1 function, equal to zero for y near ∞; integration by parts yields

$$\int_a^\infty \left(u_y - \frac{u}{2y} \right)^2 dy = \int_a^\infty \left(u_y^2 - \frac{uu_y}{y} + \frac{u^2}{4y^2} \right) dy$$
$$= \int_a^\infty \left(u_y^2 - \frac{u^2}{4y^2} \right) dy + \frac{1}{2a} u^2(a). \tag{43}$$

Define $\varphi(y) = (2y - a)/a$; since $\varphi(a) = 1$, $\varphi(a/2) = 0$, $\varphi' \equiv 2/a$, we can express

$$u^2(a) = \int_{a/2}^a \partial_y(\varphi u^2)\, dy = \int_{a/2}^a (\varphi' u^2 + 2\varphi u u_y)\, dy$$
$$\leq \frac{2}{a} \int_{a/2}^a u^2\, dy + 2 \left[\int_{a/2}^a u^2\, dy \int_{a/2}^a u_y^2\, dy \right]^{1/2}$$
$$\leq \frac{2}{a} \int_{a/2}^a u^2\, dy + \frac{1}{a} \int_{a/2}^a u^2\, dy + a \int_{a/2}^a u_y^2\, dy.$$

In the third step we have used the Schwarz inequality, and in the last the arithmetic-geometric inequality. Since $y \leq a$ in the range of integration, we deduce that

$$u^2(a) \leq a \int_{a/2}^a \left(\frac{3u^2}{y^2} + u_y^2 \right) dy. \tag{44}$$

Let $u(x, y)$ be a C^2 automorphic function that is zero for y near ∞. We integrate inequality (44) with respect to x over $[-\frac{1}{2}, \frac{1}{2}]$; we obtain

$$\frac{1}{2a} \int u^2(x, a)\, dx \leq \int\!\!\int_{a/2}^a \left(\frac{3}{2} \frac{u^2}{y^2} + \frac{1}{2} u_y^2 \right) dy\, dx \leq \int\!\!\int_{T_a} \left(\frac{3}{2} \frac{u^2}{y^2} + \frac{1}{2} u_y^2 \right) dx\, dy. \tag{44'}$$

Integrating (43) with respect to x, we obtain

$$\int\!\!\int_{T^a} \left(u_y^2 - \frac{u^2}{4y^2} \right) dx\, dy \geq -\frac{1}{2a} \int u^2(x, a)\, dx. \tag{43'}$$

Combine (43') and (44'):

$$\iint_{T^a} \left(u_y^2 - \frac{u^2}{4y^2} \right) dx\, dy \geq - \iint_{T_a} \left(\frac{3}{2}\frac{u^2}{y^2} + \frac{1}{2}u_y^2 \right) dx\, dy. \qquad (45)$$

Denote the integrand on the right in (42) by q, split the integral into two parts, and use inequality (45):

$$(\mathbf{L}u, u) = \iint_{T^a} q\, dx\, dy + \iint_{T_a} q\, dx\, dy$$

$$\geq \iint_{T^a} \left(u_y^2 - \frac{u^2}{4y^2} \right) dx\, dy + \iint_{T_a} q\, dx\, dy$$

$$\geq \iint_{T_a} \left(q - \frac{3}{2}\frac{u^2}{y^2} - \frac{1}{2}u_y^2 \right) dx\, dy = \iint_{T^a} \left(u_x^2 + \frac{1}{2}u_y^2 - \frac{7}{4}\frac{u^2}{y^2} \right) dx\, dy.$$

Define the quadratic functional K as

$$K(u) = \iint_{T_a} \frac{2u^2}{y^2}\, dx\, dy,$$

and add it to both sides:

$$(\mathbf{L}u, u) + K(u) \geq C(u), \qquad (46)$$

where

$$C(u) = \iint_{T_a} \left(u_x^2 + \frac{1}{2}u_y^2 + \frac{1}{4}\frac{u^2}{y^2} \right) dx\, dy.$$

It follows from Rellich's compactness theorem that for any positive ε there is a subspace of u of finite codimension on which $K(u) \leq \varepsilon C(u)$. Taking, modestly, $\varepsilon = 1$, we conclude from (46) that $(\mathbf{L}u, u) \geq 0$ on such a subspace. It follows from this and (41) that the spectral resolution of $\mathbf{L}$ on $[-\frac{1}{4}, 0]$ has finite-dimensional range; therefore the spectrum of $\mathbf{L}$ over $[-\frac{1}{4}, 0]$ consists of a finite number of eigenvalues.

Since T has finite area, $u \equiv 1$ is square integrable, and an eigenfunction of $\mathbf{L}$ with eigenvalue $-\frac{1}{4}$. There are in fact no others in $[-\frac{1}{4}, 0]$, but we skip the proof.

(ii) Both the operator $\mathbf{L}$ and the fundamental domain T are invariant under reflection across the y axis: $x \to -x$. It follows that the domain of $\mathbf{L}$ can be reduced as the direct sum of even and odd automorphic functions. For odd functions the first condition in (37) becomes $u = 0$ on the boundary; the second condition is automatically satisfied.

We will show that under the Dirichlet boundary condition $u = 0$, the resolvent $(\mathbf{L} + \mathbf{I})^{-1}$ is a compact operator. To see this, denote $(\mathbf{L} + \mathbf{I})^{-1}w = u$; this means that

$$\mathbf{L}u + u = w.$$

Take the scalar product of both sides with u, and on the left side use identity (42). Estimate the right side by the Schwarz inequality; we get

$$\iint_T (u_x^2 + u_y^2)\, dx\, dy + \tfrac{1}{4}\|u\|_T^2 \leq \tfrac{1}{2}\|w\|_T^2. \tag{47}$$

Since $u(x, y)$ vanishes at $x = \pm\tfrac{1}{2}$ for $y > 1$, by Wirtinger's inequality

$$\int u^2\, dx \leq \pi^2 \int u_x^2\, dx.$$

Integrating this with respect to y over $[Y, \infty]$ gives for $Y > 1$,

$$\int_Y^\infty \int u^2\, dx\, dy \leq \pi^2 \int_Y^\infty \int u_x^2\, dx\, dy.$$

From this we deduce that

$$\int_Y^\infty \int u^2 \frac{dx\, dy}{y^2} \leq \frac{1}{Y^2} \int_Y^\infty \int u^2 dx\, dy \leq \frac{\pi^2}{Y^2} \int_Y^\infty \int u_x^2\, dx\, dy. \tag{48}$$

We claim that the image of the ball $\|w\|_T \leq 1$ under $(\mathbf{L} + \mathbf{I})^{-1}$ is a precompact set: (47) shows that for functions u in this image the square integrals of u_x and u_y over T are uniformly bounded. We use Rellich's compactness criterion (see theorem 2 in chapter 22), applied to the compact portion T_Y of T. Combined with (48), the uniform smallness of the hyperbolic L^2-norm of u over the remainder T^Y of T shows the precompactness of the set of u in the norm $\|u\|_T$. It follows then from the spectral theory of compact symmetric operators (see chapter 28) that $(\mathbf{L}+\mathbf{I})^{-1}$ has a complete set of eigenfunctions over the space of odd functions in T, and that the eigenvalues are real, positive, and tend to zero. The corresponding eigenvalues of $\mathbf{L}$ tend to ∞.

(iii) It is in this part of theorem 14 that the renormalization of the Laplace-Beltrami operator plays an important role. We employ the hyperbolic wave equation

$$u_{tt} + \mathbf{L} u = 0, \tag{49}$$

introduced by Faddeev and Pavlov. The conservation of energy can be derived in the usual fashion by taking the scalar product of (49) with u_t. We get

$$\frac{1}{2}\frac{d}{dt}[(u_t, u_t)_T + (\mathbf{L}u, u)_T] = 0.$$

We conclude that the conserved energy is

$$E_T(u) = (u_t, u_t)_T + (\mathbf{L}u, u)_T = \iint_T \left(\frac{u_t^2}{y^2} + u_x^2 + u_y^2 - \frac{u^2}{4y^2}\right) dx\, dy; \tag{50}$$

here we have used (42).

The bilinear functional associated with the quadratic functional E_T is

$$E_T(u, v) = (u_t, v_t)_T + (\mathbf{L}u, v)_T. \tag{50'}$$

Solutions of the hyperbolic wave equation are uniquely determined by their initial data $\{u(0), u_t(0)\}$. Since the operator $\mathbf{L}$ is invariant under hyperbolic motions γ, if $u(z, t)$ is a solution of (49), defined for all z in $\mathbb{H}$, $u(\gamma(z), t)$ too is a solution. If the initial data of u are automorphic, then $u(z, t)$ and $u(\gamma(z), t)$ have the same initial data and therefore are equal. In other words, if the initial data of a solution of the hyperbolic wave equation are automorphic, the solution $u(t, z)$ is automorphic for all t.

Denote by $\mathbf{U}(t)$ the operator relating automorphic initial data $\{u(0), u_t(0)\}$ of finite energy in T to data at time t, $\{u(t), u_t(t)\}$. Since energy is conserved, $E_T(u(t)) = E_T(u(0))$.

Exercise 16. Show that $E_T(u(t), v(t)) = E_T(u(0), v(0))$ for all pairs of automorphic solutions of finite energy in T.

According to part (i) of theorem 14, the spectrum of $\mathbf{L}$ on $[-\frac{1}{4}, 0]$ consists of the single eigenvalue $-\frac{1}{4}$, with eigenfunction $\equiv 1$. It follows that if $(u, 1)_T = 0$, energy defined in (50) is positive. We claim that if the initial data of a solution u are orthogonal to 1,

$$(u(0), 1)_T = 0 = (u_t(0), 1)_T,$$

then $u(t)$ is orthogonal to 1 for all t. This follows since $(u(t), 1)_T$ satisfies the second-order equation

$$\partial_t^2(u(t), 1)_T = (u_{tt}, 1)_T = -(\mathbf{L}u, 1)_T = -(u, \mathbf{L}1)_T = \tfrac{1}{4}(u, 1)_T.$$

Denote by H the space of all automorphic initial data with finite energy in T that are orthogonal to all the eigenfunctions of $\mathbf{L}$. The operators $\mathbf{U}(t)$ map H into itself. Since energy is positive for such data, we define the square root of energy to be the norm in H. The operators $\mathbf{U}(t)$ are unitary in the energy norm.

We will construct a pair of representations of H as $L^2(\mathbb{R})$ that transmute the action of $\mathbf{U}(t)$ into translation. Let $h = \{h_1, h_2\}$ be an element of H, $u(x, y, t)$ the solution of the wave equation with initial data h. Denote by $\bar{u}(y, t)$ the x-average of u:

$$\bar{u}(y, t) = \int_{-1/2}^{1/2} u(x, y, t)\, dx.$$

Using the automorphic boundary condition (37), we get for the x-average of the wave equation (49)

$$\bar{u}_{tt} - y^2\bar{u}_{yy} - \tfrac{1}{4}\bar{u} = 0.$$

The change of variables $\bar{u} = y^{1/2}v$, $y = e^s$ turns this equation into the classical wave equation

$$v_{tt} - v_{ss} = 0.$$

This can be factored as

$$(\partial_t + \partial_s)(v_t - v_s) = 0;$$

it follows that $v_t - v_s$ is a function of $s - t$. We define

$$\sqrt{2}k_+(s) = v_s - v_t = \partial_s e^{-s/2}\overline{h}_1(e^s) - e^{-s/2}\overline{h}_2(e^s) \tag{51}$$

to be the outgoing translation representation of the initial data $\{h_1, h_2\} = h$. Clearly, the outgoing representation of $U(t)h$ is $k_+(s - t)$.

We next show that every function in $L^2(\mathbb{R})$ appears as the representer of some element in H, and that the representation is an isometry. Take any smooth function m of compact support in $1 < y$, and define w_+ on T as

$$w_+(y, t) = y^{1/2}m(\log y - t). \tag{52}$$

For $t \geq 0$, $w_+(y, t)$ satisfies the matching conditions (37) on the boundary of T, and therefore can be extended as an automorphic function to the whole hyperbolic plane. In T, the initial data h_+ of w_+ are $h_+ = \{h_1, h_2\}$, where

$$h_1 = y^{1/2}m(\log y), \quad h_2 = -y^{1/2}m'(\log y). \tag{52'}$$

For $s > 0$, the outgoing representation of $\{h_1, h_2\}$ is, according to formula (51),

$$\sqrt{2}k_+(s) = \partial_s m(s) + m'(s) = 2m'. \tag{51'}$$

Since $w_+(y, t) = 0$ for $0 < t$ and $1 \leq y \leq e^t$, the representer $k_+(s, t)$ of $\{w_+(y, t), \partial_t w_+(y, t)\}$ is zero for $0 \leq s \leq t$. Since $k_+(s, t) = k_+(s - t)$, it follows that $k_+(s) = 0$ for $s < 0$. In other words, (51') holds for all real s. Using formula (50), we calculate the energy of w_+:

$$E_T(w_+) = \int_1^\infty \left[\frac{(y^{1/2}m')^2}{y^2} + \left(\frac{1}{2}\frac{m}{y^{1/2}} + \frac{m'}{y^{1/2}} \right)^2 - \frac{ym^2}{4y^2} \right] dy$$

$$= \int \left[\frac{2m'^2}{y} + \frac{mm'}{y} \right] dy = \int \left[2m'^2 + mm' \right] ds = 2\int m'^2 \, ds.$$

Clearly, $E_T(w) = \|k_+\|^2$; in words:

For solutions w_+ of form (52), the translation representation (51) is an isometry.

This explains the need for the factor $\sqrt{2}$ in formula (51).

The initial data h_+ of w_+ do not, in general, belong to H; to make them orthogonal to the eigenfunction 1, we apply the projection $\mathbf{Q}$:

$$\mathbf{Q}\, h_+ = (h_1 - c, \ h_2 - d),$$

c and d being constants determined by the equations

$$\iint_T (h_1 - c)\frac{dx\,dy}{y^2} = 0, \qquad \iint_T (h_2 - d)\frac{dx\,dy}{y^2} = 0.$$

Using the formulas $(52')$ for h_1 and h_2, we can rewrite these equations as

$$\int_1^\infty y^{1/2} m(\log y)\frac{dy}{y^2} = cA, \qquad -\int_1^\infty y^{1/2} m'(\log y)\frac{dy}{y^2} = dA,$$

where A is the area of T. Switching to $s = \log y$ as variable of integration yields

$$\int_0^\infty e^{-s/2} m(s)\,ds = cA, \qquad -\int_0^\infty e^{-s/2} m'(s)\,ds = dA. \tag{53}$$

Integration by parts shows that $d = -c/2$.

Lemma 15.

(i) $\mathbf{Q}h_+$ *has the same energy as* h_+.

(ii) $\mathbf{Q}h_+$ *has the same translation representation as* h_+.

Proof. (i) By definition (50),

$$E_T(\mathbf{Q}h_+) = \iint_T \left(\frac{(h_2 - d)^2}{y^2} + h_{1\,y}^2 - \frac{(h_1 - c)^2}{4y^2} \right) dx\,dy$$

$$= E_T(h_+) - 2d \iint_T \frac{h_2}{y^2} dx\,dy + d^2 A + \frac{c}{2} \iint_T \frac{h_1}{y^2} dx\,dy - \frac{c^2}{4} A.$$

By formula (53), this can be rewritten as

$$E_T(h_+) - d^2 A + \frac{c^2}{4} A,$$

which is equal to $E_T(h_+)$, since $d = -c/2$.

(ii) By definition (51), the difference between the outgoing translation representation of h_+ and of $\mathbf{Q}h_+$ is

$$\sqrt{2}(\partial_s e^{-s/2} c - e^{-s/2} d) = -\sqrt{2} e^{-s/2} \left(\frac{c}{2} + d \right) = 0. \qquad \square$$

Exercise 17. Show that for $t > 0$ the solution of the hyperbolic wave equation with initial data $\mathbf{Q}h_+$ is $w_+(t) - ce^{-t/2}$.

Clearly, the functions $\mathbf{Q}h_+$ are orthogonal to all odd eigenfunctions. We claim that they also are orthogonal to all even square integrable eigenfunctions p of $\mathbf{L}$ with positive eigenvalue. To see this, take the x-average of the eigenvalue equation $\mathbf{L}p = \mu^2\bar{p}$:

$$-y^2\bar{p}_{yy} - \tfrac{1}{4}\bar{p} = \mu^2\bar{p}.$$

Solutions of this equation are linear combinations of the functions $y^{(1/2)+i\mu}$ and $y^{(1/2)-i\mu}$. Neither of these is square integrable with respect to dy/y^2 near $y = \infty$, whereas p, and therefore $\bar{p}$, are. So it follows that $\bar{p} \equiv 0$, from which we have the orthogonality of p to w_+.

The functions $k_+(s)$ that represent data in H include all functions of form $m'(s)$, m any C_0^2 function supported on $\mathbb{R}_+$ and their translates; these are dense in $L^2(\mathbb{R})$. Denote by K the subspace of the corresponding data in H. K is a closed invariant subspace for the operators $\mathbf{U}(t)$, which are represented as translations on $\mathbb{R}$. It is not hard to show that $\mathbf{L}$ has absolutely continuous spectrum on K, which covers all of $\mathbb{R}_+$.   ☐

In chapter 36, section 36.4, we introduced the concept of an outgoing subspace F_+ for a group of unitary operators $\mathbf{U}(t)$, as follows:

   (i) $\mathbf{U}(t)F_+ \subset F_+$ for $t > 0$.

   (ii) $\cap \mathbf{U}(t)F_+ = \{0\}$.

   (iii) $\cup \mathbf{U}(t)F_+ = H$.

We claim that the space formed by the initial data $\mathbf{Q}h_+$, where h_1, h_2 are given by formula (51), is such an outgoing subspace F_+. Properties (i) and (ii) are obvious. Property (iii) asserts that K, the closure of the union of $\mathbf{U}(t)F_+$, is all of H; a short proof can be found in Lax and Phillips's paper in the *Transactions of the AMS*.

It is worth pointing out that we have constructed the translation representation directly, *without* appealing to the translation representation theorem in chapter 35, section 35.5.

The theory described above for the modular group can be carried over with only trivial changes to any discrete subgroup, except that in general, one cannot determine precisely the location of the point spectrum. According to a conjecture/theorem of Phillips, Sarnak and Wolpert, there is in general, no point spectrum embedded in the continuous spectrum.

If the fundamental polygon has n vertices at infinity, the continuous spectrum of $\mathbf{L}$ has multiplicity n on $\mathbb{R}_+$. Even if a whole side of the fundamental polygon lies at infinity, the theory can be pushed through, with a continuous spectrum of infinite multiplicity.

There is an entirely analogous construction of an incoming representation; the incoming subspace F_- is formed by $\mathbf{Q}h_-$, where h_- denotes the initial data on T of incoming solutions w_- of form $w_-(y, t) = y^{1/2}n(\log y + t)$, $t < 0$, where n is supported on $\mathbb{R}_+$.

Exercise 18. Show that for $t < 0$, the solution of the hyperbolic wave equation with initial data $\mathbf{Q}h_-$ is $w_-(t) - ce^{t/2}$.

Exercise 19. Show that F_+ and F_- are orthogonal in the energy norm.

Clearly, $w_- = y^{1/2}n(\log y + t)$, $t < 0$, describes a wave arriving in T from infinity through the channel $-\frac{1}{2} \leq x \leq \frac{1}{2}$, just as w_+ is for $t > 0$ a wave traveling to infinity through the same channel. It follows from property (iii) for F_- and F_+ that everything that flows in from infinity eventually flows out to infinity. How fast is an interesting question, as we will show.

It was explained in section 37.8 that a pair of orthogonal incoming and outgoing translation representations are linked to each other by a scattering operator. The corresponding spectral representations are related via multiplication by a scattering matrix $\mathcal{M}(\lambda)$. In the present situation, where the multiplicity of the continuous spectrum is one, the scattering matrix is a scalar function. $|\mathcal{M}(\lambda)| = 1$ for λ real, and $\mathcal{M}$ has an analytic continuation into the lower halfplane $\lambda + i\eta$, $\eta < 0$, where $|\mathcal{M}(\lambda + i\eta)| \leq 1$. In chapter 38 we will meet these functions again in Beurling's theory of the arithmetic of bounded analytic functions.

Faddeev and Pavlov have determined the scalar scattering matrix that arises from the hyperbolic wave equation for solutions that are automorphic with respect to the modular group. It is, aside of inessential factors,

$$\mathcal{M}(\lambda) = \frac{\zeta(2i\lambda)}{\zeta(1 + 2i\lambda)}\mathcal{F}(\lambda),$$

where $\mathcal{F}(\lambda)$ is a product of gamma functions, and ζ is the Riemann zeta function. If the Riemann hypothesis is true, $\mathcal{M}$ has zeros in the lower half-plane on the line $-\frac{1}{4}i + \lambda$; its meromorphic continuation into the upper half-plane by Schwarz reflection has poles on the line $\frac{1}{4}i + \lambda$. We saw in section 37.8, theorems 13 and 13', that if $\lambda + i\eta$ is a zero of $\mathcal{M}$, then $\eta + i\lambda$ is an eigenvalue of $\mathbf{G}$, the generator of the semigroup $\mathbf{Z}(t)$ associated with the group of unitary operators $\mathbf{U}(t)$ and the pair of incoming and outgoing subspaces F_- and F_+. Faddeev and Pavlov point out that if one could show that $\mathbf{G}$ has no eigenvalues γ whose real part exceeds $-\frac{1}{4}$, the Riemann hypothesis would be proved. According to Phillips's spectral mapping theorem for semigroups, theorem 12 in chapter 34, if γ belongs to the spectrum of $\mathbf{G}$, $e^{\gamma t}$ belongs to the spectrum of $\mathbf{Z}(t)$. Since the spectral radius does not exceed the norm, $|e^{\gamma t}| \leq \|\mathbf{Z}(t)\|_E$. Taking logarithm and the limit $t \to \infty$, we deduce that

$$\operatorname{Re}\gamma \leq \lim_{t \to \infty} \frac{1}{t}\log\|\mathbf{Z}(t)\|_E.$$

Therefore to prove the Riemann hypothesis, it is sufficient to show that

$$\lim_{t \to \infty} \frac{1}{t} \log \|\mathbf{Z}(t)\|_E \le \frac{1}{4}. \tag{54}$$

Faddeev and Pavlov point out that (54) is necessary as well. It is not hard to show that (54) would follow if

$$\lim_t \frac{1}{t} \log \|\mathbf{Z}(t)h\|_E \le \frac{1}{4}$$

could be proved for a set of data h dense in the domain of the semigroup.

Could this formulation lead to a proof of the Riemann hypothesis? If it does, you will hear about it.

BIBLIOGRAPHY

Beardon, A. F. *The Geometry of Discrete Groups.* Graduate Texts in Mathematics, **91**. Springer-Verlag, 1983.

Birman, M. Sh. A test for the existence of the wave operators. *Dokl. Akad. Nauk. SSSR*, **147** (1962): 506–509.

Cook, J. M. Convergence to the Møller wave matrix. *J. Math. Phys.*, **36** (1957): 82–87.

Faddeev, L. D. The inverse problem in the quantum theory of scattering. *Usp. Mat. Nauk.*, **14**, 57 (1959); English translation by B. Seckler, *J. Math. Phys.*, **4** (1963): 72–104.

Faddeev, L. D. and Pavlov, B. S. Scattering theory and automorphic functions. *Seminar Steklov Math. Inst. Leningrad*, **27** (1972): 161–193.

Fröhlich, J. and Spencer, T. A rigorous approach to Anderson localisation. *Common Trends in Particle and Condensed Matter Physics.* Les Houches, 1983; *Phys. Rev.*, **103** (1984): 1–4, 9–25.

Heisenberg, W. Die beobachtbaren Grössen in der Theorie der Elementarteilchen. *Z. Physik*, **120** (1943): I, 513–538; II, 673–702.

Jauch, J. M. Theory of the scattering operator. *Helv. Phys. Acta*, **31** (1958): 127–158.

Kato, T. Wave operators and unitary equivalence. *Pacific J. Math.*, **15** (1965): 171–180.

Kato, T. Perturbation of continuous spectra by trace class operators. *Proc. Jap. Acad.*, **33** (1957): 260–264.

Kato, T. *Perturbation Theory for Linear Operators.* Grundlehren der Math. Wiss in Einzeldarstellung, **132**. Springer, Verlag, 1966.

Kuroda, S. T. On a theorem of Weyl-von Neumann. *Proc. Jap. Acad.*, **34** (1958): 11–15.

Kuroda, S. T. On the existence and the unitary property of the scattering operator. *Nuovo Cimento*, **12** (1959): 431–454.

Lax, P. D. and Phillips, R. S. *Scattering Theory.* Academic Press, New York, 1967.

Lax, P. D. and Phillips, R. S. *Scattering Theory for Automorphic Functions.* Ann. Math. Studies, Princeton University Press, Princeton, 1976.

Lax, P. D. and Phillips, R. S. Translation representation for automorphic solutions of the wave equation in non-Euclidean spaces. *CPAM*, **37** (1984): 303–328, 780–813.

Lax, P. D. and Phillips, R. S. Translation representation for automorphic solutions of the wave equation in non-Euclidean cases; the case of finite volume. *Trans. AMS*, **289** (1985): 715–735.

Møller, C. General properties of the characteristic matrix in the theory of elementary particles. *Kgl. Dansk. Videnskab, Selskab, Mat.-fys. Medd.*, **22**, 1 (1945); **23**, 10 (1946).

von Neumann, J. Characterisierung des Spectrums eines Integraloperators. *Actualités Sci. Ind.*, **229** (1935): 38-55.

Phillips, R. S. and Sarnak, P. Perturbation theory for the Laplacean on automorphic functions. *J. AMS*, **5** (1992): 1–3.

Reed, M. and Simon, B. *Scattering Theory*, Academic Press, New York, 1979.

Rellich, F. Störungstheorie der Spectralzerlegung. *Math. An.*, **113** (1937): 600–619, 677–685; **116** (1939): 555–570; **117** (1940): 356–382; **118** (1942): 462–484.

Rosenblum, M. Perturbation of the continuous spectrum and unitary equivalence. *Pacific J. Math.*, **7** (1957): 997–1010.

Shatten, R. *A Theory of Cross Spaces*. Ann. Math. Studies, **26**. Princeton University Press, Princeton, 1950.

Weyl, H. Über beschränkte quadratische Formen deren Differenz vollstatig ist. *Rend. Circ. Palermo*, **27** (1909): 373–392.

Wolpert, S. A. Disappearance of cusp forms in special families. *An. Math.*, (2), **139** (1994): 239–291.

38

A THEOREM OF BEURLING

38.1 THE HARDY SPACE

In this chapter we study the space of square integrable analytic functions in relation to the algebra of bounded analytic function.

The Hilbert space $\ell^2(\mathbb{Z}_+)$ consists of vectors $x = (a_0, a_1, \ldots)$, a_j complex numbers such that

$$\|x\|^2 = \sum |a_j|^2 < \infty. \tag{1}$$

This space can be represented as a space of analytic functions $f(z)$ in the unit disk:

$$f(z) = \sum_0^\infty a_n z^n. \tag{2}$$

In this representation the space is called Hardy space and is denoted as H_+. We have already come across this space in chapter 27, section 27.2.

The L^2-norm of f on any circle of radius $r < 1$ is

$$\int |f(re^{i\theta})|^2 \, d\theta = \sum_0^\infty |a_n|^2 r^{2n}, \qquad d\theta = \frac{d\theta}{2\pi}. \tag{3}$$

We define

$$\|f\|^2 = \sup_{r<1} \int |f(re^{i\theta})|^2 \, d\theta \tag{3'}$$

as the norm in H_+, isometric with (1). The difference $f(re^{i\theta}) - f(se^{i\theta})$ has norm

$$\|f(re^{i\theta}) - f(se^{i\theta})\|^2 = \sum |a_n|^2 |r^n - s^n|^2,$$

which shows that $f(re^{i\theta})$ converges as $r \to 1$ in the L^2 sense. This limit is the *boundary value* of the function $f(z)$ on the unit circle,

$$f(e^{i\theta}) = \sum_0^\infty a_n e^{in\theta}, \tag{4}$$

513

where the series on the right converges in the L^2 sense. Its L^2 norm is the H_+-norm of f:

$$\|f\|^2 = \int |f(e^{i\theta})|^2 d\theta. \tag{4'}$$

Denote by $\mathcal{B}$ the algebra of bounded analytic functions in the open unit disk. Define the norm $|b|$ to be the sup norm:

$$|b| = \sup_{|z|<1} |b(z)|. \tag{5}$$

Functions b in $\mathcal{B}$ have boundary values in the sense of convergence a.e. as $r \to 1$.

Theorem 1.

   (i) *If the boundary values of a function f in H_+ are bounded, then f belongs to $\mathcal{B}$.*

   (ii) *Let b denote a function $\mathcal{B}$, and denote by $\mathbf{B}$: $H_+ \to H_+$ the operation of multiplying a function f in H by b. We claim that $\mathbf{B}$ is bounded, and that*

$$\|\mathbf{B}\| = |b|. \tag{6}$$

Proof. (i) Let s_ϵ be a sequence of smooth functions approximating the δ function (see chapter 11, section 11.11). For any function f of class H_+, define f_ϵ by

$$f_\epsilon(z) = \int f(ze^{i\phi}) s_\epsilon(\phi) d\phi. \tag{7}$$

Clearly, f_ϵ is of class H_+ and tends to f in the H_+-norm as ϵ tends to zero. Furthermore $f_\epsilon(z)$ is continuous in the closed unit disk. If the boundary values of $|f|$ are bounded by, say, 1, so are the boundary values of f_ϵ. According to the maximum value principle, at any interior point z of the unit disk $|f_\epsilon(z)| \leq 1$, but then also their L^2 limit satisfies $|f(z)| \leq 1$ at all points with $|z| < 1$.

(ii) Using the definition (3′) of the norm in H_+, we deduce that for any function b in $\mathcal{B}$, $\|bf\| \leq |b| \|f\|$. It follows that

$$\|\mathbf{B}\| \leq |b|. \tag{6'}$$

To see that the sign of equality holds, we argue as follows: if not, we would have, after renormalizing b,

$$\|\mathbf{B}\| < 1 < |b|.$$

It would follow that $\|\mathbf{B}^n\| \leq \|\mathbf{B}\|^n$ tends to zero as $n \to \infty$. On the other hand, for any f in H_+,

$$\int |b^n(re^{i\theta}) f(e^{i\theta})|^2 d\theta$$

tends to ∞ as $n \to \infty$ if r is so large that $\max_\theta |b(re^{i\theta})| > 1$. This shows, by (3'), that $\|\mathbf{B}^n f\| \to \infty$, a contradiction. $\qquad\qquad\qquad\qquad\qquad\qquad\qquad\qquad$ $\square$

Exercise 1. Show that every bounded mapping $\mathbf{C}: H_+ \to H_+$ that commutes with multiplication by all functions in B is itself multiplication by some function c in B.

Exercise 2. Show that B has no divisors of zero; that is, if the product bc of two functions b and c in B is zero, then one of the factors b or c is zero.

38.2 BEURLING'S THEOREM

The basic result of this chapter, due to Arne Beurling, establishes an important relation between the Hilbert space H_+ and the algebra B:

Theorem 2. *Let N be a closed subspace of H_+ that is invariant under multiplication by functions b in B, namely that $bN \subset N$ for all b in B. Then N can be represented as*

$$N = pH_+, \tag{8}$$

where p is a function in B that has absolute value 1 on the unit circle:

$$|p(e^{i\theta})| = 1; \tag{9}$$

p is unique up to a complex constant factor of absolute value 1.

The beautiful proof below is due to Paul Halmos.

Proof. It is easy to see that every N of form (8) is invariant under multiplication by b in B. It follows from (4)' and (9) that multiplication by p is an isometry; therefore N is closed.

Conversely, consider any given closed invariant subspace N. We claim that zN is a proper subspace of N; for if not, every f in N could be written as

$$f = zf_1 = z^2 f_2 = \cdots,$$

which would show that f has a zero of infinite order at $z = 0$, impossible for an analytic function.

Multiplication by z is an isometry of H_+; therefore zN is a closed proper subspace of H_+. Denote its orthogonal complement in N by M:

$$N = M \oplus zN. \tag{10}$$

Replace N on the right by its orthogonal decomposition given by (10); k-fold repetition of this operation gives

$$N = M \oplus zM \oplus \cdots \oplus z^{k-1}M \oplus z^k N. \tag{10'}$$

Letting k tend to ∞, we deduce from $(10')$ that

$$N \supset M \oplus zM \oplus \cdots. \tag{11}$$

We claim that the right side of (11) is actually equal to N; for if not, there would be a g in N that is orthogonal to every $z^j M$. But by $(10')$, such a g would belong to $z^k N$ for every k, and thus would have a zero of infinite order at $z = 0$. This is impossible, and so

$$N = M \oplus zM \oplus \cdots. \tag{12}$$

Next we examine the space M. Let m be any function in M; it follows from $(10')$ that m is orthogonal to $z^k N$, $k \geq 1$, and so in particular, to $z^k m$.

$$(z^k m, m) = \int e^{i\theta k} |m(e^{i\theta})|^2 d\theta = 0, \qquad k = 1, 2, \ldots \tag{13}$$

Taking complex conjugates we conclude that (13) holds for $k = -1, -2, \ldots$ as well. Thus all Fourier coefficients of $|m(e^{i\theta})|^2$ except the 0th are $= 0$, which implies that $|m(e^{i\theta})|$ is constant.

We claim that M is one dimensional. To see this, let m and p be two functions in M; then $m + ap$, a any constant, belongs to M, and so by what has been shown above, for $z = e^{i\theta}$,

$$|m + ap|^2 = (m + ap)(\bar{m} + \bar{a}\bar{p}) = |m|^2 + |a|^2|p|^2 + 2 \operatorname{Re} ap\bar{m} = \text{const.}.$$

Since a is an arbitrary complex constant, $p\bar{m}$ is constant; dividing by $m\bar{m}$, we conclude that p/m is constant, that is, that p and m are proportional.

Normalize $p(e^{i\theta})$ in M to have $|p| = 1$; then all functions in M are constant multiples of p. Setting this into (12) shows that every function f in N can be decomposed as

$$f = a_0 p + z a_1 p + \cdots = p(a_0 + a_1 z + \cdots) = pg. \tag{14}$$

Since $|p(e^{i\theta})| = 1$, $|f(e^{i\theta})| = |g(e^{i\theta})|$; since f belongs to H_+, so does g. Thus (14) is the desired representation (8) of Beurling's theorem. $\square$

Exercise 3. Show that p is uniquely determined by N, up to a constant factor of absolute value 1.

A function p in the algebra $\mathcal{B}$ that has absolute value 1 on the unit circle is called an *inner function*.

Note that in the proof of theorem 2 we only used the fact that N is invariant under multiplication by z. This is no gain in generality, as shown in

Exercise 4. Show that any closed subspace N of H_+ that is invariant under multiplication by z is invariant under multiplication by any function in $\mathcal{B}$.

Beurling has shown how to use theorem 2 to factor bounded analytic functions:

Theorem 3. *Every function b in B can be factored essentially uniquely as*

$$b = pu, \tag{15}$$

where p and u are bounded analytic functions, $|p(e^{i\theta})| = 1$, and uH is dense in H.

Proof. Define N to be the closure of bH_+. Clearly, N is invariant under multiplication by z, so by Beurling's theorem N is of the form $N = pH_+$. Since b belongs to N, it is of the form pu, u in H. Since $|p(z)| = 1$ for $|z| = 1$, $|b(z)| = |u(z)|$; since b is bounded, so is u.

Denote the closure of a set S in H_+ as $\overline{S}$. By definition, $N = \overline{bH_+}$; using the factored form of b, we can write this as $N = \overline{puH}$. Since $|p| = 1$ for $|z| = 1$, multiplication by p is an isometry, so

$$N = \overline{puH} = p\overline{uH}.$$

Since N is of the form pH, we deduce from this that $\overline{uH_+} = H_+$. That p and u are unique up to constants of absolute value 1 follows from the uniqueness of p in the representation (8). $\qquad\square$

The function p is called the *inner factor* of b, u its *outer factor*.

The product of two inner factors is clearly an inner factor, and it is not hard to show that the product of two outer factors is outer. It follows from this that if b_1 and b_2 are a pair of bounded analytic functions, the inner factor of the product $b_1 b_2$ is the product of the inner factors of b_1, and b_2, and likewise for their outer factors. This shows that to establish that a bounded analytic function is divisible by another, it is sufficient to verify the divisibility of its inner and outer factors by those of the other function.

Divisibility by an outer factor is a particularly simple matter:

Theorem 4. *Let b be any function in B, u an outer function. Then b is divisible by u in the algebra B iff $b(z)/u(z)$ is a bounded function on the unit circle $|z| = 1$.*

Proof. By definition of outer factor, uH is dense in H_+. In particular, there is a sequence $\{c_n\}$ in H_+ such that

$$L^2 - \lim uc_n = 1. \tag{16}$$

Suppose b/u is bounded on the unit circle. We claim that the sequence $\{bc_n\}$ converges in H_+. To see this, we write $bc_n = buc_n/u$; since b/u is bounded on the unit circle and uc_n converges in H_+ to 1, we deduce that $\{bc_n\}$ converges in L^2 on the unit circle to a limit we denote by d:

$$L^2 - \lim bc_n = d. \tag{16'}$$

The functions bc_n belong to H_+; therefore so does their H_+ limit d.

Multiply (16′) by the bounded function u; the resulting sequence converges in H_+

$$L^2\text{-}\lim bc_n u = du. \tag{16''}$$

On the other hand, using (16), we conclude that the sequence on the left in (16″) converges to b. This proves that $b/u = d$ belongs to H_+. Since, by hypothesis, b/u is bounded on the unit circle, it follows from part (i) of theorem 1 that b/u belongs to B.   □

The simplicity of the criterion for division by an outer function shows that they may be called quasi-units of the algebra B.

We turn now to divisibility by inner functions:

Theorem 5. *An inner function p divides a function b in B iff pH_+ contains bH_+.*

Proof. Clearly, if $b = pc$, then $bH_+ = pcH_+$ is contained in pH_+. Conversely, if bH_+ is contained in pH_+, then $b = b \cdot 1$ is of the form pf, f in H_+. This shows that b/p belongs to H_+. But, since b is bounded and p has absolute value 1 on the boundary, their quotient is bounded. This shows according to theorem 1 that b/p belongs to B.   □

In this proof the closed invariant subspaces of H_+ replace the role of ideals in the algebra B; Beurling's theorem is an analogue of the *principal ideal theorem*. We will exploit this idea further.

Definition. Let b and c be two functions in B. Denote by N the closure of $bH_+ + cH_+$:

$$\overline{bH_+ + cH_+} = N. \tag{17}$$

According to theorem 2, N is of the form rH_+, r an inner function. We define r to be the *greatest common divisor* of b and c.

The g.c.d. defined above has the usual properties:

Theorem 6. *Suppose that b, c, and s belong to B, and that s is an inner function that divides both b and c in B. Then s divides the g.c.d. of b and c.*

Proof. According to theorem 5, s divides b iff sH_+ contains bH_+, and divides c iff sH_+ contains cH_+. It follows that then the linear space sH_+ contains $bH_+ + cH_+$. Since sH_+ is closed,

$$sH_+ \supset \overline{bH_+ + cH_+}, \tag{18}$$

but the right side is rH_+, r the g.c.d. of b, c. Appealing again to theorem 5, we conclude that s divides r.   □

Definition. Two functions b and c in $\mathcal{B}$ are *relatively prime* if their g.c.d. is 1. According to definition (17) this means that b and c are relatively prime iff $bH_+ + cH_+$ is dense in H_+.

Theorem 7. *Let c, d and e be three functions in $\mathcal{B}$, e relatively prime to both c and d. Then e is relatively prime to their product bc.*

Proof. By definition of relatively prime, $eH_+ + cH_+$ and $eH_+ + dH_+$ are dense in H_+. Then $d(eH_+ + cH_+)$ is dense in dH_+; therefore $eH_+ + d(eH_+ + cH_+) = eH_+ + dcH_+$ is dense in H_+. This proves that e and dc are relatively prime. $\square$

These results on divisibility in the algebra $\mathcal{B}$ can be used to develop a theory of primes in $\mathcal{B}$:

Theorem 8. *Let u be any point in the unit disk: $|u| < 1$. The function*

$$p(z) = \frac{z - u}{\bar{u}z - 1} \tag{19}$$

is a prime inner function of the algebra $\mathcal{B}$.

Proof. An elementary calculation shows that $|p(z)| = 1$ for $|z| = 1$, so p given by (19) is an inner function. To see that it is a prime, we note that since p vanishes at $z = u$, so does every function in $N = pH_+$. Conversely, if f in H_+ vanishes at $z = u$, then f/p belongs to H_+, so f belongs to $N = pH_+$. This shows that pH_+ consists exactly of those functions in H_+ that vanish at u. It follows that N has codimension 1 in H_+. This implies that p is a prime. To see this, suppose that q is an inner function that divides p; according to theorem 5, then qH_+ contains pH_+. Since pH_+ has codimension 1, qH_+ is either pH_+ or H_+; in either case q is, by theorem 3, a trivial factor of p. $\square$

We will see a little later that the functions (19) are the only primes in $\mathcal{B}$. However, $\mathcal{B}$ contains in addition *prime powers*. These are technically easier to analyze if we switch from the unit disk to the upper half-plane. The transformation

$$z = \frac{w - i}{w + i} \tag{20}$$

maps the upper half-plane $\operatorname{Im} w > 0$ onto the unit disk $|z| < 1$. The relation

$$g(w) = f\left(\frac{w - i}{w + i}\right) \tag{20'}$$

assigns a bounded analytic function g in the upper half-plane to every bounded analytic function f in the unit disk, and conversely. This relation is an isometric isomorphism of these two normed algebras. We will use the same symbol $\mathcal{B}$ for either of these two isomorphic algebras.

Theorem 9. *The function*

$$p(w) = e^{iw} \tag{21}$$

is an inner function. Furthermore it is a prime power in the sense that its only factorization by inner functions is

$$e^{iw} = e^{iaw} e^{ibw}, \qquad a, b > 0, \ a + b = 1. \tag{22}$$

Proof. Obviously $|p(w)| < 1$ for $\operatorname{Im} w > 0$, $= 1$ for $\operatorname{Im} w = 0$. Suppose that there were a nontrivial factorization

$$e^{iw} = p(w)\, q(w), \tag{22'}$$

p and q inner functions. Denote the real and imaginary parts of w as $x + iy$. Take the log of the absolute value of (22'):

$$-y = \log|p| + \log|q|. \tag{22''}$$

We define

$$h = -\log|p| \tag{23}$$

and claim:

   (i) h is harmonic in the upper half-plane.
   (ii) h is positive in the upper half-plane and is continuous down to the boundary $y = 0$, where it is equal to zero.

For it follows from (22') that p and q have no zeros in the upper half-plane; this shows that h, the real part of the analytic function $\log p$, is harmonic as asserted in (i).

Since both p and q are nontrivial inner functions, $|p|$ and $|q|$ are < 1, and so $\log|p|$ and $\log|q| < 0$ in the upper half-plane. These inequalities combined with (22'') give

$$0 < h(x, y) < y.$$

The assertions (ii) follow immediately from these inequalities.

A harmonic function that vanishes on a straight line boundary can be continued by reflection across that boundary as a regular harmonic function. Thus h can be extended to the whole plane by setting

$$h(x, -y) = -h(x, y).$$

We apply now the Poisson formula, see equation (29') in chapter 11, to h, replacing the unit disk by a disk of radius R:

$$h(x, y) = \int_0^{2\pi} \frac{R^2 - r^2}{R^2 - 2Rr\cos(\theta - \phi) + r^2} k(\phi, R)\, d\phi, \tag{24}$$

where $(x, y) = r(\cos\theta, \sin\theta)$ and

$$k(\phi, R) = h(R\cos\phi, R\sin\phi). \tag{24'}$$

Since h is an odd function of y, $k(\phi + \pi, R) = -k(\phi, R)$, so we can rewrite (24) as

$$h(x, y) = \int_0^\pi Q(R, r, \theta, \phi)\, k(\phi, R)\, d\phi,$$
$$Q = P(R, r, \theta - \phi) - P(R, r, \theta + \phi), \tag{25}$$

where P is the Poisson kernel appearing in the integral (24). Using the formula for P, we can express Q explicitly as

$$Q = \frac{(R^2 - r^2)4Rr\sin\theta\sin\phi}{(R^2 - 2Rr\cos(\theta - \phi) + r^2)(R^2 - 2Rr\cos(\theta + \phi) + r^2)}. \tag{26}$$

We write Q as

$$Q = y Q_0(R, r, \theta, \phi). \tag{26'}$$

The function Q_0 is positive for $0 < \phi < \pi$ and satisfies

$$\lim_{R\to\infty} \frac{Q_0(R, r_1, \theta_1, \phi)}{Q_0(R, r_2, \theta_2, \phi)} = 1, \tag{27}$$

uniformly in the parameter ϕ. We take now any two points (x_1, y_2) and (x_2, y_2) in the upper half-plane and use formula (25):

$$\frac{h(x_1, y_1)}{h(x_2, y_2)} = \frac{y_1}{y_2} \frac{\int Q_0(R, r_1, \theta_1, \phi)\, k(\phi, R)\, d\phi}{\int Q_0(R, r_2, \theta_2, \phi)\, k(\phi, R)\, d\phi}. \tag{28}$$

Now we let R tend to ∞; the integrands in the numerator and denominator on the right in (28) are positive, and by (27) their quotient tends to 1. We conclude that the ratio of the integrals also tends to 1. Since the left side is independent of R, it follows that

$$\frac{h(x_1, y_1)}{h(x_2, y_2)} = \frac{y_1}{y_2}.$$

This proves that $h(x, y) = ay$, $a > 0$. By (23), $|p(w)| = e^{-ay}$ and so $p(w) = e^{iaw}$; similarly $q(w) = e^{ibn}$, $b > 0$, as asserted in (22). $\square$

The analytic mapping $w \to (c - w)^{-1}$, c any real number carries the upper half-plane into itself. It follows therefore from (21) that the inner functions

$$p(w) = \exp\{i(c - w)^{-1}\} \tag{29}$$

are prime powers in the sense of theorem 9.

Next we will show that the functions (21) and (29) are the only prime powers by completing the discussion of factorization in $\mathcal{B}$. We retreat to the unit disk and consider any bounded analytic function $b(z)$. Denote by $\{u_j\}$ the set of zeros of b with multiplicity m_j, and by p_j the corresponding primes of form (19) except for a normalization that makes them real and positive at $z = 0$:

$$p_j(z) = -\frac{|u_j|}{u_j} \frac{u_j - z}{\bar{u}_j z - 1}. \tag{30}$$

We appeal now to the analogous discussion in chapter 9, section 9.2, that shows that the infinite product, called a *Blaschke product*,

$$p(z) = \prod p_j(z)^{m_j}, \tag{31}$$

converges in the unit disk, is an inner function, and divides b:

$$b = pc, \qquad c \text{ in } \mathcal{B}. \tag{32}$$

The function c is bounded, $|c| \leq |b|$; it has no zeros, because all the zeros of b were thrown into p. Suppose $|b| \leq 1$; then $-\log c$ is an analytic function whose real part is positive; for such functions we have derived in chapter 11, section 11.6, an integral representation, equation (29'):

$$-\log c(z) = \int C(z, \theta) \, dm, \qquad C = \frac{e^{i\theta} + z}{e^{i\theta} - z}, \tag{33}$$

m being a nonnegative measure of finite total mass. Decompose m into its absolutely continuous and singular parts:

$$m = m_{\text{sing}} + m_{\text{ac}}.$$

The corresponding decomposition of c is

$$c(z) = \exp\left\{-\int C dm_{\text{sing}}\right\} \exp\left\{-\int C \, dm_{\text{ac}}\right\}. \tag{33'}$$

It is not too hard to show that the first factor on the right in (33') is an inner function, the second an outer function. Setting (33') into (32) gives a factorization of b:

$$b = p \exp\left\{-\int C \, dm_{\text{sing}}\right\} \exp\left\{-\int C \, dm_{\text{ac}}\right\}. \tag{34}$$

This is the inner-outer factorization of b, the inner factor being the product of the first two factors in (34). The first of the two is a product of primes, in general infinitely many, the second is a mixture of discrete and continuous products of prime powers, smeared out with respect to a singular measure.

It follows from the representation (34) that, as claimed earlier, all primes are of the form

$$\frac{u-z}{\bar{u}z-1}, \qquad |u| < 1,$$

and that all prime powers are of form

$$\exp\frac{v+z}{v-z}, \qquad |v| = 1.$$

We conclude this section by remarking that parts of this theory, in particular, Beurling's theorem, can be carried over to vector-valued analytic functions that are square integrable, respectively, bounded in the unit disk or upper half-plane; see Halmos. The inner factors are operator valued analytic functions, unitary a.e. on the boundary, and ≤ 1 in norm in the upper half-plane. In the language of chapter 37, section 37.7, scattering matrices are operator valued inner functions.

38.3 THE TITCHMARSH CONVOLUTION THEOREM

We give now an application of theorem 9; see Lax. We consider L^1 functions on the positive axis $\xi \geq 0$. We denote the lower end of the support of such a function F by ℓ_F. That is,

$$\ell_F = \max\{\eta\colon F(\xi) = 0 \text{ for } \xi < \eta\} \tag{35}$$

An important theorem of Titchmarsh asserts:

Theorem 10. *Let A and B be two L^1 functions on $\mathbb{R}_+$, $A * B$ their convolution:*

$$(A * B)(\xi) = \int_0^\infty A(\eta)\, B(\xi - \eta)\, d\eta. \tag{36}$$

Then

$$\ell_{A*B} = \ell_A + \ell_B. \tag{37}$$

Proof. For $\xi < \ell_A + \ell_B$ the integrand on the right in (36) is zero, since at least one of the two factors is zero. Therefore the integral is zero, that is, $(A * B)(\xi) = 0$ for $\xi < \ell_A + \ell_B$. This proves that the left side of (37) is $\geq$ the right side; what remains to be shown is that equality holds. To do this, we rely on a characterization due to Paley and Wiener of the lower end of the support of an $L^1(\mathbb{R}_+)$ function F in terms of its Fourier transform

$$f(w) = \int_0^\infty F(\xi)\, e^{i\xi w}\, d\xi. \tag{38}$$

Clearly, $f(w)$ is a bounded analytic function in the upper half-plane. Next we need

Theorem 11 (Paley-Wiener). *The function F in $L^1(\mathbb{R}_+)$ vanishes on the interval $[0, \ell]$ iff its Fourier transform $f(w)$ is $\leq$ const. $e^{-\ell y}$ in absolute value in the upper half-plane* Im $w = y > 0$. *In the language of (35) we can express this as follows:*

$$\ell_F = \max\{\ell: |f(w)e^{-i\ell w}| \leq \text{const.}\}.$$

Proof. Suppose that F vanishes on the interval $[0, \ell]$; then we can write its Fourier transform (38) by setting $\xi = \sigma + \ell$ as

$$f(w) = \int_\ell^\infty F(\xi)\,e^{i\xi w}\,d\xi = e^{i\ell w}\int_0^\infty F(\sigma + \ell)\,e^{i\sigma w}\,d\sigma.$$

Clearly, $f(w)e^{-i\ell w}$ is bounded in the upper half-plane. Conversely, suppose that $f(w)e^{-i\ell w}$ is bounded in the upper half-plane; we claim that F, the Fourier inverse of f, vanishes on $[0, \ell]$. We want to prove this by showing that for any smooth function $G(\xi)$ supported on $[0, \ell - d]$, d any positive number, the scalar product $(F, G) = 0$. Denote the Fourier transform of G by g:

$$g(x) = \int_0^{\ell-d} G(\xi)\,e^{i\xi x}\,d\xi. \tag{39}$$

According to Parseval's formula

$$(F, G) = (f, g), \tag{40}$$

where

$$(f, g) = \int f(x)\,\bar{g}(x)\,dx. \tag{40'}$$

From (39) we have

$$\bar{g}(x) = \int_0^{\ell-d} \overline{G}(\xi)\,e^{-i\xi x}\,d\xi. \tag{39'}$$

This formula shows that $\bar{g}(x)$ can be extended as an entire function $h(w)$ to the whole complex plane

$$h(w) = \int_0^{\ell-d} \overline{G}(\xi)\,e^{-i\xi w}\,d\xi.$$

Since G is smooth, $h(w)$ is bounded in the upper half-plane Im $w = y > 0$ by $e^{(\ell-d)y}/(1 + |w|^2)$. Since f is analytic in the upper half-plane, we can by Cauchy's theorem shift the line of integration in (40') from the real axis to the line Im $w = y > 0$:

$$(f, g) = \int f(x)\,h(x)\,dx = \int f(x + iy)\,h(x + iy)\,dx. \tag{41}$$

By construction,

$$|h(x+iy)| \leq \frac{e^{(\ell-d)y}}{1+x^2+y^2},$$ (42)

and by hypothesis,

$$|f(x+iy)| \leq \text{const. } e^{-\ell y}.$$ (43)

Using the estimates (42), (43) on the right in (41) shows that the right side of (41) tends to zero as y tends to ∞. Since the left side is independent of y, it is zero. It follows from Parseval's formula (40) that $(F, G) = 0$ for all smooth functions G supported on some subinterval of $[0, \ell]$. This proves that F vanishes on $[0, \ell]$. □

We can restate the Paley-Wiener theorem in the language of division in the algebra $\mathcal{B}$:

The lower end ℓ_F of the support of a function F in $L^2(\mathbb{R}_+)$ is the highest power of e^{iw} that divides in $\mathcal{B}$ the Fourier transform of F.

We return to Titchmarsh's convolution theorem 10. Denote the Fourier transform of A and B by $a(w)$ and $b(w)$; these are bounded analytic functions in the upper half-plane. The Fourier transform of the convolution $A * B$ is the product ab. Using the Paley-Wiener theorem, we can reformulate the statement (37) as follows:

Let ℓ_A and ℓ_B denote the highest powers of e^{iw} that divide the functions $a(w)$ and $b(w)$, respectively. The highest power of e^{iw} that divides their product ab is $\ell_A + \ell_B$.

To prove this, we factor a and b as $a = e^{i\ell_A w}c$, $b = e^{i\ell_B w}d$, c and d in $\mathcal{B}$. The functions c and d are relatively prime to e^{iw}. Indeed, according to theorem 9, the only divisors of e^{iw} are of the form e^{ikw}, $k > 0$; on the other hand, c and d have no divisors of that form, because then a and b would be divisible by a higher power of e^{iw} than stipulated. We appeal now to theorem 7 to conclude that the product cd is relatively prime to e^{iw}. This proves that $ab = e^{i(\ell_A+\ell_B)w} cd$ is not divisible by a power of e^{iw} higher than $\ell_A + \ell_B$. □

NOTE. It is a curious fact that Titchmarsh's convolution theorem is a result about functions of real variables, yet Titchmarsh's original proof in 1924 used complex variable theory. A real variables proof has been furnished only in 1952 by Ryll-Nardzewski, see Mikusinski. The proof described in this chapter shows that the approach through complex variables is not unnatural.

HISTORICAL NOTE. During the Second World War the British Secret Service had broken the "Enigma" code of the German armed forces. The advance information gained through intercepts was of decisive help in many battles. It is less well known that the Swedes also had broken the "Enigma" code; the mathematician who has led this effort was Arne Beurling.

The mathematician leading the British code breaker was the great logician Alan Turing. After the war he was prosecuted for homosexuality and hounded into suicide.

BIBLIOGRAPHY

Beurling, A. On two problems concerning linear transformations in Hilbert space. *Acta Math.*, **18** (1949): 239–255.

Halmos, P. Shifts on Hilbert spaces. *Crelles J.*, **208** (1961): 102–112.

Lax, P. D. Translation invariant spaces. *Acta Math.*, **101** (1959): 163–178.

Lax, P. D. Translation invariant spaces. in *Proc. Int. Symp. on Linear Spaces.* Israeli Acad. Press, Jerusalem, 1961, pp. 299–306.

Paley, R. E. A. C. and Wiener, N. Fourier transforms in the complex domain. *AMS Coll. Publ.*, **19**. American Mathematical Society, New York, 1934.

Mikusinski, J. Operational Calculus. *Int. Series Monographs in Pure and Applied Math*, **8**. Pergamon Press, New York, 1959.

Titchmarsh, E. C. The zeros of certain integral functions. *Proc. London Math. Soc.*, **25** (1926): 283–302.

TEXTS

Banach, S., *Théorie des opérations linéaires*, Monografje Matematyczne, Warsaw, 1932, Chelsea, 1955.

Brezis, H., *Analyse fonctionelle*, Théorie et application, Masson, 1983.

Conway, J. B., *A Course in Functional Analysis*, Springer Verlag, 1985.

Day, M. M., *Normed Linear Spaces*, Springer Verlag, 1962.

Douglas, R. G., *Banach Algebra Techniques in Operator Theory*, Graduate texts in mathematics Vol. 179, Springer Verlag, 2nd ed. 1997.

Dunford, N. and Schwartz, J. T., *Linear Operators, Part I: General Theory*, (1958), *Part II: Spectral Theory*, (1963), Wiley-Interscience Series on Pure and Applied Mathematics, John Wiley and Sons.

Edwards, R. E., *Functional Analysis: Theory and Applications*, Holt, Rinehart and Winston, 1965.

Hille, E. and Phillips, R. S., *Functional Analysis and Semi-groups*, Colloquium Publ. AMS, 1957.

Johnson, W. B. and Lindenstrauss, Y., eds., *Handbook on the Geometry of Banach Spaces*, North Holland, to appear.

Lindenstrauss, J. and Tzafriri, L., *Classical Banach Spaces*, I (1977), II (1979), Springer Verlag.

Morrison, T. J., *Functional Analysis: An Introduction to Banach Space Theory*, Wiley-Interscience Series on Pure and Applied Mathematics, John Wiley and Sons, 2001.

Reed, M. and Simon, B., *Methods of Modern Mathematical Physics, I: Functional Analysis*, Academic Press, 1972.

Riesz, F. and Sz. Nagy, B., *Leçons d' analyse fonctionelle*, Akadémiai Kiadó, 1952, *Functional Analysis*, F. Ungar, 1955.

Rudin, W., *Functional Analysis*, McGraw-Hill, 1973.

Schechter, M., *Principles of Functional Analysis*, Academic Press, 1971.

Taylor, A.E. and Lay, D. C., *Introduction to Functional Analysis*, John Wiley and Sons, 1980.

Yosida, K., *Functional Analysis*, 1st ed. (1964), 6th ed. (1980), Springer Verlag.

Zeidler, E., *Nonlinear Functional Analysis and its Applications*, Springer Verlag, 1985.

A

RIESZ-KAKUTANI
REPRESENTATION THEOREM

There is a conundrum in mathematical analysis similar to the chicken or the egg question: Which comes first, the Lebesgue integral or the Lebesgue measure? My answer is, neither; first comes the space L^1. The traditional approaches enlarge the class of continuous functions, and then show this class to be sufficiently large, that is, complete in the L^1-norm. The approach described in the pages to follow puts the horse before the cart. The object of our desire, L^1, is defined as an abstract space, the completion in the L^1-norm of the space of continuous functions. Then each element of L^1 is identified as a down-to-earth function, defined almost everywhere.

The theorem in the title states that every bounded, linear functional ℓ on the space $C(Q)$ of continuous functions c on a *compact* Hausdorff space Q can be represented as an integral with respect to a signed finite measure on the σ-algebra of the Borel sets of Q:

$$\ell(c) = \int c\,dm.$$

In this note we use functional analysis to give a simple and natural proof of this basic proposition in the case where Q is a compact metric space.

A.1 POSITIVE LINEAR FUNCTIONALS

We will study bounded linear functionals ℓ on the linear space $C(Q)$ of continuous real-valued functions c on a compact metric space. Boundedness means that

$$\ell(c) \leq \text{const.} \, |c|_{\max}$$

for all continuous functions c. A linear functional is called *positive* if $\ell(c) \geq 0$ for all nonnegative functions c. Note that a positive linear functional is *monotone*: $c_1 \leq c_2$ implies that $\ell(c_1) \leq \ell(c_2)$. It follows that a positive linear functional is bounded. This is because every c in $C(Q)$ satisfies $c \leq |c|_{\max}u$, where u is the unit function on $Q: u(q) = 1$ for all q in Q; therefore, by monotonicity, $\ell(c) \leq |c|_{\max}\ell(u)$.

It is a standard result, and not hard to show, that every bounded linear functional on $C(Q)$ can be decomposed as the difference of two positive linear functionals. Therefore it suffices to prove the representation theorem for positive linear functionals; in this case the representing measure is positive.

Given a positive linear functional ℓ on $C(Q)$, we define the ℓ-norm of a continuous functions as follows:

$$|c|_\ell = \ell(|c|), \tag{1}$$

where $|c|$ is the absolute value of the function c : $|c|(q) = |c(q)|$ for every point q of Q.

The quantity $|c|_\ell$ is a semi-norm on the space $C(Q)$. If we identify the continuous functions whose difference has ℓ-norm 0, we get a genuine norm on the quotient space. We denote by L the *completion* in the ℓ-norm of this quotient space. We recall that the elements of the completion are Cauchy sequences in the ℓ-norm of continuous functions; two Cauchy sequences are lumped together if their difference is a null sequence in the ℓ-norm.

It follows from definition (1) of the ℓ-norm and monotonicity that $|\ell(c)| \leq |c|_\ell$. It follows that the functional ℓ can be extended by continuity to the whole space L. If $\{c_n\}$ is a Cauchy sequence tending to f in the ℓ-norm, we define $\lim \ell(c_n) = \ell(f)$; ℓ is a bounded linear functional on L:

$$\ell(f) \leq |f|_\ell. \tag{2}$$

Next we show that the elements of L have some functionlike attributes.

Theorem 1. *Let ϕ be a Lipschitz continuous function $\mathbb{R} \to \mathbb{R}$:*

$$|\phi(x) - \phi(y)| \leq k|x - y|. \tag{3}$$

Let f be any element of L; then $\phi(f)$ can be defined as an element of L. Furthermore, for any pair of elements f and g of L,

$$|\phi(f) - \phi(g)|_\ell \leq k|f - g|_\ell. \tag{4}$$

Proof. Let $\{c_n\}$ be a Cauchy sequence in the ℓ-norm of continuous functions; denote their limit in L by f. It follows from the definition of the ℓ-norm, the monotonicity of ℓ, and (3), that

$$|\phi(c_n) - \phi(c_m)|_\ell \leq k|c_n - c_m|_\ell.$$

Thus $\{\phi(c_n)\}$ too is a Cauchy sequence; denote its limit in L by the symbol $\phi(f)$.

Concerning the second part: let $\{d_n\}$ be a sequence of continuous functions tending in the ℓ-norm to g. By (3),

$$|\phi(c_n) - \phi(d_n)| \leq k|c_n - d_n|$$

holds at every point q of Q. Since ℓ is monotonic,

$$\ell(|\phi(c_n) - \phi(d_n)|) \leq k\ell(|c_n - d_n|);$$

by definition of the ℓ-norm, this can be expressed as

$$|\phi(c_n) - \phi(d_n)|_\ell \leq k|c_n - d_n|_\ell.$$

As n tends to ∞, the left and right sides tend to the left and rights of (4).   □

The usual rules of functional calculus hold: Here are two important examples:

Example 1.

$$\phi_+(x) = \begin{cases} 0 & \text{for } x \leq 0 \\ x & \text{for } 0 \leq x. \end{cases}$$

We denote $\phi_+(f)$ as f_+, and call it the *positive part* of f.

Definition 1. f in L is called *nonnegative*, denoted as $f \geq 0$, if $f_+ = f$.

Note that f in L is nonnegative iff it is the limit in the ℓ-norm of a sequence of nonnegative continuous functions. It follows that if $f \geq 0$, $|f|_\ell = \ell(f)$. Note that if $f \geq 0, g \geq 0$, then $f + g \geq 0$.

We say that $f \leq g$ if $g - f \geq 0$; clearly this relation is transitive in L.

Example 2.

$$\phi_a^b(x) = \begin{cases} a & \text{for } x \leq a \\ x & \text{for } a \leq x \leq b \\ b & \text{for } b \leq x \end{cases}$$

for f in L we denote $\phi_a^b(f)$ as f_a^b.

Definition 2. f in L is said to be *bounded* if $f = f_a^b$ for some a and b.

Note that f in L is bounded iff it is the limit in the ℓ-norm of a sequence of bounded continuous functions.

Note that if f in L is bounded, and ϕ is Lipschitz continuous on the interval $[a, b]$, then $\phi(f)$ can be defined as an element of L.

If $\phi(x, y)$ is a Lipschitz continuous function of two variables, then $\phi(f, g)$ can be defined for any pair of elements f, g in L. We will use only a special case of this. Suppose that f and g are elements of L that are both bounded. Then we can define their product fg as an element of L that is bounded, as follows:

Let $\{c_n\}$ and $\{d_n\}$ be Cauchy sequences of bounded continuous functions tending in the ℓ-norm to f and g, respectively. It is easy to show that $\{c_n d_n\}$ is a Cauchy sequence in the ℓ-norm; its limit in L is defined as the product fg. Note that if $f \geq 0$, $g \geq 0$, the $fg \geq 0$.

The following result, although simple, is very useful.

Theorem 2 (Monotone Convergence Theorem). *Let $\{fn\}$ be a monotone sequence of elements in L, say increasing, $f_n \leq f_{n+1}$. Suppose that the sequence of numbers $\ell(f_n)$ is bounded; then f_n converges in the ℓ-norm to a limit f in L.*

Proof. Since the functional ℓ is monotone, $\ell(f_n)$ is an increasing sequence of real numbers. Being bounded, it has a limit. We claim that $\{f_n\}$ is Cauchy sequence, for, since $f_n - f_m$ is nonnegative for $n > m$,

$$|f_n - f_m|_\ell = \ell(f_n - f_m) = \ell(f_n) - \ell(f_m).$$

Since $\ell(f_n)$ is a convergent sequence, the right side tends to zero as n, m tend to ∞.

Having shown that $\{f_n\}$ is a Cauchy sequence, it follows the completeness of L that f_n converges to a limit in L. $\qquad\qquad\square$

Theorem 3. *For any f in L*

$$\ell - \lim_{\substack{b \to \infty \\ a \to -\infty}} f_a^b = f.$$

Proof. Let $\{c_n\}$ be a Cauchy sequence of continuous functions whose ℓ-limit is f. It follows that for any $\epsilon > 0$ there is an N such that

$$|f - c_N|_\ell < \epsilon. \tag{5}$$

Since c_N is a continuous function on a compact space, it is bounded: $a \leq c_N(q) \leq b$ for all q in Q, for any a and b exceeding the upper and lower bounds of c_N. Define ϕ_a^b as in example 2; by inequality (4),

$$|\phi_a^b(f) - \phi_a^b(c_N)|_\ell \leq |f - c_N|_\ell < \epsilon.$$

Since the values of c_N lie in the interval $[a, b]$, $\phi_a^b(c_N) = c_N$, so we can rewrite the inequality above as

$$|f_a^b - c_N|_\ell < \epsilon. \tag{6}$$

Using the triangle inequality and the estimates (5) and (6), we get

$$|f - f_a^b|_\ell = |f - c_N + c_N - f_a^b|_\ell \leq |f - c_N|_\ell + |c_N - f_a^b|_\ell < 2\epsilon. \quad\square$$

A.2 VOLUME

In this section we show how to use the positive functional ℓ to define the volume of any open set G in Q.

Definition 3. A continuous function c is called *admissible* for the open set G if

   (i) the support of c is contained in G.

   (ii) $c(q) \le 1$ for all q in Q.

The *volume* $V(G)$ with respect to ℓ of an open set G is defined as

$$V(G) = \sup_{c \text{ admissible}} \ell(c) \tag{7}$$

Theorem 4.

   (i) *The volume of the empty set is zero.*

   (ii) *V is a monotonic set function: if $G \subset H$, then $V(G) \le V(H)$.*

   (iii) *V is countably subadditive: $V(\cup_1^\infty G_n) \le \sum_1^\infty V(G_n)$.*

   (iv) *V is countably additive, that is, if G_n is a collection of pairwise disjoint open sets,*

$$V(\cup G_n) = \sum V(G_n).$$

Proof. Parts (i) and (ii) are obvious. To prove (iii), take any continuous function c admissible for $\cup G_n$. The support of c is a closed set; since Q is compact, so is the support of c. Being admissible, it is covered by $\cup G_n$; therefore by compactness it is covered by a finite subcollection $\cup_1^N G_n$. We show now that c can be decomposed as

$$c = \sum_1^N c_n, \tag{8}$$

where c_n is admissible for $G_n, n = 1, \ldots, N$. (iii) would follow from this; for, applying ℓ gives

$$\ell(c) = \sum_1^N \ell(c_n);$$

since by definition of volume $\ell(c_n) \le V(G_n)$,

$$\ell(c) \le \sum_1^\infty V(G_n).$$

Taking the supremum of over all admissible c yields (iii).

   We turn now to the decomposition (8). Each q in the support of c belongs to at least one of the sets $G_n, n = 1, \ldots, N$. There exists a continuous function b_q with the following properties:

(i) b_q is nonnegative.

(ii) $b_q(q) > 0$.

(iii) The support of b_q lies in one of the sets G_n.

The set where b_q is positive is an open set O_q which contains q; the union of these open sets O_q contains the support of c. Since the support of c is compact, it is covered by a finite number of these open sets. The sum of the corresponding functions b_q is positive on the support of c:

$$\sum b_{q_i} > 0 \qquad \text{on supp } c.$$

Assign to each b_{q_i} an open set G_n that contains its support. Denote by b_n the sum of those b_{q_i} assigned to G_n. Clearly, $\sum_1^N b_n = \sum b_{q_i}$; now define

$$c_n = \frac{b_n}{\sum b_n} c, \qquad n = 1, \ldots, N.$$

Clearly, $\sum c_n = c$, and each c_n is admissible for G_n. This completes the construction of the decomposition (8).

(iv) Given any N, we can choose admissible functions c_n for $G_n, n \leq N$, so that

$$V(G_n) - \frac{1}{N^2} \leq \ell(c_n).$$

Since the G_n are pairwise disjoint, $\sum_1^N c_n$ is admissible for $\cup_1^N G_n$, and therefore

$$\ell \left(\sum_1^N c_n \right) \leq V(\cup_1^N G_n)$$

Using the linearity of ℓ and the inequalities above, we get

$$\sum_1^N V(G) - \frac{1}{N} \leq V(\cup_1^N G_n).$$

Letting $N \to \infty$, we get

$$\sum V(G_n) \leq V(\cup G_n).$$

By countable subadditivity the opposite inequality hold; so equality must prevail.

$\square$

The following simple estimate is useful:

Theorem 5. *Let h be a continuous nonnegative function defined on Q, a some nonnegative number, G_a the open set $G_a = \{q : h(q) > a\}$. Then*

$$V(G_a) \leq \frac{1}{a}\ell(h). \tag{9}$$

Proof. We show that any continuous function c_a admissible for G_a satisfies for every q,

$$c_a(q) \leq \frac{1}{a}h(q).$$

Clearly, this is true for q not in G_a, for then $c_a(q) = 0$, and $h(q) \geq 0$. It is equally true for q in G_a, for then by definiton $h(q)/a$ is greater than 1, while $c_a(q) \leq 1$. Since ℓ is monotonic,

$$\ell(c_a) \leq \frac{1}{a}\ell(h).$$

Taking the supremum of this inequality over all admissible functions c_a yields inequality (9).   □

Definition 4. A subset S of Q is called *negligible* if it can be covered by an open set of arbitrarily small volume.

It follows from the countable subadditivity of volume [see part (iii) of theorem 4] that the countable union of negligible sets is negligible.

When a relation holds for all points of q with the exception of a negligible set, we say it holds *almost everywhere*, abbreviated as a.e.

A.3 *L* AS A SPACE OF FUNCTIONS

In this section we show how to associate to any element f in L a function $f(q)$ on Q, up to negligible sets of points.

Definition 5. A Cauchy sequence $\{c_n\}$ is called *rapidly converging* if

$$|c_n - c_{n+1}|_\ell \leq \frac{k}{n^4} \tag{10}$$

for all n and some constant k.

Note that every Cauchy sequence contains rapidly convergent subsequences. Clearly, a sequence satisfying (10) is a Cauchy sequence.

Theorem 6. A rapidly converging Cauchy sequence $\{c_n\}$ of continuous functions converges almost everywhere, that is except for a negligible subset of Q.

Proof. Denote by G_n the open set of points satisfying

$$G_n = \left\{ q : |c_n(q) - c_{n+1}(q)| > \frac{1}{n^2} \right\}. \tag{11}$$

Combining inequality (9) for $h = |c_n - c_{n+1}|$ and $a = 1/n^2$ and (10), we have

$$V(G_n) \le n^2 \ell(|c_n - c_{n+1}|) \le \frac{k}{n^2}. \tag{12}$$

It follows from (11) that for q not in G_n, $|c_n(q) - c_{n+1}(q)| \le 1/n^2$; therefore the sequence $c_n(q)$, written as a sum

$$c_n = \sum_1^n (c_k - c_{k-1}),$$

converges at every point q that belongs to only a finite number of the sets G_n. The exceptional set of points, where $\{c_n(q)\}$ fails to converge, can thus be enclosed in the union $\cup_N^\infty G_n$. By countable subadditivity, the volume of this set is bounded by

$$V(\cup_N^\infty G_n) \le \sum_N^\infty V(G_n)$$

which according to (12) is $\le \sum_N^\infty k/n^2$ and thus tends to zero as N tends to ∞. Thus the exceptional set can be covered by open sets of arbitrarily small volume, and therefore form a negligible set. $\square$

It is important to note that the sequence $c_n(q)$ converges *uniformly* outside the open set $\cup_N^\infty G_n$, N arbitrary.

Given any element f in L, take any Cauchy sequence $\{c_n\}$ of continuous functions that converges to f. It is easy to see we can select a subsequence that converges rapidly; the pointwise limit of this sequence is called a *realization* of f.

We claim that realizations obtained from any two Cauchy sequences converging rapidly to f are equal a.e., for, we can select a single rapidly convergent Cauchy sequence that contains infinitely many terms from both sequences.

The following theorem characterizes the relation of f in L to its realization.

Theorem 7.

   (i) *For any Lipschitz continuous* ϕ, $\phi(f)(q) = \phi(f(q))$.
   (ii) $(f \pm g)(q) = f(q) \pm g(q)$.
   (iii) *If f and g are bounded elements of L,* $(fg)(q) = f(q)g(q)$.
   (iv) *The realization of elements of L by functions is faithful, that is, if $f(q) = g(q)$ a.e., then $f = g$ in L.*
   (v) *Suppose that ℓ-limit $f_n = f$ and that $f_n(q)$ converges a.e. Then the point-wise limit of $f_n(q)$ is a realization of the ℓ-limit f.*

Proof. Parts (i), (ii), and (iii) are obvious. To prove (iv), it suffices to show that if f in L is represented by a function that is zero a.e., then $f = 0$ in L. We argue indirectly, supposing that $f \neq 0$. So we may assume that f is nonnegative. Since according to theorem 3, f^b tends to f in the ℓ-norm, $f^b \neq 0$ for b large enough. It follows that it suffices to consider nonnegative, bounded elements of L; we may take the upper bound to be 1.

Let c_n be a rapidly converging Cauchy sequence of continuous functions, each of which lies between 0 and 1, and tends in the ℓ-norm to f, and to zero a.e. As noted at the conclusion of theorem 6, for any $\epsilon > 0$ there is an open set G_ϵ of volume less than ϵ, such that outside G_ϵ the sequence $c_n(q)$ converges uniformly to $f(q)$. Since $f(q)$ is assumed to be zero a.e. for all N large enough

$$c_N(q) < \epsilon \text{ on the complement of } G_\epsilon, \tag{13}$$

$$|f - c_N|_\ell < \epsilon. \tag{13'}$$

We decompose c_N as follows:

$$c_N = c_N - 2\epsilon u + 2\epsilon u \leq (c_N - 2\epsilon)_+ + 2\epsilon u, \tag{14}$$

where u is the unit function on Q. Since ℓ is monotonic, we deduce from (14) that

$$\ell(c_N) \leq \ell((c_N - 2\epsilon u)_+) + 2\epsilon \ell(u). \tag{15}$$

It follows from (13) and the fact that $c_N(q) \leq 1$ for all q, that the function $(c_N - 2\epsilon u)_+$ is admissible for G_ϵ. Therefore

$$\ell((c_N - 2\epsilon u)_+) \leq V(G_\epsilon).$$

Setting this into (15), and using the fact that $V(G_\epsilon) < \epsilon$, we obtain

$$\ell(c_N) \leq (1 + 2\ell(u))\epsilon.$$

Using this inequality and (13′) combined with (2), we obtain

$$\ell(f) = \ell(f - c_N) + \ell(c_N) \leq |f - c_N|_\ell + (1 + 2\ell(u))\epsilon < 2(1 + \ell(u))\epsilon.$$

Since f is a nonnegative element of L, $\ell(f) = |f|_\ell$; since ϵ is arbitrary, we conclude that $|f|_\ell = 0$. This contradicts the presumption that $f \neq 0$ in L, and completes the demonstration of (iv).

To prove (v), select from $\{f_n\}$ a rapidly converging subsequence, also denoted as $\{f_n\}$, in the sense of inequality (10). For each n choose a continuous function c_n so that

(a) $|f_n - c_n|_\ell \leq 1/n^4$,

(b) $|f_n(q) - c_n(q)| \leq 1/n^2$, except on an open set of volume $\leq 1/n^2$.

It follows from (a) and (10) for $\{f_n\}$ that $\{c_n\}$ converges rapidly in the ℓ-norm, and that its ℓ-limit is f. Therefore $c_n(q)$ tends to $f(q)$ a.e. On the other hand, it follows from (b) that $\lim c_n(q) = \lim f_n(q)$ a.e. $\qquad\square$

Theorem 8. *Let f be an element of L whose realization $f(q)$ is ≥ 0 a.e. Then f is ≥ 0 as an element of L.*

Proof. According to part (i) of theorem 7, $f_+(q) = f(q)_+$. If $f(q) \geq 0$ a.e., it follows that $f_+(q) = f(q)$ a.e. It follows then from part (iv) of theorem 7 that $f_+ = f$. but this is the definition of $f \geq 0$. ☐

Corollary. *Two elements f and g in L satisfy $f \leq g$ iff $f(q) \leq g(q)$ a.e.*

Proof. The relation $f \leq g$ means that $g = f + p$, p nonnegative. Now apply theorem 8 to p. ☐

A.4 MEASURABLE SETS AND MEASURE

Definition 6. A set S in Q is *measurable* if its characteristic function c_S:

$$c_S(q) = \begin{cases} 1 & \text{when } q \text{ in } S \\ 0 & \text{when } q \text{ not in } S \end{cases} \tag{16}$$

is the realization of some element f_S in L. The *measure* of S is defined to be

$$m(S) = \ell(f_S). \tag{17}$$

Note that the notion of measurability, and the value of measure, depends on the linear functional ℓ.

We show first that this notion is not vacuous.

Theorem 9. *Every open set is measurable, and its measure is its volume.*

Proof. We will use the function $d(q, G^c)$, the distance of q from the complement of G. Set

$$\phi_n(x) = \begin{cases} 0 & \text{for } x \leq 1/2n \\ \text{linear} & \text{in between} \\ 1 & \text{for } 1/n \leq x. \end{cases}$$

Define the continuous functions $c_n(q) = \phi_n(d(q, G^c))$. Clearly, c_n is an increasing sequence of functions, and equally clearly, $\ell(c_n)$ is bounded by $\ell(u)$. Therefore, by theorem 2, the monotone convergence theorem, the sequence $\{c_n\}$ converges in the ℓ-norm to a limit which we denote as f_G. On the other hand, $\{c_n(q)\}$ converges for every q to the characteristic function $c_G(q)$ of G. According to part (v) of theorem 7, $c_G(q)$ is the realization of f_G; this proves that G is measurable.

To determine the measure of G, we note that since c_n is ℓ-convergent to f_G, $\ell(c_n)$ tends to $\ell(f_G)$. Since each c_n is admissible for G, $\ell(c_n) \leq V(G)$, and so $\ell(f_G) = \lim \ell(c_n) \leq V(G)$. On the other hand, given any admissible c, we can choose n so large that $c \leq c_n$. Since $V(G) = \sup \ell(c)$, it follows that $\ell(f_G) = \lim \ell(c_n) \geq V(G)$. Combining the two inequalities shows that $\ell(f_G) = V(G)$.   □

Theorem 10. *The collection of sets measurable in the sense of (16) is a σ-algebra, and the set function $m(S)$ defined by (17) is a measure. That is,*

    (i) the complement of a measurable set is measurable.

    (ii) the intersection of two measurable sets is measurable.

    (iii) the denumerable union of measurable sets is measurable.

    (iv) $m(S)$ is a countably additive set function.

Proof. (i) S is measurable if its characteristic function c_S is the realization of an element of f_S in L. But then the realization of $u - f_S$ is $1 - c_S$, the characteristic function of the complement of S.

(ii) If f_{S_1} and f_{S_2} are elements in L whose realizations are the characteristic functions of S_1 and S_2, then the realization of the product $f_{S_1} f_{S_2}$ is the characteristic function of $S_1 \cap S_2$.

(iii) Let $\{S_n\}$ be a denumerable collection of measurable sets. We can replace this collection by another one, $\{T_n\}$, where $T_n = S_n \cap (S_1 \cup \ldots \cup S_{n-1})^c$. Clearly, $T_1 \cup \ldots \cup T_n = S_1 \cup \ldots \cup S_n$, and therefore $\cup T_n = \cup S_n$. The sets T_n are pairwise disjoint. Consider now the series $\sum_1^\infty f_{T_n}$; since each f_{T_n} is nonnegative, the partial sums $\sum_1^N f_{T_n}$ of this series are increasing. Furthermore, since the T_n are pairwise disjoint, $\sum_1^N f_{T_n} \leq u$; therefore $\ell(\sum_1^N f_{T_n}) \leq \ell(u)$. We appeal now to the monotone convergence theorem to conclude that the partial sums converge in the ℓ-norm to a limit that we denote by f_T:

$$\sum_1^\infty f_{T_n} = f_T \tag{18}$$

On the other hand, the realization of the partial sums is $\sum_1^N c_{T_n}(q)$, which converge pointwise to $c_T(q)$, where $T = \cup T_n$. By part (v) of theorem 7, c_T is the realization of f_T; therefore $T = \cup T_n = \cup S_n$ is a measurable set.

(iv) Since the partial sums of (18) converges in the ℓ-norm to f_T, we can apply ℓ to obtain $\sum_1^\infty \ell(f_{T_n}) = \ell(f_T)$. Using the definition (17) on both sides, we obtain $\sum_1^n m(T_n) = m(T)$.   □

Theorem 11. *A realization $f(q)$ of any f in L is a measurable function, that is, the sets K_a:*

$$K_a = \{q : f(q) \geq a\} \tag{19}$$

a any real number, are measurable sets.

Proof. We may assume that a is positive, for we can always add a constant to f. We may further assume that $a = 1$, for we can always divide f by a. Recall that f_0^1

denotes the truncation of f from below by 0, and from above by 1. The sequence $f_n = (f_0^1)^n$ is decreasing, since $f_n - f_{n+1} = (f_0^1)^n(1 - f_0^1)$ is the product of two nonnegative elements of L. Since each f_n is nonnegative, all the $\ell(f_n)$ are bounded from below by zero. So by the monotone convergence theorem, the sequence f_n converges in the ℓ-norm to a limit. On the other hand, the sequence of realizations

$$f_n(q) = (f_0^1(q))^n$$

converges to 1 at all points q where $f(q) \geq 1$, and to 0 at all points where $f(q) < 1$. By the part (v) of theorem 7, this shows that the characteristic function of the set K_1 defined by (19) is a realization of the ℓ-limit of $(f_0^1)^n$.   □

Replacing f by $-f$, and taking complements, we can show that all sets of the form

$$\{q : f(q) < a\}, \quad \{q : f(q) \leq a\}, \quad \{q : f(q) > a\}$$

are measurable as well.

We are ready to show that the measure m we have constructed in (17) yields the linear functional ℓ,

$$\ell(c) = \int c \, dm \tag{20}$$

for all continuous functions c. To see this, we form the sets $K_{j,\epsilon}$:

$$K_{j,\epsilon} = \{q : j\epsilon \leq c(q) \leq (j+1)\epsilon\}. \tag{21}$$

Denote the characteristic function of $K_{j,\epsilon}$ by $k_{j,\epsilon}$. It follows from (21) that

$$\sum j\epsilon k_{j,\epsilon}(q) \leq c(q) \leq \sum ((j+1)\epsilon k_{j,\epsilon}(q) \tag{22}$$

holds at each point q. Since c is bounded, these sums are finite. According to the corollary to theorem 8, we deduce from (22) that

$$\sum j\epsilon k_{j,\epsilon} \leq c \leq \sum (j+1)\epsilon k_{j,\epsilon} \tag{23}$$

holds in L. Since ℓ is a positive functional, we deduce that

$$\sum j\epsilon \ell(k_{j,\epsilon}) \leq \ell(c) \leq \sum (j+1)\epsilon \ell(k_{j,\epsilon}).$$

Using the definition (17) of measure for the sets $K_{j,\epsilon}$, we can rewrite this as

$$\sum j\epsilon m(k_{j,\epsilon}) \leq \ell(c) \leq \sum (j+1)\epsilon m(k_{j,\epsilon}). \tag{24}$$

The left and right side in (24) are lower and upper sums for the integral $\int c \, dm$ in (20). Their difference,

$$\epsilon \sum m(k_{j,\epsilon}) = \epsilon m(\cup k_{j,\epsilon}) = \epsilon m(Q),$$

tends to zero as ϵ does. Thus (24) shows that $\ell(c)$ is the only number $\geq$ all lower sums and $\leq$ all upper sums. This proves the representation formula (20). □

We remark that the representation formula (20) holds as well for all f in L. In this case the sums in (23) may be infinite. Taking f to be nonnegative, the convergence of the infinite series in (23) follows from the monotone convergence theorem. Since any f in L can be written as the difference of two nonnegative elements, the general case follows.

We close this section with the converse of theorem 11.

Theorem 12. *Every function $g(q)$ defined on Q that is measurable and integrable is the realizaiton of an element g of L.*

Proof. Measurability means that the sets $H_a = \{q : g(q) < a\}$ are measurable. Define the function $n_g, \mathbb{R} \to \mathbb{R}$, by

$$n_g(a) = m(H_a).$$

The function g is called integrable if

$$\int |a| dn_g < \infty.$$

Define g_ϵ in L by

$$g_\epsilon = \sum j\epsilon k_{j,\epsilon},$$

where $k_{j,\epsilon}$ is defined after equation (21). As ϵ tends to zero, the realizations $g_\epsilon(q)$ tend to $g(q)$. It is easy to show that if we set $\epsilon = 2^{-n}$, we get a Cauchy sequence in the ℓ-norm, whose limit g is realized by $g(q)$. □

The uniqueness of the representing measure is a standard fact of measure theory.

A.5 THE LEBESGUE MEASURE AND INTEGRAL

If we take Q to be a Euclidean multitorus, and ℓ to be the Riemann integral, our construction gives the Lebesgue measure and integration. I consider this approach more natural than the traditional ones; for the most important object in the Lebesgue theory is the *complete* space L^1 (as well as the spaces L^p). In the traditional approaches the completeness of L^1 is the *last* item to emerge; in the present approach it is the *first*.

B

THEORY OF DISTRIBUTIONS

In his formulation of quantum mechanics Dirac treated the continuous spectrum by employing a function on $\mathbb{R}$, denoted as δ—and ever since called *Dirac's delta function*—that is zero everywhere except at $x = 0$, where it is so large that the integral of δ over $\mathbb{R}$ equals 1. Of course there is no such function. Von Neumann, in his book on quantum mechanics (1932), warns against basing a theory on such a fiction; he knew very well how to treat the continuous spectrum rigorously.

The δ-function can be given new life as a *generalized function*. The need for such generalized functions was keenly felt in the 1920s and 1930s. Bochner introduced—rigorously—such notions in the context of the Fourier transform, and Sobolev in the context of partial differential equation. Hadamard's use of the "finite part" of an integral in his formula for solutions of hyperbolic equations foreshadowed the need for generalized functions, as did Wiener's justification of the Heaviside calculus. L. C. Young's idea of "generalized curves" is a step in a similar direction. But it was Laurent Schwartz, in the 1940s, who came up with the notion of distributions that was general enough and supple enough to serve most purposes of both the theory of partial differential equations and harmonic analysis. Harald Bohr, a leading mathematician of his time (and brother of Niels Bohr), was among the first to recognize the value of Schwartz's ideas. The world soon followed; at the International Congress of Mathematician in 1950 Laurent Schwartz was honored by a Fields Medal; pockets of resistance eventually faded.

This appendix presents the bare bones of the theory of distributions, with a few scraps of meat thrown in. For a fuller presentation I recommend Robert Strichartz's book on the subject.

B.1 DEFINITIONS AND EXAMPLES

We denote by C_0^∞ the space of all complex-valued infinitely differentiable functions in $\mathbb{R}^n$ of compact support. We say that a *sequence* $\{u_k\}$ of C_0^∞ functions *converges* to u if the support of all u_k is contained in the *same compact set* K, and if for each multi-index $\alpha = \alpha_1, \ldots, \alpha_n$, $D^\alpha u_k = D_1^{\alpha_1} \ldots D_n^{\alpha_n} u_k$ converges *uniformly* to $D^\alpha u$, where D_i denotes partial differentiation with respect to x_i.

Definition. A *distribution* is an element in the *dual* of C_0^∞, that is, a complex-valued *linear functional* ℓ on C_0^∞ that is continuous under sequential convergence as defined above:

$$\ell(u_k) \to \ell(u) \qquad \text{if } u_k \to u.$$

This kind of continuity is equivalent to the following apparently more stringent one.

Theorem 1. *Given a distribution ℓ, to each compact set K there is a positive integer $N(K)$ and a positive number $c(K)$ such that for all C_0^∞ functions u whose support is contained in K*

$$|\ell(u)| \le c|u|_N, \qquad \text{where } |u|_N = \max_{|a| \le N} |D^a u(x)|, \ |a| = \sum \alpha_j. \qquad (1)$$

Proof. Given K, suppose that (1) is false for all c and all N; that means that for each N there is a C_0^∞ function u_n with support in K such that $\ell(u_n) = 1$, $|u_n|_N < 1/N$. Clearly, $u_n \to 0$ in the sense defined above; therefore $\ell(u_n)$ ought to tend to $\ell(0) = 0$, which contradicts $\ell(u_n) = 1$. $\qquad\square$

We call $|u|_N$ the C^N norm. We *embed* C_0^∞ in the space of distributions by assigning to each v in C_0^∞ the linear functional

$$\ell(u) = \int uv\, dx = (u, v). \qquad (2)$$

Clearly, (2) defines a continuous linear functional, and clearly different v's define different functionals. Thus C_0^∞ functions are special kinds of distributions.

By the same token, distributions may be regarded as generalized functions. In what follows we will often write

$$\ell(u) = (u, \ell).$$

In (2) we have examples of distributions; we now give others. In each case we leave it to the reader to verify that the given functional depends continuously on u in the sense of sequential convergence.

Example 1. $\ell(u) = \int uv\, dx$, v any continuous function.

Example 2. $\delta(u) = u(0)$, the Dirac delta "function."

Example 3. $\ell(u) = \int (D^a u)v\, dx$, v any integrable function, α any multi-index.

Example 4. p a C_0^∞ function in $\mathbb{R}^1$ whose zeros are simple: if $p(y) = 0$, $p'(y) \ne 0$.

Define $\ell(u)$ as the *principal value* integral

$$\ell(u) = PV \int \frac{u}{p}\, dx = \lim_{\varepsilon \to 0} \int_{|p(x)|>\varepsilon} \frac{u(x)}{p(x)}\, dx.$$

Later we will show that the most general distribution can be built up from distributions of the form given in example 3.

Definition. Let D be an open set in $\mathbb{R}^n$; denote by C_D^∞ the space of all C^∞ functions with support in D. A *distribution in* D is a linear function on C_D^∞ continuous under sequential convergence as defined above.

B.2 OPERATIONS ON DISTRIBUTIONS

In this section we show how certain operations that can be carried out for ordinary functions can also be carried out for distributions. These operations **T** are the following kind:

  (i) **T** is a linear operator that maps C_0^∞ continuously into C_0^∞.
 (ii) **T** has a transpose **T'** that also carries C_0^∞ continuously into C_0^∞. **T'** being the transpose of **T** means that for all u, v in C_0^∞,

$$(\mathbf{T}'u, v) = (u, \mathbf{T}v), \tag{3}$$

where $(\cdot, \cdot)$ is the symmetric bilinear functional used in (2) to embed C_0^∞ in its dual. Note that the transpose **T'** is uniquely determined by (3) and that the transpose of **T'** is **T** itself.

The following rules dealing with transposes are obvious and useful:

  (i) If **T** and **S** have transposes, so do $a\mathbf{T} + b\mathbf{S}$, and $(a\mathbf{T} + b\mathbf{S})' = a\mathbf{T}' + b\mathbf{S}'$.
 (ii) The transpose of **TS** is $(\mathbf{TS})' = \mathbf{S}'\mathbf{T}'$.

Theorem 2 (Extension Theorem). *Let* **T**, **T'** *be continuous linear operators mapping* C_0^∞ *into* C_0^∞, *transposes of each other. Let* ℓ *be any distribution; we define* $\mathbf{T}\ell$ *as the distributions given by*

$$(\mathbf{T}'v, \ell) = (v, \mathbf{T}\ell). \tag{4}$$

Proof. The reader can easily verify that the left side of (4) depends linearly and continuously on v, since continuity of **T'** is taken to mean that **T'** maps every convergent sequence of C_0^∞ functions into a convergent sequence of C_0^∞ functions. □

Exercise 1. Show that if **T** and **S** as mappings of C_0^∞ into C_0^∞ commute, then they also commute as mappings of distributions into distributions.

Note that if ℓ happens to be in C_0^∞, the definition of $\mathbf{T}\ell$ by (4) is the same as the original. We now give some interesting examples of linear operators with transposes.

Example 5. Let t be a C^∞ function; the operation **T** of multiplying by t clearly maps C_0^∞ linearly and continuously into C_0^∞; equally clearly, **T** is its own transpose.

Example 6. $\mathbf{T} = D_i$, meaning differentiation with respect to x_i. Clearly, **T** maps C_0^∞ into itself linearly and continuously, and $\mathbf{T'} = -\mathbf{T}$.

Example 7. $(\mathbf{T}_a u)(x) = u(x - a)$, meaning that $\mathbf{T}_a$ is translation by a. Clearly, $\mathbf{T}_a$ maps C_0^∞ into itself linearly and continuously, and $\mathbf{T}'_a$ is translation by $-a$.

Example 8. $(\mathbf{R}u)(x) = u(-x)$. Clearly, $\mathbf{R'} = \mathbf{R}$.

Example 9. Let t be a continuous function with compact support; define $\mathbf{T}u$ as the *convolution* of u with t, that is,

$$\mathbf{T}u = (t * u)(x) = \int t(y)u(x - y)\, dy. \qquad (5)$$

Clearly, **T** maps C_0^∞ into itself linearly and continuously; $\mathbf{T'}$ is convolution with $\mathbf{R}t$. It follows from the extension theorem and example 9 that the convolution of any distribution ℓ and any C_0^∞ function u is well defined as a distribution. We claim now that $\ell * u$ is a C^∞ function. To see this, we note that the classical formula (5) makes sense for distributions t, provided that the integral on the right is interpreted as in (2),

$$(\ell * u)(x) = (u_x, \ell), \qquad (6)$$

where u_y denotes the function $u(y - x)$. Clearly, u_x, as an element of C_0^∞, depends continuously and differentiably on y, from which it follows that (u_x, ℓ) is a C^∞ function of x.

Example 10. Let $\phi: \mathbb{R}^n \to \mathbb{R}^n$ be a C^∞ mapping that takes compact subsets of $\mathbb{R}^n$ into compact subsets; suppose that ϕ is invertible and that its inverse ψ has the same properties. Then the transformation **T** defined by $(\mathbf{T}u)(x) = u(\phi(x))$ maps C_0^∞ linearly and continuously into C_0^∞. The transpose of **T** is $(\mathbf{T'}v)(y) = v(\psi(y))J(y)$, where J is the Jacobian of ψ; $\mathbf{T'}$ maps C_0^∞ into itself, linearly and continuously.

Applying the extension theorem to these examples we can give meaning to the following operations with distributions:

(i) The product of a C^∞ function and a distribution is a distribution.
(ii) The derivative to any order of a distribution is a distribution.

(iii) The translate of a distribution is a distribution.

(iv) The convolution of a distribution with a C_0^∞ function is a C^∞ function.

(v) The composition of a distribution with an invertible C^∞ mapping is a distribution.

On the other hand, the product of two distributions, or the composite of a distribution with a noninvertible C^∞ mapping, cannot be defined in general. In particular, there is no reasonable way of defining $\delta^2(x)$ or $\delta(x^2)$. But there is a way of defining the product of distributions of disjoint variables:

Exercise 2. Show that if ℓ_1 and ℓ_2 are distributions in the disjoint variables $x_1, \ldots, x_m$ and $y_1, \ldots, y_n$ respectively, then their product $\ell_1 \ell_2$ can be defined as a distribution in $\mathbb{R}^{n+m}$. In particular, prove that

$$\delta(x_1)\delta(x_2)\ldots\delta(x_n) = \delta(x_1, \ldots, x_n).$$

Exercise 3. Find the first derivative of these distributions in $\mathbb{R}^1$:

(a) $\ell(u) = \int u(x)|x|\, dx.$

(b) $\ell(u) = PV \int u(x)/x\, dx.$

(c) $\ell(u) = \int_{-\infty}^{0} u(x)\, dx.$

(d) $\ell(u) = \delta(u) = u(0).$

B.3 LOCAL PROPERTIES OF DISTRIBUTIONS

We recall that the support of a continuous function f is the closure of the set of points where $f(x) \neq 0$. We will define the analogous concept for distributions; our starting point is the following equivalent characterization of the support of a function:

The complement of the support of f is the largest open set on which f is zero.

We will give meaning to the concept of a distribution being zero on an open set:

Definition. A distribution ℓ is *zero* on an open set G if $\ell(u) = 0$ for all C_0^∞ functions u whose support is contained in G.

Lemma 3. *If a distribution ℓ is zero on two open sets G_1 and G_2, then ℓ is zero on the union of G_1 and G_2.*

Proof. We have to show that $\ell(u) = 0$ for all u whose support lies in $G_1 \cup G_2$. We will accomplish this by decomposing u as $u = u_1 + u_2$ where support of u_1 is contained in G_1, that of u_2 in G_2. We proceed as follows: Since each point x in the support of u belongs to at least one of the sets G_1 or G_2, we construct a function h_x with the following properties:

(i) $h_x(y) \geq 0$ for all y.

(ii) h_x is C^∞.

(iii) $h_x(x) > 0$.

(iv) The support of h_x lies in G_1 or G_2.

The set where $h_x > 0$ is an open set containing x; the union of these cover the support of u. By compactness, a finite number of them cover the support of u.

Denote by h_1 the sum of those h_x of this finite collection whose support lies in G_1, by h_2 those whose support lies in G_2. Their sum $h_1 + h_2$ is positive on the support of u. Now set

$$u_1 = \frac{h_1}{h_1 + h_2} u, \quad u_2 = \frac{h_2}{h_1 + h_2} u$$

Clearly, support $u_1 \subset G_1$, support $u_2 \subset G_2$, u_1 and u_2 both C_0^∞, and $u_1 + u_2 = u$. Then

$$(u, \ell) = (u_1 + u_2, \ell) = (u_1, \ell) + (u_2, \ell) = 0 + 0,$$

since ℓ is assumed to be zero both in G_1 and G_2. This completes the proof of the lemma.   □

It follows by finite induction from the lemma that if ℓ is zero on the open sets $G_1, G_2, \ldots, G_n$, then u is zero on their union. We claim that the same conclusion holds for an infinite collection of open sets G_j. We have to show that if the support of u is contained in $\cup G_j$, the $\ell(u) = 0$. By compactness, the support of u is contained in the union of finitely many G_j, so the conclusion follows.

The union of all open sets on which a given distribution ℓ is zero is the *largest open set* on which ℓ is zero. The *complement* of this set is defined to be the *support* of ℓ.

The following result follows from the definition of support:

Theorem 4. *Let ℓ be a distribution, u a C_0^∞ function, and suppose that the supports of u and ℓ are disjoint; then $\ell(u) = 0$.*

Exercise 4. Show that if ℓ is a distribution with compact support, and w a C_0^∞ function, then $\ell * w$ is a C_0^∞ function.

Exercise 5. Show that if ℓ and m are two distributions and one of them has compact support, then $\ell * m$ can be defined as a distribution.

Exercise 6. Let f be a C^∞ function, ℓ a distribution such that $f\ell = 0$. Show that ℓ is zero on the open set where $f(x) \neq 0$.

Exercise 7. Show that the support of the derivative of a distribution ℓ is contained in the support of ℓ.

The support of the Dirac δ distribution is the single point $x = 0$. Therefore by exercise 7 the support of all derivatives of δ consists of $x = 0$. Conversely:

Theorem 5. *Every distribution ℓ whose support consists of a single point, say $x = 0$, is of the form $\ell = \sum_{|\alpha| \leq N} c_\alpha D^\alpha \delta$, N some positive integer, c_α complex numbers.*

Proof. We will use the following lemma:

Lemma 6. *Let ℓ be a distribution whose support consists of the origin. Then there is an integer N such that $\ell(u) = 0$ for all C_0^∞ functions u which, together with all their derivatives up to order N, are zero at the origin.*

Proof. Let f be a C^∞ function with these properties:

$$f(x) = \begin{cases} 0 & \text{for } |x| \leq 1 \\ 1 & \text{for } 2 \leq |x|. \end{cases}$$

Denote by v the function $(1 - f)u$; then since $fu = 0$ for $|x| < 1$, by theorem 4, $\ell(fu) = 0$. Therefore

$$\ell(v) = \ell(u) - \ell(fu) = \ell(u).$$

This shows that it suffices to look at functions u whose support is contained in, say, the ball $|x| < 3$. According to theorem 1, for such functions ℓ is continuous in some C^N-norm.

Take any positive real number k, define f_k as $f_k(x) = f(kx)$, and set $u_k = f_k u$. We will show that as $k \to \infty$, u_k tends to u in the C^N-norm. It follows from the definition of f that $f_k(x) = 1$ for $|x| \geq 2/k$; therefore $u_k(x) = u(x)$ for $|x| \geq 2/k$. We want to estimate u_k and its derivatives for $|x| < 2/k$. Since u and all its derivatives of order $\leq N$ are zero at $x = 0$, for $|\beta| \leq N$,

$$|D^\beta u(x)| = 0(|x|^{N+1-|\beta|});   \tag{7}$$

therefore for $|x| < 2/k$,

$$|D^\beta u(x)| \leq 0(k^{|\beta|-N-1}).   \tag{8}$$

By calculus

$$D^\alpha u_k = D^\alpha f_k u = \sum_{\beta \leq \alpha} \binom{\alpha}{\beta} D^{\alpha-\beta} f_k D^\beta u.   \tag{9}$$

Since $f_k(x) = f(kx)$, $|D^\gamma f_k| = 0(k^{|\gamma|})$; combining this with (8) and (9), we get $|D^\alpha u_k(x)| = 0(k^{|\alpha|-N-1})$ for $|x| < 2/k$. Since $u_k(x) = u(x)$ for $|x| \geq 2/k$, this completes the proof that $u_k \to u$ in the C^N-norm. According to theorem 1, it follows that $\ell(u_k)$ tends to $\ell(u)$. Since f_k, and therefore u_k, is zero in a ball around the origin, it follows from theorem 4 that $\ell(u_k) = 0$. But then so is $\ell(u) = \lim \ell(u_k)$. $\square$

Let u_1 and u_2 be two C_0^∞ functions whose values, and the values of all their derivatives up to order N, are equal at $x = 0$. It follows from lemma 6 that then $\ell(u_1) = \ell(u_2)$. In other words, $\ell(u)$ only depends on the value of u and its derivatives up to order N at $x = 0$. It follows that ℓ must be of the form

$$\ell(u) = \sum_{|a| \leq N} a_\alpha D^\alpha(u)|_{x=0}.$$

The conclusion of theorem 5 is just a restatement of this relation.   □

Theorem 7. *Every distribution ℓ of compact support can be written in the form*

$$\ell = \sum_{|a| \leq L} D^\alpha g_\alpha,$$

where the g_α are continuous functions, and L some whole number.

Proof. Let h be a C_0^∞ function that equals 1 on an open set containing the support of ℓ. It follows from theorem 4 that for all $u \in C_0^\infty$, $\ell(u) = \ell(hu)$.

Since every function of the form of hu is zero outside the support of h, it follows from theorem 1 that there is a positive constant and an integer N such that $|\ell(hu)| \leq$ const. $|hu|_N$. Clearly, $|hu|_N \leq$ const. $|u|_N$, the constant depending only on h. Combining these relations, we deduce that

$$|\ell(u)| \leq \text{const. } |u|_N \tag{10}$$

for all u, with a constant and N independent of u.

We introduce now the norms

$$\|u\|_M = \left(\sum_{|a| \leq M} \int |D^\alpha u|^2 \, dx \right)^{1/2} \tag{11}$$

and denote by H_M the completion of C_0^∞ under this norm. Since this norm comes from a scalar product that we denote as $(\,,\,)_M$, H_M is a Hilbert space. It follows from the Riesz-Frechet theorem that every continuous linear functional in H_M is of the form

$$(u, g)_M, \qquad g \in H_M. \tag{12}$$

H_M is the Sobolev space $W^{M,2}$ introduced in chapter 5.

According to an important inequality due to Sobolev (see Adams's book cited in chapter 5), for some constant depending only on the size of the support of u,

$$|u|_N \leq \text{const. } \|u\|_M \qquad \text{for } N < M - \frac{n}{2}. \tag{13}$$

It follows from (13) that every Cauchy sequence $\{u_k\}$ in the H_M-norm of C_0^∞ functions is also a Cauchy sequence in the C^N-norm. So such a sequence $\{u_k\}$ tends to a C^N function u. This mapping of H_M into C^N is one-to-one, so that this is an embedding of H_M in C^N.

Combining (10) and (13), we see that $|\ell(u)| \leq$ const. $\|u\|_M$, which implies that $\ell(u)$ is a bounded linear function in H_M. Therefore it can be represented in the form (12):

$$\ell(u) = (u, g)_M = \sum_{|a| \leq M} (D^\alpha u, D^\alpha g).$$

The right side can be rewritten as

$$\ell(u) = \sum (-1)^{|a|}(u, D^{2\alpha}g) = \left(u, \sum (-1)^{|a|} D^{2\alpha}g \right),$$

where $D^{2\alpha}g$ denotes a distribution derivative of g. Thus

$$\ell = \sum (-1)^{|a|} D^{2\alpha}g. \tag{14}$$

Since g belongs to H_M, g is a C^N function; thus (14) is the representation of ℓ stated in theorem 7, with $L = 2M - N$. $\qquad\square$

The next result shows that just like functions, distributions are determined up to a constant by their first derivatives. More precisely, we have

Theorem 8. *Let G be an open, connected subset of $\mathbb{R}^n$, ℓ a distribution in $\mathbb{R}^n$ such that all first partial derivatives $D_j\ell$ of ℓ are zero in G. Then $\ell =$ const. in G.*

Proof. $\ell =$ const. in G means that there is a constant c such that for all u whose support lies in G,

$$\ell(u) = c \int u \, dx. \tag{15}$$

We will use the following lemma.

Lemma 9. *Let $b(x)$ denote any C_0^∞ function whose support lies in the unit ball $|x| < 1$, and whose integral $\int b \, dx = 1$. Let k be any positive number, and define b_k as $b_k(x) = k^n b(kx)$. Then for any distribution ℓ, $\ell_k = b_k * \ell$ tends to ℓ in the sense that for any C_0^∞ function, $\ell_k(u)$ tends to $\ell(u)$ as k tends to ∞.*

Proof. There is no harm in supposing that b is symmetric: $b(-x) = b(x)$; then so is b_k, and so convolution with b_k, acting on C_0^∞ functions, is its own transpose. By definition of $b_k * \ell$ based on the extension theorem, for any C_0^∞ u,

$$\ell_k(u) = (b_k * \ell)(u) = (b_k * \ell, u) = (\ell, b_k * u). \tag{16}$$

It is not hard to prove (see chapter 11, section 11.1) that the sequence of C_0^∞ functions $b_k * u$ converges to u. Then $\ell(b_k * u)$ converges to $\ell(u)$, and so the lemma follows from (16). $\qquad\square$

For C_0^∞ functions, convolution and differentiation commute. Therefore, according to exercise 1, they commute also for distributions:

$$D_j(b * m) = (D_j b) * m$$

for any C_0^∞ function b and any distribution m.

Exercise 8. Show that

$$D_j(b * m) = b * D_j m$$

for every C_0^∞ function b and every distribution m.

Lemma 10. *Let b denote a C_0^∞ function whose support lies in a ball of radius r around the origin. Let m be any distribution; then every point x in the support of $b * m$ has distance $\leq r$ from some point of the support of m.*

Exercise 9. Prove lemma 10.

Define b_k as in lemma 9. It follows from exercise 8 that

$$D_j \ell_k = D_j(b_k * \ell) = b_k * D_j \ell.$$

The support of b_k is confined to a ball of radius $1/k$ around the origin, and according to the assumption in theorem 8, the support of $D_j \ell$ is contained in the complement of G. Therefore it follows from lemma 10 that $D_j \ell_k$ is zero in G_k, the set of all points of G whose distance from the boundary of G is greater than $1/k$.

Let u be any C_0^∞ function whose support S is contained in G. It is not hard to show that since G is connected, there exists a positive d such that any two points of S can be connected by a polygonal path P in G whose distance from the boundary of G is $\geq d$. For k larger than $1/d$, such a path P lies in G_k, and therefore all partial derivatives D_j of ℓ_k are zero along P. It follows that ℓ_k is a constant along P, the same constant for all points of S. Call this constant c_k; then by (2)

$$\ell_k(u) = \int \ell_k(x) u(x) \, dx = c_k \int u \, dx.$$

According to lemma 9, the left side tends to $\ell(u)$ as $k \to \infty$; therefore the right side tends to a limit. For $\int u \, dx \neq 0$ it follows that c_k tends to a limit c; it is easy to show that c is independent of u, as asserted in (15). $\square$

Theorem 11. *Suppose that g is a continuous function whose derivative in the sense of distributions $D_j g$ is a continuous function. Then $D_j g$ is the derivative of g not only in the sense of distributions but in the classical sense.*

Proof. According to lemma 9, g is the limit in the sense of distributions of the sequence of C_0^∞ functions $g_k = b_k * g$. Since differentiation commutes with convolution, $D_j g_k = b_k * D_j g$. Since both g and $D_j g$ are continuous functions, according to chapter 11, section 11.1, $g_k \to g$ and $D_j g_k \to D_j g$ uniformly on compact subsets. According to calculus

$$g_k(b) - g_k(a) = \int_a^b D_j g_k \, dx_j.$$

Letting k tend to ∞, we obtain

$$g(b) - g(a) = \int_a^b D_j g \, dx_j;$$

from this integral relation it follows that $D_j g$ is the derivative of g in the classical sense. □

Definition. A distribution ℓ is called *positive* if $\ell(u)$ is nonnegative for every nonnegative C_0^∞ function.

Examples of positive distributions abound:

Example 11. ℓ a nonnegative continuous function.

Example 12. $\ell = \delta(x - a)$.

Example 13. $\ell(u) = \int u \, dm$, m a measure.

Lemma 12. *Let ℓ be a positive distribution. To each compact set K in $\mathbb{R}^n$ there is a constant c such that $|\ell(u)| \leq c|u|_{\max}$ for every C_0^∞ function u supported in K.*

Proof. Denote by p a nonnegative C_0^∞ function that is equal to 1 on K. Since u is supported in K, for every x,

$$u(x) \leq |u|_{\max} p(x).$$

Since ℓ is a positive distribution, $\ell(u) \leq |u|_{\max} \ell(p)$. Similarly $-u(x) \leq |u|_{\max} p(x)$, so $-\ell(u) \leq |u|_{\max} \ell(p)$. This gives the inequality in lemma 12, with $c = \ell(p)$. □

Using inequality $|\ell(u)| \leq c|u|_{\max}$, we can extend, by continuity, the linear functional ℓ to all *continuous* functions u with compact support; the extended functional remains positive.

According to the Riesz-Kakutani representation theorem (see Appendix A) a positive linear functional on the space C_0 of continuous functions with compact support can be represented as an integral with respect to a measure. This proves

Theorem 13. *Every positive distribution is a measure:*

$$\ell(u) = \int u \, dm.$$

B.4 APPLICATIONS TO PARTIAL DIFFERENTIAL EQUATIONS

In chapter 9 we constructed the regular part g of Green's function for smoothly bounded domains in the plane. Given any point q in D, $g = g(p; q)$ is a harmonic function of p whose value on the boundary of D equals $\log |p - q|$. The difference $\log |p - q| - g(p; q)$ is Green's function $G(p; q)$. We want to show that Green's function satisfies, in the sense of distributions, the equation

$$\Delta G = 2\pi \delta(p - q),$$

where Δ is the Laplace operator $\Delta = D_x^2 + D_y^2$, x and y the Cartesian coordinates of p. Since the regular part of Green's function satisfies $\Delta g = 0$, it suffices to show that

$$\Delta \log \sqrt{x^2 + y^2} = 2\pi \delta(x, y); \tag{17}$$

here we have chosen $q = 0$.

Proof. According to the definition of derivatives of distributions, (17) means that for all C_0^∞ function u,

$$\int \log |p| \Delta u \, dx \, dy = 2\pi u(0). \tag{17'}$$

To see this, we write the integral on the left as the limit as $\varepsilon \to 0$ of integrals over the exterior of circular disks of radius ε around the origin:

$$\int_{|p| \geq \varepsilon} \log |p| \Delta u \, dx \, dy. \tag{18}$$

Integration by parts changes (18) into

$$\int_{|p| \geq \varepsilon} (\Delta \log |p|) u \, dx \, dy + \int_{|p| = \varepsilon} \log |p| \partial_n u \, ds - \int_{|p| = \varepsilon} (\partial_n \log |p|) u \, ds. \tag{18'}$$

Here ∂_n is the outward normal derivative on the boundary of the domain $|p| \geq \varepsilon$; that is, $\partial_n = -\partial/\partial r$ on the circle $|p| = \varepsilon$, where $r = |p|$.

Since $\log |p|$ is a regular harmonic function for $p \neq 0$, $\Delta \log |p| = 0$ for $|p| \geq \varepsilon$, so the double integral in (18)' is zero. The first line integral in (18)' is less in absolute value than const. $2\pi \varepsilon |\log \varepsilon|$, where the constant is an upper bound for $|\partial_n u|$. So as ε tends to zero, this term tends to zero. In the second line integral

$$\partial_n \log |p| = -\frac{d}{dr} \log r = -\frac{1}{r}.$$

On the circle $|p| = \varepsilon$ this equals $-1/\varepsilon$. The value of u at every point of the circle $|p| = \varepsilon$ equals $u(0) + 0(\varepsilon)$. It follows that as ε tends to zero, the last term in (18)' tends to $2\pi u(0)$; this proves formula (17). $\square$

One of the uses of distributions in the theory of partial differential equations is first to prove the existence of a distribution solution, and then to show that this object is actually a C^∞ function. Below we present an example in a simple case of how the second step is carried out; it is an extension of a classical result of Weyl.

Theorem 14. *Let ℓ be a distribution in an open set D in $\mathbb{R}_n$, which satisfies there Laplace's equation $\Delta \ell = 0$. Then ℓ is a C^∞ harmonic function in D.*

Proof. The proof is based on

Lemma 15. *Let f denote a spherically symmetric function in $\mathbb{R}^n$, $n > 2$, namely $f(x) = g(|x|)$, and suppose that $f(x) = 0$ for $|x| \geq R$. We further require that*

$$\int f(x)\, dx = 0, \qquad \int |x|^{2-n} f(x)\, dx = 0. \tag{19}$$

Then there exists a spherically symmetric C^∞ function h, $h(x) = 0$ for $|x| \geq R$, that satisfies

$$\Delta h = f. \tag{20}$$

Proof. Write h, to be determined, as $h(x) = p(|x|)$. In terms of the polar coordinate r we can write (20) as

$$\Delta h = p'' + \frac{n-1}{r} p' = g(r), \tag{20'}$$

where $'$ denotes differentiation with respect to r. Multiplying (20)$'$ by r^{n-1} gives an equation that can be written as

$$(r^{n-1} p')' = g(r) r^{n-1}.$$

Integrating this equation gives

$$r^{n-1} p'(r) = \int_0^r s^{n-1} g(s)\, ds. \tag{21}$$

We claim that the *right side* is zero for $r \geq R$. Indeed, $g(s) = f(|x|)$, so we can rewrite it as

$$\int_{|x| \leq r} f(x)\, dx.$$

According to (19) this is zero for $r \geq R$, since the support of f is contained in the ball of radius R around the origin.

Divide (21) by r^{n-1} and integrate:

$$p(r) = \int_0^r t^{1-n} \int_0^t s^{n-1} g(s)\, ds. \tag{22}$$

We claim that $p(r)$ is zero when $r \geq R$. To see this, integrate (22) by parts. Since $p'(r) = 0$ for $r \geq R$, there are no boundary terms, so we get, since $n > 2$,

$$p(r) = \frac{1}{n-2} \int_0^r t g(t) \, dt.$$

We can rewrite this integral as

$$\int_{|x| \leq R} |x|^{2-n} f(x) \, dx,$$

which is assumed in (19) to be zero because the support of f is contained in $|x| \leq R$.

Next we show that h is a C^∞ function; first we claim that p is, which is not hard to deduce from formula (21).

Setting $x = (r, 0, \ldots, 0)$ into the relation $f(x) = g(|x|)$, we get that

$$f(r, 0, \ldots, 0) = g(|r|).$$

This shows that $g(r)$ can be extended as an *even* C^∞ function to all real r. It follows from (20)' that $p(r)$ too can be extended as an even C^∞ to all real r; therefore all derivatives of p of odd order are zero at $r = 0$. From this it is not hard to deduce that $p(r) = q(r^2)$, q another C^∞ function. Then $h(x) = p(|x|) = q(|x|^2)$ is also C^∞, as claimed in the lemma.   □

Exercise 10. State and prove lemma 15 for $n = 2$.

We return to theorem 14. We will, as in lemma 9, approximate ℓ by a sequence of C^∞ functions. Take $n > 2$ and let b denote a spherically symmetric C^∞ function, supported in $|x| \leq 1$, satisfying the two conditions:

$$\int b(x) \, dx = 1, \quad \int |x|^{2-n} b(x) \, dx = 0. \tag{23}$$

We define, as in lemma 9, the functions b_k as $b_k(x) = k^n b(kx)$. Clearly, these functions, too, satisfy conditions (23):

$$\int b_k(x) \, dx = 1, \quad \int |x|^{2-n} b_k(x) = 0. \tag{23'}$$

The support of b_k is confined to a ball of radius $1/k$ around the origin. The convolution $\ell_k = b_k * \ell$ can be defined as a C^∞ function in the domain D_k defined as the set of all points of D whose distance from the boundary of D is greater than $1/k$. By formula (6), for y in D_k,

$$\ell_k(y) = (b_{k,y}, \ell),$$

where $b_{k,y} = b_k(y - x)$. Compare two of these functions:

$$\ell_k(y) - \ell_m(y) = (b_{k,y} - b_{m,y}, \ell). \tag{24}$$

The difference $b_{k,y} - b_{m,y}$ is spherically symmetric about the point y, and its support is contained in a ball of radius R around the point y, where $R = \max(1/k, 1/m)$. It follows from (23′) that each of the functions $b_{k,y} - b_{m,y}$ satisfies conditions (19). Therefore according to lemma 15, it is the Laplacean of a C^∞ function h, spherically symmetric about the point y, whose support is contained in the ball of radius R around y, satisfying $\Delta h = b_{k,y} - b_{m,y}$. So we can rewrite (24) as

$$\ell_k(y) - \ell_m(y) = (\Delta h, \ell).$$

According to theorem 2 we can rewrite the right side as $(h, \Delta \ell)$, which is zero since $\Delta \ell = 0$. So we conclude that at all points y in D, $\ell_k(y) = \ell_m(y)$ when k and m are greater than $1/d$, d the distance of y to the boundary of D. Therefore in every compact subset of D, ℓ_k does not depend on k for k large enough. According to lemma 9, the C^∞ functions ℓ_k converge in the sense of convergence of distributions to ℓ. This proves that ℓ is C^∞ in D, as claimed in theorem 14. $\qquad\square$

Theorem 14 is true for solutions of any elliptic partial differential equation with C^∞ coefficients.

Solutions of hyperbolic partial differential equations are quite different in regard of differentiability. Again we take the simplest case, the wave equation in one space dimension:

$$u_{tt} - u_{xx} = 0. \tag{25}$$

Every function of form

$$u(x,t) = f(x+t) + g(x-t), \tag{26}$$

where f and g are twice differentiable functions, is a solution of (25). Conversely, all twice differentiable solutions of (25) can be written in this form. To see this, rewrite (25) as

$$(D_t + D_x)(u_t - u_x) = 0, \quad (D_t - D_x)(u_t + u_x) = 0.$$

From these equations we conclude, respectively, that

$$u_t - u_x = a(x-t), \quad u_t + u_x = b(x+t),$$

where a and b are once differentiable functions. Adding these and integrating once yields the representation (26).

What about solutions that are distributions? We claim that for any pair of distributions ℓ and m in a single variable,

$$u = \ell(x+t) + m(x-t) \tag{26′}$$

is a solution of the wave equation in the sense of distributions. To verify this, we note that it follows from example 6 in section B.2 that $\ell(x+t)$, $m(x-t)$ can be defined as distributions in x, t, and that their partial derivatives can be calculated by the chain rule.

Exercise 11. Carry out these steps.

Furthermore it can be shown that every distribution solution of (25) is of form (26)'.

Exercise 12. Show.

What is the use of distribution solutions of the wave equation? Plenty! Take the propagation of acoustic waves, governed by the wave equation. For instance, the honking of a horn can be described as a solution of the form (26)', with ℓ and m functions equal to some constant c on an interval I, and zero outside I.

Other uses of distributions are given in section 7.2 and in Lax (1955).

B.5 THE FOURIER TRANSFORM

We saw in the previous sections that the class of C_0^∞ functions is the natural domain for some operators, such as differentiation. However, it is too narrow for other equally important operators, such as the Fourier transformation. For this operator the following larger space of functions, denoted as S, turns out to be natural:

S consists of all complex-valued C^∞ functions u defined in $\mathbb{R}^n$, which, together with all their derivatives, tend to zero faster than any power of $|x|^{-1}$ as $|x| \to \infty$. That is, a function u belongs to S iff for all multi-indexes α and all positive integers b,

$$\lim_{|x|\to\infty} |x^b D^\alpha u(x)| = 0. \tag{27}$$

Define the norms

$$|u|_{b,\alpha} = \max_x |x|^b |D^\alpha u(x)|; \tag{28}$$

we can describe S as the collection of those C^∞ functions for which all the norms (28) are finite.

Exercise 13. Show that the finiteness of the norms (28) implies (27).

Definition. A sequence $\{u_n\}$ of functions in S is said to *converge* to u if $\lim |u_n - u|_{b,\alpha} = 0$ for all b and all α.

Definition. Define the *distance* $d(u, v)$ of two functions u and v in S as

$$(u, v) = \sum \frac{1}{2^{b+|\alpha|}} \frac{\|u - v\|_{b,\alpha}}{1 + \|u - v\|_{b,\alpha}}, \tag{29}$$

the summation over all multi-indices α and positive integers b.

Exercise 14.

(a) Show that a sequence $\{u_n\}$ converges to u iff $d(u_n, u) \to 0$.

(b) Show that $d(u, v)$ as defined by (29) satisfies the triangle inequality.

(c) Show that S under the metric (29) is a complete metric space.

(d) Show that S is a linear space.

(e) Show that C_0^∞ is a dense subspace of S in the metric (29).

Definition. The dual S' of S consists of all *linear functionals* ℓ on S that are *continuous*; that is, if

$$\lim u_n = u, \quad \text{then} \quad \lim \ell(u_n) = \ell(u).$$

S' is a linear space. Since C_0^∞ is a subspace of S, ℓ acts as a linear functional on C_0^∞.

Exercise 15.

(a) Show that ℓ restricted to C_0^∞ is continuous in the sense of convergence in C_0^∞ defined in section B.1.

(b) Show that if ℓ and m are in S' and $\ell \neq m$, then $\ell \neq m$ acting on C_0^∞.

It follows from exercise 15 that elements of S' are distributions; they are called *tempered* distributions. As before, we will use the notation $\ell(u) = (u, \ell)$ for u in S, ℓ in S'.

Here are some examples of tempered distributions:

Example 14. Any distribution of compact support.

Example 15. Any function $v(x)$ that grows slower than some power of $|x|$ as $|x| \to \infty$, $\ell(u) = \int v u \, dx$.

We come now to the main topic of this section.

Definition. The *Fourier transform* of any function u of class S, denoted as $\mathbf{F}u$, and also as $\tilde{u}$, is defined in the usual fashion by the formula

$$\tilde{u}(\xi) = (\mathbf{F}u)(\xi) = \int u(x)e^{i\xi \cdot x} \, dx, \quad \text{where } dx = \frac{dx}{(2\pi)^{n/2}}. \tag{30}$$

Exercise 16. Show that $|\mathbf{F}u|_{\max} \leq (2\pi)^{-n/2}|u|_{L^1}$.

Theorem 16. $\mathbf{F}$ *maps S into S continuously.*

Proof. The integral in formula (30) can be differentiated with respect to ξ, giving

$$D^\alpha \mathbf{F} u = \mathbf{F}((ix)^\alpha u). \tag{31}$$

For u in S, $x^\alpha u$ tends to zero as $|x| \to \infty$ faster than any power of $|x|$; therefore $x^\alpha u$ belongs to L^1. It follows that for any multi-index α, $D^\alpha \mathbf{F} u$ is a continuous function. This proves that $\mathbf{F} u$ is C^∞. To show that $\mathbf{F} u$ is in class S, we deduce from (30) by integration by parts on the right that for any multi-index β,

$$\xi^\beta \mathbf{F} u = \mathbf{F}((iD)^\beta u). \tag{32}$$

Combining (31) and (32), we get $\xi^\beta D^\alpha \mathbf{F} u = i^{\alpha+\beta} \mathbf{F}(D^\beta x^\alpha u)$. This shows that $\xi^\beta D^\alpha \mathbf{F} u$ is a linear combination of the Fourier transforms of functions of the form $x^\alpha D^\delta u$. Since for u of class S these belong to L^1, it follows from exercise 16 that their Fourier transform is bounded in $\mathbb{R}^n$. This shows that $\mathbf{F} u$ belongs to class S.

$$\square$$

Exercise 17. Show that the mapping $u \to \mathbf{F} u$ is a continuous mapping of S into S.

The following theorem summarizes the relation of the Fourier transform to the usual operations in $\mathbb{R}^n$.

Theorem 17.

(i) $\mathbf{F}$ *transmutes translation in* $\mathbb{R}^n$ *into multiplication by* $e^{ia\cdot\xi}$. *That is, define* T_a *by* $(T_a u)(x) = u(x - a)$. *Then*

$$\mathbf{F} T_a u = e^{ia\cdot\xi} \mathbf{F} u \quad and \quad T_a \mathbf{F} u = \mathbf{F} e^{-ia\cdot x} u.$$

(ii) *The infinitesimal version of (i) is*

$$\mathbf{F} i D_j u = \xi_j \mathbf{F} u, \quad D_j \mathbf{F} u = \mathbf{F} i x_j u.$$

(iii) $\mathbf{F}$ *commutes with rotation around the origin, and with reflection* $\mathbf{R}$ *defined by* $(\mathbf{R}u)(x) = u(-x)$.

(iv) *Let* A *be an invertible map of* $\mathbb{R}^n$ *into* $\mathbb{R}^n$. *Denote* $u(Ax)$ *as* $u_A(x)$. *Then*

$$\mathbf{F} u_A = \frac{1}{|\det A|} (\mathbf{F} u)_B,$$

where $B = (A^{-1})'$.

Exercise 18. Prove theorem 17.

Exercise 19. Prove that the convolution of two functions u and v in S is in S, and that

$$\mathbf{F}(u * v) = (2\pi)^{n/2} (\mathbf{F} u)(\mathbf{F} v).$$

The kernel of the Fourier transform is a symmetric function of x and ξ. Therefore $\mathbf{F}$ is its own transpose $\mathbf{F}'$. So we can use theorem 2, the extension theorem, to *define* the Fourier transform of a tempered distribution

$$(\mathbf{F}v, \ell) = (v, \mathbf{F}\ell) \tag{33}$$

for all v in S.

Theorem 17′. *Theorem 19 is valid for the Fourier transform of tempered distributions.*

Exercise 20. Prove theorem 17′.

For distributions ℓ with compact support, we can use definition (30) directly to define $\mathbf{F}\ell$ as a C^∞ function by the formula

$$\tilde{\ell}(\xi) = (e(\xi), \ell), \quad e(\xi) = (2\pi)^{-n/2} e^{i\xi \cdot x}. \tag{34}$$

Exercise 21.

(a) Show that the Fourier transform of a distribution ℓ of compact support, as defined by (34), is a C^∞ function.
(b) Show that $\mathbf{F}\ell$ as defined by (34) satisfies (33).

We now present some examples of tempered distributions and their Fourier transforms; the first five are set in $\mathbb{R}^1$.

Example 16.

$$\ell(x) = \begin{cases} 1 & \text{for } |x| < 1 \\ 0 & \text{for } |x| > 1, \end{cases} \quad \tilde{\ell}(\xi) = \sqrt{\frac{2}{\pi}} \frac{\sin x}{x}.$$

Example 17.

$$\ell(x) = \begin{cases} 1 - |x| & \text{for } |x| < 1 \\ 0 & \text{for } |x| > 1, \end{cases} \quad \tilde{\ell}(\xi) = \sqrt{\frac{2}{\pi}} \frac{1 - \cos\xi}{\xi^2}.$$

Example 18.

$$\ell(x) = e^{-|x|}, \quad \tilde{\ell}(\xi) = \sqrt{\frac{2}{\pi}} \frac{1}{1 + \xi^2}.$$

Example 19.

$$\ell(x) = \frac{1}{1 + x^2}, \quad \tilde{\ell}(\xi) = \sqrt{\frac{\pi}{2}} e^{-|\xi|}.$$

Example 20.

$$u(x) = e^{-x^2/2}, \quad \tilde{u}(\xi) = e^{-\xi^2/2}.$$

Example 21.

$$\ell = \delta, \quad \tilde{\ell}(\xi) \equiv (2\pi)^{-n/2}.$$

Exercise 22. Verify, using formula (30) or (34), the Fourier transforms given in examples 16 to 21.

Theorem 18. *The Fourier transform of $\ell \equiv 1$ in $\mathbb{R}^n$ is $\tilde{\ell} = (2\pi)^{n/2}\delta$.*

Proof. Denote the Fourier transform of $\ell \equiv 1$ by d. According to part (i) of theorem $17'$, for $j = 1, \ldots, n$,

$$x_j d = x_j \mathbf{F}1 = \mathbf{F}(i D_j 1) = 0. \tag{35}$$

According to exercise 6, if f is a C^∞ function and ℓ a distribution such that $f\ell = 0$, it follows that the support of ℓ is contained in the nullset of f. Applying this to $f = x_j, j = 1, \ldots, n$, we conclude from (35) that the support of d is $x = 0$. According to theorem 5 it follows that d is of the form

$$d = \sum_{|a| \leq N} c_a D^a \delta. \tag{36}$$

It follows from (35) that for any multiindex $\beta, |\beta| > 0, x^\beta d = 0$. Combining this with (36), we conclude that for any $\beta, |\beta| > 0$,

$$x^\beta d = \sum_{|a| \leq N} c_x x^\beta D^a \delta = 0. \tag{37}$$

We want to deduce from this that $N = 0$; we need the following lemma:

Lemma 19.

$$x^\beta D^a \delta = \begin{cases} 0 & \text{if } |a| < |\beta| \\ 0 & \text{if } |a| = |\beta|, \alpha \neq \beta \\ (-1)^{|a|}\alpha! \delta & \text{if } \alpha = \beta. \end{cases}$$

Proof. Let u be any C_0^∞ function:

$$(u, x^\beta D^a \delta) = (x^\beta u, D^a \delta) = (-1)^{|a|}(D^a x^\beta u, \delta). \tag{38}$$

For $|a| < |\beta|$, and for $|a| = |\beta|$ but $\alpha \neq \beta$, the function $D^a x^\beta u$ is zero at $x = 0$. Therefore for such values of α and β, (38) is zero. For $\alpha = \beta$, $(D^a x^\alpha u, \delta) = \alpha! u(0)$ at $x = 0$. This is the assertion of lemma 19. $\square$

Suppose now that N is > 0; then there is an α, $|a| = N$, for which $c_\alpha \neq 0$. Combining (37) with lemma 19, we conclude that $c_\alpha = 0$, a contradiction. Therefore we conclude from (36) that $d = c_0 \delta$. So all that remains to prove theorem 18 is to determine the constant c_0. This is easily done; in the definition (33) of $\mathbf{F}\ell$,

$$(\mathbf{F}v, \ell) = (v, \mathbf{F}\ell),$$

set $\ell(\xi) \equiv 1$ and $v = e^{-x^2/2}$. $\mathbf{F}\ell = d = c_0\delta$, and according to exercise 5, $\mathbf{F}v = e^{-(\xi)^2/2}$:

$$(e^{-\xi^2/2}, 1) = (e^{-x^2/2}, c_0\delta).$$

The left side equals $\int_{\mathbb{R}^n} e^{-\xi^2/2}d\xi = (2\pi)^{n/2}$; the right side equals c_0, the assertion of theorem 18. $\qquad\square$

Theorem 20.

(i) $\mathbf{F}$ is an invertible mapping of S into S, and its inverse is given by

$$u(x) = \int \tilde{u}(\xi)e^{-ix\cdot\xi}\,d\xi. \tag{39}$$

(ii) $\mathbf{F}$ is an invertible mapping of S' into S', and $\mathbf{F}^{-1} = \mathbf{FR}$, where $\mathbf{R}$ is defined in part (iii) of theorem 17.

Proof.

(i) According to part (i) of theorem 17, $\mathbf{F}e^{-ia\xi} = T_a\mathbf{F}$. This implies, by theorem $17'$, that for every distribution ℓ,

$$\mathbf{F}e^{-ia\cdot\xi}\ell = T_a\mathbf{F}\ell.$$

Take $\ell \equiv 1$; we get, using theorem 18, that

$$\mathbf{F}e^{-ia\cdot\xi} = (2\pi)^{n/2}\delta(x - a).$$

Setting this into formula (33), we get for all v in S and $\ell = e^{-ia\cdot\xi}$ that

$$(\tilde{v}, e^{-ia\cdot\xi}) = (2\pi)^{n/2}(v, \delta(x - a)).$$

The left side is $\int \tilde{v}(\xi)e^{-ia\cdot\xi}d\xi$, the right side is $(2\pi)^{n/2}v(a)$. This proves (39).

(ii) In (39) replace ξ by $-\xi$ as variable of integration; we get that $\mathbf{F}^{-1} = \mathbf{FR}$. We can express this as $\mathbf{FRF} = \mathbf{I}$, the identity. So for any function v in S and any tempered distribution ℓ,

$$(v, \ell) = (\mathbf{FRF}v, \ell) = (\mathbf{RF}v, \mathbf{F}\ell) = (\mathbf{F}v, \mathbf{RF}\ell) = (v, \mathbf{FRF}\ell).$$

This proves that $\mathbf{FRF}\ell = \ell$, that is, that $\mathbf{RF}$ is a right inverse for $\mathbf{F}$ on S', and $\mathbf{FR}$ a left inverse. Since $\mathbf{F}$ and $\mathbf{R}$ commute, this proves part (ii). $\qquad\square$

Theorem 21 (Parseval's Formula). *The Fourier transform $\tilde{u}$ of every L^2 function u lies in L^2 and $\|\tilde{u}\|_{L^2} = \|u\|_{L^2}$.*

Proof. We want first to prove this for u in S. Take the complex conjugate of formula (30) defining the Fourier transform:

$$\overline{Fu} = \int \bar{u} e^{-i\xi \cdot x}\, dx.$$

The right side can be written as $\mathbf{RF}\bar{u}$, so

$$\overline{Fu} = \mathbf{RF}\bar{u}. \tag{40}$$

Since the Fourier transform is its own transpose,

$$(\mathbf{F}u, v) = (u, \mathbf{F}v).$$

Now set $v = \overline{Fu}$; using (40), we get

$$(\mathbf{F}u, \overline{Fu}) = (u, \mathbf{R}\overline{Fu}) = (u, \mathbf{FRF}\bar{u}) = (u, \bar{u}),$$

since we have shown in the course of proving theorem 20 that $\mathbf{FRF} = \mathbf{I}$. This proves the L^2 isometry of the Fourier transform acting of the class S. We get the result for any u in L^2 by approximating u with a sequence of functions of class S.   □

Exercise 23. Prove that $\tilde{u}$, defined as the L^2 limit of $\tilde{u}_n$, satisfies (33).

In words, theorem 21 says that the Fourier transform is a unitary operator that maps $L^2(\mathbb{R}^n)$ onto itself.

The next result is about distributions in $\mathbb{R}^1$.

Theorem 22. *Define for any a in $\mathbb{R}^1$*

$$p_a = \sum_m \delta(x - am), \tag{41}$$

summed over all integers m.

(i) *p_a is a tempered distribution.*

(ii) *The Fourier transform of p_a is $(\sqrt{2\pi}/|a|)p_b$, $b = 2\pi/a$.*

Proof. Part (i) is obvious. To prove part (ii), we observe that p_a is periodic with period a:

$$T_a p_a - p_a = 0.$$

We take the Fourier transform of this relation; using the relation of Fourier transform to translation (see theorem 17'), we get

$$(e^{ia\xi} - 1)\tilde{p}_a = 0. \tag{42}$$

The function e^{-ibx}, $b = 2\pi/a$, is $= 1$ at all points am, m an integer. Therefore it follows from the definition (41) of p_a that

$$e^{-ibx} p_a = p_a.$$

Take the Fourier transform of this relation; using again theorem 17, we get

$$T_b \tilde{p}_a = \tilde{p}_a, \tag{43}$$

that is, that $\tilde{p}_a$ is periodic with period b.

Let $u(\xi)$ be any function of class S that is zero at all integer multiples of b:

$$u(nb) = 0, \qquad n \text{ in } \mathbb{Z}.$$

Such a function is divisible by the function $e^{ia\xi} - 1$:

$$u(\xi) = e^{ia\xi - 1} v(\xi), \tag{44}$$

v in S. Therefore for such a u

$$(u, \tilde{p}_a) = ((e^{ia\xi} - 1)v, \tilde{p}_a) = (v, (e^{ia\xi} - 1)\tilde{p}_a) = 0, \tag{45}$$

where in the last step we have used (42).

It follows from (45) that if $u(nb) = 0$ for $|n| > N$, then the value of $(u, \tilde{p}_a)$ depends only on the values of $u(nb)$, $|n| \leq N$. Since this dependence is linear,

$$(u, \tilde{p}_a) = \sum_{|n| \leq N} c_n u(nb). \tag{46}$$

Since, by (43), $\tilde{p}_a$ is b-periodic, it follows that all the c_n are equal; we denote their common value by c:

$$(u, \tilde{p}_a) = c \sum u(nb), \tag{47}$$

where $c = c(a)$ depends on a.

Any u in S can be approximated in the topology of S by a sequence of functions u_N such that $u_N(nb) = 0$ for $|n| > N$. Setting $u = u_N$ in (47) and letting $N \to \infty$ we conclude that (47) holds for all u in S. We can restate this result simply in the notation of (41) as

$$\tilde{p}_a = c(a) p_b, \qquad b = \frac{2\pi}{a}. \tag{48}$$

Take the Fourier transform of both sides. Since P_a is an even function, the Fourier transform of $\tilde{p}_a$ is P_a, and so we get

$$P_a = c(a) \tilde{p}_b. \tag{48'}$$

Interchanging a and b in (48), $\tilde{p}_b = c(b) P_a$. From this and (48)$'$ we deduce that $c(a)c(b) = 1$ for $ab = 2\pi$. For $a = b = \sqrt{2\pi}$ we get $c(\sqrt{2\pi})^2 = 1$, so $c(\sqrt{2\pi})$ is either 1 or -1.

The definition of the Fourier transform of a tempered distribution such as P_a is that for all functions u of class S, $(\tilde{u}, P_a) = (u, \tilde{p}_a)$. Using the definition (41) of P_a and formula (47) for $\tilde{p}_a$, we get, with $b = 2\pi/a$,

$$\sum_m \tilde{u}(am) = c(a) \sum_n u(bn). \tag{50}$$

Given u and any number r, define u_r by $u_r(x) = u(rx)$. The Fourier transform of u_r is

$$\tilde{u}_r(\xi) = \int u(rx)e^{i\xi x}\, dx = \int u(y)e^{i\frac{\xi}{r}y}\frac{dy}{|r|} = \frac{1}{|r|}\tilde{u}\left(\frac{\xi}{r}\right).$$

Setting u_r into (49) in place of u gives

$$\frac{1}{|r|}\sum_m \tilde{u}\left(\frac{a}{r}m\right) = c(a)\sum_n u(rbn). \tag{49$'$}$$

We can use (49) with a replaced by a/r, and b replaced by rb, to express the left side of (49$'$) as

$$\frac{c(a/r)}{|r|}\sum u(rbn).$$

Since this equals the right side of (49$'$), we deduce that $(1/|r|)c(a/r) = c(a)$. Setting $r = a/\sqrt{2\pi}$ we get $(\sqrt{2\pi}/|a|)c(\sqrt{2\pi}) = c(a)$. Since we have shown that $c(\sqrt{2\pi}) = \pm 1$, we deduce that

$$c(a) = \pm\frac{\sqrt{2\pi}}{|a|}.$$

We claim that the correct sign is the positive one. For take any positive, even function v of class S, and define u as $v * v$. Then u is positive, and since the Fourier transform of the even function v is real, $\tilde{u} = \sqrt{2\pi}\tilde{v}^2$ too is positive. So it follows from formula (49) that $c(a)$ is positive as well. This completes the proof of theorem 22.   □

Setting $c = \sqrt{2\pi}/|a|$ into (49) yields

Poisson Summation Formula. *For every function u of class S and for all real a,*

$$\sum_m \tilde{u}(am) = \frac{\sqrt{2\pi}}{|a|}\sum_n u\left(\frac{2\pi}{a}n\right). \tag{50}$$

Poisson's formula is valid for a much wider class of functions than S.

NOTE. In chapter 30, section 30.6, we have derived Poisson's formula as a special case of the trace formula applied to the convolution operator. We used there a different normalization for the Fourier transform.

We indicate now how to extend theorem 22 to $\mathbb{R}^n$. Instead of all integer multiples of a, we consider all points of *a lattice L* defined as follows:

Definition. A *lattice* in $\mathbb{R}^n$ is a set L of vectors in $\mathbb{R}^n$ with the following properties:

   (i) The sum and difference of vectors in L belongs to L.
   (ii) The set L has no point of accumulation in $\mathbb{R}^n$.
   (iii) The vectors in L span $\mathbb{R}^n$.

Here are some examples.

Example 22. The set E of all vectors with integer coefficients form a lattice.

Example 23. The image of any lattice under an invertible linear map of $\mathbb{R}^n \to \mathbb{R}^n$ is a lattice.

Lemma 23.

   (i) *Every lattice L can be represented as*

$$L = AE, \tag{51}$$

   where A is an invertible linear map of $\mathbb{R}^n \to \mathbb{R}^n$, and E is the integer lattice described in example 1.
   (ii) *The representation of L in form (51) is not unique. But in all such representations $|\det A|$ has the same value.*

Proof. For a proof of part (i) we refer the reader to appendix 5 of a splendid text on linear algebra by the author of these pages.

   (ii) Let $L = A_1 E$ and $L = A_2 E$ be two representations of L of form (51). It follows that $A_2^{-1} A_1$ maps E onto E. We want to show now that a linear map M that maps E onto itself has integer entries and determinant ± 1. Clearly, if M had a noninteger entry m_{ij}, then the ith component of Me_j, where e_j is the jth unit vector, is m_{ij}, contrary to the assumption that M maps every vector in E into E.

   Since M^{-1} also maps E onto E, its entries, too, are integers. Since $MM^{-1} = I$, $(\det M)(\det M^{-1}) = 1$. Since the entries of M and M^{-1} are integers, their determinants are integers; therefore $\det M$ can have no other value than ± 1. Applying this to $M = A_2^{-1} A_1$, we conclude that $\det M = (\det A_2)^{-1}(\det A_1) = \pm 1$. $\qquad\square$

Definition. The *dual L'* of a lattice L consists of all vectors b such that $a \cdot b$ is an integer for all a in L.

Exercise 24.

(a) Show that the dual of a lattice is a lattice.
(b) Show that $L'' = L$.
(c) Show that for every lattice L and every invertible linear map A of $\mathbb{R}^n \to \mathbb{R}^n$, $(AL)' = BL'$, where $B = (A^{-1})'$.

Let L be a lattice; denote by p_L the tempered distribution

$$p_L = \sum_{a \text{ in } L} \delta(x - a). \tag{52}$$

Theorem 24. *The Fourier transform of p_L is*

$$\tilde{p}_L = c(L)p_{2\pi L'}, \tag{53}$$

where $c(L) = (2\pi)^{n/2}/|\det A|$, and A is a matrix that appears in the representation of L of form (51).

Exercise 25. Prove theorem 24 by imitating the steps that went into the proof of theorem 22.

Exercise 26. Formulate Poisson's summation formula for lattices in $\mathbb{R}^n$.

B.6 APPLICATIONS OF THE FOURIER TRANSFORM

The Fourier transform is used in a large part of mathematics to solve, or at least reformulate, problems. Here are a few examples.

Liouville's Theorem. *Let $f(z)$ be an analytic function defined in the whole complex plane that is of polynomial growth:*

$$|f(z)| \leq \text{const.} (1 + |z|)^M.$$

Then f is a polynomial of degree $\leq M$.

Proof. An analytic function satisfies the Cauchy-Riemann equation

$$\partial_{\bar{z}} f = \tfrac{1}{2}(\partial_x - i\partial_y)f = 0. \tag{54}$$

A function f of polynomial growth is a tempered distribution; so are its derivatives. The Fourier transform of (54) is

$$(i\xi + \eta)\tilde{f}(\xi, \eta) = 0.$$

It follows from exercise 6 that the support of $\tilde{f}$ is the origin. By theorem 5, such a distribution f is of the form

$$\tilde{f} = \sum_{|a|\leq N} c_\alpha D^\alpha \delta. \tag{55}$$

According to theorem 18, the Fourier inverse of δ is constant, so, according to theorem 17', the Fourier inverse f of the right side of (55) is a polynomial in x and y. Since f is analytic, it is a polynomial in $z = x + iy$.   □

Unlike the usual proofs of Liouville's theorem, this one uses nothing of the theory of analytic functions beyond the Cauchy-Riemann equations. In fact the same proof gives the following much more general result:

Theorem 25. *Let* $P(\xi_1,\ldots,\xi_n) = P(\xi)$ *be a homogeneous elliptic polynomial, that is, whose only real zero is* $\xi = 0$. *Let* f *be a tempered distribution that is a solution in all* $\mathbb{R}^n$ *of the partial differential equation*

$$P(D_1,\ldots,D_n)f = 0.$$

Then f *is a polynomial.*

Examples of such partial differential equations are $\Delta f = 0$, $\Delta^2 f = 0$, and many others.

We give some applications of the Poisson summation formula. Set $u(x) = e^{-x^2/2}$ into formula (40). Since $\tilde{u}(\xi) = e^{-\xi^2/2}$, we get that for all real a,

$$\sum_m e^{-\frac{a^2m^2}{2}} = \frac{\sqrt{2\pi}}{|a|} \sum_n e^{-b^2n^2/2}, \qquad b = \frac{2\pi}{a}.$$

Denoting the function on the left as $Z(a)$, we can rewrite this as

$$Z(a) = \frac{\sqrt{2\pi}}{|a|} Z\left(\frac{2\pi}{a}\right),$$

which is an interesting functional equation.

Exercise 27. Set $u(x) = e^{-x^2/2+xt}$ into the Poisson summation formula, and see what you get.

B.7 FOURIER SERIES

The Fourier analysis of periodic distributions is fairly straightforward. Let u be a C^∞ function on the unit circle S^1. Its Fourier coefficients b_n are

$$b_n = \int_{S^1} e^{-in\theta} u(\theta)\, d\theta/2\pi.$$

Exercise 28.

(a) Show that for any N there is a constant such that $|b_n| \leq \text{const.} |n|^N$.

(b) Show that the partial sums of the Fourier series of u, $u_k = \sum_{-k}^{k} b_n e^{in\theta}$, converge to u in the C^N topology.

(c) Let ℓ be a distribution. Its Fourier coefficients a_n are defined by

$$a_n = (e^{-in\theta}, \ell)/2\pi.$$

Show that for any C^∞ function u,

$$(u, \ell) = \sum b_n a_{-n}.$$

(d) Let $\{a_n\}$ be a sequence of complex numbers that satisfy

$$|a_n| \leq \text{const.} |n|^N$$

for some N. Show that the a_n are the Fourier coefficients of some distribution on S^1.

•

BIBLIOGRAPHY

Bochner, S. *Vorlesungen über Fouriersche Integrale.* Akademische Verlagsgesellschaft, Leipzig, 1932.

Dirac, P. A. M. *The Principles of Quantum Mechanics.* Clarendon Press, Oxford, 1930.

Lax, P. D. On Cauchy's problem for hyperbolic equations and the differentiability of solutions of elliptic equations. *CPAM* **8** (1955): 615–633.

Lax, P. D. *Linear algebra.* Series on pure and applied mathematics. Wiley-Interscience, 1997.

Lützen, J. *The Prehistory of the Theory of Distributions.* Springer Verlag, 1982.

von Neumann, J. *Matematische Grundlagen der Quantenmechanic.* Die Grundlehren der mathematischen Wissenschaften, **38**. Springer, Berlin, 1932.

Schwartz, L. *Théorie des distributions.* Hermann, Paris, 1950–1951.

Sobolev, S. L. Méthode nouvelle à résoudre le probleme de Cauchy pour les équation linéaires hyperboliques. *Mat. Sb.*, **1** (1936): 39–71.

Strichartz, R. *Guide to Distribution and Fourier Transform.* Studies in Advanced Mathematics. CRC Press, Boca Raton, 1994.

Weyl, H. The method of orthogonal projection in potential theory, *Duke Math. J.*, **7** (1940): 411–444.

Wiener, N. The operational calculus, *Math Ann.* **95** (1926): 557–584.

C

ZORN'S LEMMA

Zorn's lemma is a theorem in the Zermelo-Fraenkel system of set theory. It is equivalent logically with the axiom of choice. Thus its use introduces a highly nonconstructive step; therein lies its power.

Zorn's lemma deals with *partially ordered sets*, nonempty sets where an *order relation* $a \leq b$ is defined for some pairs of elements in the set, which satisfies

(i) transitivity: if $a \leq b$ and $b \leq c$, then $a \leq c$.
(ii) reflexivity: $a \leq a$ for all a in the set.

A subset of a partially ordered set is called *totally ordered* if for every pair x, y of elements in it, either $x \leq y$ or $y \leq x$.

An element u of a partially ordered set is said to be an *upper bound* of a subset if $x \leq u$ for every element x in the subset.

An element m of a partially ordered set is called *maximal* if every element b of the set satisfies $b \leq m$.

Zorn's lemma. *If every totally ordered subset of a partially ordered set has an upper bound, then the partially ordered set has a maximal element.*

AUTHOR INDEX

Adams, R. A., 51, 550
Agnew, R. P., 24, 28
Ahlfors, L. V., 267
Akhiezer, N. I., 159, 414
Alaoglu, L., 120, 121
Aronszajn, N., 275, 278, 282
Arzela, C., 243, 245
Ascoli, G., 243, 245
Atiyah, M., 312
Atkinson, F. V., 305, 313

Banach, S., 19, 28, 168, 172, 260, 527
Beardon, A. F., 511
Berger, C., 312, 313
Bernstein, A. R., 282
Bernstein, S., 89, 138
Beurling, A., 221, 225, 282, 513, 515, 516, 517, 523, 525, 526
Birkhoff, G., 151, 152, 156, 159, 251, 252
Birman, M. Sh., 490, 511
Blaschke, W., 89, 522
Bochner, S., 141, 144, 146, 149, 543
Bohnenblust, H. F., 27, 28
Bohr, H., 543
Bohr, N., 459
Boltzmann, L., 446
Boutet de Monvel, L., 312, 313
de Branges, L., 126, 132
Brezis, H., 527
Brodsky, M. S., 280, 283
Buskes, G., 28

Calderon, A. P., 470, 475
Calkin, J. W., 234, 244
Carathéodory, 125, 128, 141, 142, 143, 149, 152, 159
Carleman, T., 393, 470, 475
Carleson, L., 215, 225
Cartan, E., 172

Chernoff, P., 121, 429, 438
Choquet, G., 128, 130, 131, 133, 135, 146, 150, 151, 157
Clarkson, J. A., 46, 51
Coburn, L. A., 312, 313
Conway, J. B., 527
Cook, J. M., 482, 511
Courant, R., 41, 246, 252, 318, 319, 322, 328, 345

David, G., 191
Davidson, K. R., 279, 283
Day, M. M., 51, 527
Diestel, J., 131, 132
Dieudonné, J., 303, 305, 313
Dirac, P.A.M., 108, 543, 566
Donoghue, W. F., 280, 283
Doob, J. L., 378, 393
Douglas, R. G., 311, 313, 527
Dunford, N., 112, 121, 527
Dyson, F. J., 271, 273, 274

Eberlein, W. F., 105, 107, 121
Edwards, R. E., 527
Enflo, P., 282, 283

Faddeev, L., 492, 505, 510, 511
de Finetti, B., 156, 159
Fischer, E., 318, 319, 322, 328
Frayn, M., 459
Frechet, M., 57, 62
Fredholm, I., 260, 268, 274, 279, 342
Friedrichs, K. O., 115, 117, 402, 412, 414, 432, 502
Frölich, J., 491, 511

Garabedian, P. R., 71, 98, 94
Gelfand, I. M., 159, 195, 202, 208–210, 221, 222, 224, 271, 273, 274, 371, 376, 436, 437

573

SUBJECT INDEX

Absolute value, 330
Almost orthogonal bases, 251–252
Analytic function
 entire, 265
 positive real part, 115
 resolvent, 195, 382
 strongly, 111
 weakly, 111
Anderson localization, 491
Annihilator, 76, 244
Approximation
 by powers, 89–90
 by weighted polynomials, 88
 of the δ function, 108
Arzela–Ascoli theorem, 243

B^* algebra, 222
Baire category principle, 169
Banach algebra, 192, 435
Banach limits, 31
Banach space, 38
 reflexive, 78, 82
 uniformly convex, 45
Bessel's inequality, 60
Blaschke product, 89, 522
Borel subsets, 354, 361
Bounded linear maps, *see* Linear maps

Capacity, 239
Cauchy transform, 381
Cayley transform, 389, 400
Closed graph theorem, 170–171, 377
Compact, 43
 map, 233
 noncompact, 43
Completely monotone functions, 137
Convexity, 5
 closest point, 45, 54
 convex combination, 5

convex function, 135, 337
convex hull, 6
 extreme point, 6, 124
 extreme subset, 6, 124
 Riesz theorem, 177
 subset, 45
 uniform, 45
Convolution operator, 216, 323, 348
Cross norm, 479

Deficiency index, 400
Discrete subgroup, 501
 automorphic, 501
 fundamental group, 501
 fundamental polygon, 501
Determinant, 341–342. *See also* Fredholm
  integral equation
Dimension, codimension, 4
Dirichlet's problem, 65, 94, 112, 327, 405, 462
Distribution, 113, 543
 delta function, 108, 544
 Fourier transform, 558–559
 support, 547–548
 tempered, 559
Dual, 72
 of $C(Q)$, 82
 of L^p, 79
 variational problems, 76–77, 86

Eigenvalue, 229, 238, 248, 253, 266, 287
 algebraic multiplicity, 267, 278
Eigenvector, 229, 248
Elliptic PDE, 249, 258, 286, 461–462
Energy, 447, 454
 decay, 473
Equivalence theorem, 427
Ergodic mapping, 157–158, 443
Extreme point, *see* Convexity

PURE AND APPLIED MATHEMATICS

A Wiley-Interscience Series of Texts, Monographs, and Tracts

Founded by RICHARD COURANT
Editors: MYRON B. ALLEN III, DAVID A. COX, PETER LAX
Editors Emeriti: PETER HILTON, HARRY HOCHSTADT, JOHN TOLAND

*Now available in a lower priced paperback edition in the Wiley Classics Library.
†Now available in paperback.

*Now available in a lower priced paperback edition in the Wiley Classics Library.
†Now available in paperback.

Printed in the United States
By Bookmasters